D1598813

Organophosphates

Organophosphates

Chemistry, Fate, and Effects

Editors

Janice E. Chambers

College of Veterinary Medicine
Mississippi State University
Mississippi State, Mississippi

Patricia E. Levi

Department of Toxicology
North Carolina State University
Raleigh, North Carolina

Academic Press

Harcourt Brace Jovanovich, Publishers

San Diego New York Boston London Sydney Tokyo Toronto

This book is printed on acid-free paper. ♾

Academic Press, Inc.
San Diego, California 92101

United Kingdom Edition published by
Academic Press Limited
24–28 Oval Road, London NW1 7DX

Library of Congress Cataloging-in-Publication Data

Organophosphates : chemistry, fate, and effects / [editors] Janice E. Chambers, Patricia E. Levi.
p. cm.
Includes index.
ISBN 0-12-167345-6
1. Organophosphorus compounds--Toxicology. 2. Insecticides--Toxicology. I. Chambers, Janice E. II. Levi, Patricia E.
[DNLM: 1.Cholinesterase Inhibitors--pharmacology. 2. Insecticides, Organophosphate--metabolism. 3. Insecticides, Organophosphate--pharmacology. WA 240 06803]
RA1242.P56074 1992
615.9'517--dc20
DNLM/DLC
for Library of Congress 91-25947
CIP

PRINTED IN THE UNITED STATES OF AMERICA
91 92 93 94 9 8 7 6 5 4 3 2 1

Contents

V
Summary and Conclusions

Contributors

Numbers in parentheses indicate the pages on which the authors' contributions begin.

Mohamed B. Abou-Donia (327), Department of Pharmacology, Duke University Medical Center, Durham, North Carolina 27710

Zoltan Annau (419), Department of Environmental Health Sciences, The Johns Hopkins University, Baltimore, Maryland 21205

Howard W. Chambers (3), Department of Entomology, Mississippi State University, Mississippi State, Mississippi 39762

Janice E. Chambers (229, 435), College of Veterinary Medicine, Mississippi State University, Mississippi State, Mississippi 39762

Shao-Kuang Chang (241), Cutaneous Pharmacology and Toxicology Center, North Carolina State University, Raleigh, North Carolina 27606

Lucio G. Costa (271), Department of Environmental Health, SC-34, University of Washington, Seattle, Washington 98195

Walter C. Dauterman (169), Department of Toxicology, North Carolina State University, Raleigh, North Carolina 27695

Amira T. Eldefrawi (257), Department of Pharmacology and Experimental Therapeutics, School of Medicine, University of Maryland at Baltimore, Baltimore, Maryland 21201

Mohyee E. Eldefrawi (257), Department of Pharmacology and Experimental Therapeutics, School of Medicine, University of Maryland at Baltimore, Baltimore, Maryland 21201

Mark E. Hansen (107), Department of Avian Sciences and Environmental Toxicology, University of California, Davis, Davis, California 95616

Ing K. Ho (285), Department of Pharmacology and Toxicology, University of Mississippi Medical Center, Jackson, Mississippi 39216

Ernest Hodgson (141), Department of Toxicology, North Carolina State University, Raleigh, North Carolina 27695

Michael J. Hooper (107), Institute of Wildlife, and Department of Environmental Toxicology, Clemson University, Clemson, South Carolina 29631

Beth Hoskins (285), Department of Pharmacology and Toxicology, University of Mississippi Medical Center, Jackson, Mississippi 39216

David Jett (257), Department of Pharmacology and Experimental Therapeutics, School of Medicine, University of Maryland, Baltimore, Maryland 21201

Yutaka Kasai (169), Kao Corporation, Tochigi Research Labs, 2606 Akabane Ihikaimachi Haga, Tochigi 321–34, Japan

Paul A. Kitos (387), Department of Biochemistry, The University of Kansas, Lawrence, Kansas 66045

Takamichi Konno (169), Nikon Nohyako Company Ltd., Biological Research Center, 4-31 Hondacho, Kawachi-Nagano, Osaka 586, Japan

Patricia E. Levi (141), Department of Toxicology, North Carolina State University, Raleigh, North Carolina 27695

Donald M. Maxwell (183), U.S. Army Medical Research Institute of Chemical Defense, Biochemical Pharmacology Branch, Aberdeen Proving Ground, Maryland 21010

Tsutomu Nakatsugawa (201), College of Environmental Science and Forestry, State University of New York, Syracuse, Syracuse, New York 13210

Pamela S. Nieberg (107), Department of Avian Sciences and Environmental Toxicology, University of California, Davis, Davis, California 95616

Stephanie Padilla (353), Health Effects Research Laboratories, Neurotoxicology Unit, Cellular and Molecular Toxicity Branch, U.S. Environmental Protection Agency, Research Triangle Park, North Carolina 27711

Stephen B. Pruett (367), Department of Biological Sciences, Mississippi State University, Mississippi State, Mississippi 39762

Kenneth D. Racke (47), Environmental Chemistry Laboratory, DowElanco, Midland, Michigan 48641

Rudy J. Richardson (299), Toxicology Program, The University of Michigan, Ann Arbor, Michigan 48109

J. Edmund Riviere (241), Cutaneous Pharmacology and Toxicology Center, North Carolina State University, Raleigh, North Carolina 27606

Lester G. Sultatos (155), Department of Pharmacology/Toxicology, University of Medicine and Dentistry of New Jersey, Newark, New Jersey 07103

Oranart Suntornwat (387), Department of Biochemistry, The University of Kansas, Lawrence, Kansas 66045

Charles M. Thompson (19), Department of Chemistry, Loyola University of Chicago, Chicago, Illinois 60626

Bellina Veronesi (353), Health Effects Research Laboratories, Neurotoxicology Unit, Cellular and Molecular Toxicity Branch, U.S. Environmental Protection Agency, Research Triangle Park, North Carolina 27711

Kendall B. Wallace (79), Department of Pharmacology, School of Medicine, University of Minnesota, Duluth, Minnesota 55812

Barry W. Wilson (107), Department of Avian Sciences and Environmental Toxicology, University of California, Davis, Davis, California 95616

Preface

Organophosphorus insecticides have been an important part of global agricultural chemistry for at least three decades, are still widely used today, and undoubtedly will have utility in the future. Consequently, these compounds are the subject of significant research by scientists in academia, industry, and government. In addition, the organophosphorus nerve agents constitute a current military threat, as has been highlighted in recent years in the Middle East.

The concept behind this book is to bring together overviews of the diverse areas of current research on organophosphorus anticholinesterases, with primary, although not exclusive, focus on the insecticidal organophosphorus (OP) compounds. Much of the biochemistry and toxicology associated with OP compounds is oriented toward their anticholinesterase properties. However, since the emphasis is on recent developments in OP toxicology, the book includes information on effects probably not resulting directly from acetylcholinesterase inhibition, such as teratogenicity, delayed neuropathy, and immunotoxicity. The authors include many of the current researchers in the field of OP research, and their chapters cover the most pertinent areas that must be understood in considering the spectrum of reactions in which OP compounds can participate.

As the subtitle "Chemistry, Fate, and Effects" suggests, the book presents a synthesis of several important aspects of the science of OP compounds. Although these topics are diverse, the focus of each section is on the reactions occurring between the organophosphorus compound and the organism, both in terms of toxicity and possible defenses. The editors have taken license in the title by using the popular but less accurate term "organophosphates" instead of the cumbersome but more accurate term "organophosphorus compound."

This book was developed as an expansion of a symposium of the same name presented at the Agrochemicals Division of the American Chemical Society meeting in Boston in April, 1990. In that connection, the Public Health Service participated in the support of this meeting under conference grant R13 ES05375 from the National Institute of Environmental Health Sciences. The editors are also grateful for the generous support of the symposium from E. I. duPont de Nemours and Company, Haskell Laboratory, and American Cyanamid Corporation.

We also express much appreciation to the contributors for their articles and their participation in the symposium. Many thanks are due to Carol Majors of Publications Unlimited, Raleigh, North Carolina, for patience and diligence in the preparation of chemical structures for the book.

I
Introduction

1

Organophosphorus Compounds: An Overview

Howard W. Chambers
Department of Entomology
Mississippi State University
Mississippi State, Mississippi

This initial chapter is intended as an overview, quite general and somewhat simplified, to acquaint the reader with the broad scope of the subject of organophosphorus compounds, or OP compounds, as they will be referred to in the book. For more detailed information, complete with literature citations, the reader is referred to the short bibliography at the end of this chapter and, of course, to subsequent chapters of this book.

I. Introduction and Historical Development

Organophosphorus compounds are a massive and highly diverse family of organic chemicals, with many uses, not the least of which is the control of

Organophosphates: Chemistry, Fate, and Effects

pests of plant, animal, and human health importance. Though OP compounds with herbicidal and fungicidal properties are known and used, far more are used as insecticides, acaricides, nematocides, and helminthicides. As will be discussed in this and other chapters, the acute lethality of the latter groups of OP pesticides can be attributed primarily, if not entirely, to their ability to inhibit acetylcholinesterase (AChE), an enzyme vital to normal nerve function. This property has led also to development of several important pharmaceuticals and, less fortunately, to chemical warfare agents.

The history of organic phosphorus chemistry apparently began about 1820 when Lassaigne reacted ethanol with phosphoric acid to obtain triethyl phosphate. Tetraethyl pyrophosphate was first synthesized by de Clermont in 1854 by reaction of the silver salt of pyrophosphoric acid with ethyl iodide. Some 80 years would pass, however, before the insecticidal properties of this chemical, known today as TEPP, would be discovered by Gerhard Schrader in Germany. Other early reactions that would be of significance later in the development of OP insecticides include (1) discovery of thiophosphoric esters by Cleoz in 1847, (2) conversion of trialkyl phosphites to dialkyl alkylphosphonates by Michaelis and Becker in 1897, and (3) synthesis of phosphoramidodichloridates from phosphorus trichloride by Michaelis in 1903.

The high toxicity of certain OP compounds was noted first in 1932 by Lange and Krueger, who were studying the alkylation of silver salts of monofluorophosphoric acid. This discovery led to the synthesis of tabun and sarin by Schrader in 1937, diisopropyl phosphoro fluoridate (DFP) by Saunders in 1941, and soman by Riser in 1944.

Though Schrader synthesized two important *nerve gases,* his primary interest was in insecticides. His research led to dimefox and schradan (OMPA, octamethyl pyrophosphoramide) in 1941, an improved practical synthesis of TEPP in 1943, and parathion in 1944. He is generally recognized today as the "father" of OP insecticides.

Shortly after the discovery of parathion, methyl parathion and EPN were recognized as insecticides by Bayer and du Pont, respectively, and the groundwork was laid for the subsequently rapid development of OP compounds as a major class of insecticides. A common characteristic of all the early OP insecticides was high, acute mammalian toxicity. The introduction of malathion by American Cyanamid in 1950 represented a breakthrough in combining good insecticidal activity with low mammalian toxicity. The discovery of additional low-toxicity compounds, combined with growing problems of pest resistance to organochlorine pesticides, led to OP compounds becoming the dominant class of insecticides worldwide, with annual production exceeding 100 million pounds by 1970. Though the introduction of the modern synthetic pyrethroids appreciably decreased the use of OP compounds, these insecticides remain even today an important weapon in our arsenal for insect control.

II. Chemistry

A. Classification and Nomenclature

Organophosphorus compounds are a large and diverse family of chemicals. The nomenclature is complex, and classification may follow various schemes, depending on the portion of the structure used in the classification system. A common scheme, for example, considers the nature of the most reactive substituent (commonly called the leaving group, for reasons that will become clear later). Thus, OP compounds may be classified as anhydrides, aliphatics, aromatics, heterocyclics, etc. For the purposes of this chapter, only the nature of the atoms immediately surrounding the central phosphorus atom will be considered. In all OP insecticides in current use, four atoms are directly attached to the P atom, usually three by single bonds and one by a coordinate covalent bond (commonly shown as a double bond, though this is not truly accurate).

The large majority of OP insecticides may be regarded as derivatives of phosphoric acid. The partial structures of the 12 types of OP insecticides known, along with two additional classes of OP compounds, are shown in Fig. 1, with structures of representative OP insecticides in Fig. 2. Though no phosphinate-type OP compounds are in use, several interesting compounds have been described.

The *true* phosphates, triesters of phosphoric acid, may be regarded as the prototypes of the entire family of OP compounds. All four atoms surrounding the phosphorus are oxygen. These often highly reactive materials are particularly useful where short residual activity is desirable, such as on vegetables near the time of harvest and on or near dairy cattle.

Far more numerous are the sulfur-containing OP compounds, especially those with a P=S moiety. The phosphorothionates include such important insecticides as parathion and its methyl homolog (methyl parathion), diazinon, chlorpyrifos, and many others. Phosphorothiolates are usually considerably more toxic and are used largely as soil insecticides or soil-applied plant systemics. Demeton (II-isomer), oxydemetonmethyl, and omethoate are representative examples.

Another large subclass is that of the phosphorthionothiolates, to which most phosphorodithioates belong. In the subclass, one S atom is as P=S, and the other, as a thioester. While the thioester linkage is to an alkyl substituent in a few OP compounds, such as sulprofos, more commonly it is the leaving group, which is attached via the S atom. This large and diverse group includes highly toxic materials such as phorate and terbufos, along with the relatively safe compounds malathion and dimethoate. Only two phosphorodithiolates are presently in production, ethoprop and ebufos. The latter is used outside the United States, but is not as yet registered within the United States.

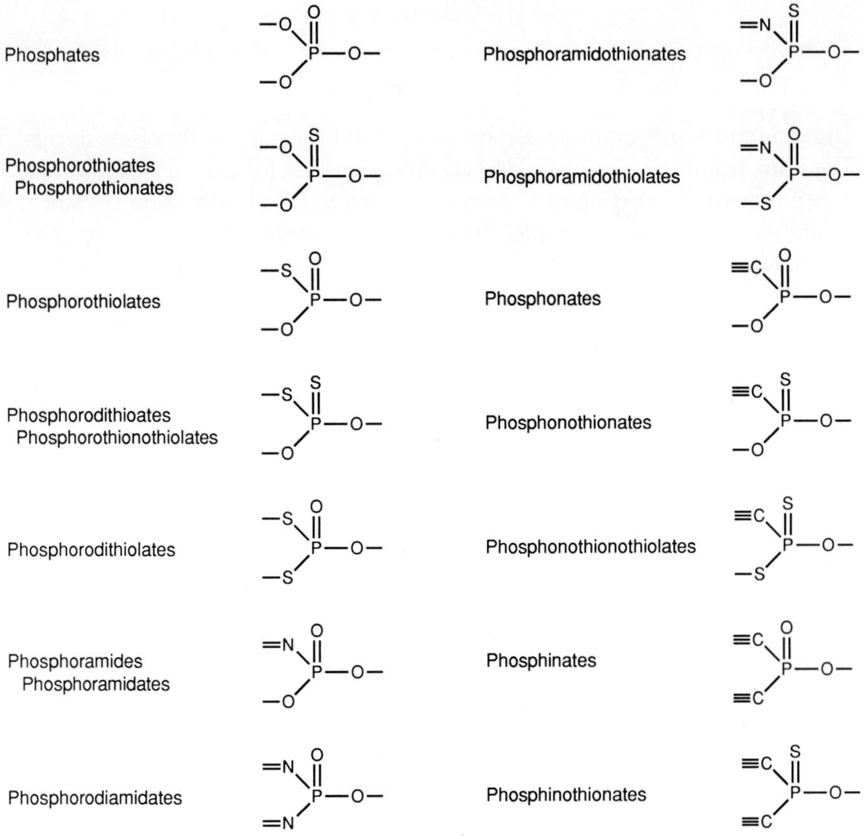

Phosphates

Phosphorothioates
Phosphorothionates

Phosphorothiolates

Phosphorodithioates
Phosphorothionothiolates

Phosphorodithiolates

Phosphoramides
Phosphoramidates

Phosphorodiamidates

Phosphoramidothionates

Phosphoramidothiolates

Phosphonates

Phosphonothionates

Phosphonothionothiolates

Phosphinates

Phosphinothionates

Figure 1 Structures of organophosphorus esters.

As with carboxylic acids, the phosphorus acids form amides as well as esters. Though the first OP insecticides described by Schrader were phosphoramides, the class has remained quite small. Of the three phosphorodiamidates once used, none is in current production. Of the seven phosphoramides in use today, three are phosphoramidates (fenamiphos, phosfolan, and mephosfolan), two are phosphoramidothionates (propetamphos and isofenphos), and two are phosphoramidothiolates (methamidophos and acephate).

Phosphonates and related OP compounds have one substituent, which is attached by a phosphorus–carbon bond. Never a large class of OP compounds in terms of number of individual insecticides, only two are currently used. These are trichlorfon, a phosphonate, and fonofos, a phosphonothionothiolate. Although the four phosphonothionates once produced have all been discontinued, EPN was a major insecticide for nearly 40 years.

No phosphinates or phosphinothionates have been registered. Though several have appeared promising in lab bioassays, results under practical application conditions were disappointing. The U.S. Army Department of Chemical Defense has recently investigated phosphinates as prophylactic agents against nerve gas poisoning because of the rapid spontaneous recovery and ease of oxime reactivation of phosphinylated cholinesterase.

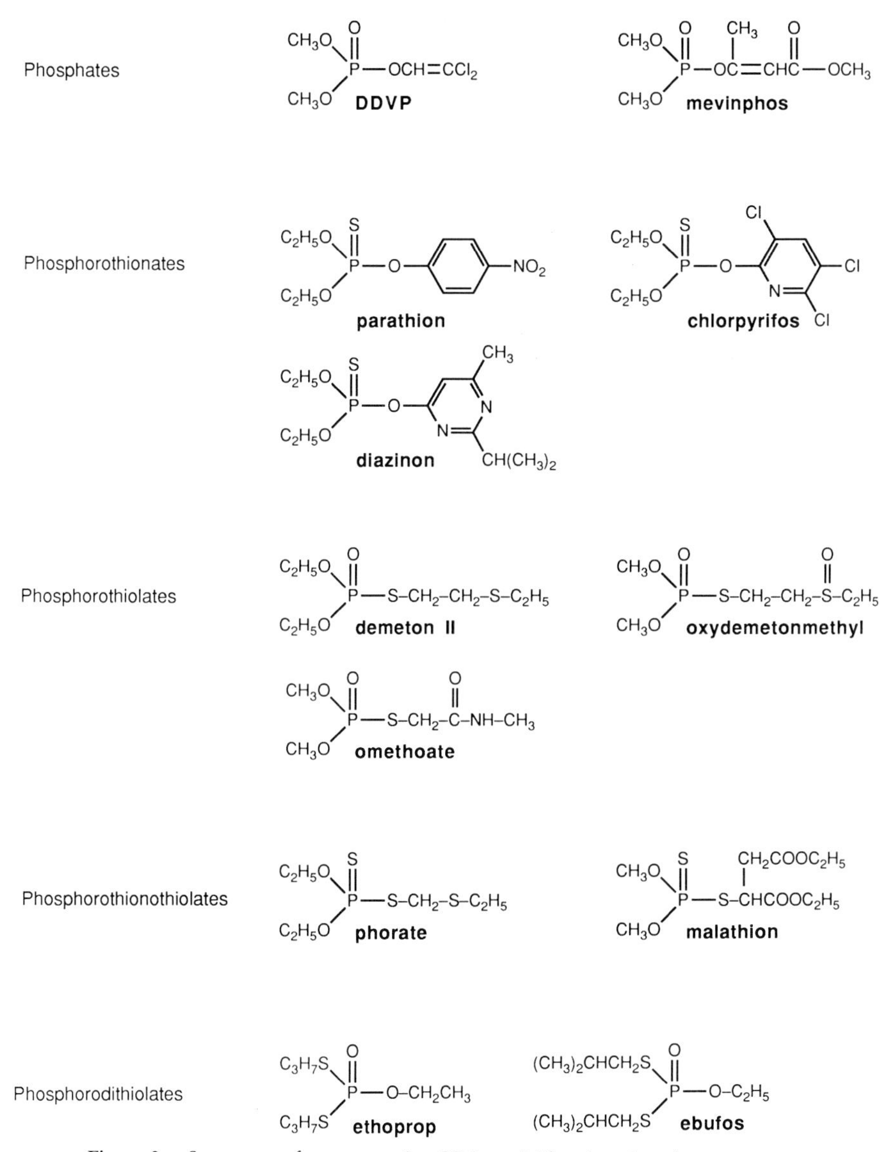

Figure 2 Structures of representative OP insecticides. *(continued on next page)*

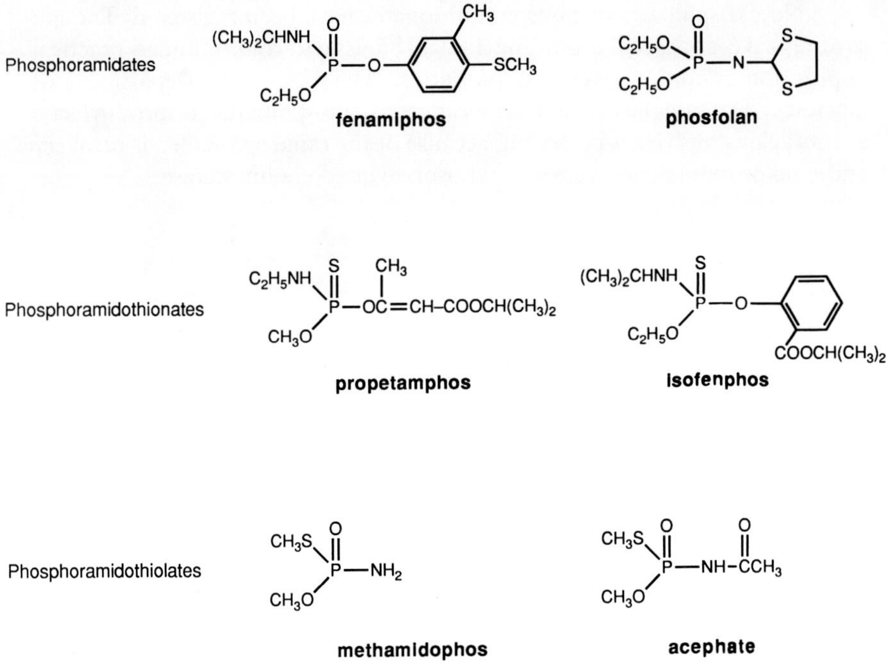

Phosphoramidates

fenamiphos phosfolan

Phosphoramidothionates

propetamphos isofenphos

Phosphoramidothiolates

methamidophos acephate

Figure 2. *continued*

B. Major Synthetic Reactions

Though the OP compounds are described above as derivatives of phosphoric acid, the acid itself is not the usual starting material for synthesis. Commercial preparations generally begin with elemental phosphorus and proceed through phosphorus pentasulfide (P_2S_5) or phosphorus trichloride (PCl_3). The former is converted into dialkyl dithiophate by reaction with the appropriate alcohol.

$$2\ P + 5\ S \rightarrow P_2S_5 \tag{1}$$

$$P_2S_5 + 4\ R\text{—}OH \rightarrow 2\ (R\text{—}O)_2P(:S)\text{—}SH \tag{2}$$

PCl_3 is used to prepare several useful intermediates as shown.

$$2\ P + 3\ Cl_2 \rightarrow 2\ PCl_3 \tag{3}$$

$$PCl_3 + 0.50_2 \rightarrow P(:0)Cl_3 \tag{4}$$

$$+\ S \rightarrow P(:S)Cl_3 \tag{5}$$

$$+\ 3\ R\text{—}OH \xrightarrow{\text{base}} (R\text{—}O)_3P \tag{6}$$

$$+\ 3\ R\text{—}OH \xrightarrow{\text{no base}} (R\text{—}O)_2P(:O)\text{—}H \tag{7}$$

$$+ R—Mg—Cl \rightarrow R—PCl_2 \quad (8)$$

Phosphates, then, are prepared from either phosphoryl chloride from reaction (4) or dialkyl phosphites from (7). The former is particularly useful when all three substituents are different, i.e., asymmetrical esters and diester-monoamides. Dialkyl phosphites are easily and cleanly converted to phosphates by reaction with alcohol or phenol with base (e.g., triethyl amine) in carbon tetrachloride. The reaction is possible because the phosphite is converted *in situ* to the phosphorochloridate by reaction with CCl_4.

$$(R—O)_2P(:O)—H + R'—OH \xrightarrow{\text{base } CCl_4} (R—O)_2P(:O)—O—R' \quad (9)$$

Phosphorothionates are generally synthesized from dialkyl phosphorochloridothionate. These are obtained by chlorination of dialkyl phosphorodithioates (usually with Cl_2), or by reaction of thiophosphoryl chloride, from (5) with two equivalents of appropriate alcohol. The third substituent is added by standard esterification.

Trialkyl phosphites, from reaction (6), have proven particularly useful in synthesis of vinyl phosphates, which would otherwise be difficult to prepare. These materials react directly with α-chloroaldeheydes and ketones. Interaction between the P and carbonyl O atoms with simultaneous dealkylation and dechlorination produces the vinyl phosphate and alkyl chloride. Synthesis of DDVP serves as an example.

$$(CH_3—O)_3P + HC(:O)CCl_3 \rightarrow (CH_3—O)_2P(:O)—O—CH{=}CCl_2 \quad (10)$$

Interestingly, the same aldehyde reacts with dimethyl phosphite by a rather different mechanism to yield the phosphonate trichlorfon.

$$(CH_3—O)_2P(:O)—H + HC(:O)CCl_3 \rightarrow (CH_3—O)_2P(:O)—CH(OH)CCl_3 \quad (11)$$

Since mercaptans do not readily react with dialkyl phosphorochloridates to form thioesters, but rather attack the alkyl ester bond, phosphorothionothiolates are generally prepared directly from the dialkyl phosphorodithioic acid [from reaction (2)], or its salts. The salts react with alkyl halides to produce the thioester. The acid adds across carbon–carbon double bonds, a reaction used in the preparation of malathion from diethyl maleate. A widely used reaction of the dithioic acid is with formaldehyde and a thiol or cyclic amine to produce insecticides containing $P—S—CH_2—X$-groupings, where X is S or N.

$$(R—O)_2P(:S)—SH + H_2C(:O) + HS—R' \rightarrow (R—O)_2P(:S)—S—CH_2—S—R' \quad (12)$$

$$+ HN{=}R' \rightarrow (R—O)_2P(:S)—S—CH_2—N{=}R' \quad (13)$$

Finally, alkyldichlorophosphines from reaction (8) may be reacted with O or S to produce phosphonodichloridates and thionates, respectively, for further processing into phosphonates, phosphonothionates, and phosphonothionothiolates. While these reactions do not encompass all processes used, the large majority of our commercial insecticides are synthesized by these or closely related reactions.

C. Chemical Reactions

While OP compounds may undergo a wide variety of chemical reactions, many involving the substituents, the subsequent discussion will consider only those involving the P atom or the atoms immediately surrounding it. Of the three reactions discussed, two are of importance also as biochemical processes and will be seen again later.

Hydrolysis is a reaction common to all esters. As with other esters, the rate of hydrolysis is a function of the nature of the acid and alcohol moieties, pH, and temperature. With most OP compounds, hydrolysis occurs at the leaving group and increases with higher pH, though acid-catalyzed hydrolysis is important in mineral acids at very low pH. Hydrolysis increases with temperature and is often enhanced by UV light. Hydrolysis is appreciably slower for phosphorothionates than for the corresponding phosphates, and hydrolysis of diesters (products of initial reaction) is usually very slow in bases.

An interesting oxidation occurs with phosphorothionates in which the sulfur of the P=S group is replaced by oxygen. This so-called desulfuration utilizes O_2 and is enhanced by UV. As will be discussed, desulfuration is also an extremely important biochemical process.

The alkyl substituents may act as alkylating agents, especially methoxy groups. Of particular interest is *self-alkylation* in which one molecule of phosphorothionate alkylates the sulfur (P=S) of another producing the *S*-alkyl phosphorothiolate. Though the product appears to be one of an intramolecular rearrangement, it is more likely that the intermolecular reaction described is the mechanism involved. Usually a very minor reaction, significant amounts of phosphorothiolate may be formed if OP compounds in storage are exposed to high temperatures for extended periods. The chemistry of some of these compounds is discussed in greater detail in Chapter 2 by Thompson, this volume.

III. Inhibition of Acetylcholinesterase

While OP compounds are well known to react or interact with a number of proteins including a variety of enzymes, there seems to be little doubt that the acute lethality may be attributed almost entirely to inhibition of AChE, an enzyme critical to the normal function of the nervous system. As discussed in Chapter 13 by Eldefrawi *et al.*, this volume, OP compounds are also known to react with cholinergic receptors, at least *in vitro*. Toxic effects not attributable to AChE inhibition are also known, however, and will be considered in a subsequent section.

A. Phosphorylation of AChE

Inhibition of AChE by OP compounds is an irreversible process best described as a transesterification, or transphosphorylation, reaction. It should be noted that the term irreversible is applied not because the enzyme inhibition itself is irreversible but because the intact OP molecule is not recovered upon recovery of the enzyme activity. During the hydrolysis of acetylcholine (ACh), the normal substrate of AChE, the serine–hydroxyl group at the catalytic center of the enzyme is acetylated, releasing the choline moiety. Subsequent hydrolysis of acetyl-AChE completes the reaction. During inhibition, the OP molecule *mimics* ACh, and phosphorylation of the serine-hydroxyl occurs. Hydrolysis of the phosphoryl-AChE, however, is exceedingly slow, and the catalytic activity of the enzyme molecule is lost. The half-life for recovery of diethyl phosphoryl-AChE, for example, is greater that 60 hr, compared to the deacetylation half-life of about 0.15 milliseconds (see chapters 4 by Wallace and 5 by Wilson (this volume) for discussions of binding and reactivation).

The potency of OP compounds as anti-AChE agents, then, depends largely on the degree to which phosphorylation occurs. This is governed, of course, by the nature of the substituents on the P atom itself. Numerous structure–activity relationship studies have contributed a great deal to our knowledge of the requirements for anti-AChE activity. Although these factors will not be covered in detail, a few basic principles will be discussed.

Since the transphosphorylation reaction relies on interaction of the P atom with an unshared pair of electrons of the oxygen of the serine-hydroxyl group, electrophilicity of the P atom is critical. The presence of electron-donating atoms or groups in the molecule greatly diminishes activity. Diesters, with one acidic (and ionic at physiological pH) group, are inactive against AChE. With few exceptions, the sulfur atom of phosphorothionates is sufficiently electron donating to yield extremely poor inhibitors. While an amido-NH_2 group either on the phosphorus or within the leaving group is acceptable, an amino-NH_2 strongly deactivates the molecule, as do hydroxyl and carboxylic acid moieties. On the other hand, halogens, nitro and cyano groups, ketones and carboxylic esters represent activating groups commonly found in OP insecticides. While a thiono-sulfur atom deactivates, replacement of the oxygen in the ester linkage to the leaving group with sulfur usually enhances activity.

Though the effects of the alkyl substituents are usually less profound, they are certainly not absent. For example, while the di-*n*-propyl and di-*n*-butyl phosphates are commonly as effective against mammalian AChE as are the methyl and ethyl homologs, they are often appreciably less effective against insect AChE. With both enzymes, branched-chain alkyl groups decrease anti-AChE potency, probably by steric effects (see Chapter 4 by Wallace, this volume, for species-related differences).

Clearly, metabolic alteration of an OP compound *in vivo* would alter the toxicity compared to that expected from *in vitro* data. Metabolism, for the most part, will be covered in a later section, but one extremely important reaction should be considered here. As noted, phosphorothionates are very poor AChE inhibitors, though they are commonly used as commercial insecticides. Their effectiveness is attributable to their rapid desulfuration to the corresponding *oxon* by the cytochrome P450-dependent monooxygenases. The enzyme-catalyzed reaction proceeds much more rapidly than the chemical reaction, and probably accounts for virtually all of the active agent involved in the poisoning following application of a phosphorothionate.

B. Signs of Poisoning

The specific symptomology following exposure to an OP compound will vary with species of animal, dosage, route of exposure, and the chemical involved. Observed differences, however, are undoubtedly the result of the complexity of the nervous system and not of any significant actions of OP compounds, other than inhibition of AChE. In the most simple terms, OP poisoning leads to the abnormal persistence and accumulation of the neurotransmitter ACh in cholinergic nerve synapses and neuromuscular junctions. The result, then, is hyperexcitability associated with these junctions, leading to multiple postsynaptic impulses generated by single presynaptic stimuli. In the somatic nervous system, i.e., that which controls voluntary skeletal muscle, hyperstimulation produces isolated muscle twitches, tremors, tonic and/or clonic convulsions, and occasionally tetanic paralysis. In the autonomic system, however, the situation is more complex. Both the sympathetic and parasympathetic branches of this system are cholinergic in their ganglionic synapses. At the neuroeffector sites, however, they often exert opposing effects on the tissue receiving input. Heart rate, for example, is increased by sympathetic stimulation and decreased by parasympathetic stimulation. Depending on the relative effects on the two branches, OP compounds may produce tachycardia, bradycardia, fibrillation, or cardiac arrest. The picture is further complicated by the fact that at very high concentrations of ACh, some cholinergic transmission is blocked. This so-called desensitization or depolarizing blockade is most notable in ganglia and at the nicotinic receptors of skeletal muscle. The summation of these multiple actions gives the clinical signs of poisoning. Listed are the observed effects of a lethal (presumably accidental) oral dose of parathion in man.

1. Nausea, vomiting, abdominal cramps, diarrhea;
2. Excessive salivation, rhinorrhea;
3. Headache, vertigo;
4. Fixed pinpoint pupils, blurred vision, ocular pain;
5. Muscle twitches, especially of face, tongue, and neck;
6. Difficulty in breathing, primarily due to excessive secretions and bronchoconstriction;
7. Random jerky movements, convulsions; and
8. Respiratory paralysis and death.

C. Reactions of Phosphorylated AChE

Though usually rather slow, phosphorylated AChE undergoes further chemical reactions. Two primary processes are of interest here, recovery and *aging*. Recovery is the hydrolytic removal of the phosphoryl moiety that regenerates the active enzyme. The rate of recovery varies considerably, depending on the nature of the alkyl substituents. Among the more common OP compounds, the dimethyl compounds recover fastest with a half-life of about 2 hr for human red blood cell AChE. As noted previously, the half-life for diethyl phosphates is greater than 2 days. Diisopropyl OP compounds yield inhibition complexes for which recovery cannot be detected. The rate of dephosphorylation can be greatly enhanced by certain chemicals that act as acceptors for a second transphosphorylation. First observed with hydroxylamine and later with hydroxamic acids, the chemically induced reactivation of phosphorylated AChE is of great interest as a tool in chemotherapy of OP poisoning. Today, the most-studied reactivators are the oximes (R—CH=N—OH), with the most active being quaternized pyridine-2-aldoxime. The best known in the United States is 2-PAM or praldoxime (1-methylpyridinium-2-aldoxime iodide), though several others are under intense investigation. An expanded discussion of oximes and aging can be found in Chapter 5 by Wilson, in this volume.

The phenomenon known as aging was first observed in studies of AChE reactivators. It was found that freshly inhibited AChE was readily reactivated, but inhibited enzyme held overnight in the refrigerator could not be reactivated fully. Further investigation revealed that some of the phosphoryl-AChE had undergone dealkylation, and it was this monoalkyl phosphoryl-AChE that was refractory to chemical reactivation. Subsequent studies have shown that methoxy groups age more rapidly than do ethoxy groups, and n-propoxy shows almost no aging. Branched-chain substituents such as isopropoxy and the highly branched 3,3-dimethyl-2-butoxy group of soman age much more rapidly. Despite considerable effort, no effective reactivator for aged AChE has been found. It is generally assumed that the ultimate recovery of aged AChE is, in fact, replacement by *de novo* enzyme synthesis rather than by dephosphorylation.

IV. Toxic Effects Not Related to AChE Inhibition

Not surprisingly, the highly reactive nature of OP compounds may lead to toxic actions other than the acute lethality attributable to AChE inhibition. Of the three best-known types of poisoning, none is fully understood.

Certain byproducts of OP synthesis, which may remain in the formulated product, have been shown to produce a delayed lethality. Among the most active are *O,O,S*-trimethyl phosphorothiolate and *O,S,S*-trimethyl phosphorodithiolate, both initially isolated from technical malathion (see Chapter 3 by Racke, this volume). Symptoms of poisoning do not resemble those of typical AChE inhibitors and are best described as a *wasting away,* with death often delayed a week or more following treatment. The specific biochemical lesion has not been found.

Several OP compounds are known teratogens. The effects are best known in chickens following injection of the OP compound into the egg, and in frogs following exposure of eggs to the OP compound in water. In chicks, teratogenesis is accompanied by a decrease in nicotinamide adenine dinucleotide (NAD) and can be reversed, at least in part, by treatment with NAD or precursors. It has been suggested that decreased kynurenin formamidase activity is involved in the process. In frog eggs and tadpoles, however, no correlation between teratogenicity and NAD levels was found. Rather decreased collagen synthesis and decreased lysyl oxidase activity appeared to correlate best with observed effects. Teratogenic effects of OP compounds are discussed further by Kitos *et al.* in Chapter 20, this volume.

The most intensively studied effect of the non-anti-AChE actions of OP compounds is organophosphate-induced delayed neuropathy (OPIDN). First described as a demyelination syndrome, OPIDN is characterized by a dying back of long myelinated nerve axons, particularly in the sciatic nerve and within the spinal cord. Appreciable onset of symptoms is delayed for 10 days to several weeks following treatment. Though histological examination reveals damage at earlier times, there is still a delay of several days. The damage to individual axons is apparently irreversible, and the *recovery* observed in mild poisonings is undoubtedly by compensation for the lost nerve fibers. The sequence of events leading to the axonal degeneration remains largely unknown. A primary target has been described but has not yet been universally accepted. The proposed target is a nerve protein termed NTE (for neurotoxic esterase or neuropathy target enzyme). Though the protein has esterase activity against phenyl valerate and certain other esters, it has been suggested that its primary function is not as an esterase. The working definition of NTE activity is that component of esterase activity inhibited by mipafox (50 μ*M*) but not by paraoxon (40 μ*M*). Results of numerous structure–activity studies may be summarized by the conclusion that OP compounds that inhibit NTE, and further undergo aging, are capable of producing OPIDN. Aging is a critical component as evidenced by findings that nonaging inhibitors (certain phosphinates, carbamates, and sulfonyl fluorides) fail to elicit neuropathy. Indeed, nonaging inhibitors may be used prophylactically to prevent OPIDN resulting from challenge with a known active agent. Clearly, much valuable research remains to be done on this intriguing phenomenon. Various aspects of OPIDN are considered in Chapters 16 by Richardson, 17 by Abou-Donin, and 18 by Veronesi and Padilla.

V. Metabolism of Organophosphorus Compounds

A. Cytochrome P450

Cytochrome P450 is the terminal oxidase of the cytochrome P450-dependent monooxygenase system (P450), formerly, and occasionally currently called mixed function oxidases (MFO's). Capable of catalyzing a variety of oxidative reactions, P450 is very important in the metabolism of OP compounds. The desulfuration of phosphorthionates has been mentioned. A seemingly associated reaction is the P450-dependent dearylation of the same group of OP compounds, a reaction that has on occasion been referred to as oxidative hydrolysis. It has been postulated that both processes share a common intermediate, a phosphooxathiiran. This three-membered ring containing P, O, and S could be formed from P═S by a reaction parallel to epoxidation of a C═C moiety. Elimination of the S would yield the oxon. The dearylation is less obvious since the final products are those expected from simple hydrolysis. A possible, but unconfirmed, explanation is that hydrolysis of the P-O-aryl bond occurs with the phosphooxathiiran, and O rather than S is eliminated from the acidic product. Interestingly, different P450 isozymes may yield different ratios of products.

Dealkylation of dimethyl and diethyl OP compounds has been described. Formation of an unstable hydroxyalkyl ester and subsequent cleavage produces the dealkylated OP and the aldehyde. The latter distinguishes the process from hydrolysis or glutathione-dependent alkyl transfer.

In addition, other oxidations may occur within the leaving group, depending on groups that may serve as substrates (i.e., sulfoxidation, ring hydroxylation, *N*-dealkylations). It is beyond the scope of this overview to consider the many reactions catalyzed by P450. Monooxygenase metabolism is discussed further in Chapters 6 by Levi and Hodgson and 13 by Eldefrawi *et al.*

B. Glutathione Transferases

Reduced glutathione is known to be an acceptor for several alkyl and aryl transfer reactions. Methyl iodide and dichloronitrobenzene are commonly utilized as model substrates for the glutathione transferase enzymes. Glutathione also adds to epoxides such as styrene oxide. Certain OP compounds have also been found to be substrates for glutathione-dependent metabolism. For alkyl transfer (dealkylation of the OP compound), dimethyl phosphates and phosphorthionates are generally the best substrates, with activity decreasing with increasing alkyl chain length. In some cases, the leaving group may be involved. Dearylation of parathion and diazinon, for example, has been reported. Though a few OP compounds such as fenitrothion have been extensively investigated, little is known about the relative importance of glutathione-dependent metabolism for most compounds. Further discussion of glutathione-dependent metabolism can be found in Chapter 7 by Sultatos, this volume.

C. Esterases

Two types of esterases have been proposed to be important in detoxication of OP compounds. The A-esterases, such as paraoxonase, actively hydrolyze some OP compounds, but their spectrum of activity appears to be somewhat limited. Phosphorothionates are not substrates. Based on *in vitro* studies, paraoxon itself is a good substrate for paraoxonase only at concentrations of 100 μM or higher. Since paraoxon inhibits AChE at sub-micromolar concentrations, there is some question as to whether paraoxonase contributes appreciably in *in vivo* poisoning. Again, much remains to be learned. See Chapter 8 for additional discussion by Kasai *et al.* of these esterases.

The B-esterases, such as aliesterases (carboxylesterases), are thought to contribute to OP detoxication by acting as alternative phosphorylation sites. Though occurring at the highest concentrations in the liver, aliesterases are found in many tissues. Commonly more sensitive to inhibition than AChE, aliesterases become readily available targets for OP compounds, and since their inhibition is of no apparent acute ill effect, they may protect the more vital AChE. Several *in vitro* and *in vivo* studies support this hypothesis. This phenomenon is considered further in Chapter 9 by Maxwell.

Finally, aliesterases have been shown to actively degrade OP compounds containing carboxylic ester moieties within the leaving group. The most notable of these is malathion, which owes most of its low mammalian toxicity to rapid hydrolysis of carboxylester groups in the molecule by liver esterases. Presence of such an ester, however, does not ensure low mammalian toxicity, as evidenced by mevinphos. In this case, presumably, mevinphos inhibits the aliesterases that otherwise detoxify the compound.

In summary, the OP compounds are a large and diverse family of chemicals with several distinct, known toxic actions and complex patterns of metabolic activation and detoxication. Despite several decades of research and hundreds of published articles, there may be more we do not know about these chemicals than the amount we know.

References

Doull, J., Klaassen, C. D., and Amdur, M. O. (ed.). (1986). "Toxicology: The Basic Science of Poisons" 3rd Ed. Macmillan, New York.

Eto, E. (1961). "Organophosphorus Pesticides: Organic and Biological Chemistry," CRC Press, Cleveland, OH.

Fest, C., and Schmidt, K.-J. (1982). "The Chemistry of Organophosphorus Pesticides" 2nd Ed. Springer-Verlag, Berlin, Heidelberg, New York.

Gilman, A. G., Goodman, L. S., and Gilman, A. (eds.). (1980). "The Pharmacological Basis of Therapeutics." Macmillan, New York.

Heath, D. F. (1961). "Organophosphorus Poisons." Pergamon Press, Oxford, London, New York, Paris.

Hodgson, E., and Levi, P. E. (eds.). "Introduction to Biochemical Toxicology" 2nd Ed. Elsevier, New York-in press.

O'Brien, R. D. (1960). "Toxic Phosphorus Esters." Academic Press, New York, London.

Farm Chemicals Handbook, Vol. 75. (1989). Meister Publishing, Willoughby, Ohio.

2

Preparation, Analysis, and Toxicity of Phosphorothiolates

Charles M. Thompson
Department of Chemistry
Loyola University of Chicago
Chicago, Illinois

I. Introduction

A high percentage of organophosphorus (OP) insecticides contain the sulfur atom in an array of atomic arrangements, oxidation states, and composition. These varied structural arrangements result in marked differences in chemical reactivity, metabolism, and biological activity. The aim of this chapter is to review a particular class of sulfur-containing insecticides: the phosphorothiolates. The varied atomic arrangements of sulfur, however, warrant a brief clarification of structure so as to enable us to specifically identify and separate the class of phosphorothiolates from other OP compounds and thus permit a

Organophosphates: Chemistry, Fate, and Effects

more focused coverage. The more common phosphorothionate and phosphate structures are discussed in Chapter 1 by Chambers, this volume.

A. Nomenclature and Structure

To remove any ambiguity in this particular article, representative structures are provided (Fig. 1) for phosphorothionate **1** (P═S), phosphorothiolate **2** (P—S), and phosphorothiolothionate **3** (P(S)—S:phosphorodithioate) esters. Relatively speaking, the most common sulfur-containing OP insecticides are those that possess the phosphorothionate **1** arrangement, as found in parathion, fenitrothion, diazinon, and related materials. Equally as important as the thionates, but with higher insect:mammal toxicity ratio, are the phosphorothiolothionates **3** or *dithioates* (e.g., malathion, guthion, and dimethoate), which possess the thionate (P═S) and thiolate (P—S) sulfur linkages. Worth mentioning as a subclass of dithioates are the phosphorodithiolates [RO-P(O)$(SR)_2$], although these materials have been commercially less exploited. Finally, a less frequently used group of sulfur-containing OP insecticides and the focus of this report is the class of compounds known as phosphorothiolates **2** (acephate, demeton-S, iprobenphos). This report will restrict coverage to phosphorothiolates **2**, where the substituents X and Y are alkoxy, aryloxy, and amino groups (i.e., phosphoramidothiolates), and R groups are defined as alkyl or aryl. Compounds containing the phosphorus-carbon bond (phosphono- and phosphinothiolates: no intervening heteroatom) are not included in this review.

B. Chemical Character and Reactivity

The prominent atomic features that affect the reactivity of phosphorothiolates are the phosphorus-sulfur sigma bond and a P═O linkage, the latter of which imparts a high electrophilicity at phosphorus; thus, the thiolates would be expected to be more hydrolytically unstable than the corresponding dithioates or the phosphorothionate compounds. However, the specific contribution of a sigma-bonded sulfur moiety to hydrolysis at phosphorus also deserves attention. For example, bond cleavage of phosphorothiolates is illustrated in the comparative hydrolysis of paraoxon **4** and isoparathion **5** (the *S*-ethyl isomer),

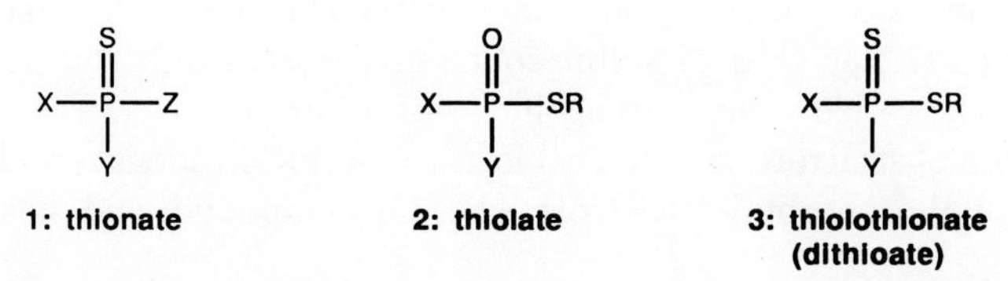

Figure 1 Representative structures for phosphorothionate (P═S), phosphorothiolate (P—S), and phosphorothiolothionate [P(S)-S] esters.

where the latter material is hydrolyzed at the *p*-nitrophenoxy group 470 times faster (Scheme 1) (Heath, 1961). This enhanced hydrolytic susceptibility may be explained, in part, by decreased pπ–dπ overlap between the sulfur and phosphorus atom (as compared to the oxygen analog) and increased polarizability by the sulfur atom. These factors contribute significantly to charge

SCHEME 1

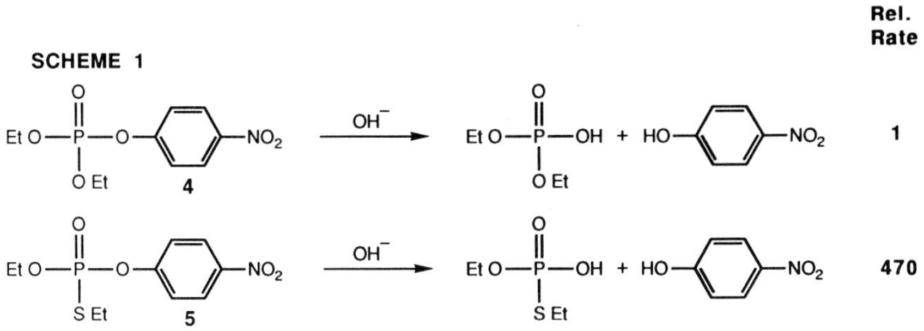

			Rel. Rate
$EtO-P(=O)(OEt)-O-C_6H_4-NO_2$ (4)	$\xrightarrow{OH^-}$	$EtO-P(=O)(OEt)-OH + HO-C_6H_4-NO_2$	1
$EtO-P(=O)(SEt)-O-C_6H_4-NO_2$ (5)	$\xrightarrow{OH^-}$	$EtO-P(=O)(SEt)-OH + HO-C_6H_4-NO_2$	470

delocalization during attack of OH^- thus lowering the transition state energy (E_a) and enhancing reaction rate. Under slightly acidic conditions, however, thiolates **2** and thiolothionates **3** have been reported to have very similar hydrolytic half-lives (Ruzicka *et al.*, 1967). The role of the thiolate linkage as a reactive (leaving) group during hydrolysis presents an additional factor. When the hydrolysis of methamidophos **6** was examined, a methoxy ligand was preferentially cleaved over a thiomethyl under certain aqueous basic conditions (Fahmy *et al.*, 1972), yet the biochemical mechanism differed (Scheme 2) (Thompson and Fukuto, 1982). Where specific hydrolysis of the

SCHEME 2

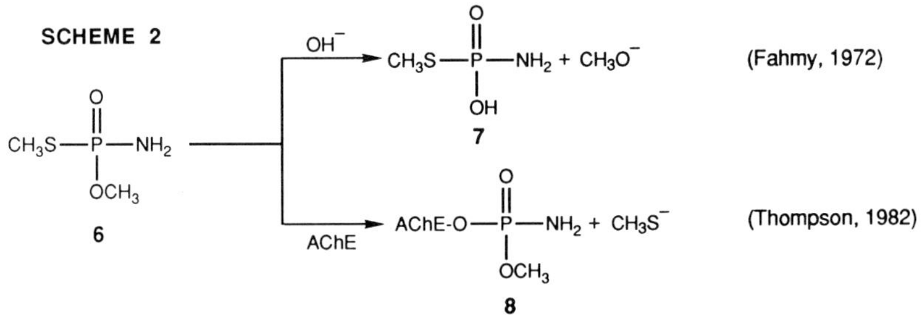

$CH_3S-P(=O)(OCH_3)-NH_2$ (**6**)

$\xrightarrow{OH^-}$ $CH_3S-P(=O)(OH)-NH_2 + CH_3O^-$ (**7**) (Fahmy, 1972)

$\xrightarrow{AChE}$ $AChE\text{-}O-P(=O)(OCH_3)-NH_2 + CH_3S^-$ (**8**) (Thompson, 1982)

P-S-alkyl linkage is expected, the sulfur-phosphorus bond is usually cleaved owing to the greater carbon-sulfur bond strength. Consistent with this reactivity, the phosphorothiolate compounds are less likely to be alkylating agents than are the corresponding phosphates. Overall, the unique reactivity bestowed on an OP compound by the thiolate linkage should be considered seriously in any study of mechanisms.

C. Convergence of Organophosphorus Insecticide Structural Type

Phosphorothiolates are the isomerization contaminants of thionate insecticides and are also the bioactive form of the corresponding dithioates (via oxidative desulfuration) (Scheme 3). As a result, all three compound classes may be considered united through a variety of pathways, for one reason or another, by a P(O)-S linkage. Furthermore, this convergence of structure type provides the impetus for attempting to better understand this multiderived compound class.

SCHEME 3

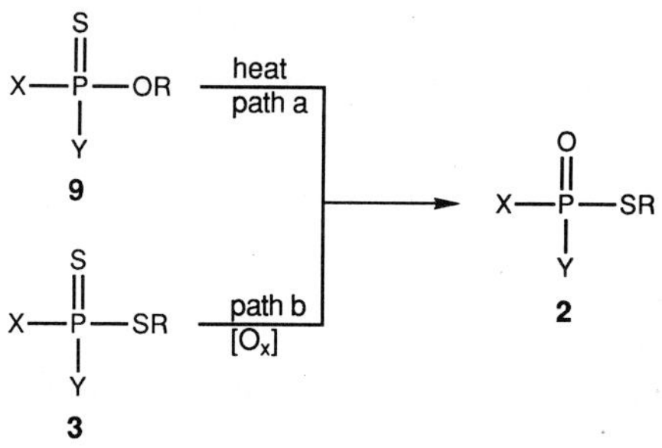

$$X{-}P({=}S)(Y){-}OR \;(\mathbf{9}) \xrightarrow[\text{path a}]{\text{heat}} X{-}P({=}O)(Y){-}SR \;(\mathbf{2}) \xleftarrow[{[O_x]}]{\text{path b}} X{-}P({=}S)(Y){-}SR \;(\mathbf{3})$$

The formation of phosphorothiolates from other OP compounds deserves closer examination. The extent to which this transformation occurs depends greatly on the OP compound and the experimental or environmental conditions under consideration. Oxidative desulfuration (Scheme 3: path b) of thiolothionates **3** to the corresponding oxon (P═O) is observed *in vivo* and *in vitro*. Malathion [**3**:X═Y═OCH_3, R═$CH(CH_2CO_2Et)CH_2CO_2Et$], for example, is converted to malaoxon (a phosphorothiolate) *in vivo* by insects and *in vitro* by crude mammalian liver homogenates, or oxidized chemically in organic solvents by peracids (*meta*-chloroperoxybenzoic acid; *m*-CPBA) and related oxidants. The P═S to P═O reaction is quite facile, and concomitant oxidation of the thioalkyl moiety is generally not observed until forcing conditions are imposed. Thioalkyl oxidation (P—S) before the thionate linkage is generally not observed. When a sulfur atom is remotely positioned from the phosphorus (e.g., phorate; a thioether), such oxidations do, however, proceed readily.

Whereas the thiolothionate-to-thiolate conversion is primarily a metabolic path, the thionate-to-thiolate alteration (Scheme 3; path a) is manifested primarily in the manufacture of the thionate (Metcalf and March, 1953). Continuing studies on the composition of certain commercial thionate OP formulations have revealed the presence of thiolate impurities in significant quantities (Rengasamy and Parmer, 1988, and references therein). Such *isom-*

erizations are probably induced thermally or by halide-promoted dealkylation–realkylation processes. Phosphorothiolates are produced also when the corresponding thionates are exposed to ultraviolet radiation (Chukwadebe *et al.*,1989). Although only small quantities of mixed thiolates were produced in this latter study, it is certainly prudent to consider their contribution to the overall toxicity profile.

This chapter will accentuate the importance of phosphorothiolate compounds not only as insecticides but also as impurities in related insecticides and as metabolic intermediates. Specifically, studies involving the preparation, analysis, and toxicological profile of phosphorothiolates are summarized.

II. Preparation of Phosphorothiolates

A. Synthesis of Phosphorothiolates—General Considerations

This section of the chapter deals only with selected synthetic preparations of phosphorothiolates. The reactions outlined were chosen to represent a variety of approaches for investigators requiring a high degree of purity and yield as well as flexibility in isotope incorporation. The starting materials (Kosolapoff, 1950) and reagents for most of these transformations are available either commercially or through standard preparations. For further details and scope of the reactions discussed herein, more comprehensive texts are available (Fest and Schmidt, 1973; Eto, 1974).

During synthesis of phosphorothiolates, it is essential that proper techniques of manipulation, handling, and storage of toxic OP compounds be used. In addition to substantial toxicity, many phosphorothiolates have a noticeably high vapor pressure and therefore should be refrigerated when not in use. When required, phosphorothiolates may be destroyed by reaction with 2*N* NaOH overnight (hood). Bleach will destroy any remaining material or odors.

1. Isomerization of Phosphorothionates/Alkylation of Phosphorothioic Acid and Its Derivatives

Phosphorothionates may be isomerized to the corresponding thiolates by a variety of conditions and/or reagents (Scheme 4). This general conversion is important not only from a synthetic standpoint but also because it is the primary cause of contamination of certain OP thionates. The thermal isomerization of phosphorothionates, Eq. (1), may be interpreted as an alkylation of one thionate by another, leading to an ionic pair that rapidly reorganizes to afford thiolate. While thermally induced transformations become more sluggish as the alkyl group size increases, alkyl halides, Eq. (2), aprotic polar solvents and/or amino groups (Burn and Cadogen, 1961), and Lewis acids

SCHEME 4

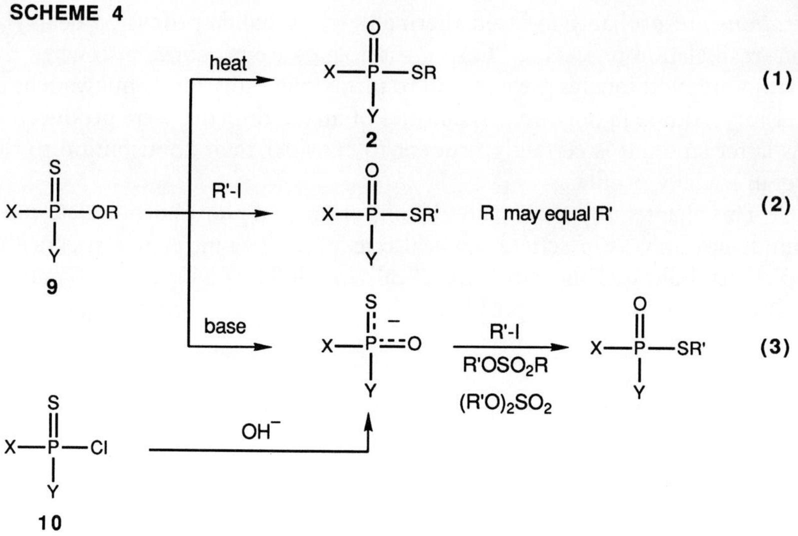

X—P(=S)(Y)—OR (9) —heat→ X—P(=O)(Y)—SR (2) (1)

X—P(=S)(Y)—OR (9) —R'-I→ X—P(=O)(Y)—SR' R may equal R' (2)

X—P(=S)(Y)—OR (9) —base→ X—P(S)(Y)⋯O$^-$ —R'-I, R'OSO$_2$R, (R'O)$_2$SO$_2$→ X—P(=O)(Y)—SR' (3)

X—P(=S)(Y)—Cl (10) —OH^-→ X—P(S)(Y)⋯O$^-$

base = OH^-, R_3N, RS^-, PEX, etc.

(Hilgetag and Teichman, 1965) promote the reaction. The scope of the thermal isomerization is quite limited; aryl groups do not migrate. However, the conversion of *O*-methyl-phosphorochloridothionate (**9**: X=Y=Cl, R=Me) to its *S*-methyl isomer (**2**: X=Y=Cl, R=Me), which was used in the radiolabeled synthesis of methamidophos (Lubkowitz *et al.*,1974), is a noteworthy utilization of the isomerization. Further, parathion and methyl parathion are converted to the thiolo isomers in 85 to 90% yield at 150°C in 24 hr and 4 hr, respectively (Metcalf and March, 1953). Similarly, amiton is prepared by thermolysis of its thionate analog (Eto, 1974).

Phosphorothioate salts may be converted to thiolates by reaction with alkyl halides, sulfonates, and sulfates, Eq. (3), in a reaction sequence that is mechanistically akin to the thermal isomerization. The phosphorothioate salts are most easily prepared by dealkylation of a thionate **9** or hydrolysis of a chloridothionate **10** (Mastin *et al.*,1945; Fest and Schmidt, 1973). Representative dealkylation reagents are mercaptans (Fest and Schmidt, 1973), amines (Hilgetag and Lehmand, 1960), and xanthates (Umetsu *et al.*, 1977). Potassium ethyl xanthate (PEX) has recently been used in the dealkylation–realkylation of thionates to prepare thiolates bearing a variety of other ligands in good yield (Thompson *et al.*,1989). A recent clever variation on this theme, using O-alkylisoureas (derived from carbodiimides) as the alkylating agent to promote phosphorothiolate formation, has been reported (Nowicki *et al.*, 1986). The coupling of alcohols and thioic acids via Mitsunobu type chemistry is also a worthy procedure, especially in the construction of branched thiolate compounds (Mlotkowska and Wartalewska-Graczek, 1987).

2. Oxidation of Phosphorothiolothionates (Dithioates)

The oxidative desulfuration of phosphorothiolothionates to form phosphorothiolates represents both synthetic and metabolic routes. Dithioates **3**, available as insecticides or prepared by reaction of dithioic acids **11** with alkyl halides, Eq. (4) are easily converted to the corresponding oxon by equimolar treatment with peracids (*m*-CPBA). The stereochemical course of thionate oxidation using *m*-CPBA has been found to proceed with retention of configuration (Lee *et al.*, 1978). Given the ready availability of dithioates, oxidative paths will continue to be of importance in the preparation of thiolates.

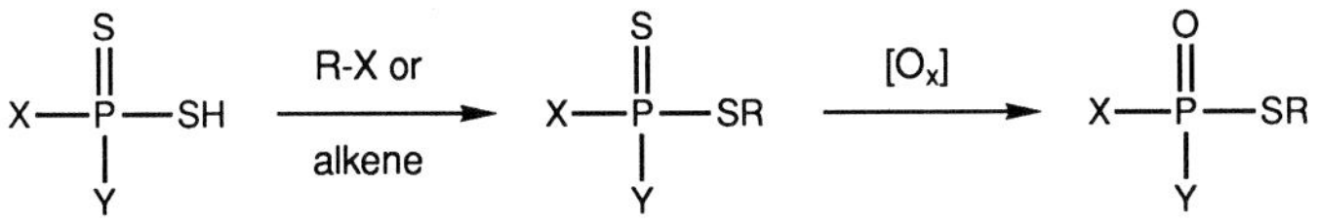

$$X{-}P(=S)(Y){-}SH \xrightarrow[\text{alkene}]{\text{R-X or}} X{-}P(=S)(Y){-}SR \xrightarrow{[O_x]} X{-}P(=O)(Y){-}SR \quad (4)$$

11 **3** **2**

$[O_x]$ = m–CPBA, N_2O_4, CF_3CO_3H

3. Reaction of Mercaptans with Phosphites

Nucleophilic phosphites **12** react by substitution at suitably substituted bivalent sulfur compounds **13** to form phosphorothiolates, Eq. (5). Sulfenyl chlo-

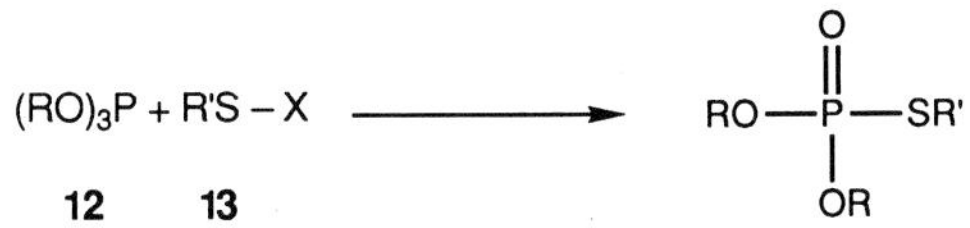

$$(RO)_3P + R'S-X \longrightarrow RO{-}P(=O)(OR){-}SR' \quad (5)$$

12 **13** **14**

X = Cl, SO_2CH_3, SR', N-maleimido, N-phthalimido

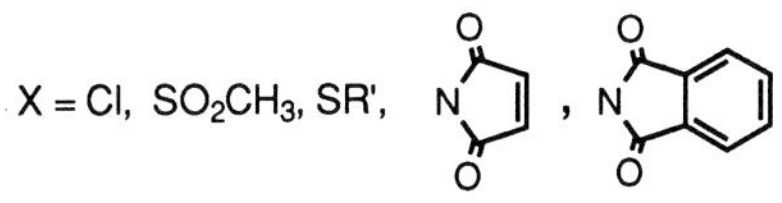

rides (Morrison, 1955), thiosulfonates (Michalski *et al.*, 1960), disulfides (Ailman, 1965) and, more recently, thioimides (Muller and Roth, 1990) all afford the thiolates upon reaction with phosphites. Inherent limitations in the commercial availability of mixed phosphites make this route less desirable as a general preparation of phosphorothiolates.

4. Reaction of Phosphoridochloridates with Mercaptans

The reaction of phosphoridochloridates **15** with mercaptans would also seem a logical choice for the preparation of dialkyl phosphorothiolates **16**, Eq. (6). However, undesired dealkylation reactions that afford **17** and **18** frequently interfere, thereby limiting this process to substrates that bear only ligands that cannot be dealkylated by RS^-. Moreover, tetraethylpyrophosphate was found to contaminate certain thiolate preparations by this method, thereby drastically altering the biological data (Gazzard *et al.*, 1974).

$$\underset{\mathbf{15}}{CH_3O{-}P(=O)(Y){-}Cl} \xrightarrow[\text{base}]{RSH} \underset{\mathbf{16}}{CH_3O{-}P(=O)(Y){-}SR} + \underset{\mathbf{17}}{{}^{-}O{-}P(=O)(Y){-}Cl} + \underset{\mathbf{18}}{RSCH_3} \qquad (6)$$

B. Synthesis of Chiral Phosphorothiolates

The isomerization of symmetrical *O,O*-dialkylphosphorothionates to *O,S*-dialkylphosphorothiolates results in the creation of a center of asymmetry at the phosphorus atom. The preparation of phosphorothiolate enantiomers will help to define the importance of stereochemical features in the mechanism of action. Thus, reliable synthetic pathways will assist in the overall toxicological assessment and, therefore, the individual contribution of each stereoisomer. Three general preparative routes (Valentine, 1984) to chiral phosphorothiolates are briefly summarized here.

1. Via Phosphorus-Containing Heterocyclic Intermediates

Chiral β-amino alcohols such as (-)ephedrine (Hall and Inch, 1979a, 1981) and serine (Thompson *et al.*, 1990) react with phosphoryl or thiophosphoryl di(tri)halides to afford cyclic covalent diastereomers, which may be separated and further manipulated to provide the chiral phosphorus compounds. A chiral dialkoxy phosphorothioic acid (Hall and Inch, 1979b) and phosphoramidothioic acid (Hall and Inch, 1979a) both prepared from (-)-ephedrine access the synthesis of several chiral thiolates through simple alkylation, Eq. (3).

2. Via Acyclic Intermediates

Chiral phosphorothiolates have been prepared via the acyclic diastereomeric *l*-proline phosphoramidates (e.g., Hirashima *et al.*, 1984; Ryu *et al.*, 1991a). This process is essentially the same as that developed for chiral phenylphosphate and phenylphosphonate materials (Koizumi *et al.*, 1978). The key transformation involves acidic alcoholysis of the phosphoramidate P-N bond to furnish the chiral target via inversion of configuration. This method places some restrictions on the alkoxy group installed to the extent that only methyl and ethyl alcohols have been used. Even with this limitation, the method remains attractive for most chiral phosphorothiolate preparations.

3. Resolution of Phosphorothioic Acid Derivatives

Traditional chemical resolution of phosphorus thioacids by chiral amine agents such as quinine, α-phenethylamine, and strychnine also has been applied to the synthesis of optically active phosphorothiolates. In fact, one of the earliest reports of chiral thiolates utilized strychnine to resolve O-methyl, O-*p*-nitrophenoxyphosphorothioic acid (Hilgetag and Lehman, 1959) in route to the antipodes of isoparathion-methyl.

C. Summary and Isotopic Labeling Considerations

Since it is usually desirable to incorporate labeled moieties at the latest stage possible in the sequence, the alkylation of phosphorothioic acid salts (isotopically enriched alkylating agents are also among the cheapest available) would be the preferred route. Alkoxy and aryloxy ligands of phosphorothiolates are most easily labeled by direct introduction of the moiety by displacement of corresponding phosphoryl halide or by oxidation of the corresponding isotopically labeled thionate.

In summary, there are several good methods for the preparation of phosphorothiolates, some of which are amenable to isotopic incorporation. To a somewhat lesser extent, methods have become successful for obtaining optically active materials, although much work remains to be done in this area.

III. Analysis of Phosphorothiolates

A. Chromatographic Techniques

1. Gas Chromatography

Although the gas chromatography (GC) of OP compounds has been reviewed (Sherma and Zweig, 1972), a brief discussion of recent advances and potential applications to phosphorothiolates is warranted. Despite relatively low molecular weights, the phosphorothiolates are fairly high boiling materials. Since phosphorothiolates tend to decompose at elevated temperatures, higher carrier gas flow rates with lower column temperature may be beneficial to analysis.

Detection of phosphorothiolates (as with OP compounds in general) is usually conducted by either alkaline flame or thermionic methods, which optimize both detection limits and selectivity. Recent advances in capillary gas chromatographic procedures have led to even lower detection levels and unique separations. Tandem GC–mass spectral analyses have further advanced residue analysis and structure identification.

The aforementioned thermal instability of phosphorothiolates has been circumvented nicely by their conversion to phosphorofluoridates after reaction with silver fluoride (Tingfa, 1986). The resulting fluoridates have greater volatility and much lower boiling points. An additional benefit of this procedure is that now thiolates may be linked chemically to the nerve gas agents, some of which contain a fluoride, for which far more chiral gas chromatographic research has been completed (Benschop and DeJong, 1988).

Surface-modified supports utilizing EGSS-X (6%) or EGSP-Z (Applied Science Labs, PA) on Chromosorb W were successful in the detection and analysis of several thionates and thiolates at 0.008 to 0.70 pmol levels (Chukwadebe *et al.*, 1989).

One of the more exciting recent advances in gas chromatographic analysis is chiral separation (Souter, 1987). For example, the Chirasil-Val capillary column has been used in the chiral separation of nerve gases (Benschop and DeJong, 1988), one of which (tabun) is chemically similar to the phosphorothiolates. In one instance, separation on a Chirasil-Val column was not observed, and lanthanide shift reagents were doped onto methyl silicone (wide-bore column) and found to resolve the enantiomeric pair. Such modifications should be useful in phosphorothiolate separations.

2. High Performance Liquid Chromatography

With the one noted limitation to GC (i.e., thermal instability), high-performance liquid chromatography (HPLC) would seem ideally suited to complement phosphorothiolate analysis, especially in those cases in which the thiolate contains a chromophore. Unfortunately, many phosphorothiolates must be monitored either at the low end of UV detection (i.e., 230–240 nm) or by the far less sensitive refractive index detection. Regardless, HPLC has been used extensively in phosphorothiolate detection and purity analysis. Both normal (organic mobile phase) and reverse phase (aqueous solvents) HPLC have been widely employed (Rengasamy and Parmer, 1988; Thompson *et al.*, 1989) for phosphorothiolate analytical and preparative separation and reaction monitoring. By comparison, phosphorothiolates are more polar than their thiolothionate counterparts, and large differences in the eluotropic properties have been recorded (Table I). A relative sense of the polarity of such similar structures can also be invoked. For example, fenchlorphos is far more nonpolar (elutes slower on reversed phase) than the other thionates. This would suggest that this particular compound, although more polar than its corresponding thionate, might still retain significant lipid solubility.

A major void in the HPLC analysis of phosphorothiolates, as in OP compounds in general is enantiomer separation. The separation of chiral carbon molecules by HPLC has become almost routine, whereas separation of

TABLE I

Comparative Organophosphate HPLC Data[a]

Compound	Retention time (min.) Methanol:water 70:30	65:35
Parathion-methyl	7.9	11.2
S-Methyl-parathion (isoparathion-methyl)	7.3	9.9
Fenitrothion	9.5	14.4
S-Methyl fenitrothion (isofenitrothion)	6.7	9.0
Fenchlorphos	27.8	53.4
S-Methyl fenchlorphos (isofenchlorphos)	10.8	15.8
Malathion	8.5	12.1
S-Methyl malathion (isomalathion)	5.9	7.4
Azinphos methyl	8.4	12.0
S-Methyl azinphos (isoazinphos)	5.8	7.1

[a] From Thompson *et al.*, (1989).

enantiomeric OP compounds has witnessed far less development (Armstrong, 1984; Souter, 1987). In preliminary experiments, we have been unsuccessful in separating certain simple racemic phosphorothiolates by either amino acid based or Pirkle-type HPLC columns. This is presumably because of a lack of complementary functional group(s) on the thiolate needed for discriminatory antipode binding.

3. Thin-Layer Chromatography

As mentioned, phosphorothiolates are generally more polar than the corresponding thionates or thiolothionates and may be easily separated from each other by silica gel thin-layer chromatography (TLC). Separation by reverse phase TLC (RPTLC; in which the thiolate migrates more quickly) is also possible, but the longer elution times may cause hydrolysis. In addition, indicating reagents stain the reverse phase binders as well as the target compound, making identification more difficult. Therefore, for routine use, silica-based TLC probably remains the more attractive at-the-bench method, although in certain instances (e.g., metabolic studies), RPTLC may be superior.

Detection of phosphorothiolates on TLC may be done by one of three methods: (1) ultraviolet lamp; (2) iodine crystals; and (3) spray or dipping reagents. Obvious limitations to UV detection frequently necessitate the use of chemical identification. Iodine stains many OP compounds, although simple phosphorothiolates are reluctant to develop. As such, it is worth highlighting those reagent stains that reliably detect thionate and thiolate OP compounds. DBQ (dibromoquininone-4-chlorimide; 5% in ethyl ether) was used in our studies of thionate-to-thiolate conversion (Thompson *et al.*,

1989), since the former stain red, and the latter stain faint yellow. Most staining reagents are designed primarily for nondiscriminate detection of OP compounds (i.e., sulfur containing and non–sulfur containing). Such reagents include palladium II chloride, anisaldehyde, phosphomolybdic acid, and potassium hexacyanoferrate. A recent report demonstrates the use of potassium iodate as a TLC reagent for sulfur-containing OP compounds at 1 μg detection levels (Patil *et al.*, 1987). Although this reagent is primarily for thionates, thiolates also stain.

B. Spectral Analyses

Several excellent texts and reviews provide essential foundation reading in the nuclear magnetic resonance (NMR) (Gorenstein, 1983) and mass spectrometry (MS) (McLafferty, 1980; Desmarchelier *et al.*, 1976) characteristics of OP compounds. For the most part, specific spectral highlights of phosphorothiolates are presented herein as a part of this larger body of information, but complemented by some specific reports. It is also the intention of this section to condense those NMR and MS features that will help to distinguish the thiolate linkage from other (related) OP compounds.

1. Nuclear Magnetic Resonance

The simplest approach to evaluating the NMR characteristics of phosphorothiolates is to organize each individual NMR nucleus in terms of general features, chemical shift, and spin–spin coupling. Because of coupling between phosphorus and other spin-active nuclei, these additional features must also be examined.

The proton NMR chemical shift of methyl group protons attached to thiolate carbons [P(O)—S—CH_3] resonate at or near 2.5 ppm (δ), which is roughly what one would expect for thioethers affected slightly by an adjacent phosphoryl moiety. Additional substituents on the thioalkyl carbon moiety may affect this chemcial shift dramatically (e.g., Me_2CH—S, 3.8 ppm; $PhCH_2$—S, 4.4 ppm; i$PrCH_2$-S, 3.3 ppm: Nowicki *et al.*, 1986), but far less change is observed by varying the phosphorus ligands. For example, the methyl group of *S*-methyl-phosphoryldichloride shows a resonance centered at 2.6 ppm, but the analogous dimethoxy and amino methoxy ligand patterns afford a thiolate resonance centered at 2.25 ppm.

Whereas carbon thiolesters [C(O)—S—R] show straightforward coupling patterns (methyl, singlet; ethyl, quartet, triplet; etc.), the phosphorothiolates have the added feature of phosphorus-proton vicinal coupling (P—S—C—H or $J_{P—S—C—H}$). Hence, a thiomethyl group produces a doublet at about 2.5 ppm with a coupling constant varying from 12 to 17 Hz depending on the phosphorus substituents appended (Thompson *et al.*, 1989; Fahmy *et al.*, 1972). Likewise, a thioethyl group frequently shows a doublet ($J_{P—S—C—H}$) of quartets ($J_{H—C—C—H}$).

In general, the carbon-13 NMR chemical shift (relative to the $CDCl_3$ triplet) of an alkane-carbon juxtaposed to a phosphorothiolate sulfur range from 12 to 14 ppm (Thompson *et al.*, 1984, 1989). As in the proton spectra, substituents directly attached to this methyl group will greatly influence the chemical shift. A second similarity to the proton spectra is the geminal (J_{P-S-C}) phosphorus coupling. Although few studies reported this coupling constant, we have found a range of 3 to 6 Hz for some representative thiolates. By contrast, carbon absorbencies from a methoxy group appear at approximately 53–55 ppm, also with a similar coupling constant (6–8 Hz).

The P-31 nucleus has many convenient properties (natural abundance, 100%, spin, ½, moderate relaxation time, etc.) that make OP analysis accessible. Until very recently, however, few reports (Greenhalgh and Schoolery, 1978) capitalized on the method. An abridged list of P-31 chemical shifts (Table II) provides an indication of the range for the phosphorothiolate moiety. In general, *S*-alkyl phosphorothiolates that bear alkoxy or aryloxy ligands resonate between 25 and 35 ppm. Slight shift differences are noted in varying the solvent in which the spectrum is recorded. *S*-Aryl phosphorothiolates appear slightly upfield (smaller value) relative to their alkyl counterparts. By contrast, the phosphorothionates appear farther downfield (approximately 65 ppm). This 40 ppm upfield shift trend was found to be general in the thionate-to-thiolate isomerization, Eq.(1) (Thompson *et al.*, 1989). Inasmuch as P-31 NMR provides a rapid and diagnostic tool for the straightforward identification of phosphorothiolates, it has also been impressively utilized in analyzing oxidative bioactivation intermediates (Bielawski and Casida, 1988).

2. Mass Spectrometry

The mass spectral characteristics of phosphorothiolates have not been specifically investigated, but some examples permit limited conclusions to be made concerning their fragmentation properties. The aim of this section is to diagnose briefly the mass spectral characteristics that are associated with the P—S-alkyl linkage of phosphorothiolates.

Many of the high-intensity fragment ions observed for phosphorothiolates are the same as for OP esters in general (Desmarchelier *et al.*, 1976). However, the thiolate group influences some interesting fragmentation patterns worthy of discussion.

Compounds containing the R′O—P(O)—SR arrangement are cleaved generally at the P—S rather than P—O linkage, and whereas ethoxy groups fragment to ethylene, ethylthio groups fragment by P—S fission to provide $C_2H_5S\bullet$ and C_2H_4S. Contrary to the fact that methylthio and ethylthio groups fragment by this process, the larger alkyl groups (i.e., Pr, iPr, Bu) afford other side-chain fissions (Tashma *et al.*, 1973).

It is also instructive to compare isomerically related compounds [RO—P(S)—versus RS—P(O)—] to further distinguish the mass spectral properties

TABLE II

Phosphorus-31 Chemical Shift Data for Representative Phosphorothiolates (Structure 2)[a]

X	Y	R	δ(ppm) CDCl$_3$	δ(ppm) d$_6$-acetone
MeO	MeO	Me	32.1[b]	36.6[h]
MeO	3-Me-4-NO$_2$-phenoxy	Me	27.0[b]/27.6[d]	—
EtO	4-SMe-phenoxy	n-Pr	25.9[c]	24.7[c]
EtO	4-S(O)Me-phenoxy	n-Pr	26.0[c]	24.8[c]
EtO	4-SO$_2$Me-phenoxy	n-Pr	26.1[c]	24.8[c]
EtO	2-Cl,4-Br-phenoxy	n-Pr	26.4[c]	31.2[h]
MeO	2-Cl,4-Br-phenoxy	n-Pr	28.2[c]	—
EtO	Cl$_5$-phenoxy	n-Pr	27.2[c]	—
MeO	Cl$_5$-phenoxy	n-Pr	29.0[c]	—
MeO	4-NO$_2$-phenoxy	Me	27.8[d]	—
MeO	2,4,5-Cl-phenoxy	Me	28.1[d]	—
MeO	MeO	Ph	26.4[e]	—
MeO	MeO	CH$_2$Ph	30.7[e]	—
MeO	MeO	n-C$_{18}$H$_{37}$	31.9[e]	—
EtO	EtO	Me	26.0[f]	29.7[g]
EtO	EtO	Et	27.9[j]	
EtO	EtO	n-Pr	27.3[j]	
EtO	EtO	i-Pr	26.1[f]/27.1[j]	32.8[g]
EtO	E tO	CH$_2$Ph	25.5[f]/27.9[j]	—
EtO	EtO	sec-Bu	26.8[f]/26.8[j]	—
EtO	EtO	CH$_2$ C(Cl)=CH$_2$	—	31.4[h]
EtO	EtO	(*cis*)CH$_2$C(Cl)=CHCl	—	30.6[h]
EtO	EtO	(*trans*)CH$_2$C(Cl)=CHCl	—	29.8[h]
MeO	NH$_2$	Me	—	40.4[h]
MeO	NHC(O)Me	Me	—	32.4[h]
MeO	NH$_2$	CH$_2$C=CH$_2$	—	37.8[h]
EtO	EtO	SPh	23. 3[i]	21.9[i]
EtO	EtO	S(O)Ph	14.2[i]	13.3[i]
EtO	EtO	SO$_2$Ph	9.3[i]	8.4[i]
EtO	EtO	H	64.3[i]	66.4[i]
EtO	PhO	H	59.6[i]	61.7[i]
EtO	EtO	OMe	24.7[i]	23.6[i]

[a] Chemical shifts made relative to H$_3$PO$_4$.
[b] Greenhalgh and Schoolery (1978); conducted in d$_6$-benzene.
[c] Hirashima *et al.* (1984).
[d] Thompson *et al.* (1989).
[e] Muller and Roth (1990).
[f] Nowicki *et al.* (1986).
[g] Segall and Casida (1983).
[h] Segall and Casida (1982).
[i] Bielawski and Casida (1988).
[j] Mlotkowska and Wartlalowska-Graczek (1987); conducted in CCl$_4$.

of phosphorothiolates. Methamidophos **6** and O,O-dimethyl-phosphoramidothionate **19** differ in their base peaks (major fragmentation route); m/e = 94 (M—SCH_2) and m/e = 78 (M—OCH_2 then —S), respectively (Scheme 5). Similarly, O,O,*S*-trimethylphosphorothiolate **20** shows a base peak at m/e = 110 (M—SCH_2) and O,O,O-trimethylphosphorothionate **21** shows a base peak at m/e = 93 (M—OCH_2 then —S). In both instances, the phosphorothiolate isomer ruptures predominantly at the P—S bond, whereas the thionate is cleaved at the P—O linkage.

SCHEME 5: Predominant MS Bond Scission (dotted line)

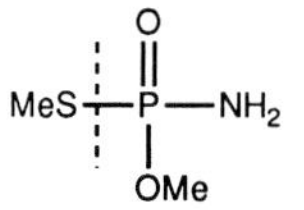

6: methamidophos 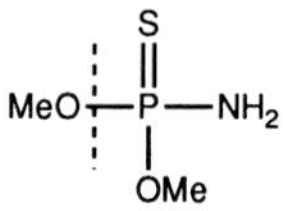 **19**

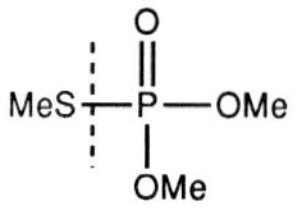

20 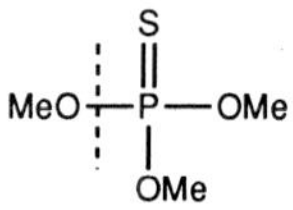 **21**

In a study designed specifically to evaluate differences in thioalkyl versus oxaalkyl ligand scission, methamidophos was independently C-13 labeled at both methyl groups. Results from this study indicated that the neutral methyl radical was lost only from the *S*-methyl moiety, and the methyl cation was predominantly formed from the O-methyl group (Thompson *et al.*, 1983) (Scheme 6). These pathways were attributed to the difference in bond energies between P—O and P—S bond (i.e., inductive effect) although more work needs to be done to generalize this observation for phosphorothiolates.

SCHEME 6: Preferred Fragmentation of Methyl Groups in Methamidophos

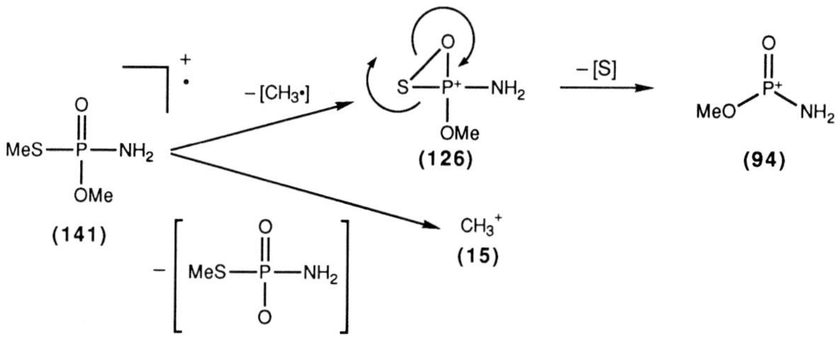

(141) **(126)** **(94)** **(15)**

IV. Toxicity of Phosphorothiolates

Phosphorothiolates, like OP insecticides in general, have been implicated in a wide variety of toxicological manifestations, many of which result from their phosphorylating ability. Within this particular mode of action, phosphorothiolates act primarily as anticholinesterases capable of eliciting the customary symptoms of OP poisoning as described in Chapter 1, by Chambers (Ecobichon and Joy, 1982). The extent to which these compounds inhibit cholinesterases and their ability to do so vary greatly, characteristics which, in many cases, may be attributable to the thiolate linkage. Phosphorothiolates have been held responsible also as inhibitors of neurotoxic esterase (NTE) (see Chapters 1 and 16) (Clothier and Johnson, 1980), potentiators of toxicity of other OP compounds (Umetsu *et al.*, 1977), delayed toxins (cf. Aldridge *et al.*, 1985), and other biochemical lesions. The goal of this section is to provide a summary of the multifarious toxicological properties of phosphorothiolates and, in certain instances, elaborate upon relevant biochemical mechanisms.

A. Phosphorothiolates as Anticholinesterases

Doubtless, the major biochemical pathway of phosphorothiolate intoxication is the phosphorylation of acetylcholinesterase (AChE) and, in this regard, these compounds may be grouped with OP insecticides in general. The *in vitro* anticholinesterase potency is conveniently determined by kinetic analysis of an appropriate AChE enzyme source, with most studies reporting either an I_{50} (molar concentration for 50% inhibition) or k_i (bimolecular inhibition constant) value. These two determinants are linked through the expression $I_{50} = 0.694/k_i t$, where t equals the time of inhibition. This expression has served to establish continuity in tabulating some relative anticholinesterase values for a variety of phosphorothiolates (Table III, Structure 2).

1. Inhibitory Potential of Phosphorothiolates

For the most part, trialkyl phosphorothiolates are only modest inhibitors of cholinesterases *in vitro*, with potency comparable to that of thionates (i.e., $k_i \sim 10^2$). In contrast, phosphorothiolates that bear a phenoxy-type leaving group appended to phosphorus are good-to-excellent inhibitors of AChE, some comparable to paraoxon in phosphorylating capability ($k_i \sim 10^7$). A most interesting feature of phosphorothiolates is by comparison to the oxo-analogs (i.e., phosphates), *S*-alkyl(aryl) OP compounds are 10–50,000 times better anticholinesterases by virtue of the thiolo effect (Eto, 1974). Thus, the contribution of a sulfur atom may exert a powerful influence on the inhibitory potency.

Few systematic studies have been conducted to determine the contribution of the thioalkyl (or thioaryl) moiety to the phosphorylating proficiency (Bracha and O'Brien, 1968a,b; Ali and Fukuto, 1982; Sanborn and Fukuto,

1972; Eto, 1974). For trialkyl phosphorothiolates, there is a slight blossoming of anti-AChE action as the thioalkyl moiety increases in size to a six-carbon chain (Bracha and O'Brien, 1968b). However, this trend was not so apparent when a 2,4-dichlorophenoxy group was held constant with k_i ranging from 23 to 320 $M^{-1}min^{-1}$ against fly head AChE (Kimura *et al.*, 1984). A similar minor steric/electronic influence was observed in the *S*-alkyl (Quistad *et al.*, 1970) and *S*-aryl (Sanborn and Fukuto, 1972) phosphoramidothiolate series with the trend certainly present in the former. Thus, when the thiolate ligand serves as the leaving group, a slight enhancement in anticholinesterase activity is expected as the chain length increases. However, when another leaving group is present, other factors appear to dominate the inhibitory potency.

In some cases, the phosphorothiolate sulfur ligand may contain a chiral center, which introduces the question of stereochemical preference in the phosphorylation of AChE (Jarv, 1984). In two separate studies examining the influence of a carbon stereocenter juxtaposed to the thiolate sulfur, the *d*-isomers of malaoxon and papoxon were found to be three to four times the more potent anticholinesterase (Hassan and Dauterman, 1968; Ohkawa *et al.*, 1976). Stereochemistry at the phosphorus of phosphorothiolates has also been shown to affect the anticholinesterase potency in the case of sulprophos and related compounds (Hirashima *et al.*, 1984: Hilgetag and Lehman, 1959). The (−)-isomer of chiral isoparathion methyl was shown to be a 4- to 15-fold better anticholinesterase than the (+)-isomer when tested against four different sources of the enzyme (Ryu *et al.*, 1991b).

As mentioned in Section I,C, phosphorothiolates are the isomerization contaminants of thionate-based insecticides, and it is therefore worthwhile to determine the relative anti-AChE potency of these impurities. The isomeride contaminants, which now bear both a good leaving group and thiolate moiety, are approximately 1000 times more potent anticholinesterases of rat brain AChE than are their thionate counterparts (Thompson *et al.*, 1989) (Table III).

2. Mechanism of Action of Representative Phosphorothiolates

With such a wide range of anticholinesterase potency, it is probable that phosphorothiolates do not share a unitary mechanism of action. In addition, Table III reveals that, in many cases, there is no clear choice of a leaving group. Therefore, it is advisable at this juncture to examine possible atomic features that contribute to the mechanism of action. Cholinesterases interact with OP compounds by reaction at an activated serine hydroxyl, which results in the expulsion of one of three possible ligands from phosphorus. When the phosphorus atom bears a phenoxy-type moiety (e.g., parathion or fenitrothion) this ligand is generally the one removed regardless of the other ligands present (Eto, 1974). To an extent, this is also true of phosphorothiolates (e.g., malaoxon, papoxon, isoparathion-methyl) although few studies have unambiguously established this fact. Certain phosphorothiolates, however, do not subscribe to such a simple case, and in an extremely adroit investigation, a

TABLE III

Bimolecular Inhibition Constants for Representative Phosphorothiolates (Structure 2)

X	Y	R	k_i ($M^{-1}min^{-1}$) BAChE[m]	RAChE	MAChE	FHAChE	Reference
MeO	MeO	Me	1–30	< 10 (p)			*a,d*
MeO	MeO	Et, n-Pr, n-Bu	< 1	< 10 (p)			*a*
MeO	MeO	i-Pr		24–40 (p,b)			*g*
EtO	EtO	Me	100	65 (p)			*a*
EtO	EtO	Et	46	370 (p)			*d,f*
EtO	EtO	n-Pr	90	78 (p)	1.5×10^3		*a,b*
EtO	EtO	i-Pr	140	100–250 (p,b)			*a,g*
EtO	EtO	i-Bu, n-C_5H_{11}	1.5×10^3	14 (p)	8.2×10^3		*a,b*
EtO	EtO	n-C_6H_{13}			3.3×10^4		*b*
EtO	EtO	$CH(Et)_2$			1.1×10^4		*c*
EtO	EtO	$(CH_2)_2$ $CH(Et)_2$			2.3×10^5		*c*
EtO	EtO	CH_2CH_2 $N(Et)_2$			1.4×10^6		*c*
MeO	4-NO_2PhO	Me		7.1×10^4 (b)			*e*
MeO	3-Me,4-NO_2PhO	Me		7.0×10^4 (b)			*e*
MeO	2,4,5-Cl_3-PhO	Me		8.0×10^3 (b)			*e*
EtO	EtO	d-$CH(CH_2CO_2Et)CO_2Et$	2.8×10^4				*h*
EtO	EtO	l-$CH(CH_2CO_2Et)CO_2Et$	6.3×10^3				*h*
EtO	EtO	dl-$CH(CH_2CO_2Et)CO_2Et$	1.5×10^4				*h*
MeO	MeO	d-$CH(Ph)CO_2Et$	4.1×10^6			4.1×10^7	*i*
MeO	MeO	l-$CH(Ph)CO_2Et$	1.4×10^6			7.4×10^7	*i*
MeO	MeO	dl-$CH(Ph)CO_2Et$	2.3×10^6			5.3×10^7	*i*

EtO	2,4-Cl_2-PhO	Me				320	*j*
EtO	2,4-Cl_2-PhO	Et				44	*j*
EtO	2,4-Cl_2-PhO	n-Pr				130	*j*
EtO	2,4-Cl_2-PhO	n-Bu				< 23	*j*
MeO	NH_2	Me		1.8×10^3	1×10^3 (b,p)	9.2×10^2	*k,m*
EtO	Nh_2	E t				1.5×10^3	*k*
n-PrO	NH_2	n-Pr				5.0×10^3	*k*
MeO	NHMe	Me				< 35	*k*
EtO	NHMe	Et				2.1×10^2	*k*
EtO	NMe_2	Et				5.0×10^3	*k*
NH_2	NH_2	Me				5.6×10^2	*k*
EtO	NH_2	ph				2.1×10^5	*l*
EtO	NH_2	4-Cl-Ph				4.3×10^5	*l*
EtO	NH_2	4-Br-Ph				5.6×10^5	*l*
EtO	NH_2	4-Et-Ph				5.9×10^5	*l*
EtO	NHMe	4-Cl-Ph				2.6×10^4	*l*
EtO	NMe_2	4-Cl-Ph				9.2×10^3	*l*

[a] Ali and Fukuto (1982).
[b] Bracha and O'Brien (1968b).
[c] Bracha and O'Brien (1968a).
[d] Clothier *et al.* (1981).
[e] Thompson *el al.* (1989).
[f] Aldridge *et al.* (1985).
[g] Ali and Fukuto (1983).
[h] Hassan and Dauterman (1968).
[i] Ohkawa *et al.* (1976).
[j] Kimura *et al.* (1984).
[k] Quistad *et al.* (1970).
[l] Sanborn and Fukuto (1972).
[m] BAChE, bovine erythrocyte; RAChE, Rat plasma (p) or brain (b); MACHE, mouse brain; FHAChE, fly head.

change of leaving group from the expected O-aryl to an *S*-propyl was found to occur (Wing *et al.*, 1984) (Scheme 7). Presumably, the thiolate moiety undergoes metabolic activation (oxidation; *vide infra*) thus converting the sulfur-containing ligand into a more superior leaving group than phenoxy. This unique feature of phosphorothiolates renders the prediction of phosphorylation mechanism more difficult.

SCHEME 7: Metabolic Activation Leading to Leaving Group Reversal (Wing, 1984)

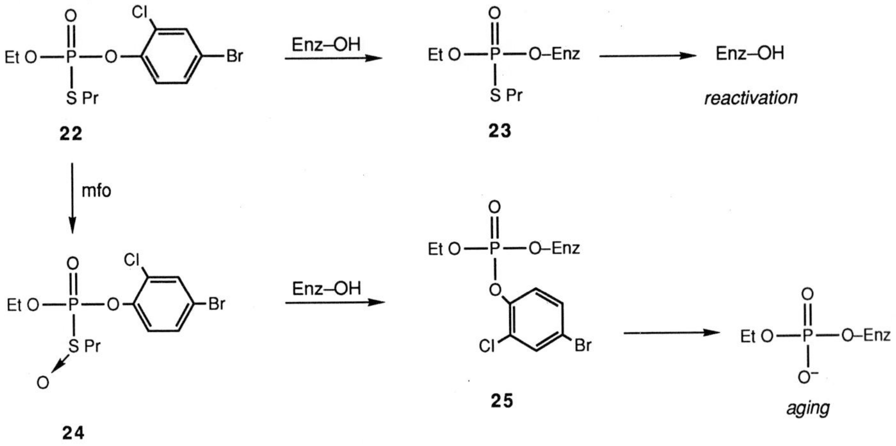

Et O—P(=O)(S Pr)—O—C₆H₃(Cl)Br **22** —Enz–OH→ Et O—P(=O)(S Pr)—O–Enz **23** ——→ Enz–OH *reactivation*

22 —mfo→ **24** (S-oxide) —Enz–OH→ Et O—P(=O)(O-aryl)—O–Enz **25** ——→ Et O—P(=O)(O⁻)—O–Enz *aging*

In some cases the choice of leaving group may be delimited to two as in the case of methamidophos **6**. Here, the amino group is not expected to be expelled during phosphorylation, rendering the thiomethyl and methoxy as candidates. To establish the nature of the leaving group, the alkyl ligands were independently radiolabeled with C-14, incubated with electric eel acetylcholinesterase (EEAChE) to complete inhibition, and the protein-inhibitor mixture was separated by gel filtration. It was determined that the thiomethyl group was lost, as evidenced by radioisotope incorporation onto the protein in the case of EEAChE inhibition by $^{14}CH_3O$-**6** (Thompson and Fukuto, 1982) (Scheme 2). This finding suggested that even in the absence of an oxidizing system, the thiomethyl group was the preferred leaving group over a methoxy moiety. This result was further supported by *in vivo* investigations with $^{14}CH_3S$-**6**, which captured a $^{14}CO_2$ quantity consistent with the anticholinesterase inhibition (Gray *et al.*, 1982). In a similar fashion, the thiomethyl group was the preferred leaving group *in vivo* for *O*,*O*,*S*-trimethylphosphorothiolate **20** (Gray and Fukuto, 1984) and the corresponding thioethyl in the case of *O*, *O*,*S*-triethylphosphorothiolate **32** (Clothier *et al.*, 1981).

3. Aging and Reactivation

OP inhibited AChE may undergo further chemical reactions, which result in either reactivation or aging of the enzyme, as described in Chapters 1 and 5 by Chambers and Wilson *et al.* Aging and reactivation phenomena imparted by the class of phosphorothiolates is not completely understood owing to the sheer diversity of compounds, but a few representative studies will serve to highlight important mechanisms.

O,O,S-Trimethyl and *O,O,S*-triethyl phosphorothiolate were found to reactivate (against bovine erythrocyte AChE; BAChE) at rates of $1.2 \times 10^6 min^{-1}$ and $2.4 \times 10^4 min^{-1}$ and to age at rates of $1.3 \times 10^5 min^{-1}$ and $2 \times 10^5 min^{-1}$, respectively (Clothier *et al.*, 1981). These values were found to be somewhat slower than the corresponding *O,S,S*-analogs and overall, the methyl congeners within a series reactivated 60 times faster than the ethyl derivatives. The study concluded that in light of these reactivation and aging parameters, oxime antidotes would be of little value if not administered immediately after exposure. Since trialkyl phosphorothiolates are relatively poor inhibitors of AChE ($k_i \sim 30-100\ M^{-1}min^{-1}$), it would be instructive to now contrast these observations with the more potent anti-ChE phosphorothiolates.

Profenphos **22** yields inhibited AChE, which undergoes reactivation via **23**. However, profenphos-inhibited AChE ages rapidly after incubation with mixed-function oxidase (MFO) (Wing *et al.*, 1984). It is likely that *S*-oxidation occurs, resulting in the *O*-aryl-*O*-ethyl phosphorylated enzyme **25** which, by virtue of a second good leaving group (*O*-aryl) is more prone to age (Scheme 7). Without oxidation, the *S*-propyl, *O*-ethyl phosphorylated enzyme reactivates with no readily hydrolyzable phosphorus-linked moiety present, other than the protein. This landmark study served to demonstrate the influential role of the thiolate linkage in reactivation–aging phenomena. Leaving group role reversal is limited to thioalkyl groups that are larger than ethyl as the oxidase apparently does not recognize the smaller chains (Hirashima *et al.*, 1984; Kimura *et al.*, 1984).

B. Oxidative Activation

Any molecule that bears a divalent sulfur atom is subject to oxidation (*in vitro* and *in vivo*); OP compounds are no exception. Traditionally, the thiono (P=S) to oxon (P=O) sulfoxidation has achieved the most attention since this particularly facile mechanism results in activation of an insecticide to a potent anticholinesterase, as described by Chambers in Chapter 1. Unlike thionate oxidation, thiolate oxidation is not so clearly resolved. Phosphorothiolates contain a P—S sigma bond, which when oxidized, Eq. (7), results in an α-sulfinyl (or *S*-oxide) phosphoryl **27a**, which rearranges to **27b** and is further oxidized to a sulfinate ester. Despite isolation problems of **27a**, *S*-oxidation of phosphorothiolates continues to be an attractive hypothesis to explain the anomalous toxicity behavior of these compounds (Casida, 1983; Schuphan and Casida, 1983).

$$R_1-P(=O)(R_2)-S-R_3 \xrightarrow{[O]} R_1-P(=O)(R_2)-S(\rightarrow O)-R_3 \rightleftharpoons R_1-P(=O)(R_2)-O-S-R_3 \xrightarrow{[O]} R_1-P(=O)(R_2)-O-S(\rightarrow O)-R_3 \quad (7)$$

26 **26** **26** **26**

One of the first phosphorothiolates to be suggested as oxidatively activated to the corresponding *S*-oxide was methamidophos **6**. While methamidophos shows poor anticholinesterase activity *in vitro*, animals exposed to it exhibit typical cholinergic symptoms, advancing the idea that **6** was converted *in vivo* to the potentially more potent *S*-oxide. Preincubation of **6** with *m*-CPBA resulted in a potent anticholinesterase suggested to be the *S*-oxide (Eto *et al.*, 1977). C-13 NMR studies in support (Thompson, 1984) and P-31 NMR investigations against (Bielawski and Casida, 1988) *S*-oxide formation have been reported. Recent work (Yang *et al.*, 1990; Leslie *et al.*, 1991) also suggests a sulfenate ester as a probable intermediate in the oxidation of phosphorothiolates.

Unquestionably, the most intensive research efforts dedicated to establishing the nature of *S*-oxide formation in phosphorothiolates have been conducted in Casida's laboratories (Bielawski and Casida, 1988; Wing *et al.*, 1984; Segall and Casida, 1983; Segall and Casida, 1982; Bellet and Casida, 1974). A brief summary of their findings will assist in recognizing which types of thiolate linkages are capable of undergoing the transformation. In consort with other investigations (Kimura *et al.*, 1984), it can be stated that there is a clear dependence between the toxic oxidative activation of phosphorothiolate and the *S*-alkyl chain length. Thioalkyl groups represented by propyl and butyl are activated to toxic materials, whereas methyl and ethyl are not. Apparently, the methyl and ethyl congeners are too hydrolytically unstable as *S*-oxides to effect phosphorylation. Further, the *n*-propyl group seems to be ideal for metabolic activation, with significant increases in anticholinesterase activity noted (Wing *et al.*, 1984; Kimura *et al.*, 1984).

The clever use of P-31 NMR to investigate possible *S*-oxide formation from *S*-aryl phosphorothiolates **29** was the subject of a report (Bielawski and Casida, 1988). P-31 chemical shifts were correlated with oxidation intermediates to aid the interpretation of mechanism. Results suggested the formation of a new possible contributor to oxidative activation induced toxicity, phosphoxathiiranes **30b**, Eq. (8).

$$R-P(=O)(R)-SPh \xrightarrow{[O]} \left[R-P(=O)(R)-S(\rightarrow O)-Ph \rightleftharpoons R_1-P(R)(O^-)\text{(cyclic }S^+(Ph)\text{–O)} \right] \longrightarrow \text{further oxidation} \quad (8)$$

29 **30a** **30b: phosphooxathiiran**

What was a suggestion 10 years ago has now been found to be a reality. It is likely that many phosphorothiolates are indeed capable of being *S*-oxidized to highly reactive intermediates within cells or organs. Although no α-sulfinyl phosphoryl has been isolated, the identification of specific rearrangement and/or reaction products of certain transient species firmly establishes their existence.

C. Delayed Toxicity

Doubtless, the primary biochemical lesion from exposure to OP compounds is inhibition of AChE. However, the phosphorothiolates have been examined as delayed toxicants with biochemical modes of action not attributable to AChE inhibition. This delayed toxicity may be separated into two deleterious manifestations: (1) OP-induced delayed neuropathy (OPIDN) which may be the result of inhibition of neuropathy target esterase (NTE, neurotoxic esterase) (Johnson, 1980), and (2) delayed pulmonary toxicity (DPT) (Aldridge *et al.*, 1985). In both instances administration of an acute dose results in toxic effects, which are not seen for several days although non-lethal symptoms resulting from AChE inhibition are frequently apparent. Whereas mammals generally recover from acute anti-AChE OP poisoning episodes, these two delayed maladies generally result in either paralysis (OPIDN) or death (DPT). Although these two areas of investigation have not been so thoroughly researched as AChE biochemistry, there is significant evidence gathered at this time to suspect phosphorothiolates as compounds that may impart their toxic action by more than one (anti-AChE) mechanism. It is the intent of this section to provide a concise review of those phosphorothiolates that may participate as delayed toxins. For additional discussions of OPIDN and neurotoxic esterases, the reader is referred to Chapters 16, 17, and 18.

1. Phosphorothiolates as Neurotoxic Esterase Inhibitors

Organophosphorus-induced delayed neurotoxicity may be attributable to inhibition of NTE but not inhibition of AChE. Yet, measurements of both NTE and AChE inactivation in the hen brain are needed to provide a guide for the complete evaluation of OPIDN. Several OP compounds [e.g., ethyl *p*-nitrophenylthiobenzene phosphonate (EPN)] have been found to cause delayed toxicity, and some general structure–activity relationships have been forwarded. It is clear from the many compounds screened that one mechanism assuredly occurs: the phosphorylated NTE must age. Therefore, OP compounds that bear two (potential) leaving groups would be expected to be the most probable candidates.

Methamidophos **6** and its congeners (Vilanova *et al.*, 1987) have been shown to inhibit NTE with one particular case of methamidophos poisoning being implicated in humans. It is interesting that the aging step in this case presumably occurs by loss of the methoxy moiety, since the thioalkyl group

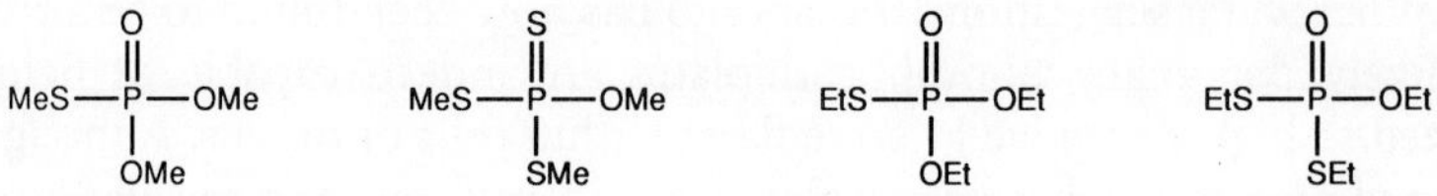

MeS—P(=O)—OMe, OMe (**20**); MeS—P(=S)—OMe, SMe (**31**); EtS—P(=O)—OEt, OEt (**32**); EtS—P(=O)—OEt, SEt (**33**)

Figure 2 Examples of phosophorothiolate compounds that exhibit delayed toxicity.

is the ligand first lost during phosphorylation (Thompson and Fukuto, 1982). To date, very few systematic studies of phosphorothiolates have been conducted for anti-NTE action, although it is reasonable to speculate that *S*-propyl and *S*-butyl phosphorothiolates bearing a phenoxy-type ligand would be suitable substrates for the bioactivation mechanism previously described (Wing *et al.*, 1984).

2. Unusual Delayed Toxicity of Trialkylphosphorothiolates

The past decade has witnessed explosive research growth into the toxicological manifestations caused by impurities in OP insecticides. For the most part, these impurities are simple trialkylphosphorothiolates or dithiolates (Fig. 2), which arise largely from the manufacture of the parent materials. These small OP molecules are generally poor to modest inhibitors of AChE (cf. Table III; Clothier *et al.*, 1981; Ali and Fukuto, 1982) that lack the obvious leaving group needed for phosphorylation. During studies designed to probe the potentiation effects of such compounds to the toxicity of phosphorothionate-type insecticides, the observation was made that animals dosed with only the trialkylphosphorothiolate were not completely recovered one day after administration and, moreover, their health steadily declined for days until they died. Subsequent investigations have tied this *delayed toxicity* to changes in the lung (Imamura *et al.*, 1983; Aldridge and Nemery, 1984). To the present time, most of the investigations have concentrated on the trimethyl **20/31** and triethyl **32/33** analogs of these phosphorothioates, since they would be the most relevant impurity types found in the commercial samples.

The biochemical mechanism of trialkylphosphorothiolate pulmonary toxicity is not fully understood but presumably results, in part, from metabolic activation of the parent compound via cytochrome P450 monooxygenases (Imamura and Hasegawa, 1984). The ultimate cause of death in phosphorothioate-treated animals is hypoxia/anoxia resulting from insufficient uptake of oxygen across an altered blood–air barrier, which resulted from gross morphological changes in the lung (Imamura and Gandy, 1988 and references therein).

V. Summary

Despite limited utility as insecticides, the class of phosphorothiolates have achieved significant attention as components in commercial insecticides and as biochemical intermediates. From both a research laboratory and a public health standpoint, these compounds must be recognized because of their unique modes of action with resultant toxic implications.

Acknowledgments

The author would like to express his appreciation for kind financial support from the National Institute of Environmental Health Sciences (ESO4434) and Loyola University of Chicago. Thanks are also due to T. Roy Fukuto (U.C. Riverside) for the many years of OP research guidance and to my group for their continued enthusiasm for OP chemistry.

References

Ailman, D. E. (1965). Synthesis of *O,O*-dimethyl *S*-(1,2-dicarboethoxy) ethyl phosphorothioate (malaoxon) and related compounds from trialkyl phosphites and organic disulfides. *J. Org. Chem.* **30**, 1074–1077.

Aldridge, W. N., and Nemery, B. (1984). Toxicology of trialkylphosphorothioates with particular reference to lung toxicity. *Fundam. Appl. Toxicol.* **4**, S215– S223.

Aldridge, W. N., Dinsdale, D., Nemery, B., and Verschoyle, R.D. (1985). Some aspects of the toxicology of trimethyl and triethyl phosphorothioates. *Fundam. Appl. Toxicol.* **5**, S47–S60.

Ali, F. A. F., and Fukuto, T. R. (1982). Toxicity of *O,O,S*-trialkyl phosphorothioates to the rat. *J. Agric. Food Chem.* **30**, 126–130.

Ali, F. A. F., and Fukuto, T. R. (1983). Toxicological properties of *O,O,S*-trialkyl phosphorothioates. *J. Toxicol. Environ. Health* **12**, 591–598.

Armstrong, D. W. (1984). Chiral stationary phases for HPLC separation of enantiomers. A review. *J. Liquid Chromatogr.* **7**, 353–376.

Bellet, E. M., and Casida, J. E. (1974). Products of peracid oxidation of organothiophosphorus compounds. *J. Agric. Food Chem.* **22**, 207–211.

Benschop, H. P., and DeJong, L. P. A. (1988). Nerve gas stereoisomers: Analysis, isolation, and toxicology. *Accts. Chem. Res.* **21**, 368–374.

Bielawski, J., and Casida, J. E. (1988). Phosphorylating intermediates in the peracid oxidation of phosphorothionates, phosphorothiolates, and phosphorodithioates. *J. Agric. Food Chem.* **36**, 610–615.

Bracha, P., and O'Brien, R. D. (1968a). Trialkyl phosphate and phosphorothiolate anticholinesterases. I. Amiton analogues. *Biochemistry* **7**, 1545–1554.

Bracha, P., and O'Brien, R. D. (1968b). Trialkyl phosphate and phosphorothiolate anticholinesterases. II. Effect of chain length on potency. *Biochemistry* **7**, 1555–1559.

Burn, A. J., and Cadogen, J. I. G. (1961). The reactivity of organophosphorus compounds. Part IX. The reaction of thionates with alkyl iodides. *J. Chem. Soc.* 5532–5541.

Casida, J. E. (1983). Propesticides: Bioactivation in pesticide design and toxicological evaluation. *In* "Pesticide Chemistry, Human Welfare, and the Environment" (J. Miyamoto and P.C. Kearney, eds.), Vol. 3, pp. 239–246. Pergamon Press, Oxford.

Chukwudebe, A., March, R. B., Othman, M., and Fukuto, T. R. (1989). Formation of trialkyl phosphorothioate esters from organophosphorus insecticides after exposure to either ultraviolet light or sunlight. *J. Agric. Food Chem.* **37**, 539–545.

Clothier, B., and Johnson, M. K. (1980). Reactivation and aging of neurotoxic esterase inhibited by a variety of organophosphorus esters. *Biochem. J.* **185**, 739–747.

Clothier, B., Johnson, M. K., and Reiner, E. (1981). Interaction of some trialkyl phosphorothiolates with acetylcholinesterase. Characterization of inhibition, aging and reactivation. *Biochim. Biophys. Acta* **660**, 306–316.

Desmarchelier, J. M., Wustner, D. A., and Fukuto, T. R. (1976). Mass spectra of organophosphorus esters and their alteration products. *Residue Rev.* **63**, 77–185.

Ecobichon, D. J., and Joy, R. M. (1982). "Pesticides and Neurological Diseases." CRC Press, Boca Raton, Florida.

Eto, M. (1974). "Organophosphorus Pesticides: Organic and Biological Chemistry." CRC Press, Cleveland, Ohio.

Eto, M., Okabe, S., Ozoe, Y., and Maekawa, K. (1977). Oxidative activation of *O,S*,-dimethyl phosphoramidothiolate. *Pestic. Biochem. Physiol.* **7**, 367–377.

Fahmy, M. A. H., Khasawinah, A., and Fukuto, T.R. (1972). Alkaline hydrolysis of phosphoramidothioate esters. *J. Org. Chem.* **37**, 617–625.

Fest, C., and Schmidt, K.-J. (1973). "The Chemistry of Organophosphorus Pesticides." Springer-Verlag, New York.

Gazzard, M. F., Sainsbury, G. L., Swanston, D. W., Sellers, D., and Watts, P. (1974). Anticholinesterase ability of diethyl S-*n*-propyl phosphorothiolate: Errors caused by the presence of an active impurity. *Biochem. Pharmacol.* **23**, 751–752.

Gorenstein, D. G. (1983). Non-biological aspects of phosphorus-31 NMR spectroscopy. *Prog. NMR Spectroscopy* **16**, 1–98.

Gray, A. J., Thompson, C. M., and Fukuto, T. R. (1982). Distribution and excretion of [$^{14}CH_3S$]methamidophos after intravenous administration of a toxic dose and the relationship with anticholinesterase activity. *Pestic. Biochem. Physiol.* **12**, 17–26.

Gray, A. J., and Fukuto, T. R. (1984). Metabolism of O,O,S-trimethylphosphorothioate in rats after oral administration of a toxic dose and the effect of coadministration of its antagonistic thionate isomer. *Pestic. Biochem. Physiol.* **22**, 295–311.

Greenhalgh, R., and Schoolery, J. N. (1978). Analysis of organophosphorus insecticide and formulations for contaminants by phosphorus-31 fourier transform nuclear magnetic resonance. *Analyt. Chem.* **50**, 2039–2042.

Hall, C. R., and Inch, T. D. (1979a). Preparation of some chiral alkyl phosphorothioates and stereochemical studies of their conversion into other chiral organophosphates. *J.C.S. Perkin Trans I.* 1104–1111.

Hall, C. R., and Inch. T. D. (1979b). Chiral O,S-dialkyl phosphoramidothioates: Their preparation, absolute configuration and stereochemistry of their reaction in acid and base. *J.C.S. Perkin Trans I.* 1646–1655.

Hall, C. R., and Inch, T. D. (1981). Stereochemistry of alkaline hydrolyses of 1,3,2-oxazaphospholidine-2-thiones and related reactions. *J.C.S. Perkin Trans I.* 2368–2373.

Hassan, A., and Dauterman, W. C. (1968). Studies on the optically active isomers of O,O-diethyl malathion and O,O-diethyl malaoxon. *Biochem. Pharmacol.* **17**, 1431–1439.

Heath, D. F. (1961). "Organophosphorus Poisons, Anticholinesterases and Related Compounds." Pergamon, Oxford.

Hilgetag, G., and Lehman, G. (1959). Beitrag zur chemie die thiophosphate. VII. Optisch aktive thiophosphate. *J. Prakt. Chem.* **8**, 224–234.

Hilgetag, G., and Lehman, G. (1960). Uber den mechanismus der isomerisierung von thionophosphate. *J. Prakt. Chem.* **4**, 5–10.

Hilgetag, G., and Teichman, H. (1965). The alkylating properties of alkyl thiophosphates. *Angew. Chem. Int. Ed. Engl.* **4**, 914–916.

Hirashima, A., Leader, H., Holden, I., and Casida, J. E. (1984). Resolution and stereoselective action of sulprofos and related S-propyl phosphorothiolates. *J. Agric. Food Chem.* **32**, 1302–1307.

Imamura, T., Gandy, J., Fukuto, T. R., and Talbot, P. (1983). An impurity in malathion alters the morphology of rat lung bronchiolar epithelium. *Toxicology* **26**, 73–79.

Imamura, T., and Hasegawa, L. (1984). Role of metabolic activation, covalent binding, and glutathione depletion in pulmonary toxicity produced by an impurity in malathion. *Toxicol. Appl. Pharmacol.* **72**, 476–483.

Imamura, T., and Gandy, J. (1988). Pulmonary toxicity in phosphorothioate impurities found in organophosphate insecticides. *Pharmacol. Ther.* **38**, 419–427.

Jarv, J. (1984). Stereochemical aspects of cholinesterase catalysis. *Bioorg. Chem.* **12**, 259–278.

Johnson, M. K. (1980). Delayed neurotoxicity induced by organophosphorus compounds; areas of understanding and ignorance. *In* "Mechanism of Toxicity and Hazard Evaluation" (B. Holmstedt, ed.), Elsevier Press, New York.

Kimura, S., Toeda, K., Miyamoto T., and Yamamoto, I. (1984). Activation and detoxication of S-alkyl phosphorothiolate insecticides. *J. Pestic. Sci.* **9**, 137–142.

Koizumi, T., Kobayashi, Y., Amitani, H., and Yoshii, E. (1977). A practical method of preparing optically active dialkyl phenyl phosphates. *J. Org. Chem.* **42**, 3459–3460.

Koizumi, T., Amitani, H., and Yoshii, E. (1978). A new method of preparing optically active alkyl phenyl phosphonates. *Tetrahedron Lett.* **39**, 3741–3742.

Kosolapoff, G. M. (1950). "Organophosphorus Compounds." John Wiley, New York.

Lee, P. W., Allahyari, R., and Fukuto, T.R. (1978). Studies on the chiral isomers of fonofos and fonofos oxon. *Pestic. Biochem. Physiol.* **9**, 23–32.

Leslie, D. R., Beaudry, W. T., Szafraniec, L. L., and Rohrbaugh, D.K. (1991). Mechanistic implications of pyrophosphate formation in the oxidation of *O,S*-dimethyl phosphoramidothioate. *J. Org. Chem.* **56**, 3459–3462.

Lubkowitz, J. A., Revilla, A. P., and Baruel, J. (1974). Synthesis of C-14 labeled O,S-dimethyl phosphoramidothioate. *J. Agric. Food Chem.* **22**, 151–152.

Mastin, T. W., Norman, G. R., and Weilmuenster, E. A. (1945). Chemistry of the aliphatic esters of thiophosphoric acids. I. *J. Am. Chem. Soc.* **67**, 1662–1664.

McLafferty, F. (1980) "Interpretation of Mass Spectra," 3rd. Ed. University Science Books, Mill Valley, California.

Metcalf, R. L., and March, R.B. (1953). The isomerization of organic thionophosphate insecticides. *J. Econ. Entomol.* **46**, 288–294.

Michalski, L. L., Modro, T., and Wieczorkowski, J. (1960). Reaction of organic thiosulfonates with trialkyl phosphites and dialkyl phosphites. *J. Chem. Soc.* 1665–1669.

Mlotkowska, B., and Wartalowska-Graczek, M. (1987). Direct conversion of alcohols into phosphorothiolates. *J. Prakt. Chem.* **329**, 735–740.

Morrison, D. C. (1955). The reaction of sulfenyl chlorides with trialkyl phosphites. *J. Am. Chem. Soc.* 77, 181–184.

Muller, C. E., and Roth, H. J. (1990). A new synthesis of thiophosphoric acid esters with a C—S—P bond. *Tetrahedron Lett.* **31**, 501–502.

Nowicki, T., Markowska, A., Kielbasinski, P., and Mikolajczyk, M. (1986). A new, efficient synthesis of thiolesters. *Synthesis* 305–308.

Ohkawa, H., Mikami, N., Kasamatsu, K., and Miyamoto, J. (1976). Stereoselectivity in toxicity and acetylcholinesterase inhibition by the optical isomers of papthion and papoxon. *Agric. Biol. Chem.* **40**, 1857–1861.

Patil, V., Padalikar, S., and Kawale, G. (1987). Use of potassium iodate for thin-layer chromatographic detection of sulfur-containing organophosphorus insecticides. *Analyst* **112**, 1765–1766.

Quistad, G. B., Fukuto, T. R., and Metcalf, R. L. (1970). Insecticidal, anticholinesterase, and hydrolytic properties of phosphoramidothiolates. *J. Agric. Food Chem.* **18**, 189–194.

Rengasamy, S., and Parmar, B. S. (1988). Investigation of some factors influencing isomalathion formation in malathion products. *J. Agric. Food Chem.* **36**, 1025–1030.

Ruzicka, J. H., Thomson, J., and Wheals, B. B. (1967). The gas chromatographic determination of organophosphorus pesticides. A comparative study of hydrolysis rates. *J. Chromatogr.* **31**, 37–44.

Ryu, S., Jackson, J. A., and Thompson, C. M. (1991a). Methanolysis of phosphoramidates with boron trifluoride-methanol complex. *J. Org. Chem.* **56**, 4999–5002.

Ryu, S., Lin, J., and Thompson, C. M.. (1991b). Comparative anticholinesterase potency of chiral isoparathion methyl. *Chem. Res. Toxicol.* **4**, 517–520.

Sanborn, J. R., and Fukuto, T. R. (1972). Insecticidal, anticholinesterase, and hydrolytic properties of S-aryl phosphoramidothioates. *J. Agric. Food Chem.* **20**, 926–930.

Schuphan, I., and Casida, J. E. (1983). Metabolism and degradation of pesticides and xenobiotics: Bioactivations involving sulfur-containing substituents. In "Pesticide Chemistry, Human Welfare, and the Environment" J. (Miyamoto and P.C. Kearney, eds.), Vol. 3, pp. 287–294. Pergamon Press, Oxford, U.K.

Segall, Y., and Casida, J. E. (1982). Oxidative conversion of phosphorothiolates to phosphinyl-oxysulfonates probably via phosphorothiolate S-oxides. *Tetrahedron Lett.* **23**, 139–142.

Segall, Y., and Casida, J. E. (1983). Reaction of proposed phosphorothiolate S-oxide intermediates with alcohols. *Phosphorus Sulfur* **18**, 209–212.

Sherma, J., and Zweig, G. (1972). Gas chromatography. *In* "Analytical Methods for Pesticides and Plant Growth Regulators" Vol. II, (G. Zweig, ed.) Academic Press, New York.

Souter, R. W. (1987). "Chromatographic Separation of Stereoisomers." CRC Press, Boca Raton, Florida.

Tashma, Z., Katzhendler, J., and Deutsch, J. (1973). Single or double hydrogen rearrangements in the mass spectrum of O-ethyl, S-alkyl phosphorothioates. *Org. Mass Spectrom.* **7**, 955–959.

Thompson, C. M., and Fukuto, T. R. (1982). Mechanism of cholinesterase inhibition by methamidophos. *J. Agric. Food Chem.* **30**, 282–284.

Thompson, C. M., Nishioka, T., and Fukuto, T. R. (1983). Determination of the electron impact fragmentation pattern of methamidophos via stable isotope enrichment. *J. Agric. Food Chem.* **31**, 696–700.

Thompson, C. M., Castellino, S., and Fukuto, T. R. (1984). ^{13}C-NMR studies on an organophosphate. Formation and characterization of methamidophos S-oxide. *J. Org. Chem.* **49**, 1696–1699.

Thompson, C. M., Frick, J. A., Natke, B., and Hansen, L. (1989). Preparation, analysis and anticholinesterase properties of O,O-dimethyl phosphorothiolate isomerides. *Chem. Res. Toxicol.* **2**, 386–391.

Thompson, C. M., Frick, J. A., and Green, D. L. C. (1990). Synthesis, configuration, and chemical shift correlations of chiral 1,3,2-oxazaphospholidin-2-ones derived from serine. *J. Org. Chem.* **55**, 111–116.

Tingfa, D. (1986). Gas chromatographic determination of O-ethyl-S-(*N*,*N*-diisopropylamino) ethyl methylphosphonothiolate and O,O-diisopropyl S-benzyl phosphorothiolate as corresponding phosphonofluoridate and phosphorofluoridate. *Int. J. Environ. Anal. Chem.* **27**, 151–158.

Umetsu, N., Grose, F. H., Allahyari, R., Abu-El-Haj, S., and Fukuto, T. R. (1977). Effect of impurities on the mammalian toxicity of technical malathion and acephate. *J. Agric. Food Chem.* **25**, 946–953.

Valentine, D. (1984). Preparation of enantiomers of compounds containing chiral phosphorus centers. *In* "Asymmetric Synthesis" (J.D. Morrison and J.W. Scott, eds.), Vol. 4, pp. 263–312. Academic Press, Orlando, Florida.

Vilanova, E., Johnson, M. K., and Vicedo, J. L. (1987). Interaction of some unsubstituted phosphoramidate analogs of methamidophos with acetylcholinesterase and neuropathy target esterase. *Pestic. Biochem. Physiol.* **28**, 224–238.

Wing, K. D., Glickman, A. H., and Casida, J. E. (1984). Phosphorothiolate pesticides and related compounds: Oxidative bioactivation and aging of the inhibited acetylcholinesterase. *Pestic. Biochem. Physiol.* **21**, 22–30.

Yang, Y-C., Szafraniec, L. L., Beaudry, W. T., and Rohrbaugh, D. K. (1990). Oxidative detoxification of phosphorothiolates. *J. Am. Chem. Soc.* **112**, 6621–6627.

3

Degradation of Organophosphorus Insecticides in Environmental Matrices

Kenneth D. Racke
Environmental Chemistry Laboratory
DowElanco
Midland, Michigan

I. Introduction

For several reasons, it is critical that discussions related to either toxicological or environmental issues of agrochemicals be conducted in light of information from both of these complementary disciplines. This is certainly true for consideration of organophosphorus (OP) insecticides.

First, it is only through the interaction of toxicology and environmental chemistry that realistic estimates of toxicological risk can be obtained. For example, toxicological information may be interpreted in a vastly different manner depending on the persistence of the compound under consideration. If a moderately toxic pesticide is nonpersistent in the environment, toxicological considerations might best be focused on exposure resulting from handling concentrated material during application, storage, or disposal rather than focusing on dietary exposure. Distorted perspectives of toxicological risk

Organophosphates: Chemistry, Fate, and Effects

can arise if environmental information is either neglected or unavailable. This is typified by the adverse publicity when disclosure of information suggests that exposure of laboratory rats to some pesticide at extremely high dosages can result in chronic toxic effects.

Second, from a systems perspective, organisms constitute just one of many environmental compartments. Thus, transport between abiotic and biotic compartments can be an important toxicological consideration (e.g., bioconcentration). In addition, it is not surprising that many of the same chemical transformations of a given pesticide occur in a variety of environmental matrices (Fig. 1). For example, hydrolysis of a phosphorothioate ester to a common metabolite may be catalyzed by the ultraviolet radiation present

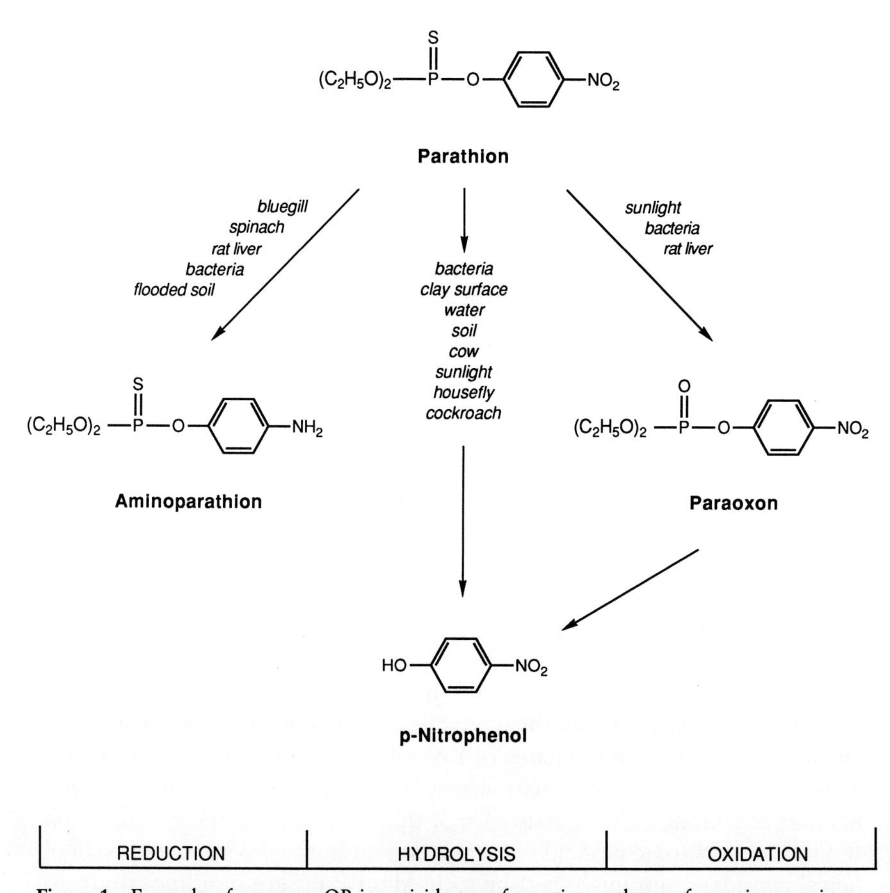

Figure 1 Example of common OP insecticide transformation pathways for various environmental matrices [Pankaskie *et al.* (1952); Grunwell and Erickson (1973); Lichtenstein *et al.* (1973); Paris and Lewis (1973); Suzuki and Uchiyama (1975); Munnecke and Hsieh (1976); Willis and McDowell (1987)].

in sunlight, by reactive surfaces of clay minerals in soil, or by enzymes associated with metabolic pathways in micro- or macro-organisms. Similar pathways of transformation of a given pesticide may operate to some extent in pond water, on the surface of a leaf, in soil, and in the liver of a vole. Therefore, consideration of the environmental transformations a pesticide may undergo can confer valuable insight into reconstructing likely pathways of metabolism in isolated biotic systems.

Third, environmental chemistry cannot be appropriately conducted in a vacuum, devoid of biological considerations. The discipline of environmental chemistry becomes relevant when it interfaces with the biological disciplines associated with the effect of chemicals on target organisms (e.g., insect pests) and nontarget organisms (e.g., humans). Thus, it is as important for environmental chemists to interact with toxicologists as it is for them to interact with entomologists and agronomists.

The purpose of the present chapter is to give an overview of the environmental behavior of OP insecticides. This will include an introduction to the uses, transport, and transformation of OP compounds in the environment. Rather than attempt a comprehensive treatment of the subject, for which an entire volume or series would be necessary to do the OP class justice, the approach here will be to cover the material in a cursory, exemplary fashion. Consideration of the environmental characteristics of compounds within the OP class should provide an appropriate perspective from which to view the toxiciological information in the succeeding chapters.

II. Organophosphorus Insecticides in the Environment

A. Diversity of Uses

Since the introduction of the first commercial product in the 1940s, the OP esters have become the most widely used class of insecticides in the world. Although more than 100,000 OP insecticides have been synthesized and screened, slightly fewer than 100 are currently being marketed as commercial products (Hassall, 1982). The *Agrochemicals Handbook* (Hartley and Hamish, 1987) summarized information on 81 OP compounds out of a total of 179 insecticides. *Commercial and Experimental Organic Insecticides* (Larson *et al.*, 1985) lists 88 OP insecticides. A World Health Organization summary provides information on 77 different OP insecticides (WHO, 1986). An indication of the prevalence of OP insecticide use is the fact that in 1988 roughly $2.3 billion out of a total world insecticide market of $6.1 billion was accounted for by this popular class.

The versatility of the OP insecticides is demonstrated by their use in virtually every insecticide, acaricide, and nematocide market imaginable. This domination has permeated agricultural, horticultural, industrial, and public health pest management systems. A few typical examples are given in Table I.

As an example of a typical use scenario, corn agriculture in the United States is heavily reliant on the use of soil-applied insecticides to control the depredations of larval corn rootworms (*Diabrotica* spp.). Having replaced the environmentally persistent chlorinated hydrocarbons (e.g., aldrin), the OP insecticides have come to dominate this important market. In the Midwestern U.S. corn belt approximately 23 million of 61 million acres of corn were treated with soil insecticides in 1989, representing a market value of over $200 million. Of these, 80% were members of the OP class, primarily terbufos, chlorpyrifos, and fonofos. In Iowa alone, approximately 4.6 million acres of corn were treated in 1985 with OP insecticides (Wintersteen and Hartzler, 1987). Statistics for a number of other pest control scenarios could be cited to corroborate this evidence of the importance of the OP insecticides.

B. Environmental Entry

The multitude of manners in which a given OP insecticide can enter the environment may be demonstrated by considering the case of one widely used product. Chlorpyrifos is formulated in a variety of manners for a broad range of insect pests of agricultural, horticultural, industrial, and medical importance. As a granular formulation, chlorpyrifos is applied to soil to control insect pests of corn, sugarbeets, and peanuts. As a foliar spray it is applied to various fruits and field crops. As an emulsifiable concentrate chlorpyrifos is injected around building foundations to control termites. Chlorpyrifos can be

TABLE I

Examples of Typical Organophosphorus Insecticide Uses

Insecticide	Crop or System	Formulation	Pest
Methyl parathion	Cotton	Microencap. spray	Bollworms
Diazinon	Turf	Granular	Grubs
Dimethoate	Apple	Emulsifiable conc.	Codling moths
Terbufos	Corn (soil)	Granular	Corn rootworms
Malathion	Stored grain	Emulsifiable conc.	Granary weevils
Chlorpyrifos	Structure (soil)	Emulsifiable conc.	Termites
Temephos	Standing water	Emulsifiable conc.	Mosquitoe larvae
Chlorpyrifos	Structure	Emulsifiable conc.	Cockroaches, fleas
Dichlorvos	Structure	No-Pest Strip	Flies, mosquitoes
Coumaphos	Cattle	Dip vat solution	Cattle ticks

aerially applied as a fog for control of mosquito pests. Wood can be treated with chlorpyrifos to discourage wood-destroying insects. A microencapsulated form of chlorpyrifos can be applied to fire ant mounds. The inside of buildings can be treated with chlorpyrifos for control of household pests such as cockroaches or fleas.

There are several routes of directed, intentional application of OP insecticides to the environment. Soil and foliar application represent major uses for these products, with much foliar-directed material reaching soil eventually, during or shortly after application. In a few cases OP insecticides are applied directly to water systems. Some uses may involve the application of aerosol fogs into the atmosphere. In some cases disposal of pesticide wastes and/or washwaters into pesticide disposal pits have been proposed or implemented.

Although not common, indirect or unintentional means of OP insecticide environmental entry include redeposition of residues resulting from air (spray drift, volatilization) or water (leaching, runoff) movement. There is also the possibility that entry may occur from spills. This latter mode of entry may be most likely to occur during storage, transport, or equipment loading/cleaning. Contemporary experience indicates that, because of the management practices that accompany use of OP insecticides, environmental entry of significant quantities of OP insecticide residues by indirect means does not generally accompany their use.

C. Classes and Environmentally Significant Properties

Although the OP family by definition centers around the phosphorus atom, variations on the bonds with this *molecular hub* divide the family into several classes. Examples of such classes would be phosphorothionates (parathion, chlorpyrifos), phosphorodithioates (terbufos, malathion), phosphoramidates (isofenphos), phosphonates (fonofos), and phosphates (dichlorvos). In some cases there may be rather distinctive degradative reactions associated with a particular class, for example, phosphate formation (oxon) from phosphorothionates. But it should be noted that owing to the tremendous diversity of aromatic and aliphatic groups that can be present, generalizations can be hard to make. This will become clear as we discuss examples of some of the degradative processes that operate for OP insecticides.

The diversity within the OP insecticide family is immediately evident when chemical and physical properties of selected examples are examined (Table II). For instance, although many OP compounds have fairly low water solubilities (1–40 ppm), some are quite soluble, an example of which would be dimethoate (25,000 ppm). Vapor pressures of OP insecticides also vary greatly, with some compounds being virtually nonvolatile [e.g., ethyl *p*-ni-

TABLE II

Environmentally Significant Properties of Organophosphorus Insecticides[a]

Compound	Water solubility (ppm)	Vapor pressure (mm Hg)	Soil adsorption (K_{oc9}, ml/g)
Parathion	12	4×10^{-5}	4800
Diazinon	40	7×10^{-7}	251
Chlorpyrifos	1	2×10^{-5}	8753
Chlorpyrifos-methyl	3	4×10^{-5}	3300
Fonofos	16	2×10^{-4}	5105
Fonofos oxon	> 2600	—	—
EPN	—	1×10^{-8}	1327
Malathion	143	4×10^{-5}	280
Dimethoate	25,000	8×10^{-6}	27
Terbufos	6	3×10^{-4}	842
Terbufos sulfoxide	>1100	—	—
Terbufos sulfone	408	—	—
Isofenphos	22	4×10^{-6}	—
Dichlorvos	10,000	1×10^{-2}	—
Ethoprophos	750	3×10^{-4}	26

[a]From Bowman and Sans (1979, 1983); Chakrabarty and Gennerich (1987); Kanazawa (1989); McCall (1987); Gustafson (1989); Felsot and Dahm (1979); Packard (1987); Kenaga (1976); Hartley and Hamish (1987); Matsumura (1985).

trophenylthiobenzene phosphonate (EPN)], and others exhibiting great volatility (dichlorvos). To a great extent the physical and chemical characteristics of OP insecticides determine their susceptibility to transformation and transport processes that in turn interact to determine environmental fate. For example, OP insecticides differ somewhat in the strength of their sorption to soil (Table II). Thus, a compound such as parathion with a large sorption partitioning factor (4800 ml/g) will tend to be relatively immobile in soil, whereas a compound such as dimethoate with a small partitioning factor (27 ml/g) will tend to be much more mobile in soil. This mobility, coupled with such factors as application rate, persistence, and rainfall, acts to determine the risk that may exist for such a compound to leach in significant quantities to groundwater. The point to emphasize here is that the somewhat unique set of physical and chemical properties possessed by a given OP insecticide indicates that, within limits, substantial variation in the environmental behavior of these compounds is to be expected. It should also be pointed out that in many cases environmentally significant metabolites can possess very different properties from those of the parent compound. For example, the oxidative metabolites of terbufos (sulfoxide, sulfone) and fonofos (oxon) are much more water soluble than the respective parents and may differ greatly in their environmental behavior (Table II).

III. Environmental Transport and Transformational Processes

A number of important processes for transport and degradation of organophosphorus insecticides interact to determine their environmental fate. The individual processes will first be discussed on an isolated basis to focus on the mechanisms involved.

A. Transport Processes

1. Volatilization

Volatilization can be an important process of dissipation of some OP insecticides from surfaces of moist soil, foliage, or water. The significance of volatility depends on several factors, including vapor pressure, solubility, adsorptive behavior, and persistence of the compound, and such environmental characteristics as temperature, moisture, and air movement. OP insecticides display a wide range of vapor pressures (Table II), and thus volatility can be an important (e.g., ethoprophos) or a relatively unimportant (e.g., EPN) process. In some instances fumigant action is required for effective insect control. Such is the case with the extremely volatile dichlorvos (vp 1×10^{-2} mm Hg), which is impregnated into No-Pest Strips for control of nuisance flying insects in buildings.

For OP insecticides of intermediate volatility, environmental factors greatly modulate the kinetics of volatilization. Volatilization rates from surfaces such as glass plates or leaves are often much greater than those from soil surfaces. For example, the volatilization rate of methyl parathion (vp 2×10^{-5} mm Hg) from a glass surface and a moist soil column was 0.44 μg/cm²/hr and 0.03 μg/cm²/hr, respectively (Spencer *et al.*, 1979). Only in the latter case did volatilization rate decrease rapidly with time resulting from surface depletion and degradation, and only 2.6% of the applied methyl parathion volatilized over a 29-day period. In contrast, the volatilization of chlorpyrifos (vp 2×10^{-5} mm Hg) was quite rapid from treated corn leaves (Fig. 2). In a study conducted in a laboratory growth chamber, approximately 80% of the foliarly applied chlorpyrifos volatilized within 48 hr at 30°C with a simulated windspeed of 0.8 km/hr (McCall *et al.*, 1985). A field study of chlorpyrifos dissipation from foliage confirmed the fairly rapid rate of volatilization, with an observed half-life of approximately 1.5 days on corn and soybean foliage (McCall *et al.*, 1984).

In general the environmental significance of volatilized OP insecticide residues appears to be low. This is because of the tremendous dilution that occurs as the pesticides are volatilized into the atmosphere and through dissipation by such degradative mechanisms as photodegradation. For ex-

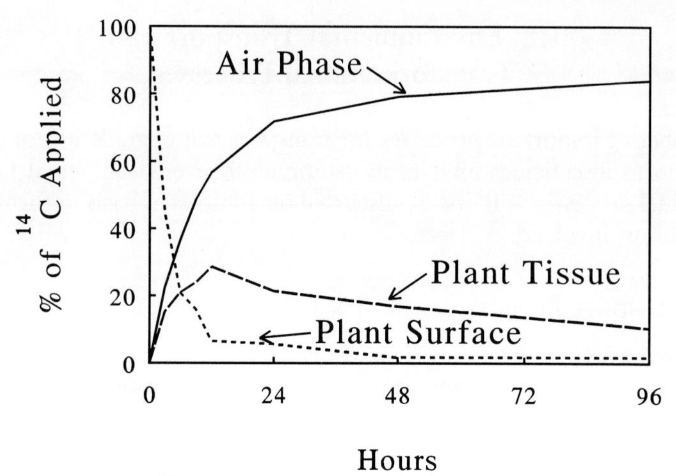

Figure 2 Distribution of ^{14}C-residues within corn agroecosystem after application of ^{14}C-chlorpyrifos to corn leaves. Adapted from McCall *et al.* (1985).

ample, the photodegradation half-life of chlorpyrifos vapor is approximately 2.6 days (Fontaine and Teeter, 1987). Sampling of air or fog has in some cases demonstrated detectable, albeit low, levels of OP insecticides such as methyl parathion or parathion present in the atmosphere (Arthur *et al.*, 1976; Glotfelty *et al.*, 1987; Seiber *et al.*, 1989). The source of the trace atmospheric levels detected has not been conclusively determined, and may result from either volatilization or spray drift of aerosol particles.

2. Leaching

Leaching involves the movement of pesticide residues into the soil profile and potentially to groundwater via percolating water. Exposure of populations to pesticides in groundwater may be a concern depending on the toxicity of the material and the concentrations observed. Pesticides partition between soil-sorbed and soil-water phases, and this latter material can move by diffusion or by mass flow through micropores or macropores. Factors that can affect leaching potential include chemical variables such as sorptive partitioning behavior and persistence as well as environmental variables such as rainfall and soil porosity (Cohen *et al.*, 1984).

Most OP insecticides tend to sorb fairly strongly to soil surfaces, and their strong affinity for soil organic matter is indicated by the rather high sorption coefficients (K_{oc}) many possess (Table II). Both laboratory and field studies that have considered the vertical mobility of OP insecticides generally find very little leaching movement occurring in soil (Edwards *et al.*, 1971; Agnihotri *et al.*, 1981; Chapman *et al.*, 1984). Somewhat typical of the OP class is the behavior of chlorpyrifos, a compound fairly strongly sorbed to soil organic matter (K_{oc} = 8753). Laboratory column leaching studies revealed that all the surface-applied residues of chlorpyrifos were confined to the upper 5 cm of several soils after elution with 20 cm of water (McCall, 1985). Field studies have confirmed this lack of mobility, and chlorpyrifos residues were confined to the upper 12 inches of soils in trials at several locations (Oliver *et al.*, 1987; Fontaine *et al.*, 1987). Even OP insecticides which have much lower affinities for sorption by soil tend to be only moderately mobile in soil. Ethoprophos, an example of such a compound, has a much lower affinity for sorption (K_{oc} = 26). In a field soil column study, nearly all of the ethoprophos residues were present in the upper 15 cm of soil during the season after application (Fig. 3), with minor traces detected in the 15 to 20-cm soil layer (Smelt *et al.*, 1977). The minor leaching of even fairly mobile OP insecticides observed in most studies can be attributed to the short persistence that characterizes their behavior in soil.

Analysis of groundwater from many locations has revealed few detections of OP insecticides (Hallberg, 1989; Leistra and Boesten, 1989). A summary of U.S. survey information demonstrates the lack of propensity for this class of pesticides to move to groundwater (Table III). The few detections that occur may represent point source pollution resulting from spills at mixing/loading sites. This is in contrast to compounds such as the carbamate insecticide aldicarb (587 detects/4004 wells) and the triazine herbicide atrazine (771 detects/5569 wells), which are commonly detected in groundwater (Parsons and Witt, 1988). The Soil Conservation Service of the U.S. Department of Agriculture has ranked the mobility of pesticides, and not surprisingly

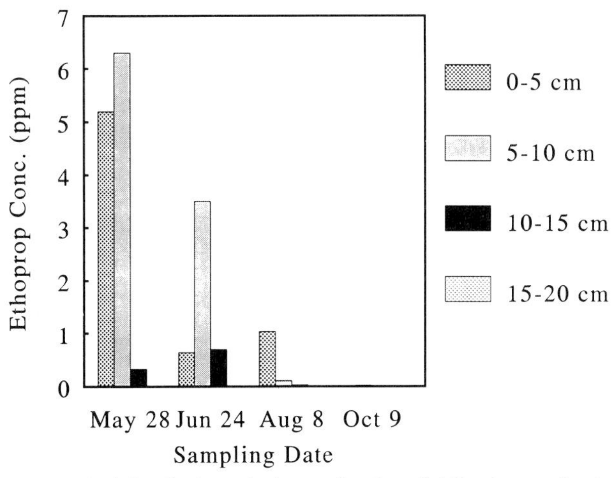

Figure 3 Season vertical distribution of ethoprophos in soil following application to field soil columns. Adapted from Smelt *et al.* (1977).

TABLE III

Summary of Well Water Analysis Information and Leaching Potential of Selected Organophosphorus Insecticides

Compound	No. of wells tested[a]	No. of detects	SCS leaching potential[b]
Azinophos-methyl			Small
Chlorpyrifos	981	3	Small
Diazinon	1473	8	Large
Dimethoate	916	2	Medium
Disulfoton	659	1	Small
Ethoprophos	—	—	Large
Fonofos	2276	1	Medium
Malathion	1347	2	Small
Methamidophos	536	14	Small
Methyl parathion	1275	6	—
Phorate	—	—	Medium
Terbufos	794	2	Small

[a]From Parsons and Witt (1988).
[b]From Wauchope (1988).

most OP insecticides are listed as possessing only small to medium leaching potential (Wauchope, 1988). In general, the environmental significance of leaching of OP insecticides appears to be of minor concern.

3. Runoff

Transport over the surface of treated fields with moving water and/or sediment is another potential route of movement of OP insecticides (Table IV). This is of concern because of considerations of aquatic non–target organism exposure. In general, pesticides with water solubilities greater than 10 ppm will move mainly in the solution phase, and less-soluble pesticides will move mainly sorbed to eroding soil particles (Wauchope, 1978). Movement of fairly soluble compounds will be governed mainly by compound persistence and environmental factors such as rainfall and hydraulic conductivity of the soil. Movement of more highly sorbed chemicals will occur mainly as an erosion-linked process and thus will be heavily dependent on erosion-management practices in addition to the preceding factors.

Although runoff of pesticides in general has not been as well studied as runoff of nutrients, information from field studies conducted under natural precipitation conditions indicates that runoff of most pesticides is of minor environmental significance (Weber and Miller, 1989). For the majority of pesticides, runoff represents typically less than 0.5% of the amounts applied (Wauchope, 1978). For example, a study conducted in an Iowa cornfield watershed reported that approximately 0.003% of 3 applications of chlorpyrifos was transported via runoff to a pond within the watershed (McCall *et al.*, 1984). Similarly, less than 0.1% of a granular diazinon application was present in runoff from cornfields (Ritter *et al.*, 1974). Studies with other OP insecticides under natural rainfall conditions reveal similar results (Edwards *et al.*, 1971; Sheets *et al.*, 1972). Information from irrigated environments such as turf indicates that because of the lack of erosion of soil particles, strongly sorbed OP insecticides such as chlorpyrifos are not transported via runoff (Watschke and Mumma, 1989).

It is only when severe erosion resulting from high-intensity storm events occurs soon after application that runoff of OP insecticides assumes potential significance. For example, when water equal to a 15-to-50-year storm event was applied to a cornfield (7–11% slope) within 2 days of terbufos application, between 0.4 and 7.0% of the applied dose was present in runoff, mainly represented as the major metabolite terbufos sulfoxide (Felsot *et al.*, 1990). The levels in runoff reported in this study were found to be extremely dependent on tillage practice. Similarly, a study of fonofos runoff from corn reported between 1 and 6% of applied material present in runoff from plots when 20 cm of simulated precipitation was applied within 24 hr of pesticide application (Baker *et al.*, 1976). Not only is storm severity critical in modulat-

TABLE IV

Runoff of Organophosphorus Insecticides in Field Studies with Natural or Simulated Rainfall

Compound	Study conditions	Rainfall	% pesticide runoff	g/ha
Methyl parathion[a]	EC on cotton 2–4% slope 17m^2	Natural 2–10 cm runoff	0.008–0.25	1–33
Diazinon[b]	GR on corn 10–15% slope 1–4 acres	Natural	≤ 0.1	0.5
Fonofos[c]	GR on corn 33 m^2	Simulated 20 cm rain/24 hr 8–12 cm runoff	1–6	—
Chlorfenvinphos[d]	EC on barley 27% slope 1.8 m strips	Natural 13.5 cm/17 wk	0.3–0.6	—
Chlorpyrifos[e]	GR on corn 6% slope 1.4 m^2	Simulated 7 and 14 cm 1st, 14 cm/hr	0.004–0.3	—
Chlorpyrifos[f]	EC on corn 4–10% slope/0.6 acres 1% slope/2.7 acres	Natural 0.4 cm runoff	0.0002–0.003	0.01–0.2

[a]Sheets *et al.* (1972).
[b]Ritter *et al.* (1974).
[c]Baker *et al.* (1976).
[d]Edwards *et al.* (1971).
[e]Sauer and Daniel (1987).
[f]McCall *et al.* (1984).

ing the quantity of organophosphorus insecticides present in runoff, but the timing of the storm event is of key importance. This is demonstrated by a study of chlorpyrifos runoff from surface application of granular formulation in corn (6% slope) (Sauer and Daniel, 1987). When a 100-year simulated rainfall event (13.6 cm) occurred less than a week after pesticide application, between 0.10 and 0.29% of the applied chlorpyrifos was present in runoff. However, if the rainfall event was delayed until 3 weeks after application, runoff represented only 0.04–0.08% of applied chlorpyrifos. Because of the harsh conditions present on the soil surface that result in rapid dissipation of many OP insecticides, high-intensity storm events that occur very shortly after pesticide application, however rare, will be of most concern.

The relative significance of runoff for OP insecticide fate can be placed in perspective by information from surveys of agricultural watersheds. These surveys reveal that although some commonly used herbicides are frequently detected in these surface waters, in very few instances are detectable quantities of OP insecticides present (Braun and Frank, 1980; Spalding and Snow, 1989; Baker and Richards, 1989).

B. Transformation Processes

In many ways OP insecticides represent an attempt to maximize insecticidal activity and minimize environmental persistence. A number of abiotic and biological degradation processes mediate OP insecticide dissipation.

1. Hydrolysis

Hydrolytic transformation is an extremely important pathway for dissipation of OP insecticides in the environment, because OP esters are very susceptible to hydrolysis. In fact, their mechanism of toxicity results from this property. Thus, from an environmental viewpoint the phosphate ester bond can be considered as a "weak link" in the molecule that is prone to cleavage resulting in detoxication. Although in some cases hydrolysis can occur at several locations in a given OP insecticide, the most common reaction involves cleavage at the phosphate ester linkage as a result of base-catalysis. A key variable in the susceptibility of OP insecticides to hydrolysis is the electron deficiency of the phosphorus atom. Thus, the electron-withdrawing properties of the substituents on the phosphorus atom can greatly modify the hydrolytic stability of the phosphate ester.

Some examples of hydrolysis half-lives for OP insecticides are listed in Table V. From these representative values it is obvious that OP insecticides vary greatly in their susceptibility to hydrolysis. The pH dependence of hydrolysis is also evident. For example, chlorpyrifos exhibits hydrolysis half-lives of 72 and 16 days at pH values of 7 and 9, respectively. In general, for compounds susceptible to base-catalyzed hydrolysis, hydrolysis rate increases

TABLE V

Hydrolysis Kinetics for Selected Organophosphorus Insecticides

Compound	Condition	Hydrolysis half-life (days)	Reference
Parathion	pH 5	301	Chapman and Cole (1982)
	pH 7	168	
	pH 8	105	
Chlorpyrifos	pH 5	72	McCall (1986a)
	pH 7	72	
	pH 9	16	
Chlorpyrifos oxon	pH 5	85	Kenaga (1971)
	pH 7	6	
	pH 9	<1	
Diazinon	pH 5	2	Sumner *et al.* (1987)
	pH 7	23	
	pH 9	12	
Phorate[a]	pH 7	4	Chapman and Cole (1982)
Phorate sulfoxide	pH 7	175	
Phorate sulfone	pH 7	56	
Chlorpyrifos	120 ppm Cu^{+2}	<<2	Chapman and Harris (1984)
	Moist soil	25	McCall *et al.* (1984)
	Air-dry soil	4	
	Air-dry clay	<<2	Getzin (1981b)

[a]Sulfoxidation.

with increasing pH, although in many cases hydrolysis rate constants tend to be pH independent below pH 7 and very pH dependent above pH 7 (Macalady and Wolfe, 1983). For compounds such as diazinon that are also susceptible to acid-catalyzed hydrolysis, hydrolysis can occur quite rapidly under acidic conditions (Table V). It is worth noting that in many cases pesticide metabolites exhibit much different hydrolytic properties than do the parent compounds. Because oxygen is more electron withdrawing than sulfur, the phosphate metabolites (oxons) of phosphorothionates and phosphorodithioates are much less hydrolytically stable than the parent pesticides. Thus, although chlorpyrifos exhibits a hydrolysis half-life of 72 days at pH 7, chlorpyrifos-oxon has a half-life of only 6 days under the same conditions.

When hydrolytic transformations in the environment are considered, it is important that the factors modifying hydrolysis rates and in some cases catalyzing hydrolysis be considered. Although hydrolysis appears to be a fairly straightforward transformation, caution must be exercised in extrapolating hydrolysis rates determined in distilled water in the laboratory to the field environment. For example, in multiphase systems such as soil or sediment, sorption of insecticides can greatly influence hydrolysis rates. In one set of experiments the rate of alkaline hydrolysis of organophosphorothioate esters in the sorbed state was approximately 10 times slower than in the bulk solution (Macalady and Wolfe, 1985). In another set of experiments no correlation was noted between pH and soil degradation half-lives of 5 OP insecticides (Chapman and Cole, 1982). A number of factors have been found to catalyze the hydrolysis of OP insecticides. Metal ions such as Cu^{+2} can catalyze the very rapid hydrolysis of some OP insecticides in solution (Chapman and Harris, 1984). Under the dry conditions characteristic of the soil surface, clay minerals present in soil can catalyze the rapid hydrolysis of compounds such as chlorpyrifos (Getzin, 1981b). In addition to inorganic mechanisms, OP insecticides can also be hydrolyzed in soil through the action of immobilized enzymes, which in some cases can persist sorbed to clay or organic fractions for rather extensive periods (Skujins, 1976).

2. Redox

Formation of oxidized or reduced metabolites of OP insecticides can occur resulting from the action of sunlight, interaction with soil components, or as an indirect consequence of microbial activities. For example, oxidation of phosphorothionates such as parathion or chlorpyrifos results in formation of the oxon metabolites, which usually are much less stable than the parent compounds. This is not true for some phosphoramidates, however, such as isofenphos, in which cases the oxons formed may be fairly persistent in soil (Felsot, 1984). Thioether oxidation readily occurs for OP insecticides such as phorate and terbufos (Racke and Coats, 1988). An important consideration about many of these processes, however, is their reversibility. Thus, phorate sulfoxide may form under aerobic conditions, but the reaction may be reversed under anaerobic conditions in lake mud (Walter-Echols and Lichtenstein, 1977). Some compounds, such as parathion, can be either reduced (amino-parathion) or oxidized (paraoxon) (Katan and Lichtenstein, 1977). Reductive dechlorinations can also occur under anaerobic conditions, as is the case with coumaphos and tetrachlorvinphos (Beynon and Wright, 1969; Shelton and Karns, 1988). The environmental significance of oxidation of OP insecticides is that metabolites produced (e.g., oxons from phosphorothionates) may be more toxic than the parent insecticides as discussed by Chambers in Chapter 1. However, this toxicological fact is often rendered a moot point because of the inherently unstable nature of oxidative metabolites such as paraoxon in most environmental compartments.

3. Photolysis

Exposure of OP insecticides to sunlight can result in photolytic degradation. This can occur via direct photolysis, in which the pesticide actually absorbs the

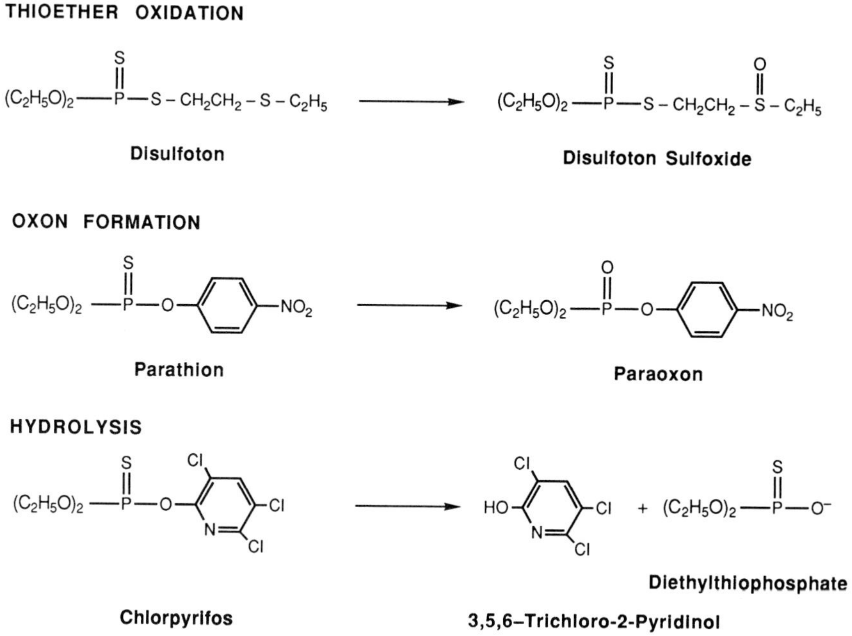

THIOETHER OXIDATION

$(C_2H_5O)_2-P(=S)-S-CH_2CH_2-S-C_2H_5 \longrightarrow (C_2H_5O)_2-P(=S)-S-CH_2CH_2-S(=O)-C_2H_5$

Disulfoton → Disulfoton Sulfoxide

OXON FORMATION

$(C_2H_5O)_2-P(=S)-O-C_6H_4-NO_2 \longrightarrow (C_2H_5O)_2-P(=O)-O-C_6H_4-NO_2$

Parathion → Paraoxon

HYDROLYSIS

Chlorpyrifos → 3,5,6-Trichloro-2-Pyridinol + Diethylthiophosphate

$(C_2H_5O)_2-P(=S)-O^-$

Figure 4 Examples of photolytic transformation of OP insecticides [Smith (1968); Grunwell and Erickson (1973); Gohre and Miller (1986)].

ultraviolet radiation and then interacts with environmental reactants or itself to undergo transformation. Some OP compounds absorb fairly strongly in the range of ultraviolet radiation that reaches the earth. More commonly, indirect photolysis occurs, in which sunlight is absorbed by humic or inorganic substances, and activated forms interact with the pesticide or produce oxygen radicals or peroxides that interact. The most common pathway of OP photolysis involves photooxidation (Miller *et al.*, 1989).

Some examples of photolytic reactions are the oxidation of thioethers, oxon formation from phosphorothionates, and hydrolysis (Fig. 4). Photolysis can affect OP insecticides in the vapor phase, in solution, or on surfaces. Photolysis on the soil surface is often limited to the top 1 or 2 mm of soil (Hebert and Miller, 1990), and it is generally believed that photolysis is most significant for OP insecticides in the vapor phase, in solution, and on the exposed surfaces of plant leaves. For exposed residues, photolytic degradation can occur at significant rates. For example, the degradative half-life for chlorpyrifos in irradiated air and water was reported as 2.6 and 11–52 days, respectively (McCall, 1986b; Fontaine and Teeter, 1987). Because of quenching of ultraviolet radiation by soil solids, photolysis degradation kinetics in soil tend not to follow first-order kinetics, but degradation often slows considerably as the most exposed pesticide residues on the extreme surface are dissipated (Hebert and Miller, 1990).

4. Microbiological Degradation

OP insecticides present in soil and aquatic systems are subject to the degradative activities of microorganisms. Microbial degradation can be a very important process for dissipation of these and other xenobiotic compounds from environmental compartments, and evidence of the pervasive importance of biodegradation is provided by the numerous studies in which degradation proceeds at a much slower pace in sterilized systems or in systems in which microbial activity is inhibited by low temperature, low moisture, or high substrate concentration. One of the major types of microbial degradative mechanisms involves hydrolysis, although other reactions such as oxidation and hydroxylation may also be employed. Microorganisms in some cases can completely mineralize pesticides to inorganic compounds.

a. Cometabolism There are two classes of microbial pesticide degradation. In the first, cometabolism, there is often an incomplete degradation of the pesticide by the microorganisms, with the transient accumulation of certain metabolite(s) (Alexander, 1981). The microorganism that initiates the degradation process obtains no benefit from its activities, and thus this process is often termed *incidental degradation.* An example is provided by parathion, which in mixed cultures of soil microorganisms was almost quantitatively transformed to aminoparathion within 12 hr (Katan and Lichtenstein, 1977). In natural systems, rather than accumulating, these metabolites are often degraded by other types of microorganisms, which may or may not obtain any benefit. If additional growth substrates are available, and microbial enzymatic activity is increased, pesticides may also be degraded fairly rapidly under laboratory or special environmental conditions. For example, in cultures of microorganisms isolated from industrial effluents, malathion was completely transformed to an unknown metabolite after 28 hr only in the presence of added ethanol carbon source; without the ethanol, no significant degradation occurred (Singh and Seth, 1989). In natural systems cometabolism often results in slow to moderate rates of degradation that slowly decrease with time as pesticide is depleted. Many OP insecticides commonly appear to undergo this form of microbial degradation in soil and water systems (Paris and Lewis, 1973; Laveglia and Dahm, 1977). For example, degradation half-lives in sterile and nonsterile sandy loam soil for chlorpyrifos were 22 and 3 weeks, respectively, and for chlorfenvinphos were > 24 and 2 weeks, respectively (Miles *et al.*, 1983).

b. Catabolism The second class of microbial metabolism is that of catabolism, in which the pesticide or a primary metabolite is completely degraded with concomitant benefit to the microbe, which utilizes the compound as a carbon, energy, or nutrient (e.g., phosphorus) source (Sethunathan and Yoshida, 1973; Cook *et al.*, 1978; Chaudhry *et al.*, 1988). In many cases this type of degradation results in complete mineralization of the compound with production of carbon dioxide. Classic examples of catabolism are provided by parathion and diazinon. Microorganisms that can degrade one or both of these common OP insecticides and utilize them as sole carbon/energy sources have been isolated from soil, aquatic systems, rice paddies, and sewage (Sethunathan and Yoshida, 1973; Munnecke and Hsieh, 1976; Spain *et al.*, 1980; Nelson, 1982). The process has been quite well characterized, and one of the enzymes involved (parathion hydrolase) has been isolated and found to be encoded by plasmid-borne genes (Mulbry *et al.*, 1986).

The ability of microorganisms to catabolize pesticides can result in an interesting adaptive phenomenon. It has been noted that repeated exposure of soil or aquatic microorganisms to some OP insecticides results in a proliferation of pesticide-degrading microorganisms, primarily bacteria. This phenomenon, *enhanced degradation*, has been found to result in such rapid degradation that the use of certain soil-applied OP insecticides has been rendered ineffective (Racke and Coats, 1990). An example is provided by the case of isofenphos, which is very persistent upon first application to soil (soil half-life 31–84 days) (Chapman and Harris, 1982; Felsot 1984). However, repeated application results in a tremendous increase in numbers of degraders in soil, and subsequent doses are rapidly mineralized (Fig. 5) (Racke and Coats, 1987). Some OP compounds appear to be susceptible to this type of microbial

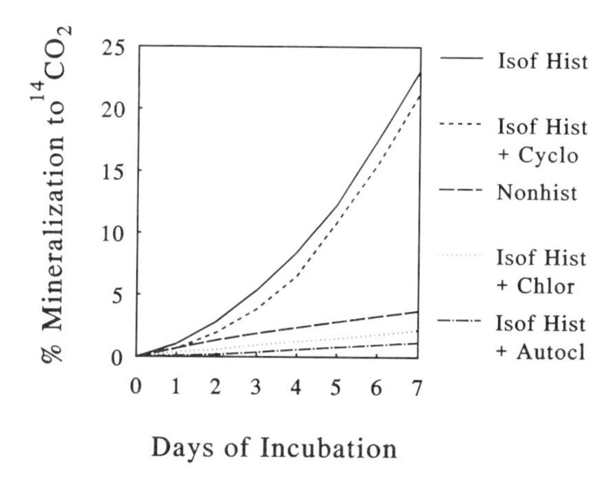

Figure 5 Enhanced microbial degradation: mineralization of ^{14}C-isofenphos in soils with and without a history of isofenphos use as affected by sterilization (autoclaving) or addition of antibacterial (chloramphenicol) or antifungal (cycloheximide) compounds [Adapted from Racke and Coats (1987)].

adaptation (diazinon, coumaphos, fensulfothion, chlorfenvinphos), whereas other compounds appear to be fairly resistant to the phenomenon (chlorpyrifos, terbufos, phorate) (Racke and Coats, 1990).

IV. Fate in Terrestrial and Aquatic Systems

The processes of transport and transformation discussed earlier are highly interactive. These interactions determine how these insecticides will behave under field conditions, and what types of toxicologically significant exposure of biota to OP insecticides may occur. Direct application of OP insecticides to plant foliage and to soil are both major uses of marketed products. In addition, the presence of OP insecticides in aquatic ecosystems is of interest because of the sensitivity of the aquatic biota to insecticidal compounds in general. A brief discussion of the fate of organophosphorus insecticides in each of these environmental compartments will provide a more comprehensive idea of how these compounds behave in the environment.

A. Fate in Soil Systems

Many OP compounds are applied to soil to control soil-dwelling pests, and other OP insecticides reach the soil surface as spray drift from foliar applications. Typical concentrations present in soil range between 1 and 10 μg/g. In soil these compounds are caught up in the dissipation processes that operate in this great recycling ecosystem. In general, these degradative forces result in fairly short persistence of most OP insecticides when they are applied to the soil. As shown in Table VI, degradative half-lives for OP insecticides in soil can be measured in terms of several weeks in most cases. Striking differences in the behavior of different members of the OP class are evident from comparative studies conducted under identical conditions. For example, in one study the degradative half-lives of dimethoate and trichloronate were 11 and 141 days, respectively (Bro-Rasmussen *et al.*, 1970). Although the relative contributions of the various degradative forces at work in the soil environment are hard to rank, microbial degradation, hydrolysis, volatilization, and photodegradation have all been shown to be important (Racke and Coats, 1990). The differences observed in the behavior of individual members of the OP insecticides are the result of differential susceptibility to the degradative forces at work in the soil environment.

It is worth noting that a great deal of variability in the behavior of a given OP insecticide can be observed under different study conditions. As shown in Table VI, soil half-lives for chlorfenvinphos in several studies were 7, 63, and 135 days (Miles *et al.*, 1979; Suett, 1975; Bro-Rasmussen *et al.*, 1970). One major difference that can be observed is the result of the formula-

TABLE VI

Comparative Degradation Half-Lives for Selected Organophosphorus Insecticides in Soil[a]

Compound	Muck[b]	Sandy loam[c]	Silty clay loam[d]	Loam[e]	Clay loam[f]	Unknown[g]
Parathion	11					
Diazinon	14			22		
Chlorpyrifos	18				14	19
Fensulfothion	7					
Pirimiphos-ethyl		70				
Isofenphos					32	
Fonofos	28		35		18	46
Trichloronate	28			141		
Dimethoate				11		
Chlormephos		28				
Ethion	56					
Phorate		7	9		8	5
Disulfoton		7			7	
Terbufos			9		14	7
Chlorfenvinphos	7	63		135		
Ethoprophos						11

[a]Reported in number of days.
[b]Miles *et al.* (1979).
[c]Suett (1975).
[d]Ahmad *et al.* (1979).
[e]Bro-Rasmussen *et al.* (1970).
[f]Harris *et al.* (1988).
[g]Kaufman (1987).

tion vehicle with which the OP insecticide enters the environment. Major differences are often noted between the behavior of technical grade pesticide (often used in laboratory studies), emulsifiable concentrates, and granular formulations. For example, recoveries of chlorpyrifos from a muck soil 4 weeks after application of an emulsifiable concentrate or a granular formulation were approximately 20 and 60% of initial levels, respectively (Chapman and Chapman, 1986). The application rate of OP insecticide to soil can also affect degradation rate, especially for compounds for which microbial degradation is important; degradation rate often decreases as initial concentrations increase (Laskowski *et al.*, 1983; Read, 1983; Racke and Lichtenstein, 1987). In addition to these factors associated with the pesticide itself, a number of environmental factors can greatly modulate the behavior of OP insecticides. These include such climatic variables as soil temperature and moisture, and management variables such as tillage practice and cropping system (Lichtenstein and Schultz, 1964; Getzin, 1981a; Abou-Assaf and Coats, 1987; Felsot *et al.*, 1990). For example, degradative half-lives of chlorpyrifos in an Ada sandy loam soil held at 25 and 75% of 0.3 Bar soil moisture tension were 61 and 37 days, respectively (McCall *et al.*, 1984). However, under air-dry conditions in this same soil, the half-life of chlorpyrifos was only 8 days. Rapid degradation of chlorpyrifos and other OP insecticides on air-dry soil surfaces is the result of clay-catalyzed hydrolysis (Getzin, 1981b; McCall *et al.*, 1984).

Even under identical conditions of application rate, temperature, and moisture, OP insecticides can exhibit strikingly different behavior in various soils. For example, a summary of chlorpyrifos soil degradation data from laboratory studies reveals that observed half-lives have been found to range between less than 10 days to greater than 120 days (Fig. 6). Such observed discrepancies are attributable to inherent soil differences of such factors as pH, organic content, clay content, and microbial populations.

In general, OP insecticides exhibit short to moderate persistence in soil. Generalizations, however, can be hard to formulate. Rather than possessing one characteristic soil behavior, OP insecticides in soil display a range of behaviors depending on characteristics of the application as well as environmental and soil variables.

Given the fairly rapid dissipation of OP insecticides in soil under many conditions, what relevant toxicological concerns may be associated with OP insecticide presence in soil? While a comprehensive treatment in the context of this chapter is not possible, several points can be made. First, acute rather than chronic exposure may be of more concern because of the relatively ephemeral nature of most OP insecticides in soil. Thus, the period during and shortly following pesticide application when maximal residues are present should be the focal point if toxicological considerations are in order. Second, potential effects of OP insecticides on the soil ecosystem may be worth

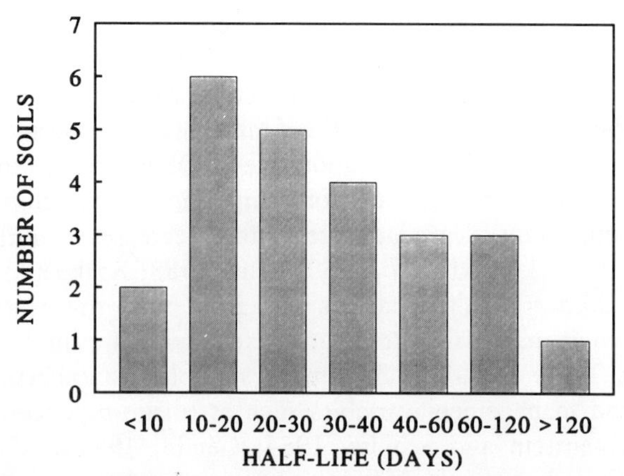

Figure 6 Distribution of laboratory-determined chlorpyrifos soil degradation half-lives [Racke *et al.* (1990)].

consideration. Although most microbial processes in soil are unaffected by the presence of OP insecticides (Tu, 1970; 1980), some perturbation of macrobiota may occur following application of certain compounds. For example, earthworm populations in soil are very little affected by application of some OP insecticides, such as chlorpyrifos. Application of other compounds, such as fensulfothion, can result in dramatic decreases in earthworm numbers (Thompson, 1971). Third, exposure of mammalian and avian systems to OP insecticides in soil may merit consideration. Although not persistent, residues present on the immediate soil surface (e.g., granules) may be available to vertebrates for a brief period following application. Potential risks vary greatly within the class, as demonstrated by an avian toxicity study estimating the number of granules necessary for bobwhite quail to ingest to reach the LD_{50} level. To illustrate the disparity within the class, this study concluded that to reach this level of intoxication, only 5 granules of a fenamiphos formulation were required while over 300 granules of a chlorpyrifos formulation were required (Hill and Camardese, 1984). Fourth, transport from the soil environment to aquatic environments is of concern because of the relative susceptibility of some aquatic organisms, especially arthropods, to certain OP insecticides. As mentioned previously, the runoff potential of most OP insecticides tends to be fairly low because of the inherently hostile nature of the soil-surface environment. Potential runoff risk would then be limited to major storm events that might occur immediately or shortly following application and would be heavily dependent on the actual quantity exposed on the erodable soil-surface layer.

B. Fate in Plant Systems

Large quantities of OP insecticides are applied to plant foliage to control insect pests of forage and field crops, turf, ornamentals, and fruit trees. Monitoring research indicates that in aerial applications between 6 and 68% of the applied dose may reach the foliar target (Willis and McDowell, 1987). In ground application, a greater quantity of applied material reaches the target foliage (14–76%). Initial residues of OP insecticides on plant foliage tend to be somewhat greater than the levels found in soil, with from 10 to 100 μg pesticide/g of tissue present after application (Sears and Chapman, 1979; McCall *et al.*, 1984; Youngman *et al.*, 1989).

One generalization that can be made for OP insecticide fate on foliage is that although initial foliar residue levels tend to be greater than those observed in soil, degradation proceeds much more rapidly on exposed plant surfaces than in soil (Table VII). For example, in a study of chlorpyrifos cornfield dissipation (McCall *et al.*, 1984), total initial residues on corn leaves of 99 to 171 ppm declined with a half-life of around 1.5 days, whereas in the soil, initial residues of 3 to 6 ppm declined with half-lives of 7 to 17 days (Fig. 7).

TABLE VII

Comparative Degradation Half-Lives and Persistence of Selected Organophosphorus Insecticides on Plant Foliage[a]

Compound	Half-life range (days)	% remaining EC 48-hr on cotton	% remaining ULV 48-hr on cotton
Parathion	0.2–7.3		
Methyl parathion	0.1–13.9	2	7
Diazinon	0.4–5.3		
Chlorpyrifos	0.7–4.0	7	12
EPN	0.6–7.0	17	10
Azinphos-methyl	1.3–17.0	26	27
Sulprofos	0.5–1.0	46	77
Methidathion	0.3–5.0	7	29
Malathion	0.3–10.9		
Dimethoate	0.9–7.2		
Ethion	2.3–10.5		
Phorate	1.4–3.6		
Phosmet	1.2–6.5		
Monocrotophos	1.3–3.4		

[a]From Willis and McDowell (1987); Ware *et al.* (1983).

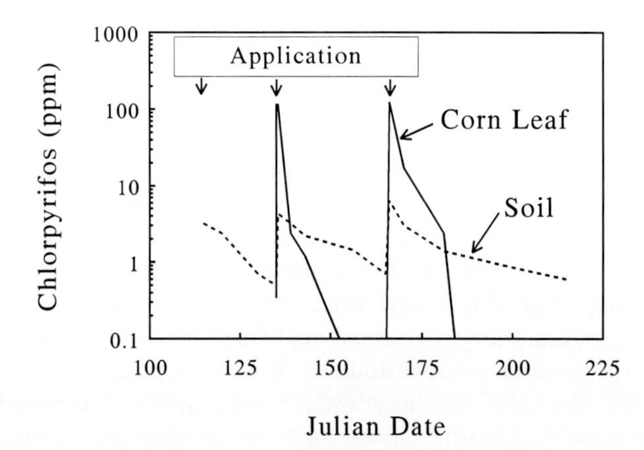

Figure 7 Dissipation of chlorpyrifos from corn foliage and surficial soil after three seasonal applications to a cornfield in Illinois. Adapted from McCall *et al.* (1985).

Similar values can be demonstrated for OP insecticides present in the turf environment. For diazinon, isofenphos, and chlorpyrifos applied to turf, degradation half-lives for dislodgeable residues on grass blades for each was less than 1 day, whereas degradation half-lives of thatch residues were 7 days or longer (Sears *et al.*, 1987). The rapid degradation of OP insecticides present on exposed leaf surfaces may be the result of the susceptibility of these residucs to such processes as volatilization and photodegradation. Additionally, insecticides may be taken up by the plant tissue and metabolized or washed off the plant surface by rainfall.

There are some differences between foliar half-lives of individual OP insecticides, with some compounds such as chlorpyrifos having fairly short persistence and others such as azinphos-methyl having greater persistence (Table VII). In addition, environmental factors such as temperature, rainfall, incident sunlight, wind, and formulation can affect foliar dissipation rates. For example, oil-based, ultra-low-volume applications may result in somewhat greater persistence than do water-based emulsifiable-concentrate formulations (Ware *et al.*, 1983).

Although a substantial treatment of the area is beyond the scope of this chapter, a few comments regarding the metabolism of OP insecticides by plants are in order. In addition to the foliar-applied pesticides that may penetrate the plant cuticle and enter plant systems, some compounds are systemic in nature and can be taken up through the roots from soil application. As an example of uptake through the plant leaf cuticle, Fig. 2 displays volatilization and plant uptake of total ^{14}C after application of ^{14}C-chlorpyrifos to the leaves of corn plants. In this case, it should be mentioned that nearly all of the ^{14}C material identified within the plant tissue was present as metabolites and less than 10% of that present within the leaf tissue was chlorpyrifos itself (Smith *et al.*, 1967). It is often difficult in these cases to determine whether uptake of metabolites that formed on the plant surface occurred or whether chloryprifos itself was absorbed by the leaf tissue and rapidly metabolized. The same distinction is often hard to make for the presence of OP insecticides and their metabolites in plant tissue as a result of soil application and systemic uptake. In some cases considerable quantities of ^{14}C can be taken up through the root systems of plants. For example, roughly 10% of the ^{14}C of an application of ^{14}C-diazinon made to rice paddy water was taken up and present in rice sprouts, of which nearly all was present as detoxified metabolite (Laanio *et al.*, 1972). Similarly, detectable quantities of metabolites of such moderately systemic compounds as ethoprophos, terbufos, and phorate can be translocated into plants following application to soil (Menzer *et al.*, 1971; Anderegg and Lichtenstein, 1984; Szeto *et al.*, 1986). In all these cases, metabolites rather than the parent OP insecticide made up most of the plant residue.

What are the potential risks associated with OP insecticides applied foliarly? First, there is the potential for consumption of plant tissues containing OP insecticide residues by humans and wildlife. Second, there is the potential for dermal exposure of workers to OP insecticide residues present on plant foliage. A factor that greatly mitigates both of these concerns is the relatively short persistence of OP compounds on exposed plant foliage as was discussed earlier in this chapter. For the first concern, field-residue testing of OP insecticide residues is performed on all commodities on which these insecticides are used to establish both appropriate application rates and suitable harvest-delays following application to allow for dissipation of residues far below toxicologically significant levels. For the second concern, similar field-residue data are generated for dislodgeable residues, which are the residues that can easily be dislodged onto clothing or skin that brushes against the treated plant surfaces, rather than total plant residue. Thus, worker reentry periods for specific crop applications may be set so as to minimize possible dermal exposure of humans to the more potent members of the class. Third, because of the method of application of most foliarly applied pesticides (spray), drift of OP insecticide spray off-target is of concern as regards exposure of aquatic systems. The use of setbacks (untreated areas) and avoidance of overspray of water with OP insecticides can minimize exposure of aquatic systems from spray drift.

C. Fate in Aquatic Systems

In a few cases OP insecticides may be directly applied to water systems to control such aquatic pests as larval mosquitoes and blackflies. Indirect entry into aquatic systems may occur in some cases from spray drift or, in rare instances, from runoff of surface residues. Typical initial water concentrations resulting from direct application of OP insecticides to water are in the range of 1 to 10 μg/liter (ppb). Measured initial values in water from drift or runoff are much lower, and generally are in the ng/liter (ppt) range. Aquatic systems in which the fate of OP insecticides have been studied include ponds, estuaries, and rice paddies.

OP insecticides applied directly to water tend to dissipate very quickly from the water column through partitioning into the sediment phase and dissipation by volatilization, photodegradation, hydrolysis, and microbial degradation. For example, a study in which small artificial ponds (2×2 m, 1 m depth) were treated with 10 μg/liter of either temephos or chlorpyrifos-methyl reported half-lives of less than 1 day for both compounds in the water column (Hughes *et al.*, 1980). As shown in Fig. 8, chlorpyrifos-methyl appeared to partition fairly rapidly into the organic debris layer, where it exhibited a degradation half-life of less than 1 week. Dissipation half-lives in the water columns of natural systems and microcosms have been demonstrated to

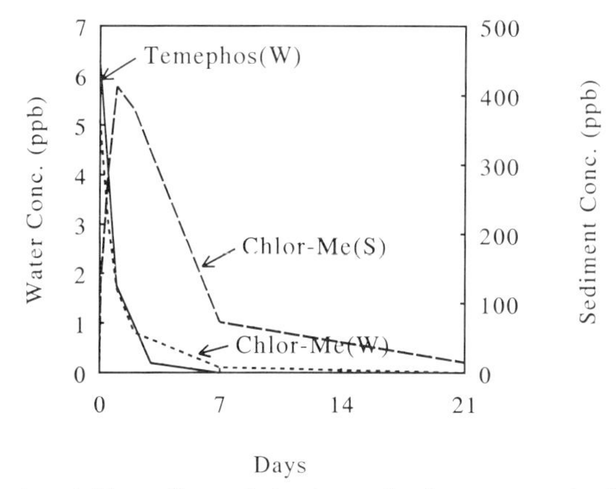

Figure 8 Dissipation of chlorpyrifos-methyl and temephos from water and sediment of artificial ponds after direct application. Adapted from Hughes *et al.* (1980).

be extremely short (< 2 days) for such compounds as fenthion, diazinon, chlorfenvinphos, malathion, and chlorpyrifos (Beynon *et al.*, 1971; Laanio *et al.*, 1972; Tagatz *et al.*, 1974; O'Neill *et al.*, 1989; Brazner *et al.*, 1989). In some cases substantial differences between OP insecticide persistence in the water versus sediment phases of aquatic systems have been noted. For example, in a littoral mesocosm study, chlorpyrifos degradation half-lives in water and sediment were reported to be < 1 day and 32–128 days, respectively (Brazner *et al.*, 1989).

Surveys of surface waters in agricultural watersheds generally result in very few detections of OP insecticides in these aquatic systems (Braun and Frank, 1980; Spalding and Snow, 1989; Baker and Richards, 1989). For example, a survey of between 725 and 949 water samples taken from 11 agricultural watersheds in Southern Ontario (1975–1977) reported detection of chlorpyrifos, ethion, and malathion only in 3, 2, and 4 cases, respectively (Braun and Frank, 1980). The concentration of these compounds was on the order of 1 to 3 pg/liter. From these studies it is apparent that OP insecticides are present in natural waters very infrequently as the result of areawide, typical use patterns. Although rare, spills may constitute the sole cases in which significant quantities of OP insecticides may enter aquatic systems. Information gathered in Ontario indicated that of 3 ponds contaminated with OP insecticides during the period 1971–1985, all resulted from accidental spills (Frank *et al.*, 1990). To place this in context, only in one pond did a fish kill occur. During the same period, 122 ponds were found to be contaminated with herbicides and 5 ponds, with fungicides. As was the case with input of low levels of OP insecticides into aquatic systems, in instances in which spills occur, dissipation from the water column rapidly follows introduction into the aquatic environment. For example, a spill of chlorpyrifos into a marine bay resulted in initial water concentrations of up to 300 μg/liter, but because of sediment sorption, dissipation and dilution, the concentration had dropped to below detectable levels within 17 days (Cowgill *et al.*, 1991).

What risks are posed by the presence of OP insecticides in aquatic systems? The most relevant concern is aquatic non–target organism toxicity. Many organophosphorus, carbamate, and pyrethroid insecticides are not only fairly potent to the target insect pest species, but are also moderately to highly toxic to aquatic arthropods and, in some cases, to fish. Within the OP insecticide class a wide range of aquatic toxicity properties has been noted. For example, the median lethal concentration (LC_{50}) (96-hr) values for 7 OP insecticides to striped bass ranged from 0.58 to 1000 μg/liter (Korn and Earnest, 1974). In light of the infrequent detection of significant quantities of OP insecticides in aquatic systems as the result of typical terrestrial use, it is not surprising that few reported cases of adverse aquatic effects have been documented under field-use conditions. One point worth noting is that direct application of OP insecticides to aquatic systems has in most cases been discontinued, and only the least toxic compounds have currently registered uses for aquatic insect control. In those rare cases in which spill accidents have occurred and resulted in reduction of aquatic community populations, the effects have usually been fairly ephemeral, and recovery of aquatic populations has occurred (Raven and George, 1989; Cowgill *et al.*, 1991).

V. Summary

In summary, the OP compounds constitute the most widely used class of insecticides. Once introduced into the environment OP insecticides are dissipated by a variety of important degradative processes, including hydrolysis, photolysis, and microbial degradation. As a result, dissipation of these compounds from the sites of application (soil, foliage, water) is fairly rapid, and degradative half-lives can usually be measured in terms of a few hours, days, or weeks. The rapid dissipation of these compounds when present in environmental matrices indicates that toxicological concerns are mainly of an acute nature, and when relevant, revolve around the handling of concentrated materials before, during, and immediately upon application. Once dispersed in the environment, the rapid dissipation characteristic of OP insecticides indicates that few chronic toxicity concerns are relevant. Although judicious use and appropriate management are required, contemporary evidence indicates that OP insecticides will continue to be popular and that their use will result in few long-term environmental effects of significant concern.

References

Abou-Assaf, N., and Coats, J. R. (1987). Degradation of [^{14}C] isofenphos in the laboratory under different soil pHs, temperatures, and moistures. *J. Environ. Sci. Health* **B22**, 285–301.

Agnihotri, N. P., Pandey, S. Y., Jain, H. K., and Srivastava, K.P. (1981). Persistence, leaching and movement of chlorfenvinphos, chlorpyriphos, disulfoton, fensulfothion, monocrotophos, and tetrachlorvinphos in soil. *Ind. J. Agric. Chem.* **14**, 27–31.

Alexander, M. (1981). Biodegradation of chemicals of environmental concern. *Science* **211**, 132–138.

Ahmad, N., Walgenbach, D. D., and Sutter, G. R. (1979). Comparative disappearance of fonofos, phorate, and terbufos soil residues under similar South Dakota field conditions. *Bull. Environ. Contam. Toxicol.* **23**, 423–429.

Anderegg, B. N., and Lichtenstein, E. P. (1984). Effects of light intensity and temperature on the uptake and metabolism of soil-applied [^{14}C]phorate by plants. *J. Agric. Food Chem.* **32**, 610–614.

Arthur, R. D., Cain, J. D., and Barrentine, B. F. (1976). Atmospheric levels of pesticides in the Mississippi delta. *Bull. Environ. Contam. Toxicol.* **15**, 129–134.

Baker, D. B., and Richards, R. P. (1989). "The Transport of Soluble Pesticides through Drainage Networks in Large Agricultural River Basins." Water Qual. Lab. Report, Heidelberg College, Tiffin, Ohio.

Baker, J. L., Johnson, H. P., and Laflen, J. M. (1976). "Effect of Tillage Systems on Runoff Losses of Pesticides: A Simulated Rainfall Study." Iowa State Water Resour. Res. Inst. Rep. ISWRRI-71, Iowa State University, Ames, Iowa.

Beynon, K. I., Edwards, M. J., Thompson, A. R., and Edwards, C. A. (1971). Persistence of chlorfenvinphos in natural waters. *Pestic. Sci.* **2**, 5–7.

Beynon, K. I., and Wright, A. N. (1969). Breakdown of the insecticide, Gardona, on plants and in soils. *J. Sci. Food Agric.* **20**, 250–257.

Bowman, B. T., and Sans, W. W. (1979). Aqueous solubility of insecticides. *J. Environ. Sci. Health* **B14**, 625–634.

Bowman, B. T., and Sans, W. W. (1983). Aqueous solubility of insecticidal compounds. *J. Environ. Sci. Health* **B18**, 221–227.

Braun, H. E., and Frank, R. (1980). Organochlorine and organophosphorus insecticides: Their use in eleven agricultural watersheds and their loss to stream waters in southern Ontario, Canada, 1975–1977. *Sci. Tot. Environ.* **15**, 169–192.

Brazner, J. C., Heinis, L. J., and Jensen, D. A. (1989). A littoral enclosure for replicated field experiments. *Environ. Toxicol. Chem.* **8**, 1209–1216.

Bro-Rasmussen, F., Noddegaard, E., and Voldum-Clausen, K. (1970). Comparison of the disappearance of eight organophosphorus insecticides from soil in laboratory and in outdoor experiments. *Pestic. Sci.* **1**, 179–182.

Chakrabarty, A., and Gennrich, S. M. (1987). "Vapor Pressure of Chlorpyrifos." Rep. ML-AL 87-40045. Dow Chemical U.S.A., Midland, Michigan.

Chapman, R. A., and Chapman, P. C. (1986). Persistence of granular and EC formulations of chlorpyrifos in a mineral and an organic soil incubated in open and closed containers. *J. Environ. Sci. Health* **B21**, 447–456.

Chapman, R. A., and Cole, C. M. (1982). Observations on the influence of water and soil pH on the persistence of insecticides. *J. Environ. Sci. Health* **B17**, 487–504.

Chapman, R. A. and Harris, C. R. (1982). Persistence of isofenphos and isazophos in a mineral and an organic soil. *J. Environ. Sci. Health* **B17**, 355–361.

Chapman, R. A., and Harris, C. R. (1984). The chemical stability of formulations of some hydrolyzable insecticides in aqueous mixtures with hydrolysis catalysts. *J. Environ. Sci. Health* **B19**, 397–407.

Chapman, R. A., Harris, C. R., Svec, H. J., and Robinson, J. R. (1984). Persistence and mobility of granular insecticides in an organic soil following furrow application for onion maggot control. *J. Environ. Sci. Health* **B19**, 259–270.

Chaudhry, G. R., Ali, A. N., and Wheeler, W. B. (1988). Isolation of methyl parathion-degrading *Pseudomonas* sp. that possesses DNA homologous to the *opd* gene from a *Flavobacterium* sp. *Appl. Environ. Microbiol.* **54**, 288–293.

Cohen, S. Z., Creeger, S. M., Carsel, R. F., and Enfield, C.G. (1984). *In* "Treatment and Disposal of Pesticide Wastes" (R. F. Krueger and J. N. Seiber eds.), pp. 297–325, ACS Symposium Series 259, American Chemical Society, Washington, D.C.

Cook, A. M., Daughton, C. G., and Alexander, M. (1978). Phosphorus-containing pesticide breakdown products: Quantitative utilization as phorphorus sources by bacteria. *Appl. Environ. Microbiol.* **36**, 668–672.

Cowgill, U. M., Gowland, R. T., Ramirez, C. A., and Fernandez, V. (1991). The history of a chlorpyrifos spill: Cartagena, Columbia. *Environ. Int.* **17**, 61–71.

Edwards, M. J., Beynon, K. I., Edwards, C. A., and Thompson, A. R. (1971). Movement of chlorfenvinphos in soil. *Pestic. Sci.* **2**, 1–4.

Felsot, A. S. (1984). Persistence of isofenphos (Amaze) soil insecticide under laboratory and field conditions and tentative identification of a stable oxygen analog metabolite by gas chromatography. *J. Environ. Sci. Health* **B19**, 13–27.

Felsot, A. S., and Dahm, P. A. (1979). Sorption of organophosphorus and carbamate insectides by soil. *J. Agric. Food Chem.* **27**, 557–563.

Felsot, A. S., Mitchell, J. K., and Kenimer, A. L. (1990). Assessment of management practices for reducing pesticide runoff from sloping cropland in Illinois. *J. Environ. Qual.* **19**, 539–545.

Fontaine, D. D., and Teeter, D. (1987). "Photodegradation of chlorpyrifos in the vapor phase." Rep. GH-C 1911. Dow Chemical U.S.A., Midland, Michigan.

Fontaine, D. D., Wetters, J. H., Weseloh, J. W., Stockdale, G. D., Young, J. R., Swanson, M. E. (1987). "Field dissipation and leaching of chlorpyrifos." Rep. GH-C 1957. Dow Chemical U.S.A., Midland, Michigan.

Frank, R., Braun, H. E., Ripley, B. D., and Clegg, B. S. (1990). Contamination of rural ponds with pesticide, 1971–1985, Ontario, Canada. *Bull. Environ. Contam. Toxicol.* **44**, 401–409.

Getzin, L. W. (1981a). Degradation of chlorpyrifos in soil: Influence of autoclaving, soil moisture, and temperature. *J. Econ. Entomol.* **74**, 158–162.

Getzin, L. W. (1981b). Dissipation of chlorpyrifos from dry soil surfaces. *J. Econ. Entomol.* **74**, 707–713.

Glotfelty, D. E., Seiber, J. N., and Lilijedahl, L. A. (1987). Pesticides in fog. *Nature* **325**, 602–605.

Gohre, K., and Miller, G. C. (1986). Photooxidation of thioether pesticides on soil surfaces. *J. Agric. Food Chem.* **34**, 709–713.

Grunwell, J. R., and Erickson, R. H. (1973). Photolysis of parathion [O,O-diethyl-O-(4-nitrophenyl)thiophosphate]. New products. *J. Agric. Food Chem.* **21**, 929–931.

Gustafson, D. I. (1989). Groundwater ubiquity score: A simple method for assessing pesticide leachability. *Environ. Toxicol. Chem.* **8**, 339–357.

Hallberg, G. R. (1989). Pesticide pollution of groundwater in the humid United States. *Agric. Ecosyst. Environ.* **26**, 299–367.

Harris, C. R., Chapman, R. A., Tolman, J. H., Moy, P., Henning, K. I., and Harris, C. (1988). A comparison of the persistence in a clay loam of single and repeated annual applications of seven granular insecticides used for corn rootworm control. *J. Environ. Sci. Health* **B23**, 1–32.

Hartley, J. J., and Hamish, H. R. (eds.). (1987). "The Agrochemicals Handbook" 2nd Ed., Royal Society of Chemistry, London, England.

Hassall, K. A. (1982). "The Chemistry of Pesticides: Their Metabolism, Mode of Action and Uses in Crop Protection." Verlag-Chemie, Weinheim, Germany.
Hebert, V. R., and Miller, G. C. (1990). Depth dependence of direct and indirect photolysis on soil surfaces. *J. Agric. Food Chem.* 38, 913–918.
Hill, E. F., and Camardese, M. B. (1984). Toxicity of anticholinesterase insecticides to birds: Technical grade versus granular formulations. *Ecotoxicol. Environ. Safety* 8, 551–563.
Hughes, D. N., Boyer, M. G., Papst, M. H., and Fowle, C. D. (1980). Persistence of three organophosphorus insecticides in artificial ponds and some biological implications. *Arch. Environ. Contam. Toxiocol.* 9, 269–279.
Kanazawa, J. (1989). Relationships between the soil sorption constants for pesticides and their physicochemical properties. *Environ. Toxicol. Chem.* 8, 477–484.
Katan, J., and Lichtenstein, E. P. (1977). Mechanism of production of soilbound residues of ^{14}C-parathion by microorganisms. *J. Agric. Food Chem.* 25, 1404–1408.
Kaufman, D. D. (1987). Accelerated biodegradation of several organophosphate pesticides. Rep. to NCR-46 Regional Technical Committee. U.S. Dept. Agric, Beltsville, Maryland.
Kenaga, E. E. (1971). Some physical, chemical and insecticidal properties of some O,O-dialkyl O-(3,5,6-trichloro-2-pyridiyl) phosphates and phosphorothioates. *Bull. World Health Org.* 44, 225–228.
Kenaga, E. E. (1976). "Environmental Evaluation of Chlorpyrifos-Methyl." Rep. GH-R 49. Dow Chemical U.S.A., Midland, Michigan.
Kenaga, E. E. (1980). Predicted bioconcentration factors and soil sorption coefficients of pesticides and other chemicals. *Ecotoxicol. Environ. Safety* 4, 26–38.
Korn, S, and Earnest, R. (1974). Acute toxicity of twenty insecticides to striped bass, *Morone saxatilis. Calif. Fish Game* 60, 128–131.
Laanio, T. L., Dupuis, G., and Esser, H. O. (1972). Fate of ^{14}C-labeled diazinon in rice, paddy soil, and pea plants. *J. Agric. Food Chem.* 20, 1213–1219.
Larson, L. L., Kenaga, E. E., and Morgan, R. W. (1985). "Commerical and Experimental Insecticides." Entomological Society of America, Hyattsville, Maryland.
Laskowski, D. A., Swann, R. L., McCall, P. J., and Bidlack, H. D. (1983). Soil degradation studies. *Residue Rev.* 85, 139–147.
Laveglia, J., and Dahm, P. A. (1977). Degradation of organophosphorus and carbamate insecticides in the soil and by soil microorganisms. *Annu. Rev. Entomol.* 22, 483–513.
Leistra, M., and Boesten, J. J. T. I. (1989). Pesticide contamination of groundwater in Western Europe. *Agric. Ecosyst. Environ.* 26, 369–389.
Lichtenstein, E. P., Fuhremann, T. W., Hochberg, A. A., Zahlten, R. N., and Stratman, F. W. (1973). *J. Agric. Food Chem.* 21, 416-424.
Lichtenstein, E. P., and Schultz, K. R. (1964). The effects of moisture and microorganisms on the persistence and metabolism of some organophosphorus insecticides in soils, with special emphasis on parathion. *J. Econ. Entomol.* 57. 618–627.
Macalady, D. L., and Wolfe, N. L. (1983). New perspectives on the hydrolytic degradation of the organophosphorothioate insecticides chlorpyrifos. *J. Agric. Food Chem.* 31, 1139–1147.
Macalady, D. L., and Wolfe, N. L. (1985). Effects of sediment sorption and abiotic hydrolyses. 1. Organophosphorus esters. *J. Agric. Food Chem.* 33, 167–173.
Matsumura, F. (1985). "Toxicology of Insecticides" 2nd Ed. Plenum Press, New York.
McCall, P. J. (1985). "Column Leaching and Sorption Studies with Chlorpyrifos." Rep. GH-C 1777. Dow Chemical U.S.A., Midland, Michigan.
McCall, P. J. (1986a). "Hydrolysis of Chlorpyrifos in Dilute Aqueous Solution." Rep. GH-C 1791. Dow Chemical U.S.A., Midland, Michigan.
McCall, P. J. (1986b). "Photodegradation of Chlorpyrifos in Aqueous Buffer." Rep. GH-C 1862. Dow Chemical U.S.A., Midland, Michigan.

McCall, P. J. (1987). "Soil Adsorption Properties of ^{14}C-chlorpyrifos." Rep. GH-C 1971. Dow Chemical U.S.A., Midland, Michigan.

McCall, P. J., Oliver, G. R., and McKellar, R. L. (1984). "Modeling the Runoff Potential and Behavior of Chlorpyrifos in a Terrestrial–Aquatic Watershed." Rep. GH-C 1694. Dow Chemical U.S.A., Midland, Michigan.

McCall, P. J., Swann, R. L., and Bauriedel, W. R. (1985). "Volatility Characteristics of Chlorpyrifos from Soil." Rep. GH-C 1782. Dow Chemical U.S.A., Midland, Michigan.

Menzer, R. E., Iqbal, Z. M., and Boyd, G. R. (1971). Metabolism of *O*-ethyl *S,S*-dipropyl phosphorodithioate (Mocap) in bean and corn plants. *J. Agric. Food Chem.* **19**, 351–356.

Miles, J. R. W., Tu, C. M., and Harris, C. R. (1979). Persistence of eight organophosphorus insecticides in sterile and nonsterile mineral and organic soil. *Bull. Environ. Contam. Toxicol.* **22**, 312–318.

Miles, J. R. W., Harris, C. R., and Tu, C. M. (1983). Influence of temperature on the persistence of chlorpyrifos and chlorfenvinphos in sterile and natural mineral and organic soils. *J. Environ. Sci. Health* **B18**, 705–712.

Miller, G. C., Hebert, V. R., and Miller, W. W. (1989). Effect of sunlight on organic contaminants at the atmosphere–soil interface. *In* "Reactions and Movements of Organic Chemicals in Soils" (B. L. Sawhney and K. Brown, eds.), pp. 99–110, SSSA Special Publication 22, Soil Science Society of America, Madison, Wisconsin.

Mulbry, W. W., Karns, J. S., Kearney, P. C., Nelson, J. O., McDaniel, C. S., and Wild, J. R. (1986). Identification of a plasmid-borne parathion hydrolase gene from *Flavobacterium* sp. by Southern hybridization with *opd* from *Pseudomonas diminuta. Appl. Environ. Microbiol.* **51**, 926–930.

Munnecke, D. M., and Hsieh, D. P. H. (1976). Pathways of microbial metabolism of parathion. *Appl. Environ. Microbiol.* **31**, 63–69.

Nelson, L. M. (1982). Biologically induced hydrolysis of parathion in soil: Isolation of hydrolyzing bacteria. *Soil Biol. Biochem.* **14**, 219–222.

Oliver, G. R., McKellar, R. L., Woodburn, K. B., Eger, J. E., McGee, G. G., and Ordiway, T. R. (1987). "Field Dissipation and Leaching Study for Chlorpyrifos in Florida Citrus." Rep. GH-C 1870. Dow Chemical U.S.A., Midland, Michigan.

O'Neill, E. J., Cripe, C. R., Mueller, L. H., Connolly, J.P., and Pritchard, P.H. (1989). Fate of fenthion in salt-marsh environments: II. Transport and biodegradation in microcosms. *Environ. Toxicol. Chem.* **8**, 759–768.

Packard, S. R. (1987). "Determination of the Water Solubility of Chlorpyrifos." Rep. ML-AL 87-71102. Dow Chemical U.S.A., Midland, Michigan.

Pankaskie, J. E., Fountaine, F. C., and Dahm, P. A. (1952). The degradation and detoxication of parathion in dairy cows. *J. Econ. Entomol.* **45**, 51–60.

Paris, D. F., and Lewis, D. L. (1973). Chemical and microbial degradation of ten selected pesticides in aquatic systems. *Residue Rev.* **45**, 95–124.

Parsons, D. W., and Witt, J. M. (1988). "Pesticides in Groundwater in the United States of America: A Report of a 1988 Survey of State Lead Agencies." Oregon State University, Corvallis, Oregon.

Racke, K. D., and Coats, J. R. (1987). Enhanced degradation of isofenphos by soil microorganisms. *J. Agric. Food Chem.* **35**, 94–99.

Racke, K. D., and Coats, J. R. (1988). Comparative degradation of organophosphorus insecticides in soil: Specificity of enhanced microbial degradation. *J. Agric. Food Chem.* **36**, 193–199.

Racke, K. D., and Coats, J. R. (1990). "Enhanced Biodegradation of Pesticides in the Environment." ACS Symposium Series 426. American Chemical Society, Washington, D.C.

Racke, K. D., Laskowski, D. A., and Schultz, M. R. (1990). Resistance of chlorpyrifos to enhanced biodegradation in soil. *J. Agric. Food Chem.* **38**, 1430–1436.

Racke, K. D., and Lichtenstein, E. P. (1987). Effects of agricultural practices on the binding and fate of ^{14}C-parathion in soil. *J. Environ. Sci. Health* **B22**, 1–14.

Raven, P. J., and George, J. J. (1989). Recovery by riffle macroinvertebrates in a river after a major accidental spillage of chlorpyrifos. *Environ. Pollut.* **59**, 55–70.

Read, D. C. (1983). Enhanced microbial degradation of carbofuran and fensulfothion after repeated applications to acid mineral soil. *Agric. Ecosyst. Environ.* **10**, 37–46.

Ritter, W. F., Johnson, H. P., Lovely, W. G., and Molnau, M. (1974). Atrazine, propachlor, and diazinon residues on small agricultural watersheds: Runoff losses, persistence, and movement. *Environ. Sci. Technol.* **8**, 38–42.

Sauer, T. J., and Daniel, T. C. (1987). Effect of tillage system on runoff losses of surface-applied pesticides. *Soil. Sci. Soc. Am. J.* **51**, 410–415.

Sears, M. K., Bowhey, C., Braun, H., and Stephenson, G. R. (1987). Dislodgeable residues of diazinon, chlorpyrifos, and isofenphos following their application to turfgrass. *Pestic. Sci.* **20**, 223–231.

Sears, M. K., and Chapman, R. A. (1979). Persistence and movement of four insecticides applied to turfgrass. *J. Econ. Entomol.* **72**, 272–274.

Seiber, J. N., McChesney, M. M., and Woodrow, J. E. (1989). Airborne residues resulting from use of methyl parathion, molinate, and thiobencarb on rice in the Sacramento Valley, California. *Environ. Toxicol. Chem.* **8**, 577–588.

Sethunathan, N., and Yoshida, T. (1973). *A Flavaobacterium* sp. that degrades diazinon and parathion. *Can. J. Microbiol.* **19**, 873–875.

Sheets, T. J., Bradley, J. R., and Jackson, M. D. (1972). "Contamination of Surface and Ground Water with Pesticides Applied to Cotton." Water Resour. Res. Inst. Rep. 60, University of North Carolina, Chapel Hill, North Carolina.

Shelton, D. R., and Karns, J. S. (1988). Coumaphos degradation in cattle-dipping vats. *J. Agric. Food Chem.* **36**, 831–834.

Singh, A. K., and Seth, P. K. (1989). Degradation of malathion by microorganisms isolated from industrial effluents. *Bull. Environ. Contam. Toxicol.* **43**, 28–35.

Skujins, J. (1976). Extracellular enzymes in soil. *CRC Crit. Rev. Microbiol.* **6**, 383–421.

Smelt, J. H., Leistra, M., and Voerman, S. (1977). Movement and rate of decomposition of ethoprophos in soil columns under field conditions. *Pestic. Sci.* **8**, 147–151.

Smith, G. N. (1968). Ultraviolet light decomposition studies with Dursban and 3,5,6-trichloro-2-pyridinol. *J. Econ. Entomol.* **61**, 793–799.

Smith, G. N., Watson, B. S., and Fischer, F. S. (1967). Investigations on Dursban insecticide. Uptake and translocation of [^{36}Cl]O,O,-diethyl O-3,5,6-trichloro-2-pyridyl phosphorotioate and [^{14}C]O,O,-diethyl O-3,5,6-trichloro-2-pyridyl phosphorotioate by beans and corn. *J. Agric. Food Chem.* **15**, 127–131.

Spain, J. C., Pritchard, P. H., and Bourquin, A. W. (1980). Effects of adaptation on biodegradation rates in sediment/water cores from estuarine and freshwater environments. *Appl. Environ. Microbiol.* **40**, 726–734.

Spalding, R. F., and Snow, D. D. (1989). Stream levels of agrichemicals during a spring discharge event. *Chemosphere* **19**, 1129–1140.

Spencer, W. F., Shoup, T. D., Cliath, M. M., Farmer, W. J., and Haque, R. (1979). Vapor pressures and relative volatility of ethyl and methyl parathion. *J. Agric. Food Chem.* **27**, 273–278.

Suett, D. L. (1975). Persistence and degradation of chlorfenvinphos, chormephos, disulfoton, phorate and pirimiphos-ethyl following spring and late-summer soil application. *Pestic. Sci.* **6**, 385–393.

Sumner, D. D., Keller, A. E., Honeycutt, R. C., and Guth, J. A. (1987). Fate of diazinon in the environment. *In* "Fate of Pesticides in the Environment." F. W. Biggar and J. M. Seiber, eds.), 109–113. Agric. Exp. Sta. Publication 3320, University of California, Oakland, California.

Suzuki, T., and Uchiyama, M. (1975). Pathway of nitro reduction of parathion by spinach homogenate. *J. Agric. Food Chem.* 23, 281–286.

Szeto, S. Y., Brown, M. J., Mackenzie, J. R., and Vernon, R. S. (1986). Degradation of terbufos in soil and its translocation into cole crops. *J. Agric. Food Chem.* 34, 876–879.

Tagatz, M. E., Borthwick, P. W., Cook, G. H., and Coppage, D. L. (1974). Effects of ground applications of malathion on salt-marsh environments in northwestern Florida. *Mosquito News* 34, 309–315.

Thompson, A. R. (1971). Effects of nine insecticides on the numbers and biomass of earthworms in pasture. *Bull. Environ. Contam. Toxicol.* 5, 577–586.

Tu, C. M. (1970). Effect of four organophosphorus insecticides on microbial activities in soil. *Appl. Microbiol.* 19, 479–484.

Tu, C. M. (1980). Influence of pesticides and some of the oxidized analogues on microbial populations, nitrification, and respiration activities in soil. *Bull. Environ. Contam. Toxicol.* 24, 13–19.

Walter-Echols, G., and Lichtenstein, E. P. (1977). Microbial reduction of phorate sulfoxide to phorate in a soil–lake mud–water microcosm. *J. Econ. Entomol.* 70, 505–509.

Ware, G. W., Buck, N. A., and Estesen, B. J. (1983). Dislodgeable insecticide residues on cotton foliage: Comparison of ULV/cottonseed oil vs. aqueous dilutions of 12 insecticides. *Bull. Environ. Contam. Toxicol.* 31, 551–558.

Watschke, T. L., and Mumma, R. O. (1989). "The Effect of Nutrients and Pesticides Applied to Turf on the Quality of Runoff and Percolating Water." Environ. Resources Res. Inst. Report ER 8904, Pennsylvania State University, University Park, Pennsylvania.

Wauchope, R. D. (1978). The pesticide content of surface water draining from agricultural fields—a review. *J. Environ. Qual.* 7, 459–472.

Wauchope, R. D. (1988). "Pesticide Properties: A Selection of Fields from the USDA-ARS Interim Pesticide Properties Database." Version 1.0. U.S. Department of Agriculture, Washington, D.C.

Weber, J. B., and Miller, C. T. (1989). Organic chemical movement over and through soil. *In* "Reactions and Movements of Organic Chemicals in Soils" (B. L. Sawhney and K. Brown, eds.), pp. 305–334. SSSA Special Publication 22, Soil Science Society of America, Madison, Wisconsin.

WHO (1986). "Organophosphorus Insecticides: A General Introduction." Environ. Health Crit. Report 63, World Health Organization, Geneva, Switzerland.

Willis, G. H., and McDowell, L. L. (1987). Pesticide persistence on foliage. *Residue Rev.* 100, 23–73.

Wintersteen, W., and Harzler, R. (1987). "Pesticides Used in Iowa in 1985." Coop. Ext. Serv. Report Pm-1288, Iowa State University, Ames, Iowa.

Youngman, R. R., Toscano, N. C., and Gaston, L. K. (1989). Degradation of methyl parathion to *p*-nitrophenol on cotton and lettuce leaves and its effects on plant growth. *J. Econ. Entomol.* 82, 1317–1322.

4

Species-Selective Toxicity of Organophosphorus Insecticides: A Pharmacodynamic Phenomenon

Kendall B. Wallace
Department of Pharmacology
School of Medicine
University of Minnesota
Duluth, Minnesota

I. Introduction

The popularity of OP compounds as broad-spectrum insecticidal agents is compromised by the high incidence of inadvertent intoxication of nontarget organisms. Their long history of use has yielded an exhaustive literature replete with reported cases of poisonings of nontarget organisms, including numerous human fatalities. Several factors contribute to this problem. Broadcast spraying of OP agents affects virtually all organisms coming in contact with the poison; selective eradication of the unwanted pests is determined by the specificity of the application procedure and by the inherent susceptibility

Organophosphates: Chemistry, Fate, and Effects

of the exposed organisms. The incidence of inadvertent poisonings can, therefore, be minimized by improving the method for selectively exposing the targeted pest, and is a function of the relative sensitivities of the exposed target and nontarget organisms to the toxic effects of the applied agent.

Comparative toxicity testing reveals a general tenet that birds are extremely susceptible, rodents intermediate, and fish and amphibians relatively resistant to acute OP intoxication (Murphy *et al.*, 1968; Benke *et al.*, 1974; Brealey *et al.*, 1980; Walker, 1983; Chattopadhyay *et al.*, 1986). The data also attest to the dilemma imposed by OP insecticides when constructing predictive models for extrapolating acute toxicity between unrelated species. Invariably, the OP insecticides represent one of the few chemical classes that fail to conform to interspecies regressions of acute toxicity for a wide array of pesticides and industrial chemicals (Bathe *et al.*, 1976; Kenaga, 1978, 1980; Janardan *et al.*, 1984; Wallace and Niemi, 1988). These observations of nonconformity have evoked considerable interest in identifying the principal determinant of the species-selectivity of OP insecticides. Resolution of this phenomenon not only will aid in the extrapolation of toxicity testing between unrelated species but also may reveal valuable insight for the development of new, more species-specific OP agents. Thus, identification of key features conferring species-selectivity can be exploited to minimize the incidence and severity of inadvertent intoxication of nontarget organisms. As a result, the overall efficacy of OP insecticidal applications may be vastly improved.

Species-related differences in sensitivity to OP intoxication may reflect differences in any one or a combination of the kinetic steps involved in the disposition of the agent in the animal. These include the absorption, distribution, metabolic transformation, and elimination of OP compounds from the body. Alternatively, species-sensitivity may be manifested at the receptor level; differences in the sensitivity of acetylcholinesterase (AChE) to inhibition by OP agents or differences in the regulation of nicotinic or muscarinic receptors by acetylcholine. This chapter provides a systematic overview of species-related differences for each step involved in the disposition and biological activity of OP insecticides. The bulk of the discussion will focus on parathion (*O,O*′-diethyl-*p*-nitro-phenylphosphorothioate) as the prototype. However, the majority of the concepts presented have been demonstrated for numerous other OP insecticides.

II. Absorption and Elimination

The pharmacodynamic responsiveness of an organism to an administered xenobiotic is a function of the steady-state concentration of the biologically active substance at the site of action, which is presumably in equilibrium with the systemic circulation. The systemic concentration is, in turn, determined by the relative rates of absorption and elimination, the rate of absorption being influenced by the route of administration. Consequently, definitive comparisons of species-related differences in pharmacokinetics are restricted to those studies in which the agent is administered via the same route to each species being compared. This complicates studies comparing rodents to aquatic organisms in that, as opposed to mammals, which most often are injected with the toxin, the majority of aquatic studies employ dissolving the toxin in the gill water and estimating the toxic dose as the median lethal concentration (LC_{50}). Consequently, correlations between fish and rodents most frequently consist of regressing the aquatic LC_{50} against the median lethal dose (LD_{50}) in mammals (Bathe *et al.*, 1976; Kenaga, 1978, 1980; Janardan *et al.*, 1984; Wallace and Niemi, 1988). Any variability between species may be ascribed to differences in the absorbed dose. The limited number of studies employing a common route of exposure (intraperitoneal injection) substantiate the relative insensitivity of sunfish and frogs to acute parathion intoxication, the selectivity ratio being between 2 and 10 when compared to mice, rats, or chickens (Benke *et al.*, 1974; Chattopadhyay *et al.*, 1986). Although definitive kinetic studies have not been performed, circumstantial evidence suggests that subtle differences in the rates of absorption and elimination are insufficient to account for the extreme insensitivity of fish compared to rodents to acute parathion intoxication (Hodson, 1985). Final resolution of the significance, or lack thereof, of species-related differences in the pharmacokinetic disposition of parathion awaits the results of more rigorous kinetic investigations.

III. Distribution

The distribution of OP insecticides in biological organisms is governed by two basic physical chemical principles; hydrophobic partitioning and nonspecific electronic interactions with various nucleophilic substituents. Accordingly, species-related differences in lipid composition or in the concentration of nonspecific binding sites, such as serum albumin or selected enzymes, may affect the concentration of free pesticide available to interact with critical biological components (Lauwerys and Murphy, 1969; Sultatos *et al.*, 1984). The association of the active metabolite of parathion (paraoxon) with nonspecific binding sites on soluble aliesterase enzymes may play a major role in limiting the toxicity of the compound (Lauwerys and Murphy, 1969; Chambers *et al.*, 1990; Chambers and Chambers, 1990). Alternatively, albumin may promote the nonenzymatic hydrolysis of OP esters rendering the compounds biologically inactive (Sultatos *et al.*, 1984). Regardless of the mechanism, these nonspecific binding sites serve as reservoirs for OP compounds, limiting the free plasma concentration and reducing the acute lethal effects of the insecticides. Species-related differences in the abundance of these nonspecific binding sites may be a significant factor in the species selectivity of OP intoxication.

Administration of tri-*o*-tolyl phosphate (TOTP), which phosphorylates nucleophilic sites and inhibits the nonspecific binding of OP agents, markedly potentiates acute paraoxon toxicity in rats (Lauwerys and Murphy, 1969). Maxwell *et al.* (1987) demonstrated that selective inhibition of carboxylesterase *in vivo* with cresylbenzodioxaphosphorin (CBDP) reduces the LD_{50} for soman in four species of rodents by 50 to 90% and eliminates the large differences in sensitivity to acute soman intoxication among these species. The results indicate that, although soman is not a substrate for carboxylesterases, interspecies differences in soman toxicity are affected by the binding of the phosphonofluoridate to nonspecific catalytic regions of the enzyme. Whether species-related differences in nonspecific binding to carboxylesterase contribute to the differences in sensitivity among nonrodent species is, however, inconclusive. Wang and Murphy (1982b) suggested that noncritical binding of diisopropyl fluorophosphate (DFP) or phosphorofluoridate to synaptosomal proteins is not a factor in distinguishing anticholinesterase activity between rats, chickens, and monkeys.

IV. Metabolism

The metabolic disposition of OP insecticides represents the concerted action of numerous esterase enzymes functioning both in series and in parallel with the oxidative desulfuration of phosphorothioates to yield complex mixtures of metabolites (see Chapters 1, and 6–10, this volume). Using parathion as an example, the metabolic disposition can be simplified by categorizing the reactions into two distinct types: metabolic activation by oxidative desulfuration of the phosphorothioate to yield the biologically active oxygen analog paraoxon (*O*,*O*′-diethyl-*p*-nitro-phenylphosphate), and detoxication by the arylesterase-catalyzed hydrolysis of either parathion or paraoxon to yield biologically inactive *p*-nitrophenyl phosphate and either diethyl thiophosphate or diethyl phosphate, respectively (see Fig. 1 of Chapter 3 and Fig. 1 of Chapter 10, this volume). In addition, the dimethyl, but not the diethyl, substituted analog of parathion is also subject to conjugation with glutathione (Benke *et al.*, 1974). A detailed description of each of these reactions is presented in Chapters 1 and 6–10, in this volume. The following paragraphs are limited to reviewing the information regarding species comparisons as they relate to differentiating the sensitivities to OP intoxication among different organisms.

A. Activation of Parathion

The fact that the acute toxicity of injected parathion is considerably less than that of paraoxon provides circumstantial evidence for the metabolic activation of the phosphorothioate (Main, 1956; Benke *et al.*, 1974; Forsyth and Chambers, 1989). *In vitro*, parathion is virtually devoid of anticholinesterase activity. However, it is converted *in vivo* to paraoxon, which is a very potent inhibitor of AChE (Diggle and Gage, 1951; Gage, 1953). The mechanism of metabolic activation of parathion to paraoxon involves cytochrome P450-mediated oxidative desulfuration and is described in detail in Chapters 1, 3, and 10, in this volume.

Comparative testing has revealed large differences in the rates of enzymatic activation of parathion among different organisms, which may be an important factor in determining species sensitivity. Virtually all species examined are capable of activating parathion to paraoxon. Potter and O'Brien (1964) compared the rates of parathion activation by liver slices from birds, fish, and amphibians with no obvious relationship between the different organisms. The activity in mammals, however, tended to be higher than that of the other species examined. Murphy (1966) observed large variability between different organisms in the rate of accumulation of paraoxon bioequivalents (measured by bioassay) in liver slices incubated with parathion at 38°C. The highest rates were observed for sparrow, sunfish, and bullhead. However, when the experiments with liver slices from sunfish were repeated at 18°C, which is closer to the normal body temperature, the accumulation of paraoxon equivalents was only 25% that observed at 38°C and was no different from that observed for rodents (Murphy, 1966). In contrast, subsequent investigations with liver homogenates revealed a twofold greater rate of parathion activation by mammalian liver, with rodents > birds > fish (Hitchcock and Murphy, 1971). The results of these investigations are based on the net accumulation of paraoxon, which is dependent on both the rate of activation of parathion and the rate of degradation of the active oxygen analog. No attempt was made in either of the studies to prevent the degradation of paraoxon once formed. Consequently, the results fail to distinguish between the generation and degradation of paraoxon and fail to provide quantitative comparisons of the rates of parathion activation among the organisms. When the assays were conducted under conditions that prohibit the metabolic degradation of paraoxon, it was found that rodents are far more capable of activating parathion than are fish (Wallace and Dargan, 1987). The desulfuration of parathion by liver tissue from rats and mice was 33- to 76-fold greater than that for trout or minnow. This lower rate of metabolic activation by fish is consistent with their resistance to acute parathion toxicity *in vivo*. However, the converse is not true; the slower rate of activation of parathion by birds compared to that of rodents is contradictory to the extreme sensitivity of birds to OP intoxication *in vivo* (Murphy, 1966; Hitchcock and Murphy, 1971).

B. Detoxication of Parathion

Parathion may be hydrolyzed by serum or tissue arylesterases to yield biologically inactive diethyl phosphorothioate and *p*-nitrophenol. The activity of these enzymes toward parathion is, however, far less than that observed for paraoxon, and thus the role of arylesterases in hydrolyzing parathion is considered to be negligible (Aldridge, 1953a). The principal mechanism responsible for the hydrolysis of parathion appears to be the same reaction involved in the activation of the compound. Rearrangement of the unstable S-oxides formed as intermediates in the cytochrome P450-catalyzed desulfuration of parathion (Fig. 1, Chapter 10, this volume) results in cleavage of the aryl-phosphate bond, oxidative dearylation (Neal, 1967; Kamataki *et al.*, 1976). Consequently, the generation of diethylphosphorothioate is thought to represent a rearrangement product of the cytochrome P450-mediated activation of parathion. This is consistent with the fact that species-related differences in parathion hydrolysis parallel the differences in the oxidative desulfuration (Wallace and Dargan, 1987).

Benke *et al.* (1974) reported that liver homogenates from sunfish incubated at 37°C in the presence of an NADPH-regenerating system catalyze the hydrolysis of parathion at approximately one half the rate catalyzed by mouse liver homogenates. A much more dramatic species difference is observed when the incubations are conducted at the physiological temperatures of the respective organisms (Wallace and Dargan, 1987). The rate of hydrolysis of parathion by liver homogenates from either rats or mice incubated at 37°C is 20- to 50-fold greater than the corresponding rates observed for fathead minnows (incubated at 25°C) or rainbow trout (11°C incubation temperature) liver fractions. This large species difference in the hydrolytic detoxication of parathion *in vitro* is, however, inconsistent with the resistance of fish to acute poisoning *in vivo* and is probably of little consequence to determining species sensitivity. As will be discussed, other parameters appear to be of far greater significance in determining species sensitivity.

C. Detoxication of Paraoxon

The hydrolysis of paraoxon is catalyzed by nonspecific arylesterases (A-esterases; EC 3.1.1.2) distributed ubiquitously in plasma and tissues of virtually all organisms. These esterases differ from the B-esterases in their rapid rate of dephosphorylation of the active site. The B-esterases, such as ali- and carboxyl esterases (EC 3.1.1.1) and AChEs (EC 3.1.1.7), reactivate very slowly and thus are subject to inhibition by OP compounds, which are suicide substrates for these enzymes. The products of paraoxon hydrolysis by A-esterases, diethyl phosphate, and *p*-nitrophenol are devoid of anticholinesterase activity, and thus the reaction represents a detoxication pathway. Although hepatic metabolism has been suggested to correlate only poorly with toxicity (see discussion of target-site activation in Chapter 11, this volume), hydrolysis of paraoxon by esterases in plasma has been invoked as the principal mechanism of detoxication *in vivo*. Butler and associates (1985) suggest that the hydrolysis of paraoxon is the major mechanism by which the compound is eliminated from the body, and that the sensitivity of an organism to intoxication is inversely proportional to serum arylesterase activity. Furthermore, it has been proposed that the bimodal distribution of plasma A-esterases in humans may account for the idiosyncrasies among individuals in their sensitivity to OP poisoning (Eckerson *et al.*, 1983; La Du and Eckerson, 1984).

Because of the implication of esterase activity as an important determinant of OP toxicity, numerous studies have been devoted to comparing the activity of plasma arylesterases among different organisms in an attempt to account for the species-selectivity observed *in vivo*. Collectively, the data provide well-documented evidence for large differences in the hydrolysis of OP esters by plasma arylesterases among various species. In contrast to insects, which lack measurable paraoxon-hydrolyzing activity (Devonshire and Moores, 1982; Mackness *et al.*, 1983; Motoyama *et al.*, 1984), plasma and tissues from numerous species of birds, mammals, fish, and amphibians contain variable amounts of this A-esterase activity. In mammals, serum arylesterase activity is proportionate to paraoxon-hydrolyzing activity (paraoxonase activity) and varies between different species (Mendoza *et al.*, 1976, 1977; Brealey *et al.*, 1980). The fact that serum paraoxon–hydrolyzing activity is lower in human sera than in any other mammalian species suggests that humans may be particularly sensitive to intoxication by OP compounds (Mendoza *et al.*, 1977; Brealey *et al.*, 1980).

The very slow rate of hydrolysis of paraoxon by liver and plasma from several species of birds correlates with the extreme sensitivity of avian species to OP intoxication (Machin *et al.*, 1976; Brealey *et al.*, 1980; Walker, 1983; Walker and Mackness, 1987). Paraoxon hydrolysis by liver homogenates and by plasma from chicken is at least 50-fold lower than that for the corresponding tissues from three species of rodents (Lauwerys and Murphy, 1969). Brealey and associates (1980) reported virtually undetectable levels of OP-hydrolyzing activity in plasma from 14 different species of birds, including chicken. This contrasts with rodents, sheep, and humans, which possess abundant plasma arylesterase activity (Brealey *et al.*, 1980). This association of low plasma and liver arylesterase activity and the extreme sensitivity to OP toxicity in birds is strongly suggestive of a causal relationship. However, the definitive experiments to test this coincidence remain to be conducted.

These species differences were observed only when paraoxon hydrolysis was assessed from the rate of liberation of *p*-nitrophenol. When measured by bioassay for anticholinesterase activity, there was virtually no difference (Lauwerys and Murphy, 1969). This lack of correlation between the results of chemical analysis versus bioassay to estimate paraoxon hydrolysis can be attributed to the interference with the bioassay by the binding to and phosphorylation of assorted B-esterases other than acetylcholinesterase.

Murphy (1966) reported that the rate of paraoxon hydrolysis by liver slices from rats or mice greatly exceeds not only that observed for sparrows and chickens, but also that for four species of fish. When analyzed by the formation of *p*-nitrophenol, paraoxon hydrolysis by liver slices from fish and birds was comparable and ranged from 10- to 300-fold less than the corresponding rates for rodents. The lower activity of paraoxon-hydrolyzing enzymes in fish compared to rats has been confirmed employing liver homogenates or plasma (Benke *et al.*, 1974; Wallace and Dargan, 1987). The rate of paraoxon hydrolysis by mouse liver homogenates is 40-fold higher than that for sunfish when all incubations are conducted at 37°C (Benke *et al.*, 1974). When the reactions are conducted at the respective physiological temperatures, the differences between rodents (rats or mice) and fish (minnows or trout) are even more dramatic (Wallace and Dargan, 1987). In contrast to birds, wherein the lower paraoxonase activity correlates with the greater susceptibility to intoxication, the lower activity of these arylesterase enzymes in fish compared to rodents is inconsistent with the resistance of fish to OP intoxication. Consequently, the relationship between paraoxon hydrolyzing activity and *in vivo* sensitivity may be purely coincidental, and other factors are more likely to be responsible for conferring the observed species selectivity.

D. Summary of Species Differences in Metabolism

To summarize the species-related differences in OP metabolism, the activities of both the activating and the detoxifying enzymes are substantially greater in rodents compared to either birds or fish. Although arguments can be made for causal relationships between the activity of any single enzyme and species-selective toxicity, the complex independence of the individual reactions nullifies any attempt to draw correlations based solely on the analysis of species-related differences in a single metabolic process. The fact that rodents are more proficient at catalyzing both the activation and the detoxication of OP compounds, but are of intermediate sensitivity compared to birds and fish, attests to the pitfalls of drawing such broad generalizations. To gain a more definitive assessment of the importance of metabolism as a determinant of species-sensitivity requires integration of the kinetic data for each of the involved reaction steps. Initial attempts at this suggest that despite large differences in specific activities of the individual enzymes, the net accumulation of the biologically active intermediate may be quite similar in fish and rodents (Wallace and Dargan, 1987). Accordingly, the importance of differences in the metabolic activation or detoxication of OP insecticides in determining species-selectivity continues to be a matter of controversy.

V. Acetylcholinesterase Inhibition

The sensitivity of acetylcholinesterase (AChE; EC 3.1.1.7) to *in vitro* inhibition is an important determinant of the acute toxicity of OP insecticides. This is illustrated by the correlations between the concentration of an OP required to inhibit 50% of the AChE activity *in vitro* (IC_{50}) and the acute LD_{50} measured *in vivo*. Such correlations have been established for a wide range of OP compounds in numerous organisms. Fukuto and Metcalf (1956, 1959) demonstrated a direct correlation between the IC_{50} for a series of diethyl-substituted phenyl phosphates measured against fly brain AChE *in vitro* and the respective topical toxicities. This provided the first indication that *in vitro* inhibition of AChE provides a reliable, and possibly predictive, measure of OP toxicity to the whole organism (Fig. 1). Furthermore, it may be implied from this correlation that the sensitivity of AChE to inhibition is a significant, if not the principal, determinant of the *in vivo* toxicity of various OP agents. Similar correlations between the *in vitro* IC_{50} and acute toxicity in insects have since been confirmed for a broad array of *p*-nitrophenyl alkylphosphonates and carbamates (Fukuto and Metcalf, 1959; Metcalf and Frederickson, 1965). Darlington and coworkers (1971) provided evidence that suggests that the *in vitro* inhibition of fly brain AChE is not only predictive of the insecticidal activity of *p*-nitrophenyl ethylphosphonates, but that this value correlates also with the acute oral toxicity in rats. Measurement of the *in vitro* inhibition constant for bovine erythrocyte AChE provides a reliable indication of both the topical LD_{50} in flies and the oral LD_{50} in rats for a series of *para*-substituted O-methyl O-phenyl enthylphosphonates (Darlington *et al.*, 1971). Becker *et al.* (1964) extended these relationships in rodents wherein the acute toxicities of alkyl and phenylalkyl phosphonates to rabbits and guinea pigs correlate with the *in vitro* inhibition of erythrocyte AChE from the respective species (Fig. 2). The regression coefficient for this correlation was 0.76. Lotti and Johnson (1978) reported a similar relationship in chickens showing that the inhibition of brain AChE *in vitro* correlates with the acute toxicity for a diverse array of unrelated OP pesticides. In addition, correlation between the inhibition of brain neurotoxic esterase (NTE) and the estimated dose that causes ataxia in hens suggests a possible and convenient method for estimating the neurotoxic dose for various OP agents (Lotti and Johnson, 1978). The authors suggest that measurement of the relative potencies of OP agents

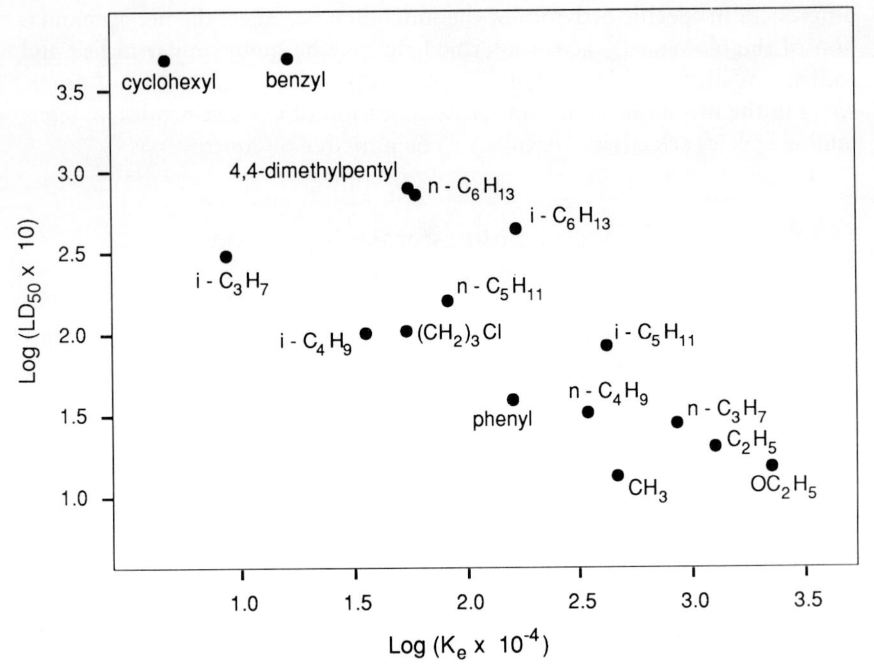

Figure 1 Correlation between the *in vitro* inhibition constant (K_e) for fly head cholinesterase and the acute topical toxicity of a series of ethyl *p*-nitrophenyl alkylphosphonates to house flies. Reproduced with permission from Fukuto, T. R., and Metcalf, R. L. (1959).

toward AChE and NTE of hens versus humans may provide a reliable means for extrapolating acute neurotoxicity data gathered for hens to predict human health hazards associated with these agents (see Chapter 16 by Richardson, in this volume, for a discussion of NTE).

Collectively, these correlations provide strong evidence supporting the theory that the sensitivity of AChE to inhibition by OP compounds is the principal determinant of acute toxicity *in vivo*. It may be implied from these relationships that species-selectivity is a pharmacodynamic phenomenon, mediated at the level of differences in the sensitivity of the AChE to inhibition by OP insecticides. Indeed the relative insensitivity of mice compared to insects to intoxication by *O,O*-dimethyl-*O*-2,2-dichlorovinyl phosphate (DDVP) is consistent with the lower sensitivity of mouse brain AChE to *in vitro* inhibition by this agent (Van Asperen and Dekhuijzen, 1958). Not only is the affinity of mouse brain AChE for DDVP lower, but once inhibited, the enzyme rapidly reactivates spontaneously. Inhibition of housefly AChE by DDVP, in contrast, is irreversible. Dauterman and O'Brien (1964) noted that isopropyl parathion is 100-fold more toxic to houseflies than to bees, and that

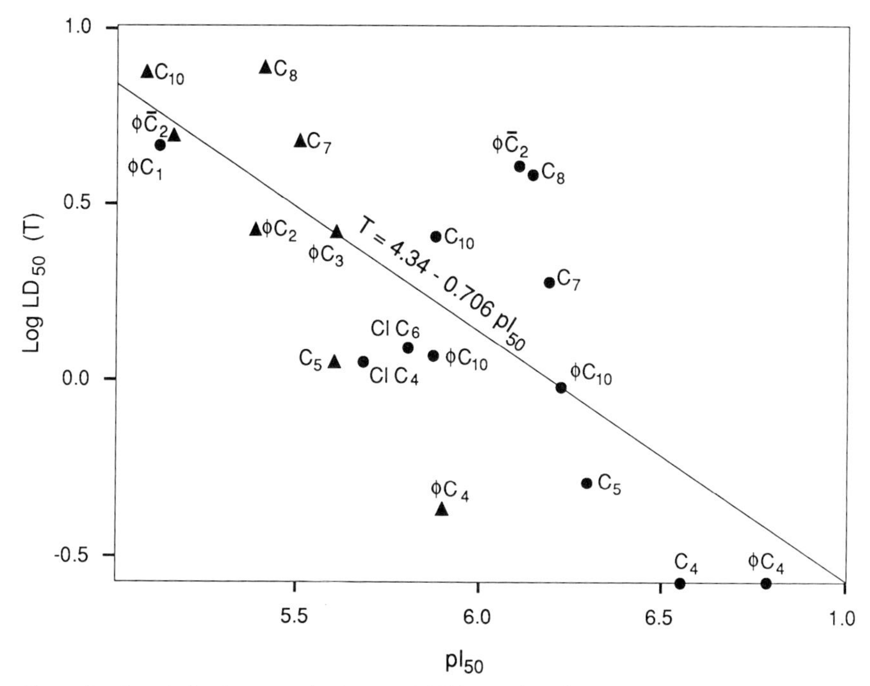

Figure 2 Correlation between the *in vitro* inhibition of erythrocyte AChE and the acute intravenous toxicity of a series of alkyl and phenyl-alkylphosphonates in rabbits and guinea pigs. Reproduced with permission from Becker, E. L., Punte, C. L., and Barbaro, J. F. (1964).

associated with this difference is a 40-fold greater sensitivity of housefly AChE to *in vitro* inhibition by isopropyl paraoxon. It has also been suggested that the resistance of certain strains of mites is conferred, to a large degree, by the insensitivity of AChE to inhibition by various OP and carbamate insecticides (Zahavi *et al.*, 1971; Anber and Overmeer, 1988).

As mentioned earlier, chickens and birds are more sensitive to acute OP poisoning than are rodents, while fish and amphibians are relatively resistant. These differences in the sensitivities of chickens, mice, sunfish, and bullheads to acute poisoning by parathion and guthion correlate well with differences in sensitivity of brain AChE from the respective species to *in vitro* inhibition by paraoxon and gutoxon (Murphy *et al.*, 1968). Chicken brain AChE is also more sensitive than rat brain AChE to inhibition by DFP, or by 2-O-(isopropyl ethyl phosphono)-acetyl-1-pyridine oxime (2-MPA-ES) or its methyl pyridinium analog 2-MPAM-ES (Andersen *et al.*, 1972). Frog brain AChE, on the other hand, is insensitive to inhibition by these agents. Similar relationships in AChE sensitivity between species (chickens > rodents > fish or amphibians) have been reported for numerous OP agents, including soman, sarin, tabun, mipafox, paraoxon and methylparaoxon, malaoxon, gutoxon and ethylgutoxon, and dimethylphosphoryl fluoride and diethylphosphoryl fluoride (Andersen *et al.*, 1977; Wang and Murphy, 1982b; Chattopadhyay *et al.*, 1986; Johnson and Wallace, 1987; Kemp and Wallace, 1990).

Although valid for brain and erythrocyte AChE, measurement of the inhibition of pseudocholinesterase in plasma does not yield a reliable indication of species differences in susceptibility to OP poisoning (Ecobichon and Comeau, 1973). Furthermore, the sensitivity of brain AChE to inhibition, rather than species-related differences in total AChE activity, is the critical determinant of species susceptibility (Chattopadhyay *et al.*, 1986; Johnson and Wallace, 1987). In addressing the obvious question—where humans fall within this spectrum of species sensitivities—it is noted that monkey brain AChE is even more sensitive than chicken brain AChE to inhibition by DFP (Wang and Murphy, 1982a) and that human and chicken brain AChE exhibit comparable (within 30%) sensitivities to inhibition by assorted OP compounds (Fig. 3).

It appears, therefore, that despite large differences in the pharmacokinetic and metabolic disposition, species-related differences in sensitivity to intoxication by OP compounds is mediated, for the most part, by differences in the sensitivity of AChE to inhibition. Inhibition of AChE is, however, a complicated process involving the association of the inhibitor with the enzyme to form a reversible complex followed by rapid phosphorylation. The inhibited enzyme subsequently undergoes either dephosphorylation or dealkylation. Figure 4 illustrates the reaction mechanism for the inhibition of AChE by OP insecticides. The active enzyme is depicted as EH, with R being an alkyl substituent and X the leaving group of the inhibitor. The association

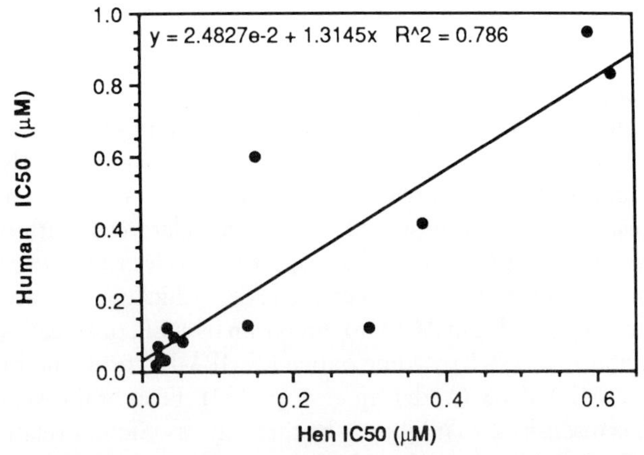

Figure 3 Correlation between the *in vitro* inhibition of human and chicken brain AChE by various OP insecticides. Data from Lotti and Johnson (1978).

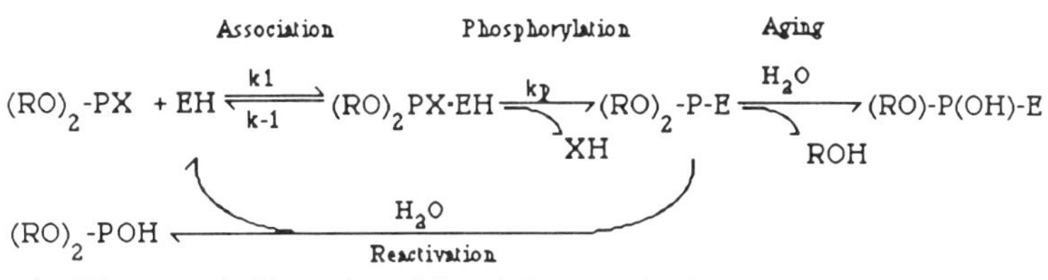

Figure 4 Diagrammatic illustration of the inhibition and subsequent spontaneous reactivation or aging of AChE.

step reflects the formation of a reversible complex between the inhibitor and the enzyme and is described by the association constant ($K_a = k_1/k_{-1}$). The ensuing phosphorylation (k_p) of the enzyme occurs rapidly and irreversibly. The phosphorylated enzyme may then undergo one of two possible fates: (1) spontaneous reactivation to regenerate the active enzyme, or (2) aging to yield an irreversibly inhibited enzyme incapable of being dephosphorylated even in the presence of strong nucleophiles.

Selected investigations have been designed to assess at which stage of the reaction species differences are most pronounced. The results have been somewhat variable; some investigators assert that solely the association or affinity of the inhibitor for the enzyme varies between species (Main and Iversen, 1966; Van Asperen and Dekhuijzen, 1958; Andersen *et al.*, 1977; Forsberg and Puu, 1984), while others suggest that a combination of both the affinity and phosphorylation constants accounts for the dramatic differences (Aldridge, 1953b; Wang and Murphy, 1982a,b; Gray and Dawson, 1987; Johnson and Wallace, 1987; Kemp and Wallace, 1990).

The rates of spontaneous reactivation and aging of phosphorylated AChE differ between species and it has been suggested and that this may contribute to species selectivity. In general, the enzyme from birds reactivates rapidly whereas that from fish or insects reactivates only slowly if at all (Davison, 1955; Mengle and O'Brien, 1960; Lee and Pickering, 1967; Wallace and Herzberg, 1988). A similar relationship between species exists for the rates of aging of phosphorylated AChE. Paraoxon- or DFP-inhibited AChE from frogs or fish ages at a much slower rate (3–10 times) than does the enzyme from mouse, rat, or chicken (Andersen *et al.*, 1972; Wallace and Herzberg, 1988). A disconcerting finding in regard to human intoxication is that following inhibition by soman, human erythrocyte AChE reactivates only very slowly, but ages rapidly compared to the bovine or rat enzyme (DeJong and Wolring, 1984, 1985). This suggests that humans are relatively susceptible to irreversible inhibition of AChE. Although not precisely defined, it appears that species-related differences in the rates of aging, and possibly reactivation, of phosphorylated AChE may contribute to the sensitivity of the different organisms to acute OP intoxication.

This marked difference in pharmacodynamic potency of OP inhibitors may reflect important differences between species in the physical or molecular properties of AChE. The mechanism of interaction of OP agents with AChE is well defined (see Chambers, Chapter 1, this volume) and provides insight into possible discriminators of enzyme inactivation. On a molecular basis, the parameters influencing species-related differences in the sensitivity of AChE to inhibition by OP compounds may be categorized according to following enzymological properties: (1) differences in the steric features of the esteratic site, molecular volume being more important than linear dimension (Hansch and Deutsch, 1966; Kemp and Wallace, 1990); (2) differences in the electronic properties of the nucleophilic center within the esteratic site; or differences in the degree of alosteric regulation of the esteratic site as conferred by (3) the electronic strength of the anionic site, and (4) the proximity of the anionic site to the esteratic site of the enzyme. The relationship between these parameters is illustrated in Fig. 5, which depicts the interaction of paraoxon (*O,O*-diethyl-*p*-nitrophenyl phosphate) with the respective domains of AChE.

A. Steric Properties of the Esteratic Site

The association of the inhibitor with AChE is limited by the finite dimensions of the active site. Rigorous analyses have revealed that a principal determinant of the differential in anticholinesterase potency for individual OP compounds is the steric exclusion of the inhibitor from the active site of AChE (Aldridge, 1953b; Fukuto and Metcalf, 1956, 1959; Fukuto *et al.*, 1959; Hansch and Deutsch, 1966; Andersen *et al.*, 1977; Mundy *et al.*, 1978; Kemp and Wal-

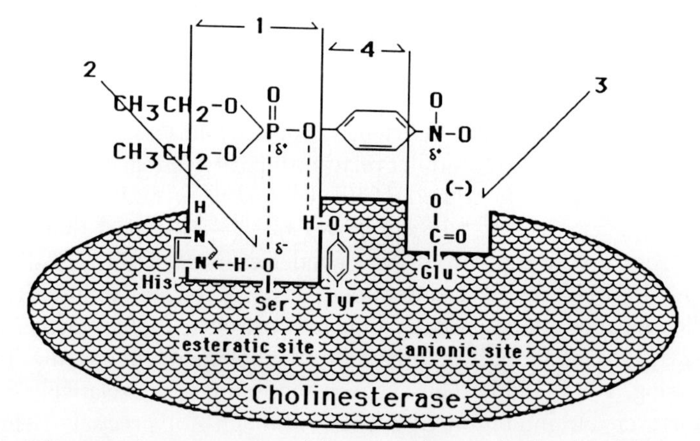

Figure 5 Diagrammatic representation of the various factors affecting the inhibition of AChE by OP agents using paraoxon as the example. Illustrated in the figure are (1) the dimensions of the esteratic site; (2) the electronic strength of the nucleophilic center within the esteratic site; (3) the presence and electronic strength of the anionic region; and (4) the distance separating the anionic from the esteratic domains of the enzyme.

lace, 1990). Rather than linear dimension, the critical factor limiting the association of the inhibitor with the enzyme is the molecular volume of the alkyl substituents on the OP (Hansch and Deutsch, 1966; Kemp and Wallace, 1990).

Since steric hindrance is dictated by the finite dimensions of the esteratic site, it may inferred that species-related differences in the sensitivity of AChE to inhibition by a selected OP compound may reflect differences in relative size of esteratic site. Insight into this possibility has been gained by comparing the kinetics of interaction of either substrates or inhibitors of varying dimensions with AChE from different sources. Dramatic differences in substrate specificity of pseudocholinesterase from 11 species are apparent from the work of Ecobichon and Comeau (1973) as illustrated in Fig. 6. The enzymes

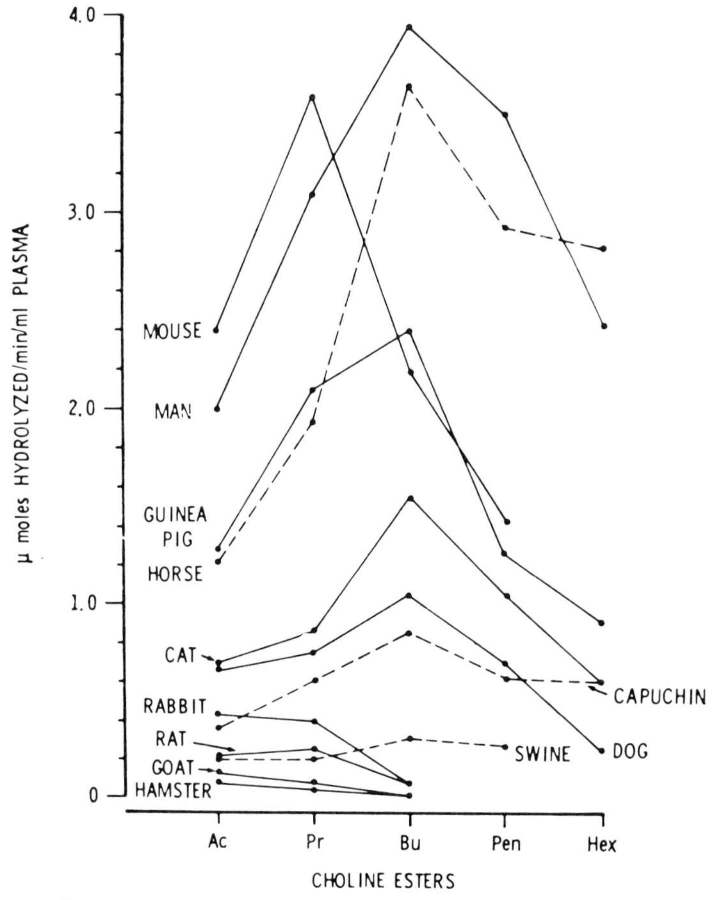

Figure 6 Substrate specificity of pseudocholinesterase from plasma of numerous mammalian species. Reprinted with permission from Ecobichon, D. J., and Comeau, A. M. (1973).

from human, horse, swine, guinea pig, cat, and dog plasma prefer butyrylcholine as substrate, suggesting a relatively large esteratic site. Pseudocholinesterase from rats and mice can be classified as propionylcholinesterases, whereas the enzyme from rabbit, goat, and hamster prefers the smaller acetylcholine substrate. In all cases, enzyme activity decreases as the length of the acylcholine ester is increased beyond a certain optimum, possibly reflecting the progressive steric exclusion of the larger substrates. Consequently, it may be inferred that in contrast to the butyrylcholinesterases, the enzymes from rabbits, goats, and hamsters possess a relatively small esteratic site.

Species-related differences in substrate specificity have also been reported for true AChE from serum and brain. Andersen and Mikalsen (1978) demonstrated that serum and brain AChE from chickens, which tend to be sensitive to acute OP poisoning, prefer propionylcholine to acetylcholine (Fig. 7). In contrast, frog AChE exhibits a strong preference for acetylcholine, which may reflect a smaller esteratic site of AChE and is consistent with the resistance of frogs to acute OP poisoning. Again, increasing the size of the acyl substituent beyond a certain optimum, which is characteristic for each enzyme, results in the progressive exclusion of the substrate from the active site of the enzyme. The optimal substrate size may, therefore, be indicative of the finite dimensions of the active site of AChE from the respective species.

Zahavi *et al.* (1971) demonstrated that AChE from resistant strains of mites prefers acetylthiocholine as substrate, whereas the enzyme from sensitive strains hydrolyzes propionylthiocholine preferentially (Fig. 8). For the resistant strain, the maximum velocity (V_{max}) for propionylthiocholine hydrolysis is only 60% that of acetylthiocholine. Brain AChE from rainbow trout, which are also resistant to acute OP poisoning, also hydrolyzes acetylthio-

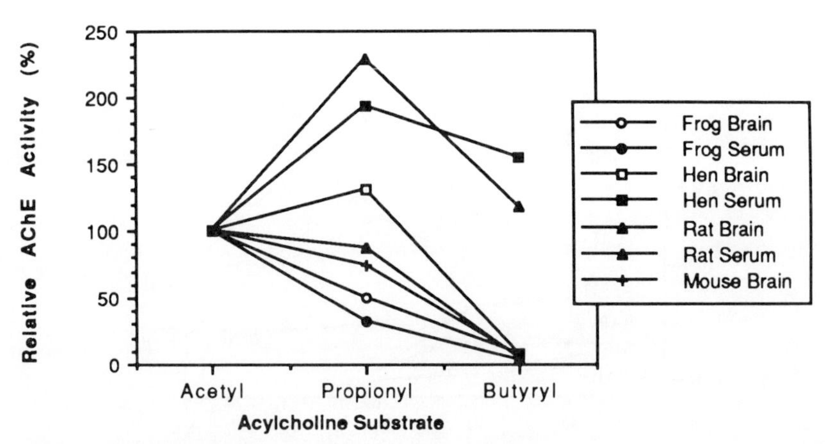

Figure 7 Substrate specificity of brain and serum AChE from various species. Data from Andersen and Mikalsen (1978).

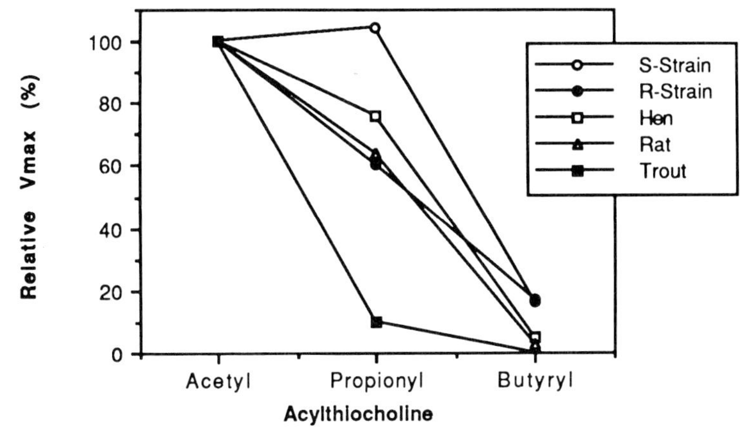

Figure 8 Substrate specificity of AChE from two strains of mites and of brain AChE from chickens, rats, and rainbow trout. Data from Zahavi *et al.* (1971) and from Kemp and Wallace (1990).

choline preferentially to propionylthiocholine (Kemp and Wallace, 1990). The 10-fold greater rate of hydrolysis of acetyl, compared to propionylthiocholine by trout AChE, contrasts with that observed for rats or hens and is consistent with the proposal that trout AChE possess a relatively restricted esteratic site. This difference in substrate specificity is also evident from the affinity constants of the enzymes for the respective substrates (Kemp and Wallace, 1990). Therefore, the ability of AChE from different sources to accommodate substrates of varying dimensions is consistent with the hypothesis that resistance to OP intoxication may be conferred, in part, by the finite dimensions of the esteratic site of AChE.

The limitations imposed by steric hindrance at the esteratic site of AChE are also evident from the data describing species-related differences in the preferential inhibition of AChE by a homologous series of unbranched dialkyl-substituted *p*-nitrophenyl phosphates (Kemp and Wallace, 1990). As illustrated in Fig. 9, inhibitor potency (pI_{50}) declines with successive methylene substitutions in all three species examined. The relative resistance of trout brain AChE is evident by the significantly lower values of pI_{50} for all four OP inhibitors. Interestingly, the pI_{50} for dimethyl and diethyl *p*-nitrophenyl phosphate were comparable for both hens and rats. Subsequent methylene substitutions, however, resulted in a progressive decrease in pI_{50}. In contrast, diethyl *p*-nitrophenyl phosphate is a much weaker inhibitor of trout AChE than is the dimethyl analog. This may reflect the limiting dimensions and greater steric exclusion in the region of the esteratic site of trout AChE. The data describing the association constants (K_a) reveal similar relationships among those species (Kemp and Wallace, 1990). Hen and rat AChE failed to

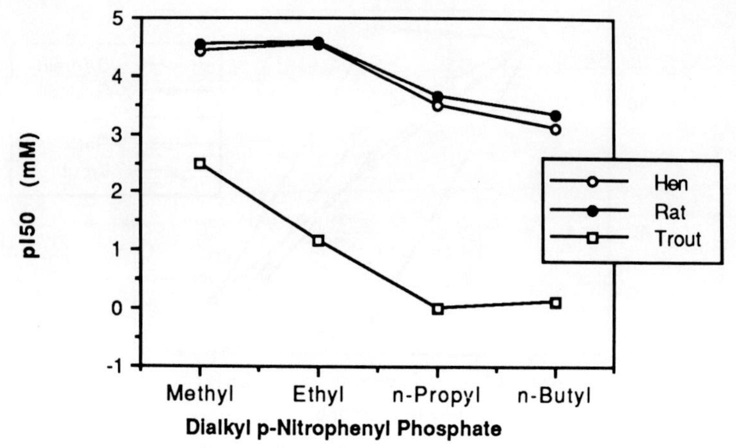

Figure 9 Inhibitor specificity of brain AChE from chickens, rats, and rainbow trout for a series of dialkyl-substituted *p*-nitrophenyl phosphates. Data from Kemp and Wallace (1990). Each point represents the mean of three independent determinations.

distinguish between dimethyl and diethyl *p*-nitrophenyl phosphate; however, K_a decreased progressively with subsequent methylene substitutions.

The data describing the association of substrates and inhibitors with AChE support the hypothesis that species-selectivity may be mediated, in part, by differences in the relative dimensions of the esteratic site. Accordingly, the apparently smaller size of the esteratic site of AChE from resistant species (rainbow trout and R-strains of mites) provides greater steric exclusion and thus relative protection from OP-induced enzyme inhibition. Steric hindrance is less pronounced for AChE from sensitive species, which apparently have a much larger esteratic site.

B. Nucleophilic Strength of the Esteratic Site

The association of the inhibitor with AChE is also governed by coulombic forces, being dependent on both the nucleophilic strength within the active site and the electrophilicity of the phosphoryl atom of the inhibitor. Charge transfers between the imidazole group of a histidine residue and serine hydroxyl group increase the nucleophilicity of the active site (Cunningham, 1957; Brestkin and Rozengart, 1965). In contrast to that observed for carbamate insecticides, hydrophobicity of the OP insecticides is inversely related to inhibitor potency (Hansch and Deutsch, 1966; Kemp and Wallace, 1990; Wallace and Kemp, 1991). Increasing the acidity of the phosphorus atom enhances the association of the inhibitor with AChE, whereas electron-releasing substituents (R or X) decrease the association constant. Rigorous analyses of various series of OP agents reveal that the electrophilicity of the respective inhibitors is a principal determinant of anticholinesterase activity (Aldridge and Davison, 1953; Fukuto and Metcalf, 1956, 1959; Fukuto *et al.*, 1959; Hansch and Deutsch, 1966; Andersen *et al.*, 1977; Mundy *et al.*, 1978).

The rates of dephosphorylation and dealkylation of phosphorylated AChE are also governed by the electronic properties of the enzyme–inhibitor interaction and are suggested to involve the imidazolium group of the active site (Michel *et al.*, 1967; Beauregard *et al.*, 1981). Consequently, differences in the nucleophilic domain may also account for species-related differences in the rates of reactivation and aging of the inhibited enzyme.

Once the reversible complex is formed, phosphorylation usually proceeds rapidly and irreversibly. This S_N2 nucleophilic attack of the serine hydroxyl group by the electron-deficient phosphorus atom is governed solely by the electronic properties of the enzyme and the inhibitor. For a given enzyme, increasing the acidity of the phosphorus atom by substituting electron-withdrawing groups on either the alkyl (R) or leaving (X) groups decreases the intramolecular P-O-X bond strength, thereby facilitating hydrolysis and enzyme phosphorylation. This relationship is illustrated by the correlation between the first-order rate constants for the nonenzymatic alkaline hydrolysis of various OP compounds and the *in vitro* inhibition constants toward AChE from a single source (Aldridge and Davison, 1953; Fukuto and Metcalf, 1956; 1959). Since the first-order rate constant for hydrolysis is proportionate to the electrophilicity of the phosphorus atom, as estimated by the Hammett's sigma constant and reflected by P-O-X bond-stretching frequencies and the alkaline hydrolysis rate constant (k_{hyd}), the *in vitro* inhibitory potency may be predicted from either sigma, k_{hyd}, infrared spectroscopy, or ^{31}P-nuclear magnetic resonance (NMR) spectroscopy (Fukuto and Metcalf, 1956; Hansch and Deutsch, 1966; Darlington *et al.*, 1971; Wallace and Kemp, 1991). The correlation between the first-order alkaline hydrolysis rate constants for a series of ethyl *p*-nitrophenyl alkylphosphonates and inhibition of housefly AChE is illustrated in Fig. 10. This relationship is, however, valid only for selected species of AChE (Wallace and Kemp, 1991).

Besides distinguishing between inhibitory potency for a series of OP agents against a specific AChE, species-related differences in OP-induced inhibition may be related also to differences in phosphorylation rate constants (Andersen *et al.*, 1977; Kemp and Wallace, 1990; Wallace and Kemp, 1991). This is true only for those enzymes possessing different properties within the nucleophilic domain of the active site of AChE. Because of the nature of the chemical mechanism, phosphorylation of AChE is governed by both the electrophilic strength of the inhibitor and the nucleophilic strength of the enzyme. Accordingly, for enzymes possessing a sufficiently weak nucleophilic center, the rate of phosphorylation is proportionate to the electronic properties of the inhibitor. In contrast, the phosphorylation of AChE enzymes

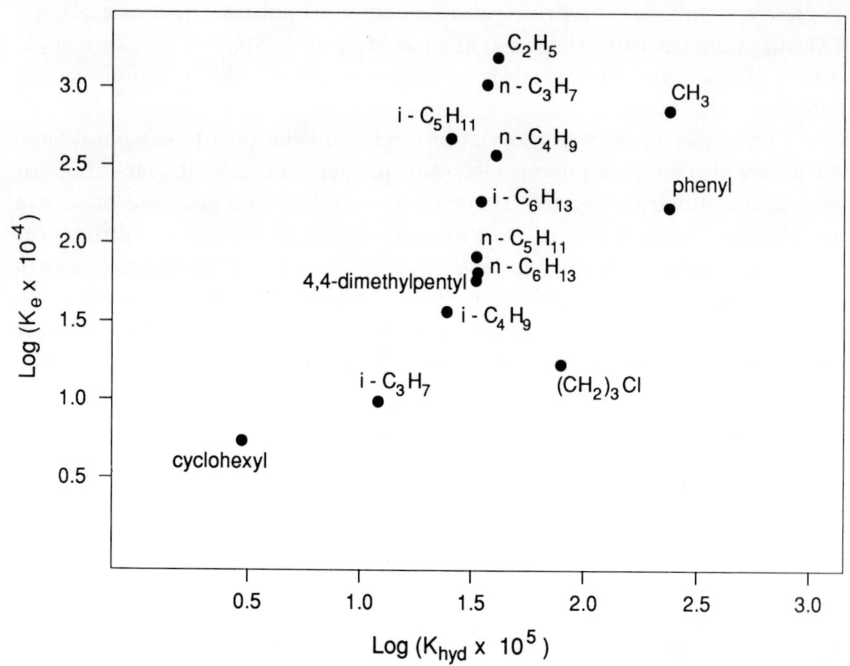

Figure 10 Correlation between the first-order alkaline hydrolysis rate constant and the *in vitro* inhibition constant (K_e) against house fly cholinesterase for a series of ethyl *p*-nitrophenyl alkylphosphonates. Reproduced with permission from Fukuto, T. R., and Metcalf, R. L. (1959).

possessing strong nucleophilic character proceeds rapidly, and the reaction is zero-order with respect to the electronic properties of the phosphorus atom. For example, the first-order phosphorylation rate constant does not vary significantly between different inhibitors of eel or bovine erythrocyte AChE, or of brain AChE isolated from monkeys, rats, mice, guinea pigs, chickens, catfish, or frogs (Chiu *et al.*, 1969; Wang and Murphy, 1982b; Gray and Dawson, 1987; Kemp and Wallace, 1990). Accordingly, physical-chemical indices of the electrophilicity of the phosphorus atom do not correlate with inhibitor potency in these species (Kemp and Wallace, 1990; Wallace and Kemp, 1991). These results suggest the presence of a strong nucleophilic center for AChE from these species with the rate of phosphorylation being independent of small changes in the electrophilic properties of the inhibitor. In the case of trout brain and housefly AChE, however, the phosphorylation rate constant (k_p) is a function of the acidity of the phosphorus atom, suggesting that these enzymes possess a weak nucleophilic center within the esteratic site (Fukuto and Metcalf, 1956, 1959; Darlington *et al.*, 1971; Kemp and Wallace, 1990; Wallace and Kemp, 1991). Accordingly, whereas physical chemical indices of the acidity of the phosphorus atom provide reasonable estimates of k_p for trout AChE (Fig. 11), there is no relationship with the rate of phosphorylation of hen or rat brain AChE (Wallace and Kemp, 1991). The weak coulombic interaction between substrates and inhibitors with the active-site of trout AChE may also be partially responsible for the low affinity of this enzyme for various OP inhibitors.

The exact nature of this difference in nucleophilic strength of the active site of AChE from different species has yet to be examined. In addition to possible differences in the actual nucleophilic substituent, it is feasible that the enzymes from trout and houseflies differ in the contribution of the adjacent tyrosine or histidine residues in modifying the pK_a of the serine hydroxyl group. For instance, in view of the importance of the imidazole group of histidine in increasing the nucleophilicity of the serine hydroxyl group, it seems reasonable to propose that the trout and fly enzyme may lack the critical histidine residue in the active site. This would also explain the slow rate of aging and reactivation of phosphorylated trout brain AChE (Brestkin and Rozengart, 1965; Beauregard *et al.*, 1981; Wallace and Herzberg, 1988).

C. Anionic Site

Although the characteristics of the anionic site are less well defined, it is thought that the association of cationic groups with this site induces a conformational change in the tertiary structure of the enzyme which is transmitted to the esteratic site. This allosteric regulation promotes both substrate

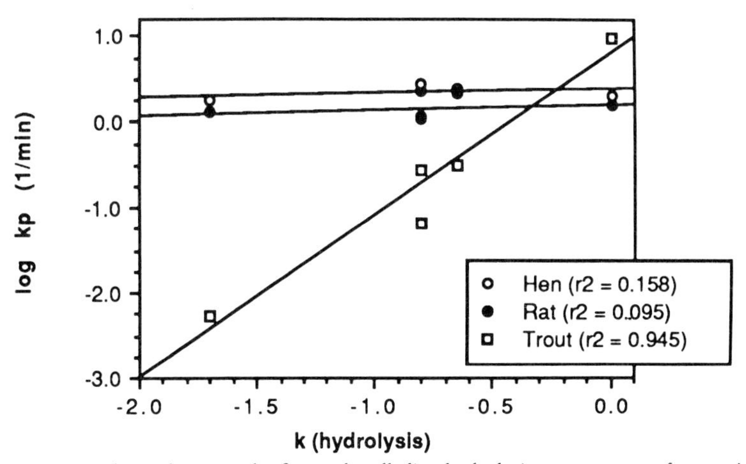

Figure 11 Correlation between the first-order alkaline hydrolysis rate constants for a series of dialkyl-substituted *p*-nitrophenyl phosphates and the *in vitro* first-order phosphorylation rate constants for brain AChE from chickens, rats, and rainbow trout. Trout AChE yielded a statistically significant correlation. Data from Wallace and Kemp (1991).

hydrolysis and enzyme inhibition and may also affect the dephosphorylation and aging of the inhibited enzyme (Green and Smith, 1958; O'Brien, 1963; Crone, 1974).

Systematic studies of species-related differences in the anionic site of AChE and its influence on species-selective toxicity of OP compounds have yet to be conducted. It may be speculated, however, that this subsite is either absent or weak in the insensitive species of AChE. Blaber and Cuthbert (1962) noted that in contrast to rat AChE, the enzyme from chickens does not react with bisquaternary substrates, suggesting possible differences between species in the allosteric regulation by the anionic site. Similarly, the failure of frog brain AChE to spontaneously reactivate has been suggested to reflect species differences in the conformational changes of the enzyme induced by the interaction of inhibitors with anionic site (Andersen *et al.*, 1972). Therefore, differences in the allosteric regulation induced by the association of the inhibitor with anionic region of AChE may also be a factor in establishing the species-related differences in AChE inhibition.

D. Proximity of the Esteratic and Anionic Sites

Critical to the allosteric regulation of the esteratic site by the anionic site is the physical proximity of these two domains of AChE. The optimal separation would be such that binding to the anionic site orients the inhibitor such that the phosphoryl group is juxtaposed to the esteratic site of the enzyme. Deviation from this optimal distance of separation decreases the cooperativity between the two sites, thereby decreasing inhibitor potency.

In view of this relationship, it may be proposed that species-related differences in the sensitivity of AChE to OP inhibition may reflect differing distances of separation between these two regions of the enzyme. O'Brien (1963) estimated a distance of 4.5 to 5.9 Å separating the anionic from the esteratic sites of fly head AChE and 4.5 Å for the enzyme from bovine erythrocytes. Hollingworth *et al.* (1967) estimated these distances to be between 4.3 and 4.7 Å and 5.0 and 5.5 Å, respectively. The distance separating the two sites in AChE from human plasma or erythrocytes is estimated to be less than 4.5 Å, and for AChE from the electric eel, 2.5 Å (O'Brien, 1963). Although the individual distances were not estimated, Moss and Fahrney (1978) suggest that even though the peripheral anionic site of fish and rat brain AChE are indistinguishable, there may be a difference in the topography of the anionic region with respect to the esteratic site of the respective enzymes. Hollingworth *et al.* (1967) advanced the hypothesis that species-related differences in distance separating the anionic and esteratic subsites is partially responsible for the differences in susceptibilities of mammalian and insect AChE to inhibition by various OP agents, as has been suggested previously (Foldes *et al.*, 1958; Wilson and Quan, 1958; Metcalf *et al.*, 1962; O'Brien, 1963; Krupka, 1965).

VI. Conclusions

Despite large differences in the pharmacokinetic and metabolic disposition of OP compounds in different organisms, it does not appear that either of these processes represents the principal determinant of the species-selective toxicity of these agents. Rather, species-specificity appears to be a pharmacodynamic phenomenon, mediated at the level of species-related differences in the susceptibility of AChE to inhibition by the phosphorylating intermediate. Molecular features of AChE that have been implicated in conferring resistance to selected species include smaller steric dimensions of the esteratic site, a weaker nucleophilic center within the active site, less effectual allosteric regulation by the anionic region, and greater distances separating the anionic from the esteratic domains of the enzyme. These distinct properties of AChE from different species caution against the employment of sentinels for monitoring the possible intoxication of diverse populations of organisms within a defined environment. However, insight is provided into factors to be considered in extrapolating OP toxicity data between unrelated species. This characterization of species-related differences in the molecular determinants of AChE sensitivity also reveals possible strategies for developing new OP compounds with specific physical-chemical properties designed to selectively inhibit AChE of the target species while sparing that of other organisms. As such, this knowledge may greatly advance the selectivity of OP insecticides and reduce the incidence of unintentional intoxication of beneficial organisms, including the inadvertent poisoning of humans.

References

Aldridge, W. N. (1953a). Serum Esterases. 2. An enzyme hydrolysing diethyl-*p*-nitrophenyl phosphate (E600) and its identity with the A-esterase of mammalian sera. *Biochem. J.* 53, 117–124.

Aldridge, W. N. (1953b). The differentiation of true and pseudo cholinesterase by organophosphorus compounds. *Biochemistry* 53, 62–67.

Aldridge, W. N., and Davison, A. N. (1953). The mechanism of inhibition of cholinesterases by organophosphorus compounds. *Biochem. J.* 55, 763–766.

Anber, H. A. I., and Overmeer, W. P. J. (1988). Resistance to organophosphates and carbamates in the predacious mite *Amblyseius potentillae* (Garman) due to insensitive acetylcholinesterase. *Pestic. Biochem. Physiol.* 31, 91–98.

Andersen, R. A., Aaraas, I., Gaare, G., and Fonnum, F. (1977). Inhibition of acetylcholinesterase from different species by organophosphorus compounds, carbamates, and methylsuphonylfluoride. *Gen. Pharmacol.* 8, 331–334.

Andersen, R. A., Laake, K., and Fonnum, F. (1972). Reactions between alkyl phosphates and acetylcholinesterase from different species. *Comp. Biochem. Physiol.* **42B**, 429–437.

Andersen, R. A., and Mikalsen, A. (1978). Substrate specificity, effect of inhibitors and electrophoretic mobility of brain and serum cholinesterase from frog, chicken, and rat. *Gen. Pharmacol.* 9, 177–181.

Bathe, R., Ullmann, L., Sachsse, K., and Hess, R. (1976). Relationship between toxicity to fish and to mammals. A comparative study under defined laboratory conditions. *Proc. Eur. Soc. Toxicol.* **17**, 351–355.

Beauregard, G., Lum, J., and Roufogalis, B.D. (1981). Effect of histidine modification on the aging of organophosphate-inhibited acetylcholinesterase. *Biochem. Pharmacol.* **30**, 2915–2920.Becker, E. L., Punte, C. L., and Barbaro, J. F. (1964). Acute toxicity of alkyl and phenylalkylphosphonates in the guinea pig and rabbit in relation to their anticholinesterase activity and their enzymatic inactivation. *Biochem. Pharmacol.* **13**, 1229–1237.

Benke, G. M., Cheever, K. L., Mirer, F. E., and Murphy, S. D. (1974). Comparative toxicity, anticholinesterase action, and metabolism of methyl parathion and parathion in sunfish and mice. *Toxicol. Appl. Pharmacol.* **28**, 97–109.

Blaber, L. C., and Cuthbert, A. W. (1962). Cholinesterases in the domestic fowl and the specificity of some reversible inhibitors. *Biochem. Pharmacol.* **11**, 113–123.

Brealey, C. J., Walker, C. H., and Baldwin, B. C. (1980). A-esterase activities in relation to the differential toxicity of pirimiphos-methyl to birds and mammals. *Pestic. Sci.* **11**, 546–554.

Brestkin, A. P., and Rozengart, E. V. (1965). Cholinesterase catalysis. *Nature (London)* **205**, 388–389.

Butler, E. G., Eckerson, H. W., and LaDu, B. N. (1985). Paraoxon hydrolysis vs. covalent binding in the elimination of paraoxon in the rabbit. *Drug Metabol. Disp.* **13**, 640–645.

Chambers, J. E., and Chambers, H. W. (1990). Time course of inhibition of acetylcholinesterase and aliesterase following parathion and paraoxon exposures in rats. *Toxicol. Appl. Pharmacol.* **103**, 420–429.

Chambers, H., Brown, B., and Chambers, J. E. (1990). Noncatalytic detoxication of six organophosphorus compounds by rat liver homogenates. *Pestic. Biochem. Physiol.* **36**, 308–315.

Chattopadhyay, D. P., Dighe, S. K., Nashikkar, A. B., and Dube, D. K. (1986). Species differences in the *in vitro* inhibition of brain acetylcholinesterase and carboxylesterase by mipafox, paraoxon, and soman. *Pestic. Biochem. Physiol.* **26**, 202–208.

Chiu, Y. C., Main, A. R., and Dauterman, W. C. (1969). Affinity and phosphorylation constants of a series of *O,O*-dialkyl malaoxons and paraoxons with acetylcholinesterase. *Biochem. Pharmacol.* **18**, 2171–2177.

Crone, H. D. (1974). Can allosteric effectors of acetylcholinesterase control the rate of aging of the phosphonylated enzyme? *Biochem. Pharmacol.* **23**, 460–463.

Cunningham, L. W. (1957). Proposed mechanism of action of hydrolytic enzymes. *Science* **125**, 1145–1146.

Darlington, W. A., Partos, R. D., and Ratts, K. W. (1971). Correlation of cholinesterase inhibition and toxicity in insects and mammals. I. Ethylphosphonates. *Toxicol. Applied Pharmacol.* **18**, 542–547.

Dauterman, W. C., and O'Brien, R. D. (1964). Cholinesterase variation as a factor in organophosphate selectivity in insects. *J. Agric. Food Chem.* **12**, 318–319.

Davison, A. N. (1955). Return of cholinesterase activity in the rat after inhibition by organophosphorus compounds. *Biochem. J.* **60**, 339–346.

DeJong, L. P. A., and Wolring, G. Z. (1984). Stereospecific reactivation by some hagedorn-oximes of acetylcholinesterases from various species including man, inhibited by soman. *Biochem. Pharmacol.* **33**, 1119–1125.

DeJong, L. P. A., and Wolring, G. Z. (1985). Aging and stereospecific reactivation of mouse erythrocyte and brain acetylcholinesterases inhibited by soman. *Biochem. Pharmacol.* **34**, 142–145.

Devonshire, A. L., and Moores, G. D. (1982). A carboxylesterase with broad substrate specificity causes organophosphorus carbamate and pyrethroid resistance in peach-potato aphids (*Mayzus persicae*). *Pestic. Biochem. Physiol.* **18**, 235–246.

Diggle, W. M., and Gage, J. C. (1951). Cholinesterase inhibition *in vitro* by *O,O*-diethyl O-p-nitrophenyl thiophosphate (parathion, E605). *Biochem. J.* **49**, 491–494.

Eckerson, H. W., Romson, J., Wyte, C., and LaDu, B. N. (1983). The human serum paraoxonase polymorphism: Identification of phenotypes by their response to salts. *Am. J. Hum. Genet.* **35**, 214–227.

Ecobichon, D. J., and Comeau, A. M. (1973). Pseudocholinesterases of mammalian plasma: Physicochemical properties and organophosphate inhibition in eleven species. *Tox. Appl. Pharmacol.* **24**, 92–100.

Foldes, F., Van Hees, G., Davis, D. L., and Shanor, S. (1958). The structure–action relationship of urethane type cholinesterase inhibitors. *J. Pharmacol. Exp. Ther.* **122**, 457–464.

Forsberg, A., and Puu, G. (1984). Kinetics for the inhibition of acetylcholinesterase from the electric eel by some organophosphates and carbamates. *Eur. J. Biochem.* **140**, 153– 156.

Forsyth, C. S., and Chambers, J. E. (1989). Activation and degradation of the phosphorothionate insecticides parathion and EPN by rat brain. *Toxicol. Appl. Pharmacol.* **38**, 1597–1603.

Fukuto, T. R., and Metcalf, R. L. (1956). Structure and insecticidal activity of some diethyl substituted phenyl phosphates. *J. Agr. Food Chem.* **4**, 930–935.

Fukuto, T. R., and Metcalf, R. L. (1959). The effect of structure on the reactivity of alkylphosphonate esters. *J. Am. Chem. Soc.* **81**, 372–377.

Fukuto, T. R., Metcalf, R. L., and Winton, M. (1959). Alkylphosphonic acid esters as insecticides. *J. Econ. Entomol.* **52**, 1121–1127.

Gage, J. C. (1953). A cholinesterase inhibitor derived from O,O-diethyl O-*p*-nitrophenyl thiophosphate *in vivo*. *Biochem. J.* **54**, 426–430.

Gray, P. J., and Dawson, R. M. (1987). Kinetic constants for the inhibition of eel and rabbit brain acetylcholinesterase by some organophosphates and carbamates of military significance. *Toxicol. Appl. Pharmacol.* **91**, 140–144.

Green, A. L., and Smith, H. J. (1958). The reactivation of cholinesterase inhibited with organophosphorus compounds. 2. Reactivation by pyridinealdoxime methiodides. *Biochem. J.* **68**, 32–35.

Hansch, C., and Deutsch, E. W. (1966). The use of substituent constants in the study of structure–activity relationships in cholinesterase inhibitors. *Biochim. Biophys. Acta* **126**, 117–128.

Hitchcock, M., and Murphy, S. D. (1971). Activation of parathion and guthion by mammalian, avian, and piscine liver homogenates and cell fractions. *Toxicol. Appl. Pharmacol.* **19**, 37–45.

Hodson, P. V. (1985). A comparison of the acute toxicity of chemicals to fish, rats, and mice. *J. Appl. Toxicol.* **5**, 220–226.

Hollingworth, R. M., Fukuto, T. R., and Metcalf, R. L. (1967). Selectivity of sumithion compared with methyl parathion. Influence of structure on anticholinesterase activity. *J. Agric. Food Chem.* **15**, 235–241.

Janardan, S. K., Olson, C. S., and Schaeffer, D. J. (1984). Quantitative comparisons of acute toxicity of organic chemicals to rat and fish. *Ecotoxicol. Environ. Safety* **8**, 531–539.

Johnson, J. A., and Wallace, K. B. (1987). Species-related differences in the inhibition of brain acetylcholinesterase by paraoxon and malaoxon. *Toxicol. Appl. Pharmacol.* **88**, 234–241.

Kamataki, T., Lee Lin, M. C. M., Belcher, D. H., and Neal, R. A. (1976). Studies of the metabolism of parathion with an apparently homogeneous preparation of rabbit liver cytochrome P450. *Drug Metab. Dispos.* **4**, 180–189.

Kemp, J. R., and Wallace, K. B. (1990). Molecular determinants of the species-selective inhibition of brain acetylcholinesterase. *Toxicol. Appl. Pharmacol.* **104**, 1–12.

Kenaga, E. E. (1978). Test organisms and methods useful for early assessment of acute toxicity of chemicals. *Environ. Sci. Technol.* **12**, 1322–1329.

Kenaga, E. E. (1980). Correlation of bioconcentration factors of chemicals in aquatic and terrestrial organisms with their physical and chemical properties. *Environ. Sci. Technol.* **14**, 553–556.

Krupka, R. M. (1965). Acetylcholinesterase: Structural requirements for blocking deacetylation. *Biochemistry* **4**, 429–435.

LaDu, B. N., and Eckerson, H. W. (1984). The polymorphic paraoxonase/arylesterase isozymes of human serum. *Fed. Proc.* **43**, 2338–2441.

Lauwerys, R. R., and Murphy, S. D. (1969). Comparison of assay methods for studying O,O-diethyl O-*p*-nitrophenyl phosphate (Paraoxon) detoxication *in vitro*. *Biochem. Pharmacol.* **18**, 789–800.

Lee, R. M., and Pickering, W. R. (1967). The toxicity of haloxon to geese, ducks, and hens, and its relationship to the stability of the di-(2-chloroethyl) phosphoryl cholinesterase derivatives. *Biochem. Pharmacol.* **16**, 941–948.

Lotti, M., and Johnson, M. K. (1978). Neurotoxicity of organophosphorus pesticides: Predictions can be based on *in vitro* studies with hen and human enzymes. *Arch. Toxicol.* **41**, 215–221.

Machin, A. F., Anderson, P. H., Quick, M. P., Waddell, D. R., Skibiniewska, K. A., and Howells, L. C. (1976). The metabolism of diazinon in the liver and blood of species of varying susceptibility to diazinon poisoning. *Xenobiotica* **7**, 104.

Mackness, M. I., Walker, C. H., Rowlands, D. G., and Price, N. R. (1983). Investigations into esterases of 3 strains of rust red flour beetle (*Tribolium castaneum*). *Comp. Biochem. Physiol.* **74C**, 65–68.

Main, A. R. (1956). The role of A-esterase in the acute toxicity of paraoxon, TEPP, and parathion. *Can. J. Biochem. Physiol.* **34**, 197–216.

Main, A. R., and Iverson, F. (1966). Measurement of the affinity and phosphorylation constants governing irreversible inhibitions of cholinesterases by di-isopropyl phosphorofluoridate. *Biochem. J.* **100**, 525–531.

Maxwell, D. M., Brecht, K. M., and O'Neill, B. L. (1987). The effect of carboxylesterase inhibition on interspecies differences in soman toxicity. *Toxicol. Lett.* **39**, 35–42.

Mendoza, C. E., Shields, J. B., and Augustinsson, K.-B. (1976). Arylesterases from various mammalian sera in relation to cholinesterases, carboxylesterases and their activity towards some pesticides. *Comp. Biochem. Physiol.* **55C**, 23–26.

Mendoza, C. E., Shields, J. B., and Greenhalgh, R. (1977). Activity of mammalian serum esterases towards malaoxon, fenitroxon, and paraoxon. *Comp. Biochem. Physiol.* **56C**, 189–191.

Mengle, D. C., and O'Brien, R. D. (1960). The spontaneous and induced recovery of fly-brain cholinesterase after inhibition by organophosphates. *Biochem. J.* **75**, 201–207.

Metcalf, R. L., and Frederickson, M. (1965). Selective insecticidal action of isopropyl parathion and analogues. *J. Econ. Entomol.* **58**, 143–147.

Metcalf, R. L., Fukuto, T. R., and Winton, M. Y. (1962). Insecticidal carbamates: Position isomerism in relation to activity of substituted-phenyl N-methylcarbamates. *J. Econ. Entomol.* **55**, 889–894.

Michel, H. O., Hackley, B. E., Berkowitz, L., List, G., Hackley, E. B., Gillilan, W., and Pankau, M. (1967). Ageing and dealkylation of soman (pinacolylmethylphosphonofluoridate)-inactivated eel cholinesterase. *Arch. Biochem. Biophys.* **121**, 29–34.

Motoyama, N., Kao, L. R., Lin, P. T., and Dauterman, W. C. (1984). Dual role of esterases in insecticide resistance in the green rice leafhopper. *Pestic. Biochem. Physiol.* **21**, 139–147.

Moss, D. E., and Fahrney, D. (1978). Kinetic analysis of differences in brain acetylcholinesterase from fish or mammalian sources. *Biochem. Pharmacol.* **27**, 2693–2698.

Mundy, R. L., Bowman, M. C., Farmer, J. H., and Haley, T. J. (1978). Quantitative structure activity study of a series of substituted O,O-dimethyl O-(*p*-nitrophenyl) phosphorothioates and O-analogs. *Arch. Toxicol.* **41**, 111–123.

Murphy, S. D. (1966). Liver metabolism of thiophosphate insecticides in mammalian, avian, and piscine species. *Proc. Soc. Exp. Biol. Med.* **123**, 392–398.

Murphy, S. D., Lauwerys, R. R., and Cheever, K. L. (1968). Comparative anticholinesterase action of organophosphorus insecticides in vertebrates. *Toxicol. Appl. Pharmacol.* **12**, 22–35.

Neal, R. A. (1967). Studies of the enzymatic mechanism of the metabolism of diethyl-4-nitrophenylphosphorothionate (parathion) by rat liver microsomes. *Biochem. J.* **105**, 289–297.

O'Brien, R. D. (1963). Binding of organophosphates to cholinesterases. *J. Agric. Food Chem.* **11**, 163–166.

Potter, J. L., and O'Brien, R. D. (1964). Parathion activation by livers of aquatic and terrestrial vertebrates. *Science* **144**, 55–56.

Sultatos, L. G., Basker, K. M., Shao, M., and Murphy, S. D. (1984). The interaction of the phosphorothioate insecticides chlorpyrifos and parathion and their oxygen analogues with bovine serum albumin. *Mol. Pharmacol.* **26**, 99–104.

Van Asperen, K., and Dekhuijzen (1958). A quantitative analysis of the kinetics of cholinesterase inhibition in tissue homogenates of mice and houseflies. *Biochim. Biophys. Acta* **28**, 603–613.

Walker, C. H. (1983). Pesticides and birds: Mechanisms of selective toxicity. *Agric. Ecosyst. Environ.* **9**, 211–216.

Walker, C. H., and Mackness, M. I. (1987). "A" esterases and their role in regulating the toxicity of organophosphates. *Arch. Toxicol.* **60**, 30–33.

Wallace, K. B., and Dargan, J. E. (1987). Intrinsic metabolic clearance of parathion and paraoxon by livers from fish and rodents. *Toxicol. Appl. Pharmacol.* **90**, 235–242.

Wallace, K. B., and Herzberg, U. (1988). Reactivation and aging of phosphorylated brain acetylcholinesterase from fish and rodents. *Toxicol. Appl. Pharmacol.* **92**, 307–314, 1988.

Wallace, K. B., and Kemp, J. R. (1991). Species-specificity in the chemical mechanisms of organophosphorus anticholinesterase activity. *Chem. Res. Toxicol.* **4**, 41–49.

Wallace, K. B., and Niemi, G. J. (1988). Structure–activity relationships of species-selectivity in acute chemical toxicity between fish and rodents. *Environ. Toxicol. Chem.* **7**, 201–212.

Wang, C., and Murphy, S. D. (1982a). Kinetic analysis of species difference in acetylcholinesterase sensitivity to organophosphate insecticides. *Toxicol. Appl. Pharmacol.* **66**, 409–419.

Wang, C., and Murphy, S. D. (1982b). The role of non-critical binding proteins in the sensitivity of acetylcholinesterase from different species to diisopropyl fluorophosphate (DFP), *in vitro*. *Life Sci.* **31**, 139–149.

Wilson, I. B., and Quan, C. (1958). Acetylcholinesterase studies on molecular complementariness. *Arch. Biochem.* **73**, 131–143.

Zahavi, M., Tahori, A. S., and Klimer, F. (1971). Insensitivity of acetylcholinesterases to organophosphorus compounds as related to size of esteratic site. *Mol. Pharmacol.* **7**, 611–619.

5

Reactivation of Organophosphorus Inhibited AChE with Oximes

Barry W. Wilson
Departments of Avian Sciences
and Environmental Toxicology
University of California
Davis, California

Michael J. Hooper
Institute of Wildlife and Department of
Environmental Toxicology
Clemson University
Clemson, South Carolina

Mark E. Hansen
Pamela S. Nieberg
Departments of Avian Sciences and
Environmental Toxicology
University of California
Davis, California

Organophosphates: Chemistry, Fate, and Effects

I. Introduction

A. Organophosphorus Compounds

Organophosphorus (OP) esters are one of the few classes of poisons for which there are specific antidotes. Acetylcholine receptor (AChR) blockers, such as atropine, inhibitors of acetylcholine (ACh) hydrolysis that are more spontaneously reactivatable than are OP esters such as physostigmine and neostigmine, and oximes that reactivate inhibited cholinesterases (ChE) have been important in developing specific prophylactic and therapeutic treatments for OP intoxication. This chapter reviews the actions of oximes, their uses in studying ChE form and function, and in treating poisonings by OP compounds and, sometimes, by organocarbamates. Emphasis is placed on agricultural chemicals whenever possible, even though much of the research on oximes has been directed toward chemical warfare agents [e.g., soman, also sarin, tabun, and *O*-ethyl *S*[2-(diisopropylamino)ethyl] methyl phosphonothionate (VX)].

B. Cholinesterases

ChEs are the primary targets of OP compounds in vertebrates. One major class is the specific acetylcholinesterases (AChEs, E.C.3.1.1.7); another is the nonspecific butyrylcholinesterases (BChEs, E.C.3.1.1.8; Silver, 1974). These enzymes are widely distributed in the nerves and muscles, and are found in both the fluid and formed elements of the blood.

Much of the current research discussed here uses the hydrolysis of ACh (e.g., the method of Johnson and Russell, 1975) or acetylthiocholine or other thiocholine substrates (e.g., the assay of Ellman *et al.*, 1961) to determine ChE activities. Multiple molecular forms of the enzymes usually are separated by sucrose gradient sedimentation (e.g., Sketelj *et al.*, 1978). Recently, ChE proteins have been purified, sequenced and antibodies to them have been produced, accelerating the study of their molecular biology (Inestrosa and Perelman, 1989).

ChEs are distributed widely within the vertebrate body. AChE forms and other esterases at synapses regulate excitation by destroying the neurotransmitter ACh. But the physiological function(s) of the ChEs in the blood [AChE in mammalian erythrocytes (RBCs), BChE and AChEs in sera], and in regions of the nervous and muscle systems outside of synapses and motor end plates, are not known. Also, ChEs are distributed differently in embryos than in adults; for example, domestic chick embryo blood contains AChE, not BChE activity, and plasma BChE activity increases after birth (Wilson *et al.*, 1973; Smucker and Wilson, 1990). High levels of AChE activity are found throughout embryo muscle fibers, and become localized at the motor end plates only after birth (Wilson *et al.*, 1973; Massoulie and Bon, 1982; Smucker and Wilson, 1990). With regard to toxicology, such distributions of ChEs, as well as the activity of OP-sensitive enzymes, like serum carboxyl esterases, OP-binding proteins, such as the A-esterases, and the rapid turnover of ChEs (one catalytic site of AChE may hydrolyze one ACh molecule per 100 sec), make it difficult to establish one-to-one relationships between OP exposure, inhibition of ChEs, and adverse physiological and behavioral effects.

Inhibitions of AChE greater than 50% are those often associated with classical symptoms of OP poisoning, such as miosis, lacrymation, salivation, and, at higher doses, convulsions, respiratory failure, and death, although there have been reports of behavioral effects from exposures to relatively low levels of OP compounds (Wolthuis and Vanwersch, Environmental Health Criteria, 1986). However, an animal may survive exposure to an OP if only a small percentage of its AChE remains active (Wolthuis and Kepner, 1978).

C. Life History of AChE

ChEs have complicated life histories; they are composed of multiple molecular forms, some of which move from site to site within cells; others are secreted into tissue fluids (Massoulie and Bon, 1982; Toutant and Massoulie, 1987). The multiple molecular forms of AChE consist of asymmetric and globular forms. Asymmetric forms tend to be localized at synapses and motor end plates (12S, 16S, or 20S); these forms have glycosylated *heads* containing the catalytic centers and collagen *tails* linked to the catalytic subunits. Globular forms (4–7S, 12S) are made up of the catalytic subunits joined together by sulfhydryl groups (Fig. 1).

A model of the life cycle of AChE (Inestrosa and Perleman, 1989; Rotundo and Fambrough, 1982) is shown in Fig. 2. AChE and BChE subunits are synthesized within excitable cells (AChE, Rotundo, 1988) and liver (BChE, Smucker and Wilson, 1990), and glycosylated within the Golgi ap-

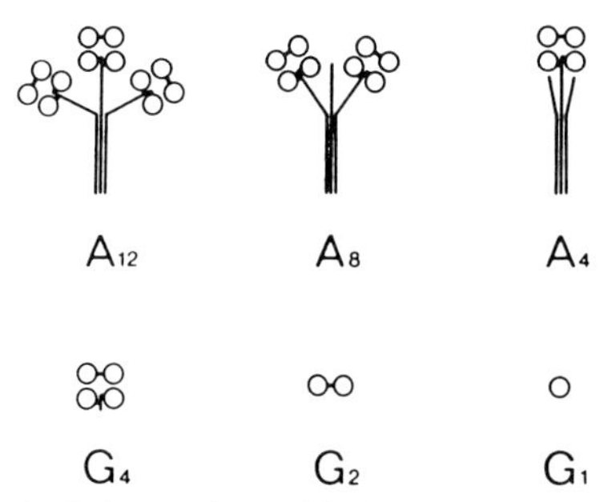

Figure 1 Structure of multiple forms of acetylcholinesterase (AChE). After Massoulie and Bon (1982).

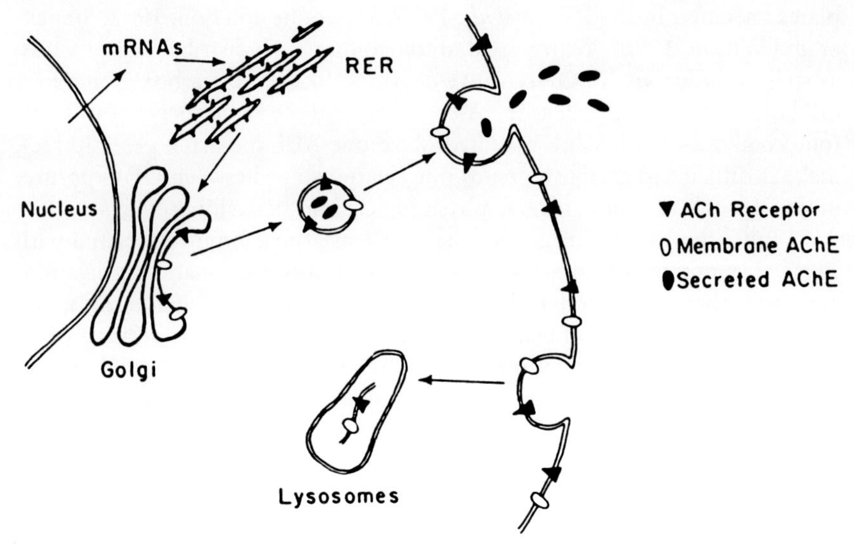

Figure 2. Model of AChE and ACh receptor metabolism in cultured muscle cells. After Rotundo and Fambrough (1982).

paratus. Once secreted from the cells, some forms become attached to the cell surface at specific binding sites (Silman and Futerman, 1987). Some globular forms are released into tissue fluids and blood, or held by ionic bonds to cell surfaces. The protein and nucleic acid sequences for several ChEs have been determined, and antibodies made to purified forms (Tsim *et al.*, 1988), including enzymes from the human (Rakonczay and Brimijoin, 1988). There are significant homologies between large sequences of the AChE forms from the electric ray (*Torpedo californica*) and the C-terminal portion of thyroglobulin, a secretory protein (Schumacher *et al.*, 1986).

II. Cholinesterase Inhibition by Organophosphorus Compounds

As discussed in Chapters 1, and 4, this volume, OP compounds inhibit ChEs by nucleophilic attack. An electronegative group, the hydroxyl of a serine at the active site, reacts with the relatively electropositive phosphorus atom of the inhibitor, resulting in an OP–ChE complex and the loss of one of the side groups on the phosphorus atom, designated the leaving group (Eto, 1974). Many factors (e.g., charge, chirality, and steric considerations) determine the inhibitory potential and the stability of the molecules. For example, resonance and inductive effects due to alkyl groups that enhance the electropositivity of the phosphorus atom will increase the reactivity of the inhibitor (Eto, 1974).

Regardless of the species and tissue sources of the ChEs, inhibitions of their activity are usually qualitatively, but not necessarily quantitatively, the same (Wallace and Herzberg, 1988; Environmental Health Criteria, 1986).

A. Recovery of Enzyme Activity

OP-inhibited ChE activity may recover by several processes *in vivo*: (1) new enzyme may be synthesized (except in RBCs and other formed elements of the blood that lack the capacity for protein synthesis); (2) spontaneous reactivation may occur as a result of dephosphorylation of the OP–enzyme complex; or (3) dephosphorylation may be chemically induced.

B. Synthesis of New Protein

Embryo muscle cell cultures, such as those from the Japanese quail (*Coturnix coturnix japonica*) are model systems for the study of AChE (Bulger *et al.*, 1982; Wilson and Nieberg, 1983; Rotundo, 1988). The differentiated cells contain both the asymmetric and the globular multiple molecular forms of AChE found in muscles of the adult bird. Such cultures are useful in studying the recovery of AChE activity after brief exposure to OP compounds.

In general, after a brief exposure to an OP compound, such as diisopropylfluorophosphate (DFP) in the experiment shown in Fig. 3, small globular forms recover first, and then the larger asymmetric forms increase in activity (Wilson and Nieberg, 1983). AChE activity appears first within the cells, and then several hours later, in the medium. Virtually no activity reappears in OP-treated cells or media when cycloheximide, an inhibitor of protein synthesis, is added.

Recovery of AChE forms after exposure to OP compounds is similar *in situ* to that occurring *in vitro*. Small forms reappear first, followed by larger forms. The time course is much longer, however, taking days instead of hours for AChE activity to return. Sung and Ruff (1987) found brain AChE activity took several weeks to recover after exposure to DFP (Fig. 4). The rates of recovery after exposure to OP compounds *in situ* tend to be biphasic, probably because both spontaneous reactivation and synthesis of new protein occur simultaneously, especially during the first week (Blaber and Creasey, 1960b). Such complex kinetics make it difficult to establish the exact contributions of dephosphorylation of preexisting enzyme and synthesis of new enzyme to the recovery of AChE activity *in situ*.

Cell-culture studies indicate that AChE molecules are rapidly degraded within cells; loss of AChE activity occurs with a half-life of 2 to 3 hr in

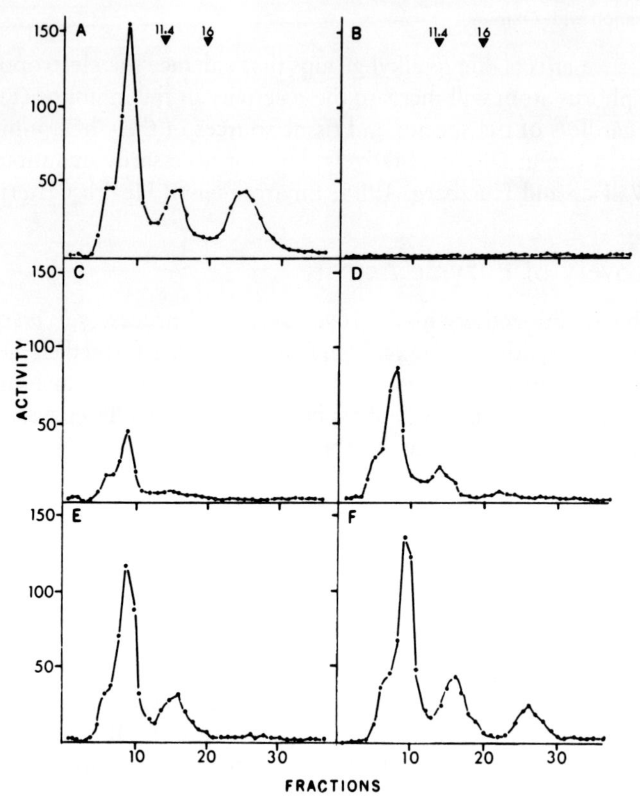

Figure 3 Recovery of multiple molecular forms of AChE in cultured muscle cells after DFP. (A) untreated 0 time controls; DFP-treated cells: (B) 0 time; (C) 2 hr; (D) 4 hr; (E) 8 hr, and (F) 24 hr after treatment with 0.1 m*M* DFP for 15 min. Arrows are catalase (11.4 s) and beta-galactosidase (16 s) standards. After Wilson and Nieberg (1983).

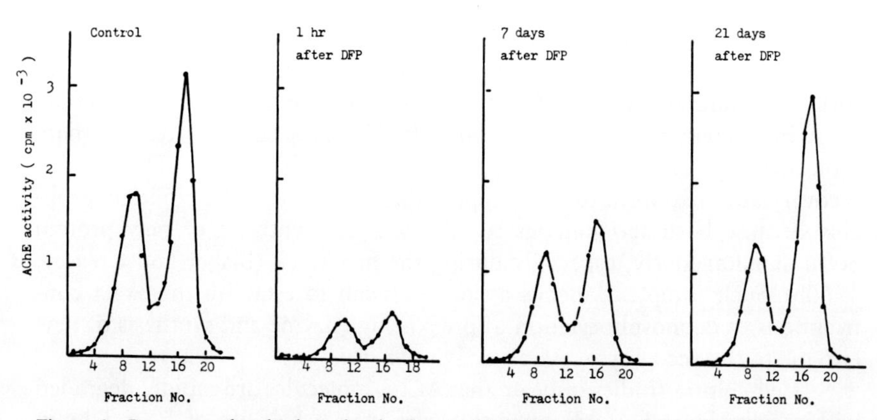

Figure 4 Recovery of multiple molecular forms of AChE in adult rat brain after DFP treatment *in vivo*. After Sung and Ruff (1987).

cultured embryo muscle cells in the absence of protein synthesis (Bulger *et al.*, 1982; Rotundo and Fambrough, 1982; Wilson and Nieberg, 1983). Once the enzyme reaches the cell surface, and thus is protected from intracellular proteases, it may remain active for days or longer.

C. Spontaneous Reactivation

Spontaneous reactivation of OP–ChE complexes usually is studied with enzymes from RBCs and brain from mammals and birds (avian RBCs do not contain ChE). It is appropriate to attribute the return of AChE activity in RBCs within hours or a few days after OP treatment to spontaneous reactivation, since these cells lack the capacity to synthesize the enzyme. However, the finding that new enzyme is rapidly synthesized in muscle after OP treatments (Leonard and Salpeter, 1979) suggests that one should not automatically attribute rapid recoveries of tissue AChE activity *in situ* to spontaneous reactivation.

Dephosphorylation of OP–ChE complexes is a nucleophilic displacement reaction. Eto (1974), in a detailed monograph on OP compounds, points out that the stability of a phosphorylated AChE tends to be similar to that of the specific OP inhibitor, noting that "AChE can be regarded as a relatively good leaving group" (p. 136). Examples of spontaneous recoveries of AChE activity are shown in Table I. In general, AChE–OP complexes from dimethoxy-substituted OPs (e.g., malathion) spontaneously dephosphorylate more rapidly than diethoxy- (e.g., parathion) or diisopropoxy- (e.g., DFP) substituted ones (Eto, 1974; Environmental Health Criteria, 1986). Such a sequence of reactivatability may be because methyl groups have less steric hindrance and greater electronegativity than do ethyl or isopropyl groups, making the phosphorus atom more electropositive, and the OP more susceptible to hydrolysis.

D. Aging

During early studies of chemically induced recovery of AChE activity after OP treatments, a progressive loss in the ability of the OP–enzyme complex to reactivate was noted (Hobbiger, 1956; Blaber and Creasey, 1960a,b). This decrease in reactivatability, designated *aging*, is considered the result of a loss of a second group, usually an alkyl group, from the phosphorus atom, stabilizing the OP–enzyme complex, and preventing both its spontaneous or chemical reactivation (Eto, 1974; see also Chapter 2, Scheme 7, by Thompson, this volume).

A number of theories concern the aging process. One proposal is that the process whereby an R-group is cleaved and released into the surrounding media, yielding a negatively charged oxygen moiety, creates an electrostatic

TABLE I

Spontaneous Reactivation Half-Life Values of Phosphorylated ChE[a]

R-groups of OP–enzyme complex	Example compound	pH	Half-life (hrs)	Enzyme	References[f]
Dimethoxy	Dichlorvos	7.4	2.5	Mouse brain AChE	1
	Dimethoate[b]	7.4	1.9	Rat brain AChE	2
	Malathion[b]	7.4	ND(a)[e]	Fly head AChE	1
		7.4	1.0	Bovine RBC AChE	3
		7.4	0.85	Human RBC AChE	2
		7.8	1.4	Rabbit RBC AChE	1,4
		7.8	1.3	Rat RBC AChE	5
		7.4	2.3	Rat plasma BChE	2
Methoxy/amido	Methamidophos	7.5	0.96	Electric eel AChE	6
Methoxy/methylthio		7.4	0.13	Bovine RBC AChE	3
Diethoxy	Chlorpyrifos[b]	7.4	58	Bovine RBC AChE	3
	Diazinon[b]	7.4	58	Human RBC AChE	7[c]
	Parathion[b]	7.8	103	Rat brain BChE	8
		7.8	2.2	Hen serum BChE	9
		7.8	200	Horse serum BChE	9
		7.8	730	Human serum BChE	9
		7.8	5.0	Rat serum BChE	8
Ethoxy/ethylthio		7.4	0.43	Bovine RBC AChE	3
Di-(2-chloroethoxy)	Haloxon	7.4	0.30	Bovine (calf) RBC AChE	10
		7.7	0.40	Sheep RBC AChE	11g
Ethoxy/dimethylamido	Tabun	7.4	ND(e)	Human RBC AChE	12
		7.4	ND(e)	Horse plasma BChE	12
		7.4	ND(e)	Human plasma BChE	12
Isopropoxy/methyl	Sarin	7.2	ND(b)	Human RBC AchE	13
		7.4	SD(e)	Purified horse serum BChE	12
		7.4	SD(e)	Human plasma BChE	12
		7.6	2.6	Electric eel AChE	14i
Diisopropoxy	DFP	7.4	ND(f)	Human plasma BChE	15
		7.4	ND(c)	Human RBC AChE	7
		7.4	ND(d)	Guinea Pig RBC	16h[d]
		7.4	ND(d)	Guinea Pig brain	16h[d]
Di-(2-chloropropoxy)		7.4	0.31	Bovine (calf) RBC AChE	10
Di-(3-chloropropoxy)		7.4	0.58	Bovine (calf) RBC AChE	10
Disecbutoxy		7.4	ND(c)	Human RBC AChE	7
Pinacoloxy/methyl	Soman	7.0	ND(b)	Electric eel AChE	17i
		7.1–7.3	ND(c)	Human RBC AChE	18i

[a]Some half-lives were calculated from $t_{1/2} = (0.693/k)$, where k is the observed spontaneous reactivation rate/hou Pinacoloxy, 1,2,2-trimethylpropoxy; AChE, acetylcholinesterase; BChE, butyrylcholinesterase or pseudocholineste ase; RBC, red blood cell or erythrocyte.

[b]Requires activation to oxon or other active metabolite.

[c]Cited in reference 19 (Reiner, 1971).

[d]Incubation *p*H not stated but probably *p*H 7.4 as listed in reference 19.

[e]ND, not detected after, SD, some detected, after (a) 5 hr; (b) 6 hr; (c) 48 hr; m(d) 72 hr; (e) 7 days; (f) unknown tim Spontaneous reactivation experiments were performed at 37°C, unless otherwise specified; (g) 36°C; (h) 30°C; (i) 25°

[f]1. Van Asperen and Dekhuijzen (1958); 2. Skrinjaric-Spoljar *et al.* (1973); 3. Clothier *et al.* (1981); 4. Aldridge (195 5. Vandekar and Heath (1957); 6. Langenberg *et al.* (1988); 7. Burgen and Hobbiger (1951); 8. Davison (1953); Davison (1955); 10. Pickering and Malone (1967); 11. Lee (1964); 12. Heilbronn (1963); 13. Davies and Green (195 14. Hovanec *et al.*(1977); 15. Hobbiger (1955); 16. Hobbiger (1951); 17. Amitai *et al.* (1980); 18. Berry and Dav (1966); 19. Reiner (1971).

barrier to nucleophilic reactivators (Harris *et al.*, 1966; Amitai *et al.*, 1982; Masson *et al.*, 1984). Another idea is that the OP residue forms an additional bond with the enzyme as the R-group is cleaved (Hobbiger, 1963). Still another is that there is a conformational change during the aging process, sinking the OP residue deeper into the active site of the anzyme (Amitai *et al.*, 1980; Amitai *et al.*, 1982).

There is evidence that a conformational change of the enzyme is required for aging to occur (Amitai *et al.*, 1982; Van der Drift, 1985; Masson *et al.*, 1984; Steinberg *et al.*, 1989). Aging does not occur if an OP-inhibited ChE is denatured, but will continue when it is renatured (Wilson, 1967).

Experiments with radiolabeled sarin (Harris *et al.*, 1966) directly demonstrated the loss of an alkyl group from the enzyme–inhibitor complex (Table II) and showed that the percentage of enzyme undergoing alkyl group cleavage was closely correlated to the percentage of the enzyme activity that was resistant to oxime reactivation.

Table III lists rates of aging of some common agricultural and other OP compounds. The rate of aging is dependent, in part, on the alkyl groups and the reactivity of the OP compound. Aging rates are often inversely related to those of spontaneous reactivation; ChE–OP complexes that rapidly age tend to spontaneously reactivate slowly. Dimethoxy-phosphorylated AChEs are exceptions; they both age and spontaneously reactivate rapidly.

Table IV shows an experiment of Clothier *et al.* (1981), in which the rates of spontaneous reactivation and aging of bovine RBC AChE, inhibited with dimethoxy-, diethoxy-, ethoxy/ethylthio- and methoxy/methylthio-substituted OP compounds were determined. AChE inhibited with dimethoxy-substituted compounds spontaneously reactivated and aged more rapidly than did AChE inhibited by the diethoxy-substituted OP compounds. Rates of spontaneous reactivation and aging were greatly increased when the oxygen

TABLE II

Loss of an Isopropyl Group from ^{32}P-Labeled Sarin after Incubation with Rat Brain AChE[a]

Time (hr)	Percentage reactivatable enzyme	Percentage ^{32}P as isopropyl methyl phosphoric acid
0.5	92	91
3	74	72
6	46	55
9	32	36
14	17	23
24	8	9

[a]2-PAM reactivation of rat brain AChE from animals treated with ^{32}P-sarin. Six pairs of brains were used for each time interval. From Harris *et al.* (1966).

TABLE III

Aging Half-Life Values of Phosphylated ChEs

R-groups of OP–enzyme complex	Example compound	Half-life (hr)	Enzyme	References[a]
Dimethoxy	Dichlorvos	2.0	Chicken brain AChE	1[d]
	Dimethoate[b]	2.0	Chicken brain AChE	1[e]
	Malathion[b]	6.7	Rat brain AChE	2
		3.9	Human RBC AChE	2
		8.9	Bovine RBC AChE	3
Methoxy/amido	Methamidophos[b]	ND(a)	Electric eel AChE	4(e)
Methoxy/methylthio		0.54	Bovine RBC AChE	3
Diethoxy	Chlorpyrifos[b]	36	Mouse brain AChE	5
	Diazinon[b]	41	Human RBC AChE	6(d)[f]
	Parathion[b]	58	Bovine RBC AChE	3
Ethoxy/ethylthio		3.6	Bovine RBC AChE	3
Di-(2-chloroethoxy)	Haloxon	ND(b)	Bovine (calf) RBC AChE	7(f)
Ethoxy/dimethylamido	Tabun	13	Human RBC AChE	8
		6.4	Human plasma BChE	8
		128	Horse plasma BChE	8
Isopropoxy/methyl	Sarin	5.8	Rat brain AChE	9
		5.8	Rat brain AChE	9[e]
		3.0	Human RBC AChE	10(c)[e]
		5.8	Human plasma BChE	8
Diisopropoxy	DFP	4.0	Mouse brain AChE	5
		4.6	Human RBC AChE	6(d)[f]
Di-(*n*-butoxy)		0.48	Bovine (calf) RBC AChE	7(f)
Di-(4-chlorobutoxy)		0.68	Bovine (calf) RBC AChE	7(f)
Pinacoloxy/methyl	Soman	0.037	Rat brain AChE	11
		0.040	Rat brain AChE	11[c]
		0.022	Human RBC AChE	12
		0.089	Dog RBC AChE	13
		0.093	Dog RBC AChE	13[e]

Some half-lives were calculated from $t\frac{1}{2} = (0.693/k)$, where k is the observed aging rate/hour. Pinacoloxy, 1,2,2-trimethylpropoxy; AChE, acetylcholinesterase; BChE, butyrylcholinesterase or pseudocholinesterase; RBC, red blood cell or erythrocyte. ND, not detected after (a) 6 hr (b) 90 min. Aging experiments were performed at 37°C and *p*H 7.3 to 7.4, unless otherwise specified; (c) *p*H 7.2, (d) *p*H 7.45, (e) *p*H 7.5, (f) *p*H not stated in paper but probably *p*H 7.4 as listed in reference 15 (Reiner, 1971).

[a]1. Witter and Gaines (1963); 2. Skrinjaric-Spoljar *et al.* (1973); 3. Clothier *et al.* (1981); 4. Langenberg *et al.* (1988); 5. Hobbiger (1957); 6. Hobbiger (1956); 7. Pickering and Malone (1967); 8. Heilbronn (1963); 9. Harris *et al.* (1966); 10. Davies and Green (1956); 11. Fleisher and Harris (1965); 12. Harris *et al.* (1978); 13. Fleisher *et al.* (1967); 14. Main (1984); 15. Reiner (1971).

[b]Requires activation to oxon or other active metabolite.

[c]Measure rate of dealkylation directly.

[d]Incubated at 40°C.

[e]*In vivo.*

[f]Cited in Reference 14 (Main, 1984).

TABLE IV

Spontaneous Reactivation and Aging of Bovine RBC AChE[a]

R_1	R_2	Reactivation	Aging
$-OCH_3$	$-OCH_3$	60	498
$-OCH_3$	$-SCH_3$	6	72
$-OC_2H_5$	$-OC_2H_5$	3480	3180
$-OC_2H_5$	$-SC_2H_5$	24	216

[a]Values given are estimated half-times in minutes. Conditions: pH 7.4; 37°C. After Clothier *et al.* (1981).

atoms bound to the methyl or ethyl groups were replaced by sulfur atoms. Langenberg *et al.* (1988), in a study of methamidophos and other phosphoramidates, found that rates of spontaneous and oxime-induced reactivation increased with increasing electronegativity of the substituted ethoxy groups.

In general, OP compounds used in agriculture (e.g., malathion and parathion) have aging rates with half-lives of hours and longer. In contrast, OP–AChE complexes of the chemical warfare agent soman age in less than 10 min (Clement, 1981).

III. Chemical Reactivation

It has been almost 40 years since I.B. Wilson and others employed nucleophiles like hydroxamic acid to study inhibition of AChE by OP compounds. The research culminated in the use of oximes to reactivate OP-inhibited AChE (Wilson and Ginsburg, 1955; Childs *et al.*, 1955). The oximes in clinical use are salts of pyridine 2-aldoxime (2-PAM). (The reader is cautioned that the terminology used to designate these oximes is not consistent. Here, we refer to pyridine 2-aldoxime methochloride as 2-PAM Cl, pyridine 2-aldoxime methiodide as 2-PAM I and pyridine 2-aldoxime methanesulphonate as P2S.)

The potency of pralidoxime has been attributed to the presence of a quaternary nitrogen that binds to the anionic site of the enzyme and to the highly nucleophilic oxime moiety situated at a specific, optimal distance from the phosphorus atom of the inhibitor, supporting the transfer of the substituted phosphate (or phosphonate) residue from the active site of the enzyme to the oxime. Hackley *et al.* (1959) demonstrated that, in addition to binding to the enzyme-bound OP residue, 2-PAM could react directly with the free OP molecule. Reactions of oximes directly with OP inhibitors, like paraoxon, and with OP-inhibited AChE, are shown in Fig. 5 (Harvey *et al.*, 1986 a,b). The extent to which such direct reactions between OP compounds and oximes occur in the body is not clear.

Other oximes, configured differently from 2-PAM, have been shown to be even more effective reactivators and antidotes to some chemical warfare agents (Fig. 6). Bisquaternary oximes such as TMB-4 [1,1′-trimethylene *bis*-(4-formyl pyridinium chloride) dioxime] and obidoxime (Toxogonin; 1,1′-oxydimethylene *bis*-(4-formyl pyridinium bromide) dioxime), and a series of oximes synthesized by Hagedorn, especially one designated HI-6 [1-(2-hydroxyiminomethyl pyridinium)-1-(4-carboxyamido-pyridinium) dimethylether dichloride], have been studied in some detail, especially with respect to chemical warfare agents. For example, Table V shows that HI-6 is a more effective prophylactic to PiMeP-Cl (O-pinacolyl methylphosphonochloridate) than are 2-PAM Cl or TMB-4 (Boskovic, 1981).

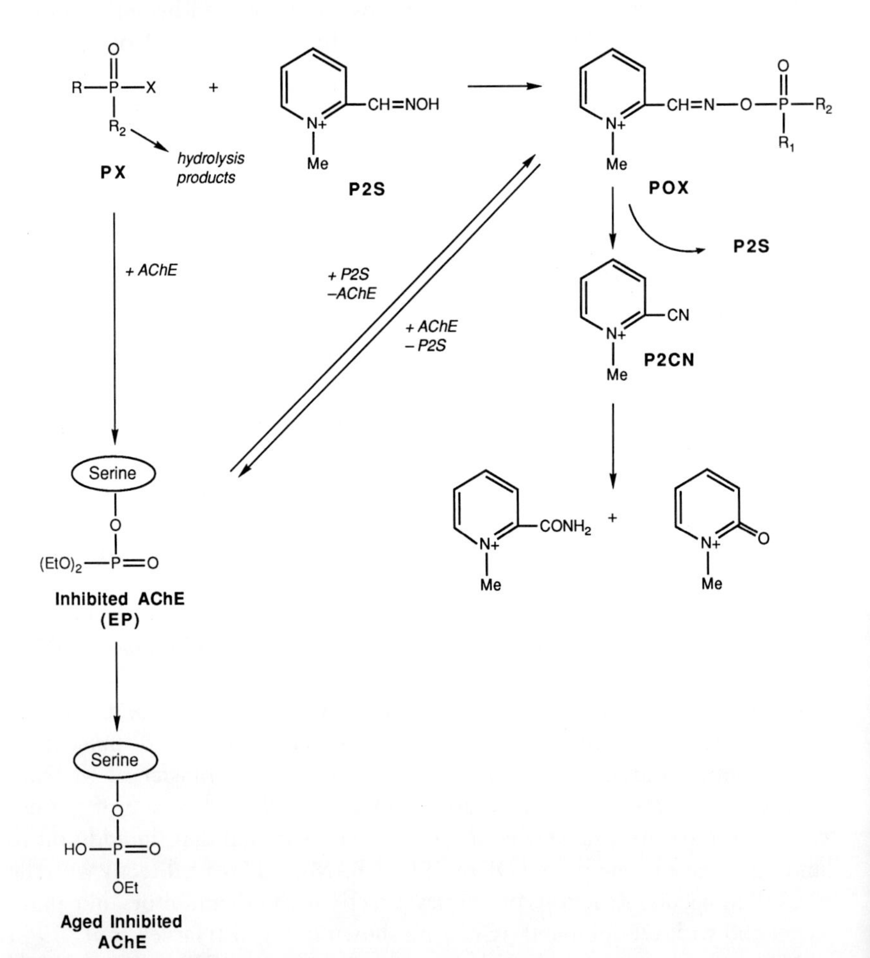

Figure 5 Reactions of 2-PAM with a diethyl OP and OP-inhibited AChE. Adapted from Harvey *et al.* (1986a).

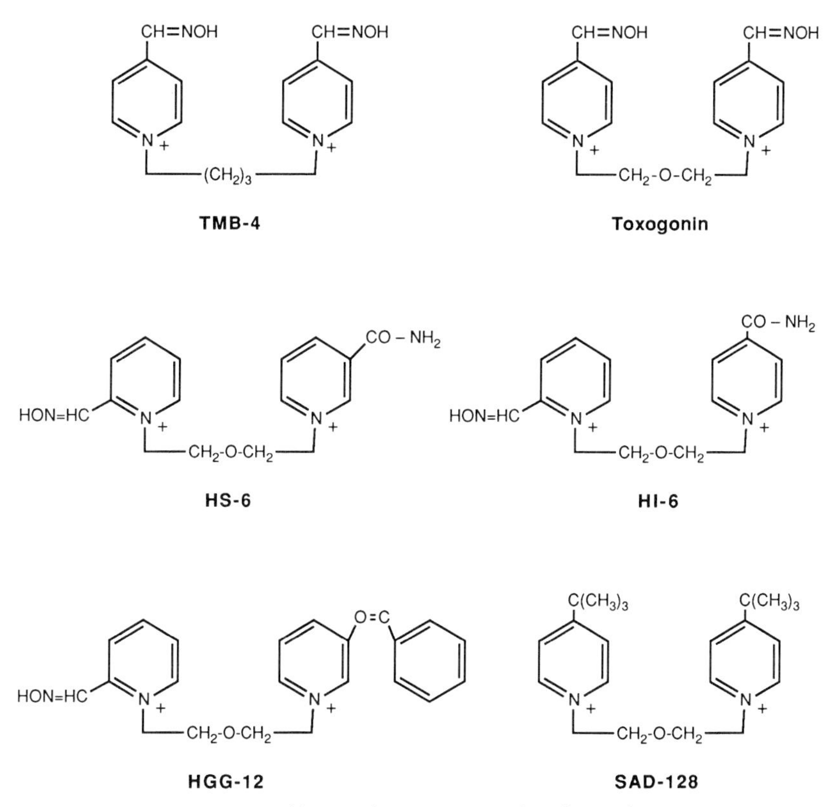

Figure 6 Structures of *bis*-pyridinium compounds. After Boskovic (1981).

There is evidence that oximes also may reduce the toxicity of soman and other chemical warfare agents by mechanisms other than reactivation. For example, HI-6 inhibits carbachol-induced contraction of chick biventer cervicis muscle (Clement, 1981), and both 2-PAM and HI-6 alter the electrical properties of the nicotinic receptor of the frog (Alkondon *et al.*, 1988). Bedford *et al.*, 1989, reported that the most potent imidazolium antidotes against soman were the poorest reactivators of soman-treated AChE. Indeed, the *bis*-pyridinium compound SAD-128 (1,1′-oxydimethylene *bis*-(4-tert-butylpyridinium chloride) lacks an oxime moiety, but is effective in protecting rodents from soman (Clement, 1981). Albuquerque *et al.* (1988), reviewing work from their laboratory, propose that "The effect on AChE appears not to be a primary mechanism in the therapeutic actions of carbamates and oximes in OP poisoning" (p. 371). They hypothesize that pharmacological actions on transmission, especially the AChR (ACh receptor)-channel blocking properties of these chemicals, was "the pivotal mechanism" for their antidotal actions.

TABLE V

Toxicity of PiMeP-Cl to Mice

Oxime[a]	Dose (mM/kg)	LD_{50} (μM/kg)	Toxicity ratio
Saline	0	580	1
2-PAM Cl	0.1	604	0.96
TMB-4	0.05	37	15.7
HI-6	0.13	926	0.61
HS-6	0.1	1107	0.5

[a]Inject oximes i.p. 5 min before s.c. (O-pinacolyl methylphosphonochloridate) (PiMeP-Cl) administration. Toxicity ratio is LD_{50} without: LD_{50} with treatment. After Boskovic (1981).

Decamethonium and SAD-128 also have been shown to retard the rate of aging and increase the rate of oxime reactivation of AChE inhibited by compounds such as sarin and soman (Harris *et al.*, 1978). One possibility is that they bind to allosteric sites on the AChE molecules, affecting the stability of the AChE–OP complex.

Much of the research directed at understanding the properties of OP antidotes has focused on chemical warfare agents. These OP compounds form OP–enzyme complexes with inhibition rate constants several orders of magnitude greater than agricultural chemicals (Eto, 1974; Ellin, 1982). Although it is probable that antidotes developed for the potent OP chemical warfare agents will be effective against OP intoxications occurring in agricultural settings, comparisons of the effectiveness of chemical warfare antidotes such as the *bis*-pyridinium compounds, HI-6 and SAD-128, and 2-PAM against agricultural chemicals are needed.

Detailed studies of the reactivation and aging of esterases other than ChEs generally are lacking. One exception is neuropathy target esterase (NTE), the enzyme associated with organophosphate-induced delayed neuropathy (OPIDN) (see Chapters 16 and 17 by Richardson and Abou-Donia, this volume). Clothier and Johnson (1981) found that oximes and potassium fluoride (KF) reactivated unaged diethylphosphinylated NTE, and used KF to study the rapid aging of NTE after inhibition by DFP (diisopropylfluorophosphate).

IV. Oxime Treatment after Organophosphorus Poisoning

2-PAM Cl is registered in the United States for use as an antidote (Protopam); its methanesulfonate salt (P2S) is used in Europe. Table VI shows an example of the efficacy of 2-PAM I as a therapeutic agent in OP poisonings. Symptoms were less severe and plasma ChE activity levels were higher in patients poisoned with parathion 30 min after treatment with the oxime (Namba, 1971). Twelve percent had plasma ChE activities depressed by 50 to 79%, 56% had ChE activities depressed by 80 to 89%, and 32% had ChE activities depressed by 90% or more of normal. None had plasma ChE activities greater than 50% of normal. (Other data cited by Namba indicate the therapeutic potential of the other salts of 2-PAM are similar.)

Aging is probably the major factor limiting the effectiveness of oxime reactivation therapy in cases of OP poisoning. The sooner treatment is begun, the greater the likelihood of success. Treatment with PAM Cl (or P2S) to regenerate AChE activity, and atropine to block muscarinic ACh receptors, have become the accepted therapy for OP intoxications (Environmental Health Criteria, 1986). Usual dosages recommended are 1 g PAM Cl, i.m. or i.v. two or three times a day, and 2 mg atropine i.v. at 15 to 30 min intervals as needed (Environmental Health Criteria, 1986).

The anionic forms of 2-PAM and other oximes are believed to be the active agents. Groups that increase the relative acidity of an oxime will lower its pK_a to the 7.5 to 8.0 range, resulting in a greater percentage of charged reactivator molecules at physiological pH. On the one hand, this should increase reactivity (Gray, 1984), but, on the other hand, it also should decrease lipophilicity and reduce ability to pass the blood–brain barrier. Even so, there are reports of reactivation of AChE within the CNS, and reduction in symptoms attributable to CNS intoxication after treatment with oximes (Lotti and Becker, 1982). The possibility of a reduction of circulating OP

TABLE VI

Actions of 2-PAM I on 25 Patients Poisoned with Parathion[a]

Symptoms, signs	Number of patients		0–10% ChE level	
	Before	After	Before	After
Nausea, vomiting	24	4	8	0
Dizziness	16	3	5	0
Pallor	18	5	8	1
Headache	15	7	3	0
Excessive saliva	15	0	6	0
Paresthesia	14	2	6	0
Muscle tremor	13	0	7	0
Dyspnea	13	2	6	0
Miosis	10	3	4	0
Impaired speech	10	2	4	0
Cramps	9	0	4	0
Bronchial secretions	6	0	2	0
Disturbed consciousness	5	0	4	0

[a]Patients with symptoms listed having 0–10% plasma ChE level before and 30 min after injection (i.v.) with 0.9 to 2.0 g pyridine 2-aldoxime methiodide (2-PAM I). Adapted from Namba (1971).

TABLE VII

LD_{50} of Oximes to Mice

Compound[a]	μmoles/kg
2-PAM Cl	695
Toxogonin	404
HS-6	880
HI-6	1430

[a]Inject i.p. in 0.9% saline, 5–6 doses; LD_{50} by probit analysis. After Clement (1981).

levels by direct chemical reaction with an oxime also deserves consideration in such cases.

Oximes themselves are toxic to animals and weak inhibitors of AChE (Table VII, Clement, 1981). In addition, the products of oxime reactivations, phosphorylated (or phosphonylated) oximes, are often as or even more reactive, but shorter-lived, than the parent OP compound (Eto, 1974; Harvey *et al.*, 1986a,b).

V. Oximes and Carbamates

The search for prophylactics to protect personnel from chemical warfare agents has stimulated research on carbamate ChE inhibitors and their reactions with oximes. The strategy is based on the idea that spontaneous decarbamylation tends to occur more readily than dephosphorylation from the active site of the enzyme. If administered before OP exposure, carbamates can prevent an OP from binding and forming its characteristic stable bond with the enzyme, thus preserving some AChE activity. While physostigmine and neostigmine have been successfully reactivated with oximes, several studies indicate that the toxicity of carbaryl, a widely used agricultural and domestic

TABLE VIII

Oximes and Toxicity of Carbamates to Rats[a]

Carbamate	Con	Obidoxime	2-PAM	Atropine	+Obidoxime	+2PAM
Carbaryl	397	101	173	2570	717	2410
Temik	3.5	8.2	5.4	18.0	> 24	22.4
NST	1.0	20.8	2.2	5.9	34.9	6.5
PST	5.3	17.9	15.9	49.3	> 100	92.6

[a]LD_{50} doses in μmol/kg. Con, carbamate treated, no antidote. Obidoxime and 2-PAM, 250 μM/kg; atropine sulfate, 17.4 mg/kg. NST, neostigmine; PST, pyridostigmine. Adapted from Natoff and Reiff (1973).

TABLE IX

Oximes and Carbamate Inhibition of Rat Brain ChE[a]

Carbamate	Obidoxime		2-PAM	
	0	1 m*M*	0	1m*M*
Carbaryl	4.57	5.71	4.23	4.81
Temik	5.43	5.15	5.43	5.15
NST	8.20	6.21	8.53	7.25
PST	7.54	6.27	7.38	6.43

[a] Log Carbamate Concentration for 50% Inhibition. Conditions as in Table VIII. Adapted from Natoff and Reiff (1973).

chemical, may be potentiated by oxime therapy. Natoff and Reiff (1973) compared the efficacy of P2S and obidoxime in reducing the toxicity of carbaryl, temik, neostigmine, and physostigmine in rats (Tables VIII and IX). The protective effects and the AChE inhibitions were dependent on the carbamate and the oxime used. Both 2-PAM and obidoxime increased rather than decreased the toxicity of carbaryl. Recently, Harris *et al.* (1989) compared the protective effects of 2-PAM Cl and HI-6 to rats exposed to physostigmine and carbaryl (Table X). The results were similar to the findings of the previous study—2-PAM increased carbaryl toxicity and reduced the effectiveness of atropine. Such data support the recommendation of Reese (1984) that 2-PAM should not be used to treat carbaryl poisoning.

TABLE X

Atropine and Oxime Therapy after Physostigmine and Carbaryl[a]

Treatment	LD_{50}	Protective ratio
	Physostigmine	
None	2.39	1.0
AT	17.2	7.2
AT + 2-PAM	21.0	8.8
At + HI-6	55.7	23.3
	Carbaryl	
None	69.9	1.0
2-PAM	89.4	0.6
AT	460	6.6
AT + 2-PAM	244	3.5
AT + HI-6	164	2.3

[a] LD_{50} for rats in mg/kg (i.p.). AT, atropine, 8 mg/kg; 2-PAM, 22 mg/kg; HI-6, 50 mg/kg. After Harris *et al.* (1989).

VI. Oxime Research

A. Cell Cultures

Oximes can be used effectively as probes to help understand regulation of cholinergic processes on the cell, tissue, and whole animal levels. The extreme toxicity of OP compounds and the fact that their target enzymes are localized within the nervous and muscular systems, make it difficult to study many of their effects *in situ* (e.g., cultured nerve and muscle cells do not die when poisoned with OP compounds such as DFP or paraoxon, the active metabolite of parathion).

One factor that has not been sufficiently investigated is the fate of the OP–AChE complex and its possible roles in regulation of AChE activity. An example of such research is a study (Wilson *et al.*, 1988b) in which 2-PAM was used to reactivate cultured quail muscle AChE after treatment with paraoxon. Results of these experiments showed that enzyme inhibited within the cells was still available for reactivation, rather than being rapidly destroyed. Highly differentiated cultures from 10-day quail embryo pectoral muscle were prepared and cultured for 10 to 14 days *in vitro*. Release of AChE into the medium was measured after inhibiting the initial ChE activity in the serum and embryo extract with DFP (Wilson and Nieberg, 1983). Cells were treated for 20 min with 1 μ*M* paraoxon at 38°C. Oxime reactivation was performed by incubating paraoxon-treated cells with 10 m*M* 2-PAM for 20 min at 38°C, a condition yielding optimal reactivation without affecting the AChE reaction itself. *Total* AChE refers to the sum of cell and medium activities; *net* AChE activity is the difference between the AChE activity of cells and medium at time t minus the AChE of the cells at time zero.

The second column of Table XI shows the total AChE activity of cells and medium in an untreated muscle cell culture over a 24-hr period. Cell AChE activity (not shown) remained relatively constant, while total AChE activity steadily increased in the medium, resulting in an increase in AChE activity from 296 to 341 mmol/ml culture. Treatment with paraoxon (column 3) inhibited virtually all AChE activity in the cells. AChE activity rapidly returned, but did not reach the level of the untreated cultures. Much of the initial AChE activity (80%) returned when paraoxon-treated cells were reactivated with 2-PAM at zero time (column 4). In another experiment (Table XII), 86% of the untreated activity of the cells returned after 2-PAM treatment. Total AChE activity tended to be higher in paraoxon-treated cultures reactivated with 2-PAM than in paraoxon or untreated cultures after 24 hr. The net increase in AChE activity was consistently higher with paraoxon-treated cultures than with untreated cultures, as we had shown earlier for DFP and paraoxon (Wilson and Walker, 1974; Cisson and Wilson, 1981). Reactivating the enzyme made little difference; net AChE activity was still higher than in untreated cultures. For example, at 24 hr, net AChE activity for the

TABLE XI

Recovery of Total AChE Activity in Muscle Cultures[a]

Hours	Untreated	Paraoxon	2-PAM
0	296 ± 14	5.0 ± 0.7	243[b]
2	259 ± 11	80 ± 4	321 ± 37
6	277 ± 10	126 ± 4	363 ± 24
12	335 ± 9	188 ± 18	407 ± 6
24	341 ± 44	214 ± 33	441 ± 117

[a]Culture in nmol/min/ml. Mean ± S.E.M. of triplicate cultures.
[b]Paired samples. Exposed to 1 μ*M* paraoxon for 20 min at zero time and subsequently to 10 m*M* 2-PAM for 20 min. Wilson, Hooper, Hansen and Nieberg, unpublished.

cultures presented in Table XI was untreated: 45 nmol/min/ml; paraoxon-treated: 209 nmol/min/ml and paraoxon/2-PAM treated: 198 nmol/min/ml. Such results suggest that treatment with the OP, and not prolonged inhibition of enzyme, is sufficient to increase the net AChE production of the cultures. It is unlikely that spontaneous reactivation played any role in the results, since, as discussed earlier, virtually no AChE activity returned in the absence of protein synthesis.

The extent to which inhibited AChE within the cells was able to reactivate was studied by adding 2-PAM to cultures 0, 2, 6, 12, 24, 48, and 72 hours after treatment with paraoxon (Table XII). Reactivatable AChE activity (2-PAM paraoxon column) decreased in the first 24 hr with a half-life of approximately 12 hr, and a half-life of about 24 hr thereafter, at least double the decrease in reactivatability that would be expected from aging of the inhibited enzyme itself. Interestingly, the sum total of AChE within para-

TABLE XII

2-PAM–Induced Recovery of Cell AChE after Treatment with Paraoxon[a]

Hours	Untreated	Paraoxon	2-PAM	Reactivated
0	256 ± 32	4 + 3	224 ± 4	220
2	270 ± 21	44 ± 3	235 ± 7	191
6	273 ± 13	81 ± 5	223 ± 14	142
12	266 ± 19	124 ± 4	236 ± 20	112
24	296 ± 43	149 ± 7	215 ± 26	66
48	294 ± 14	171 ± 7	208 ± 37	37
72	291 ± 3	213 ± 17	230 ± 17	17

[a]Culture in nmol/min/ml. Mean ± SEM of triplicate cultures; cells only. Exposed to 1 μ*M* paraoxon for 20 min at zero time, and 10 m*M* 2-PAM for 20 min at the hours indicated. Wilson, Hooper, Hansen and Nieberg, unpublished.

oxon–2-PAM treated cells remained relatively constant over time (although increasing in the medium due to secretion), as if there were a maximal AChE content permitted in the cells, regardless of whether or not the enzyme was active or inactive. Sucrose density centrifugation of cultures that had been treated with 2-PAM after exposure to paraoxon showed that all forms were reactivated, but, in several experiments, the soluble 4-7S forms did not completely recover when 2-PAM was used to reactivate AChE at zero time.

These results show that the long-term presence of phosphorylated AChE is not necessary for net AChE activity to increase after exposure to an OP compound. Logically, net AChE levels could increase owing to increased synthesis or to decreased degradation of the enzyme. Previous experiments (Wilson *et al.*, 1973; Bulger *et al.*, 1982) showed that AChE did not decrease more rapidly after cycloheximide treatment when examined several hours to a day after exposure. We have (Wilson *et al.*, 1988b) found an early decrease in the rate of AChE degradation within 2 hr following exposure that could account for at least some of the increase in net AChE activity. At best, aging of the paraoxon–AChE complex (approximately 36 to 60 hr, Table III) will not account for more than half of the progressive decrease in reactivatable AChE activity following paraoxon treatment; the remainder may have been the result of degradation of the inhibited enzyme.

Brockman *et al.* (1984) used 2-PAM reactivation, paraoxon, and short-term (2 hr) incubations of cultured rat myotubes to study the difference in relative rates of synthesis and degradation between tetrodotoxin-treated (nonfibrillating) and untreated cultures. They concluded globular forms were synthesized seven times and asymmetric forms six times faster in untreated cultures than in tetrodotoxin-blocked cultures.

Most cell-culture studies of AChE have dealt with the properties of the enzyme, using OP compounds more as chemical scalpels than as objects of inquiry in their own right. For example, the high specificity of nerve agent VX (studied under the name MPT) for AChE makes it a useful compound for examining the number and distribution of AChE molecules (Goudou and Rieger, 1983; Brockman *et al.*, 1984; Wilson *et al.*, 1988c). Studies aimed at understanding the interactions between OP compounds and AChE are often performed with RBCs. However, RBCs have only one of the many AChE forms found in the body, and are unable to synthesize AChE molecules (Massoulie and Bon, 1982; Toutant and Massoulie, 1987). Cell types such as cultured muscle and nerve possess the full repertoire of AChE regulation and afford unique opportunities to conduct toxicology experiments.

B. Motor End Plate AChE and Oximes

Several research groups have used oxime reactivation to count the number of AChE molecules at motor end plates and to study their regulation in normal and damaged muscle. Tissue is treated with unlabeled DFP; then the AChE sites are reactivated with 2-PAM, then labeled with radiolabeled DFP, and the number and location of the AChE molecules are examined with autoradiography and electron microscopy. Using this technique Porter and Barnard (1975) established that there were approximately 3000 catalytic centers of AChE per μm^2 distributed evenly over the synaptic cleft. Since each asymmetric form has 12 active centers, these sites were equivalent to 250 AChE molecules (Barnard *et al.*, 1984). Alpha-bungarotoxin labeling provided a similar estimate of 11,000 ACh receptor molecules per μm^2, localized at the tips of the clefts. Barnard *et al.*, (1984) speculated that the significance of the ratio of 1 AChE to 88 ACh receptor molecules was to provide enough space between enzyme molecules for ACh to diffuse across the synapse to bind to the receptor, and enough catalytic sites clumped together to prevent each ACh molecule from acting more than once. Salpeter's group recently used oxime reactivation, DFP, and autoradiography to study the transient damage to motor end plates and the surrounding muscle caused by OP compounds. This phenomenon has been attributed to release of calcium (Leonard and Salpeter, 1979) due to excess ACh (Dettbarn, 1984). They found that the previously established rapid recovery of motor end plate AChE activity (Leonard and Salpeter, 1979) was much slower when structural damage to the end plate by the OP treatment was prevented by atropine treatment (Kasprzak and Salpeter, 1985).

C. Cholinesterases of Wild Birds

Many of the recent studies of the actions of OP compounds on animals have been conducted with birds. Examination of the ChE levels in the blood and brains of wild birds has become one of the primary ways wildlife toxicologists determine whether exposure to OP pesticides has occurred. Workers have agreed that decreases in ChE activity greater than two standard deviations below the mean are satisfactory evidence of exposure to OP compounds (Ludke *et al.*, 1975; Hill, 1988). However, the intra- and interspecies variations in ChE levels, and the differences in methodologies between laboratories limit the usefulness of experiments in which results obtained in the field are compared to standard tables.

Oxime reactivation has been proposed as an adjunct to studying mean ChE levels (Karlog and Poulsen, 1963; Martin *et al.*, 1981). So long as aging is not advanced, an increase in AChE or BChE activities above the level of an untreated sample would suggest that OP poisoning had occurred. (In the case of tissues actively synthesizing protein, it must also be assumed or demonstrated that synthesis of new enzyme is not a factor.) Table XIII shows results of an experiment in which the reactivatability of quail brain AChE was studied from animals poisoned with parathion and kept after death in a hood at

TABLE XIII

Reactivation of Quail Brain ChE after 5 mg/kg Parathion[a]

	Dead			Live		
Days	Before 2-PAM	After 2-PAM	%	Before 2-PAM	After 2-PAM	%
1	1.91	9.75	410	4.56	8.99	97.1
2	3.08	10.4	238	5.94	8.18	37.8
4	6.16	8.67	40.7	8.24	8.50	3.15
7	4.66	4.55	−2.4	10.2	9.73	−4.6

[a]Mean μmoles/min/mg; n, 2 to 4; treated day 0 with LD_{50} dose of parathion. Reactivation with 1 m*M* 2-PAM, 60 min, 38°C. Hooper and Wilson, unpublished.

ambient temperature for several days, mimicking field conditions. There was a window of several days before aging of the diethyl phosphorylated-AChE complex became too great for exposure to be detected. Table XIV contains results of an experiment of Martin *et al.* (1981) in which there was reactivation of AChE from quail poisoned with several OP compounds but not with carbamates.

Although several laboratories presented data validating the approach, field tests of oxime reactivation were not conducted until 1988 when the blood ChEs of red-tailed hawks (*Buteo jamaicensis*) live-trapped near orchards in the Central Valley of California during winter dormant OP–spray season was examined (Wilson *et al.*, 1988a). During this time of year, many

TABLE XIV

Reactivation of Bird Brain AChE[a]

Compound	Dose	Death	24 hr	React
Control	0	22.8	15.6	16.9
Carbophenothion	170	1.1	1.0	16.4
Chlorfenvinphos	444	0.8	0.9	17.1
Dimethoate	75	2.4	1.9	5.7
Mevinphos	30	1. 3	1.7	11.5
Pirimiphosmethyl	520	4.0	3.1	6.3
Aldicarb	30	3.9	15.8	8.4
Bendiocarb	50	1.4	16.9	3.8
Methiocarb	75	3.9	15.0	9.7
Oxamyl	14	3.3	14.2	7.5
Primicarb	160	1.4	10.4	5.6
Thiofanox	3.6	3.0	16.0	8.2

[a]Japanese quail; Triton X 100 homogenates; μmol/min/g; control, N = 37; pesticide, N = 8; mevinphos, N = 7; 0.25 m*M* 2-PAM, 4°C. After Martin *et al.* (1981).

TABLE XV

Plasma Cholinesterase of Red-tailed Hawks during Dormant-Spray Season[a]

	Birds captured > ¼ mile from orchards			
		Percentage		
	Activity	Reactivated	Depressed	Inhibited
ChE	791 ± 220	5	5	10
N = 20		(1/20)	(1/20)	(2/20)
AChE	194 ± 64	10	5	15
		(2/20)	(1/20)	(3/20)
	Birds captured < ¼ mile from orchards			
		Percentage		
	Activity	Reactivated	Depressed	Inhibited
ChE	644 ± 24	27	27	38
N = 34		(9/34)	(9/34)	(13/34)
AChE	164 ± 79	29	15	35
		(10/34)	(5/34)	(12/34)

[a]Hawks sampled near Chico, California, 12/30/87 to 2/21/88. ChE, total plasma ChE. Means ± SD, nmoles/min/ml. Reactivated, 2-PAM values > 5% above control. Depressed, values 2 SD below mean of field controls (birds sampled, November and December 1987). Inhibited, sum of reactivated and depressed samples. Wilson *et al.* (1988a).

of the more than 400,000 acres of almonds in the region are sprayed with an oil–OP mixture to control San Jose scale (*Quadraspidiotus perniciosus*) and peach twig borer (*Anarsia lineatella*). The results of a pilot (Hooper *et al.*, 1989) and a larger study (Wilson *et al.*, 1988a) showed that more than one third of the birds captured within one quarter mile of the orchards exhibited blood ChE activities that were either reactivatable by 2-PAM Cl and/or were more than two standard deviations below the mean activity of controls (Table XV). At the same time, we detected OP residues on the birds and OP metabolites in their excreta (Wilson *et al.*, 1988a; Hooper *et al.*, 1989). Studies to mitigate exposure of the birds are under way, sponsored by the Almond Board of California with the collaboration of the pesticide registrants and state agencies.

D. Humans and Other Mammals

Birds do not have RBC AChE, restricting blood ChE studies to plasma, and oxime reactivations to samples in which the OP compounds present have been extracted or the samples diluted. However, oximes added to the RBCs of mammals can readily be washed away by centrifuging and resuspending the cells. One problem is that, even though the AChE activity of the RBCs is usually high, hemoglobin may interfere with both radiometric and colorimetric enzyme assays. The problem may be circumvented by hemolyzing the

50 µl Whole Blood or Washed Erythrocytes
+
900 µl "Reactivation Buffer"
(130 mM Dibasic and 195 mM Monobasic Sodium Phosphate, *p*H 8.0)

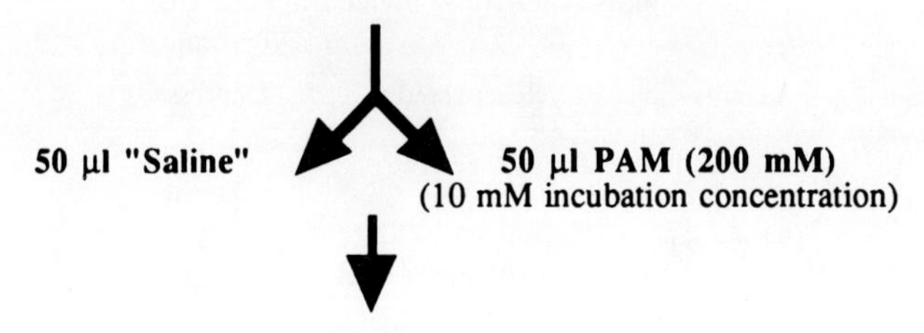

Incubate 40 to 60 min at 25°C, 40 Cycles/min

Add 5 ml "Erythrocyte Wash Buffer"
(135 mM NaCl, 20 mM Sodium Phosphate, 3 mM KCl, *p*H 8.0)
Centrifuge 15 min at 2,500*g*
Remove Supernatant

Add 5 ml "Hemolyzing Buffer"
(20 imOsm Sodium Phosphate, *p*H 7.4)

Add 1.5 ml of Sucrose Cushion
(7% Sucrose in "Hemolyzing Buffer" (w/v), *p*H 7.4)
Centrifuge 20 min at 27,800*g*
Remove Supernatant

Add "Solubilization Buffer"
(0.5% Triton X-100 in 100 mM Sodium Phosphate, *p*H 8.0)
to Yield 1.00 g of Contents
Incubate 15 min on Ice

Assay for AChE Activity

Figure 7 Flow chart for 2-PAM reactivation assay. Hansen and Wilson, unpublished.

cells, taking care to retain AChE (Dodge *et al.*., 1963). A flow sheet of the technique currently used in our laboratory is given in Fig. 7. Triplicate samples of each test permit detection of oxime reactivation with OP poisoned RBCs for several days with a 95% confidence level. Figure 8 shows the results of an experiment in which rabbits were exposed to a dermal dose of parathion for 7.5 hr, samples of blood were taken periodically for several days, and RBC AChE was subjected to reactivation with 2-PAM Cl.

We are collaborating with California Department of Food and Agriculture, the University of California at Davis Department of Occupational Medicine, and NIOSH to study the blood ChE of workers in orchards. The state of California requires workers using OP pesticides to have their blood ChE level periodically determined. The EPA is considering whether there should be nationwide regulations requiring monitoring of blood ChE levels. However, there is no single standard procedure of sampling, storage, and assay to facilitate comparisons of one study to another.

VII. Closing

Organophosphorus pesticides are one of the few classes of chemicals in use in the workplace for which antidotes are known. Although the demise of OP

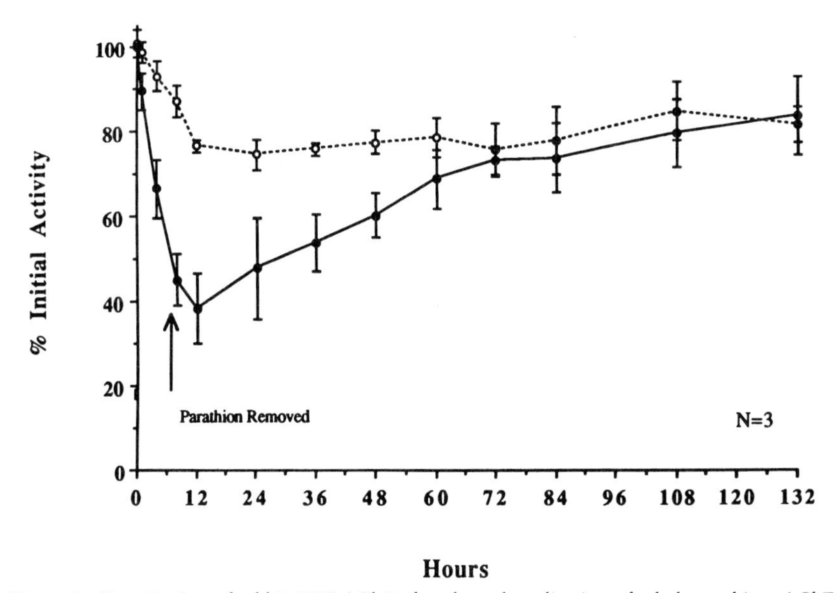

Figure 8 Reactivation of rabbit RBC AChE after dermal application of ethyl parathion. AChE activity before (solid line) and after (dotted line) 2-PAM reactivation. Reactivation as in Fig. 7. Hansen, Weisskopf, and Wilson, unpublished.

pesticides has been predicted for a number of years, they have not yet given way to a new generation of pheromones, genetically engineered microorganisms, and designer pesticides. Until then, studies of oxime reactivation of OP compounds will provide useful basic knowledge on mechanisms of OP toxicity and on exposure of wild animals and humans to these chemicals. Perhaps being able to rapidly determine blood ChE activities and test RBC AChE for oxime reactivation may make the difference between success or failure in treating an OP poisoning.

Acknowledgments

Research reported from this laboratory was supported in part by NIEHS, NIOSH, USDA, the Almond Board of California, the Endangered Species Program of the California Department of Fish and Game, the California Toxic Substances Research and Teaching Program and the UCD Agriculture Health Safety Center. The assistance of Dr. Carol P. Weisskopf and Mr. John D. Henderson is acknowledged.

References

Albuquerque, E. X., Alkondon, M., Deshpande, S. S., Cintra, W. M., Aracava, Y., and Brossi, A. (1988). The role of carbamates and oximes in reversing toxicity of organophosphorus compounds: As perspective into mechanisms. In "Molecular Basis of Drug and Pesticide Action" (G.G. Lunt, ed.), pp. 349–373, Elsevier (Biomedical Division), New York.

Aldridge, W. N. (1953). The inhibition of erythrocyte cholinesterase by tri-esters of phosphoric acid. *Biochem. J.* **54**, 442–448.

Aldridge, W. N. (1971). The nature of the reaction of organophosphorus compounds and carbamates with esterases. *Bull. W. H. O.* **44**, 25–30.

Alkondon, M., Rao, K. S., and Albuquerque, E. X. (1988). Acetylcholinesterase reactivators modify the functional properties of the nicotinic acetylcholine receptor ion channel. *J. Pharmacol. Exp. Ther.* **245**, 543–556.

Amitai, G., Ashami, Y., Shahar, A., Gafni, A., and Silman, I. (1980). Fluorescent organophosphates: Novel probes for studying aging-induced conformational changes of inhibited acetylcholinesterase and for localization of cholinesterase in nervous tissue. *Monogr. Neural Sci.* **7**, 70–84.

Amitai, G., Ashani, Y., Gafni, A., and Silman, I. (1982). Novel pyrene-containing organophosphates as fluorescent probes for studying aging-induced conformational changes in organophosphate-inhibited acetylcholinesterase. *Biochemistry* **21**, 2060–2069.

Barnard, E. A., Barnard, P. J., Jarvis, J., Jedrzejczyk, J., Lai, J., Pizzey, J. A., and Randall, W. R. (1984). Multiple molecular forms of acetylcholinesterase and their relationship to muscle function In "Cholinesterases: Fundamental and Applied Aspects" (M. Brzin, E. A. Barnard, and D. Sket, eds.), pp. 49–71. DeGruyter, Berlin.

Bedford, C. D., Harris, R. N., III, Howd, R. A., Goff, D. A., Koolpe, G. A., Petesch, M., Miller, A., Nolen, H. W., III, Musallam, H. A., and Pick, R. O. (1989). Quaternary salts of 2-[(hydroxyimino)methyl]imidazole. 2. Preparation and *in vitro* and *in vivo* evaluation of 1-(alkoxymethyl)-2-[(hydroxyimino)methyl]-3-methylimidazolium halides for reactivation of organophosphorus-inhibited acetylcholinesterases. *J. Med. Chem.* **32**, 493–503.

Benschop, H. P., and Keijer, J. H. (1966). On the mechanism of aging of phosphonylated cholinesterases. *Biochim. Biophys. Acta.* **128**, 586–588.

Berends, F., Posthumus, C. H., Van Der Sluys, I., and Deierkauf, F. A. (1959). The chemical basis of the "ageing process" of DFP-inhibited pseudocholinesterase. *Biochim. Biophys. Acta.* **34**, 576–578.

Berry, W. K., and Davies, D. R. (1966). Factors influencing the rate of "aging" of a series of alkyl methylphosphonyl-acetylcholinesterases. *Biochem. J.* **100**, 572–576.

Blaber, L. C., and Creasey, N. H. (1960a). The mode of recovery of cholinesterase activity *in vivo*after organophosphorus poisoning: 1. Erythrocyte cholinesterase. *Biochem. J.* 77, 591–596.

Blaber, J. C., and Creasey, N. H. (1960b). The mode of recovery of cholinesterase activity *in vivo* after organophosphorus poisoning: 2. Brain cholinesterase. *Biochem. J.* 77, 597–604.

Boskovic, B. (1981). The treatment of soman poisoning and its perspectives. *Fundam. Appl. Toxicol.* 1, 203–213.

Brockman, S. K., Younkin, L. H., and Younkin, S. G. (1984). The effect of spontaneous electromechanical activity on the metabolism of acetylcholinesterase in cultured embryonic rat myotubes. *J. Neurosci.* 4, 131–140.

Bulger, J. E., Randall, W. R., Nieberg, P. S., Patterson, G. T., McNamee, M.G., and Wilson, B.W. (1982). Regulation of acetylcholinesterase forms in quail and chicken muscle cultures. *Dev. Neurosci.* 5, 474–483.

Burgen, A. S. V., and Hobbiger, F. (1951). The inhibition of cholinesterases by alkylphosphates and alkylphenolphosphates. *Brit J. Pharmacol.* 6, 593–605.

Childs, A. F., Davies, D. R., Green, A. L., and Rutland, J. P. (1955). The reactivation by oximes and hydroxamic acids of acetylcholinesterase inhibited by organophosphorus compounds. *Brit J. Pharmacol.* 10, 462–465.

Cisson, C. M., and Wilson, B. W. (1981). Paraoxon increases the rate of synthesis of acetylcholinesterase in cultured muscle. *Toxicol. Lett.* **9**, 131–135.

Clement, J. G. (1981). Toxicology and pharmacology of bispyridinium oximes—Insight into the mechanism of action vs soman poisoning *in vivo. Fundam. Appl. Toxicol.* **1**, 193–202.

Clothier, B., and Johnson, M. K. (1979). Rapid aging of neurotoxic esterase after inhibition by di-isopropyl phosphorofluridate. *Biochem. J.* 177, 549–558.

Clothier, B., Johnson, M. K., and Reiner, E. (1981). Interaction of some trialkyl phosphorothiolates with acetylcholinesterase: Characterization of inhibition, aging, and reactivation. *Biochim. Biophys. Acta.* **660**, 306–316.

Coult, D. B., Marsh, D. J., and Read, G. (1966). Dealkylation studies on inhibited cholinesterase. *Biochem. J.* **98**, 869–873.

Davies, D. A., and Green, A. L. (1956). The kinetics of reactivation, by oximes, of cholinesterase inhibited by organophosphorus compounds. *Biochem. J.* 63, 529–535.

Davies, D. R., and Holland, P. (1972). Effect of oximes and atropine upon the development of delayed neurotoxic signs in chickens following poisoning by DFP and sarin. *Biochem. Pharmacol.* **51**, 3145–3151.

Davison, A. N. (1953). Return of cholinesterase activity in the rat after inhibition by organophosphorus compounds. 1. Diethyl *p*-nitrophenyl phosphate (E 600, paraoxon). *Biochem. J.* **54**, 583–590.

Davison, A. N. (1955). Return of cholinesterase activity in the rat after inhibition by organophosphorus compounds. 2. A comparative study of true and pseudocholinesterase. *Biochem. J.* **60**, 339–346.

Dettbarn, W.-D. (1984). Pesticide-induced muscle necrosis: Mechanisms and prevention. *Fundam. Appl. Toxicol.* **4**, S18–S26.

Dodge, J. T., Mitchell, C., and Hanahan, D. J. (1963). The preparation and chemical characteristics of hemoglobin-free ghosts of human erythrocytes. *Arch. Biochem. Biophys.* **100**, 119–130.

Ellman, G. L., Courtney, K. D., Andres, V., and Featherstone, R. M. (1961). A new and rapid colorimetric determination of acetylcholinesterase activity. *Biochem. Pharmacol.* 7, 88–95.

Ellin, R. I. (1982). Anomalies in theories and therapy of intoxication by potent organophosphorus anticholinesterase compounds. *Gen. Pharmacol.* **13**, 457–466.

Environmental Health Criteria 63. (1986). "Organophosphorus Insecticides: A General Introduction." Environmental Health Criteria 63. World Health Organization, Geneva, Switzerland.

Eto, M. (1974). "Organophosphorus Pesticides: Organic and Biological Chemistry." CRC press, Cleveland, Ohio.

Fleisher, J. H., and Harris, L. W. (1965). Dealkylation as a mechanism for aging of cholinesterase after poisoning with pinacoyl mehylphosphonofluoridate. *Biochen. Pharmacol.* **14**, 641–650.

Fleisher, J. H., Harris, L. W., and Murtha, E. F. (1967). Reactivation of pyridinium aldoxime methochloride (PAM) of inhibited cholinesterase activity in dogs after poisoning with pinacolyl methylphosphonofluoridate (soman). *J. Pharmacol. Exp. Ther.* **156**, 345–351.

Fleming, W. J. (1981). Recovery of brain and plasma cholinesterase activities in ducklings exposed to organophosphorus pesticides. *Arch. Environ. Contam. Toxicol.* **10**, 215–229.

Groudou, D., and Rieger, F. (1983). Recovery of acetylcholinesterase and of its multiple molecular forms in motor-end-plate-free and motor-end-plate-rich regions of mouse striated muscle, after irreversible inactivation by an organophosphorus compound (methyl phosphorothiolate). *Biol. Cell* **48**, 151–158.

Gray, A. (1984). Design and structure–activity relationships of antidotes to organophosphorus anticholinesterase agents. *Drug Metab. Rev.* **15**, 557–589.

Hackley, B. E., Jr., Steinberg, G. M., and Lamb, J. C. (1959). Formation of potent inhibitors of AChE by reaction of pyridinaldoximes with isopropyl methylphosphonofluoridate (GB). *Arch. Biochem. Biophys.* **80**, 211–214.

Harris, L. W., Fleisher, J. H., Clark, J., and Cliff, W. J. (1966). Dealkylation and loss of capacity for reactivation of cholinesterases inhibited by sarin. *Science* **154**, 404–406.

Harvey, B., Scott, R. P., Sellers, D. J., and Watts, P. (1986b). *In vitro* studies of the reactivation by oximes of phosphylated acetylcholinesterase-II: On the formation of O,O-diethyl phosphorylated AChE and O-ethyl methylphodphonylated AChE and the reactivation by P2S. *Biochem. Pharmacol.* **35**, 745–751.

Harris, L. W., Heyl, W. C., Stitcher, D. L., and Broomfield, C. A. (1978). Effects of 1,1′-oxydimethylene *bis*(4-*tert*-butylpyridinium chloride)(SAD-128) and decamethonium on reactivation of soman- and sarin-inhibited cholinesterase by oximes. *Biochem. Pharmacol.* **27**, 757–761.

Harris, L. W., Talbot, B. G., Lennox, W. J., and Anderson, D. R. (1989). The relationship between oxime-induced reactivation of carbamylated acetylcholinesterase and antidotal efficacy against carbamate intoxication. *Toxicol. Appl. Pharmacol.* **97**, 267–271.

Harvey, B., Scott, R. P., Sellers, D. J., and Watts, P. (1986a). *In vitro* studies of the reactivation by oximes of phosphylated acetylcholinesterase-I: On the reactions of P2S with various organophosphates and the properties of the resultant phosphylated oximes. *Biochem. Pharmacol.* **35**, 737–744.

Harvey, B., Scott, R. P., Sellers, D. J., and Watts, P. (1986b). *In vitro* studies of the reactivation by oximes of phosphylated acetylcholinesterase-II: On the formation of O,O-diethyl phosphorylated AChE and O-ethyl methylphodphonylated AChE and the reactivation by P2S. *Biochem. Pharmacol.* **35**, 745–751.

Heilbronn, E. (1963). *In vitro* "reactivation and aging" of tabun-inhibited blood cholinesterases:

Studies with *N*-methylpyridinium-2-aldoxime methane sulphonate and *N,N'*-trimethylene *bis*(pyridinium-4-aldoxime) dibromide. *Biochem. Pharmacol.* **12**, 25–36.

Hill, E. F. (1988). Brain cholinesterase activity of apparently normal wild birds. *J. Wildl. Dis.* **24**, 51–61.

Hobbiger, F. (1951). Inhibition of cholinesterases by irreversible inhibitors *in vitro* and *in vivo*. *Br. J. Pharmacol.* **6**, 21–30.

Hobbiger, F. (1955). Effect of nicotinhydroxamic acid methiodide on human plasma cholinesterase inhibited by organophosphates containing a dialkylphosphato group. *Br. J. Pharmacol.* **10**, 356–362.

Hobbiger, F. W. (1956). Chemical reactivation of phosphorylated human and bovine true cholinesterase. *Br. J. Pharmacol.* **11**, 295–303.

Hobbiger, F. W. (1957). Protection against the lethal effects of organophosphates by pyridine-2-aldoxime methiodide. *Br. J. Pharmacol.* **12**, 438–446.

Hobbiger, F. (1963). Reactivation of phosphorylated acetylcholinesterase. *In* "Handbuch der Experimentellien Pharmacologie: Cholinesterases and anti-cholinesterase agents, vol. 15." G. B. Koelle, ed., pp. 921–988. Springer-Verlang, Berlin.

Hooper, M. J., Nieberg, P. S., and Wilson, B. W. (1988). Oxime reactivation of OP-treated AChE in cultured muscle. *Toxicologist* **8**, 42.

Hooper, M. J., Detrich, P. J., Weisskopf, C. P., and Wilson, B. W. (1989). Organophosphate exposure in hawks inhabiting orchards during winter dormant spraying. *Bull. Environ. Contam. Toxicol.* **42**, 651–659.

Hovanec, J. W., Broomfield, C. A., Steinberg, G. M., Lanks, K. W., and Lieske, C. N. (1977). Spontaneous reactivation of acetylcholinesterase following organophosphate inhibition. *Biochim. Biophys. Acta.* **483**, 312–319.

Inestrosa, N. C., and Perelman, A. (1989). Distribution and anchoring of molecular forms of acetylcholinesterase. *TIPS* **10**, 325–329.

Johnson, C. D., and Russell, R. L. (1975). A rapid, simple radiometric assay for cholinesterase, suitable for multiple determinations. *Anal. Biochem.* **64**, 229–238.

Karlog, O., and Poulsen, E. (1963). Spontaneous and pralidoxime-induced reactivation of brain cholinesterase in the chicken after fatal nitrostigmine (parathion) poisoning. *Acta Pharmacol. Toxicol.* **20**, 174–180.

Kasprzak, H., and Salpeter, M. M. (1985). Recovery of acetylcholinesterase at intact neuromuscular junctions after *in vivo* inactivation with di-isopropylfluorophosphate. *J. Neurosci.* **5**, 951–955.

Kellner, T. P., Henderson, J. D., Higgins, R. J., and Wilson, B. W. (1988). Atropine and DFP-induced delayed neurotoxicity. *Neurotoxicology* **9**, 181–188.

Langenberg, J. P., De Jong, L. P. A. , Otto, M. F., and Benschop, H. P. (1988). Spontaneous and oxime-induced reactivation of acetylcholinesterase inhibited by phosphoramidates. *Arch. Toxicol.* **62**, 305–310.

Lee, R. M. (1964). Di-(2-chloroethyl) aryl phosphates. A study of their reaction with B-esterases, and of the genetic control of their hydrolysis in sheep. *Biochem. Pharmacol.* **13**, 1551–1568.

Leonard, J. P., and Salpeter, M. M. (1979). Agonist-induced myopathy at the neuromuscular junction is mediated by calcium. *J. Cell. Biol.* **82**, 811–819.

Lotti, M., and Becker, C. E. (1982). Treatment of organophosphate poisoning: Evidence of a direct effect on central nervous system by 2-PAM (pyridine-2-aldoxime methyl chloride). *J. Toxicol. Clin. Toxicol.* **19**, 121–127.

Lowry, O. H., Rosebrough, N. J., Farr, A. L., and Randall, R. J. (1951). Protein measurements with the Folin phenol reagent. *J. Biol. Chem.* **193**, 265–275.

Ludke, J. L., Hill, E. F., and Dieter, M.P. (1975) Cholinesterase (ChE) response and related mortality among birds fed ChE inhibitors. *Arch. Environ. Contam. Toxicol.* **3**, 1–21.

Main, A. R. (1984). Mode of action of anticholinesterases. *In* "International Encyclopedia of

Pharmacology and Therapeutics. Differential Toxicities of Insecticides and Halogenated Aromatics" (F. Matsumura, ed.), pp. 351–400. Pergamon Press, Oxford.

Martin, A. D., Norman, G., Stanley, P. I., and Westlake, G. E. (1981) Use of reactivation techniques for the differential diagnosis of organophosphorus and carbamate pesticide poisoning in birds. *Bull. Environ. Contam. Toxicol.* **26**, 775–780.

Masson, P., Marnot, B., Lombard, J. Y., and Morelis, P. (1984). Etude electrophoretique de la butyrylcholinesterase agee apres inhibition par le soman (*). *Biochimie* **66**, 235–249.

Massoulie, J., and Bon, S. (1982). The molecular forms of cholinesterase and acetylcholinesterase in vertebrates. *Annu. Rev. Neurosci.* **5**, 57–106.

Namba, T. (1971) Cholinesterase inhibition of organophosphorus compounds and its clinical effects. *Bull. W. H. O.* **44**, 289–307.

Natoff, I. L., and Reiff, B. (1973). Effects of oximes on the acute toxicity of anticholinesterase carbamates. *Toxicol. Appl. Pharmacol.* **25**, 569–575.

Newman, J. R., Virgin, J. B., Younkin, L. H., and Younkin, S. G. (1984). Turnover of acetylcholinesterase in innervated and denervated rat diaphragm. *J. Physiol.* **52**, 305–318.

O'Brien, R.D. (1967). "Insecticides: Action and Metabolism." Academic Press, New York.

Pickering, W. R., and Malone, J. C. (1967). The acute toxicity of dichloroalkyl aryl phosphates in relation to chemical structure. *Biochem. Pharmacol.* **16**, 1183–1194.

Porter, C. W., and Barnard, E. A. (1975). The density of cholinergic receptors at the endplate postsynaptic membrane: Ultrastructural studies in two mammalian species. *J. Membr. Biol.* **20**, 31–49.

Rakonczay, Z., and Brimijoin, S. (1988). Monoclonal antibodies to human brain acetylcholinesterase: Properties and applications. *Cell. Mol. Neurobiol.* **8**, 85–93.

Reese, T. V. (1984). Organophosphate poisoning. *Am. Fam. Physician* **29**, 45–47.

Reiner, E. (1971). Spontaneous reactivation of phosphorylated and carbamylated cholinesterases. *Bull. W. H. O.* **44**, 109–112.

Rotundo, R. L. (1988). Biogenesis of acetylcholinesterase molecular forms in muscle. Evidence for a rapidly turning over, catalytically inactive precursor pool. *J. Biol. Chem.* **263**, 19398–19406.

Rotundo, R. L., and Fambrough, D. M. (1982). Synthesis, transport, and fate of acetylcholinesterase and acetylcholine receptors in cultured muscle. *In* "Membranes in Growth and Development," pp. 259–286. Alan R. Liss, New York.

Schoene, K. (1978). Aging of soman-inhibited acetylcholinesterase: Inhibitors and accelerators. *Biochim. Biophys. Acta* **525**, 468–471.

Schumacher, M., Camp, S., Maulet, Y., Newton, M., MacPhee-Quigley, K., Taylor, S. S., Friedman, T., and Taylor, P. (1986). Primary structure of *Torpedo californica* acetylcholinesterase deduced from its cDNA sequence. *Nature* **319**, 407–409.

Silman, I., and Futerman, A. H. (1987). Posttranslational modification as a means of anchoring acetylcholinesterase to the cell surface. *Biopolymers* **26**, S241–S253.

Silver, A. (1974) "The Biology of Cholinesterases." Elsevier, New York.

Sketelj, J., NcNamee, M. G., and Wilson, B. W. (1978). Effect of denervation on the forms of acetylcholinesterase in normal and dystrophic chicken muscles. *Exp. Neurol.* **60**, 624–629.

Skrinjaric-Spoljar, M., Simeon, V., and Reiner, E. (1973). Spontaneous reactivation and aging of dimethylphosphorylated acetylcholinesterase and cholinesterase. *Biochim. Biophys. Acta* **315**, 363–369.

Smucker, S. J., and Wilson, B. W. (1990). Multiple molecular forms and lectin interactions of organophosphate-sensitive plasma and liver esterases during development of the chick. *Biochem. Pharmacol.* **40**, 1907–1913.

Sung, S. C., and Ruff, B. A. (1987). Intracellular distribution of molecular forms of acetylcholinesterase in rat brain and changes after diisopropylfluorophosphate treatment. *Neurochem. Res.* **12**, 15–19.

Steinberg, N., van der Drift, A. C. M., Grunwald, J., Segall, Y., Shirin, E., Haas, E., Ashani, Y., and Silman, I. (1989). Conformational differences between aged and nonaged pyrene-butyl-containing organophosphoryl conjugates of chymotrypsin as detected by optical spectroscopy. *Biochemistry* **28**, 1248–1253.

Thompson, D. F., Thompson, C. D., Greenwood, R. B., and Trammel, H. L. (1987). Therapeutic dosing of pralidoxime chloride. *Drug Intell. Clin. Pharm.* **21**, 590–593.

Toutant, J. P., and Massoulie, J. (1987). Acetylcholinesterase. *In* "Mammalian ectoenzymes" (A. J. Kenny and A. J. Turner, eds.), pp. 289–328. Elsevier Science, New York.

Tsim, K. W. K., Randall, W. R., and Barnard, E. A. (1988). Identification of a 17S asymmetric butyrylcholinesterase in chick muscle by monoclonal antibodies. *Neurosci. Lett.* **86**, 245–249.

Van Asperen, K., and Dekhuijzen, H. M. (1958). A quantitative analysis of the kinetics of cholinesterase inhibition in tissue homogenates of mice and houseflies. *Biochim. Biophys. Acta* **28**, 603–613.

Van der Drift, A. C. M. (1985). A comparative study of the aging of DFP-inhibited serine hydrolases by means of ^{31}P-NMR and mass spectrometry. *In* "Molecular Basis of Nerve Activity" (J.-P. Changeux, F. Hucho, A. Maelicke, and E. Newman, eds.), pp. 753–764. de Gruyter, Berlin.

Vandekar, M., and Heath, D. F. (1957). The reactivation of cholinesterase after inhibition *in vivo* by some dimethyl phosphate esters. *Biochem. J.* **67**, 202–208.

Wallace, K. B., and Herzberg, U. (1988). Reactivation and aging of phosphorylated brain acetylcholinesterase from fish and rodents. *Toxicol. Appl. Pharmacol.* **92**, 307–314.

Wilson, I. B. (1959). Molecular complementarity and antidotes for alkylphosphate poisoning. *Fed. Proc.* **18**, 752–758.

Wilson, I. B. (1967). Acid-transferring inhibitors of acetylcholinesterase. *In* "Drugs Affecting the Peripheral Nervous System" (A. Burger, ed.), pp. 381–397. Marcel Dekker, New York.

Wilson, I. B., and Ginsburg, S. (1955). Reactivation of acetylcholinesterase inhibited by alkylphosphates. *Arch. Biochem. Biophys.* **54**, 569–571.

Wilson, B. W., and Nieberg, P. S. (1983). Recovery of acetylcholinesterase forms in quail muscle cultures after intoxication with diisopropylfluorophosphate. *Biochem. Pharmacol.* **32**, 911–918.

Wilson, B. W., and Walker, C. R. (1974). Regulation of newly synthesized acetylcholinesterase in muscle cultures treated with diisopropylfluorophosphate. *Proc. Natl. Acad. Sci. U.S.A.* **71**, 3194–3198.

Wilson, B. W., Nieberg, P. S., Walker, C. R., Linkhart, T. A., and Fry, D. M. (1973). Production and release of acetylcholinesterase by cultured chick embryo muscle. *Dev. Biol.* **33**, 285–299.

Wilson, B. W., Hooper, M. J., and Littrell, E. E. (1988a). "Exposure of Red-Tailed Hawks to Agricultural Chemicals during Dormant-Spray Season in the Central Valley of California. California Department of Fish and Game Report, Sacramento, California.

Wilson, B. W., Nieberg, P. S., Hansen, M. E., and Hooper, M. J. (1988b). Recovery of AChE in cultured quail muscle after exposure to organophosphate esters and 2-PAM. *J. Cell Biol.* **107**, 514a.

Wilson, B. W., Henderson, J. D., Chow, E., Schreider, J., Goldman, M., Culbertson, R., and Dacre, J. C. (1988c). Toxicity of an acute dose of agent VX and other organophosphorus esters in the chicken. *J. Toxicol. Environ. Health* **23**, 103–113.

Witter, R. F., and Gaines, T. B. (1963). Rate of formation *in vivo* of the unreactivatable form of brain cholinesterase in chickens given DDVP or malathion. *Biochem. Pharmacol.* **12**, 1421–1427.

Wolthuis, O. L., and Kepner, L. A. (1978). Successful oxime therapy one hour after soman intoxication in the rat. *Eur. J. Pharmacol.* **49**, 415–425.

Wolthuis, O. L., and Vanwersch, R. A. P. (1984). Behavioral changes in the rat after low doses of cholinesterase inhibitors. *Fundam. Appl. Toxicol.* **4**, S195–S208.

II

Metabolic Fate

6

Metabolism of Organophosphorus Compounds by the Flavin-Containing Monooxygenase

Patricia E. Levi
Ernest Hodgson
Department of Toxicology
North Carolina State University
Raleigh, North Carolina

I. Introduction

The toxic action of an insecticide is dependent on its reactivity to the biochemical target (e.g., reactivity of paraoxon toward acetylcholinesterase). Toxicity *in vivo*, however, is considerably modulated by rates of absorption, metabolism, and elimination of the insecticide; thus, metabolic fate plays a pivotal role in the toxicity and persistence of insecticides. Species variation in toxicity is often attributable to differential rates or routes of biotransformation of the parent compound or one of the metabolites (see Chapters 1 and 4 by Chambers and Wallace, this volume).

Even though insecticides are subject to a wide array of phase I and phase II xenobiotic-metabolizing enzymes, the role of the monooxygenases is of primary and critical importance. The monooxygenases, in addition to detoxication reactions, produce highly reactive intermediates that play an important part in activation reactions and hence, in both acute and chronic toxicity.

Organophosphates: Chemistry, Fate, and Effects

Since the monooxygenases are enzymes involved in the metabolism of a wide variety of xenobiotics, many of which can act as inducers or inhibitors as well as substrates, the monooxygenases are a focal point for interactions between different compounds.

While the cytochrome P450 monooxygenases (P450) are probably the most important enzymes involved in the initial metabolism of insecticides and other xenobiotics, the flavin-containing monooxygenases (FMO) make a significant contribution to xenobiotic metabolism, and many compounds previously thought to be oxidized by P450 are now known to also be substrates for FMO. Table I summarizes some of the similarities and differences between the FMO and P450 monooxygenases. FMO, like P450, is located in the endoplasmic reticulum and is involved in the monooxygenation of numerous nitrogen-, sulfur-, and phosphorus-containing xenobiotics (Ziegler, 1980; Ziegler, 1984; Hodgson and Levi, 1988; Ziegler, 1988). As can be seen from Table II, there is considerable overlap between P450 and FMO in the types of reactions catalyzed and the substrates oxidized. Thus, it is of considerable importance in understanding insecticide metabolism to define the role of the FMO enzymes relative to those of the P450 system, using both microsomal and purified enzymes as well as the newer methods of molecular biology. Pesticides, such as organophosphorus (OP) compounds and carbamates containing thioether bonds, are substrates for the FMO (Hajjar and Hodgson, 1980; Hajjar and Hodgson, 1982b; Smyser *et al.*, 1985; Tynes and Hodgson, 1985a). In addition to sulfoxidation, some OP compounds, all phosphonates, will undergo oxidative desulfuration to form oxons (Hajjar and Hodgson, 1982a).

The FMO was first purified to homogeneity from pig liver microsomes (Ziegler and Poulsen, 1978) and, in our laboratory, from both pig and mouse liver (Sabourin *et al.*, 1984, Sabourin and Hodgson, 1984). Subsequently, Tynes *et al.* (1985) as well as Williams *et al.* (1984) purified from rabbit lung an FMO that was shown to be catalytically and immunologically distinct from

TABLE I

Comparison of P450 and FMO

Feature	FMO	P450
Location	Microsomes	Microsomes
Cofactors	NADPH, O_2	NADPH, O_2, reductase
Inducers	None	Many (Pb, 3-MC, PCN, EtOH)
Inhibitors	None	CO, SKF-525A, PBO
Isozymes	Few	Many
Substrates	N, S, P compounds	N, S, P, C compounds
Reactions	Oxygenation	Oxygenation, epoxidation, reduction, dealkylation

TABLE II

Summary of Major Reactions Catalyzed by FMO and P450

Reactions	Examples
P450	
Epoxidation	Aldrin, benzo(a)pyrene, aflatoxin
Hydroxylation	Nicotine, bromobenzene
N-Dealkylation	Ethylmorphine, atrazine, dimethylaniline
O-Dealkylation	*p*-Nitroanisole, chlorfenvinphos
S-Dealkylation	Methylmercaptan
S-Oxidation	Thiobenzamide, phorate, disulfoton
N-Oxidation	2-Acetylaminofluorene
P-Oxidation	Diethylphenylphosphine
Desulfuration	Parathion, fonofos, carbon disulfide
Dehalogenation	$CC1_4$, $CHC1_3$
FMO	
N-Oxygenation	Nicotine, dimethylaniline, imipramine
S-Oxygenation	Thiobenzamide, phorate, thiourea, disulfoton
P-Oxygenation	Diethylphenylphosphine
Desulfuration	Fonofos

the liver enzyme. These two enzymes have now been shown to be products of related but distinctly different genes (Lawton *et al.*, 1990). Recently Ozols (1989) reported purification of two FMO forms from rabbit liver. Thus it is now evident that there are several FMO enzymes with overlapping substrate specificities, and it is likely that the relative proportions of these isozymes vary in different tissues within and between species.

II. Catalytic Mechanism of Flavin-Containing Monooxygenases

Flavin-containing enzymes, including reductases, oxidases, and monooxygenases (FMO) occur widely in nature, catalyzing the transformation of many types of xenobiotics. Several unique features of the catalytic cycle of the FMO are important to an understanding of the mechanism of pesticide oxidation by FMO. The catalytic mechanism for the FMO has been shown to involve the formation of an enzyme bound 4a-hydroperoxy-flavin (Fig. 1) in an NADPH- and O_2-dependent reaction (Poulsen and Ziegler, 1979; Beaty and Ballou, 1980; Poulsen, 1981; Doerge and Corbett, 1984). Reduction of the flavin by NADPH occurs before binding of oxygen can occur, and activation of oxygen by the enzyme occurs in the absence of substrate by oxidizing NADPH to form $NADP^+$ and peroxide. Finally, addition of the substrate to the peroxyflavin complex is the last step before oxygenation. This is in contrast to the P450

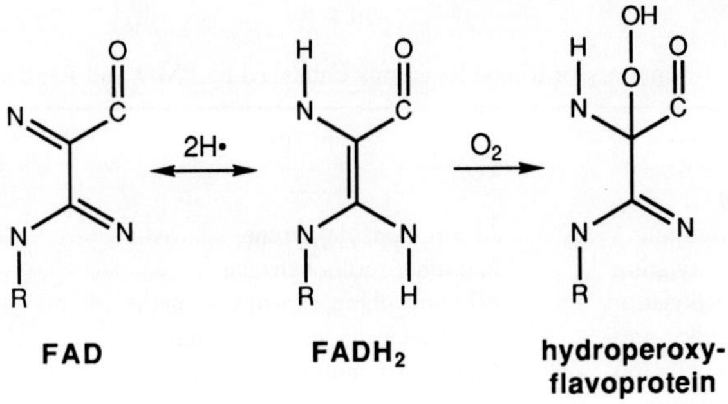

FAD $\xrightarrow{2H\bullet}$ $FADH_2$ $\xrightarrow{O_2}$ hydroperoxy-flavoprotein

Figure 1 Oxidation–reduction of flavin component of FMO and formation 4a-hydroperoxy-flavoprotein.

cycle, in which the substrate binds to the oxidized enzyme, which is then reduced.

The flavin hydroperoxide intermediate in the FMO enzyme forms a relatively stable, potent oxygenating species. Thus any nucleophile that can be oxidized by an organic peroxide and can gain access to the active site is a potential substrate for the FMO. This capability accounts for the wide substrate specificity of the FMO. Although the flavin hydroperoxide of the FMO is a strong electrophile, it exhibits a high degree of selectivity toward certain types of sulfur nucleophiles (Taylor and Ziegler, 1987; Ziegler, 1988). Compounds containing ionized carboxyl groups, which include most physiological sulfur compounds, are not substrates for the FMO, with the exception of cysteamine, which is an excellent substrate.

Reactivation of the enzyme is considered to be the rate-limiting step in the reaction (Beaty and Ballou, 1981a,b; Doerge and Corbett, 1984). Since oxidation of the substrate occurs more rapidly than the regeneration of the active enzyme, the maximal velocity (V_{max}) values are relatively similar for a variety of substrates with differing Michaelis constant (K_m) values (see kinetic constants in Table III for examples of this situation).

III. Flavin-Containing Monooxygenases as a Phosphorus Oxidase

Oxidative desulfuration of P=S compounds to form the oxon (P=O), previously considered to be entirely due to P450, is also an FMO reaction with some substrates. Desulfuration of phosphonate insecticides, such as fonofos and its analogs (Fig. 2) is catalyzed by pig liver FMO (Hajjar and Hodgson, 1982a). Structure–activity studies have demonstrated that substituents on the phosphorus atom are a critical determinant of activity (Table IV), and that at

TABLE III

Kinetic Constants for Selected Pesticides with Purified FMO from Mouse Liver[a]

Substrate	K_m (μM)	V_{max} (nmols/min/mg)
Phorate	32.2	1408
Phorate oxon	461.0	1170
Disulfoton	3.4	1693
Demeton S	110.0	1234
Demeton O	59.3	1171
Sulprofos	1.5	750
Fenthion	12.0	673
Aldicarb	607.0	1087
Thiofanox	574.0	1307
Dazomet	398.0	1409

[a]Velocities determined by following NADPH oxidation at 340 nm. Values are within 10% of the mean for three determinations. From Smyser *et al.* (1985).

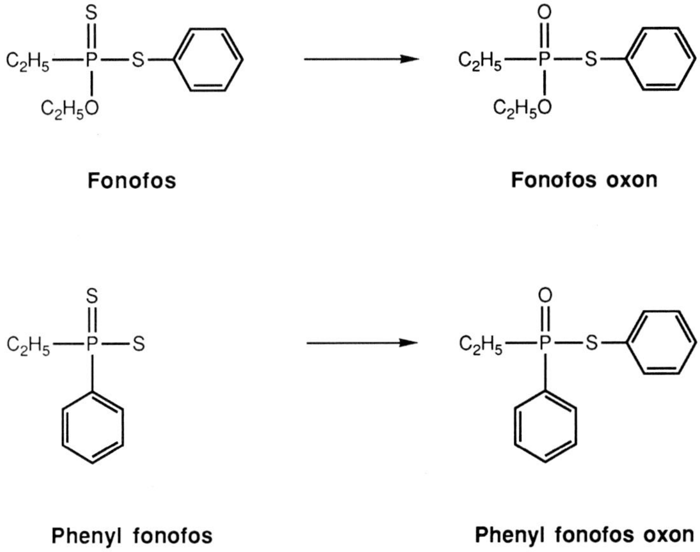

S
C_2H_5—P—S
C_2H_5O
Fonofos

O
C_2H_5—P—S
C_2H_5O
Fonofos oxon

S
C_2H_5—P—S
Phenyl fonofos

O
C_2H_5—P—S
Phenyl fonofos oxon

Figure 2 Oxidative desulfuration of fonofos and phenyl fonofos, a reaction catalyzed by both P450 and FMO.

TABLE IV

Oxidation of Fonofos and Fonofos Analogs by FMO

Substrate	Activity[a]
$(C_2H_5O)\ (C_2H_5)\ P\ (S)\ S\ (C_6H_5)$	6.45
$(C_2H_5O)\ (C_2H_5)\ P\ (O)\ S\ (C_6H_5)$	0.00
$(C_2H_5O)\ (C_2H_5)\ P\ (S)\ O\ (C_6H_5)$	0.80
$(C_2H_5O)\ (C_6H_6)\ P\ (S)\ S\ (C_6H_5)$	10.36
$(C_2H_5O)_2\ P\ (S)\ S\ (C_6H_5)$	0.00
$(C_2H_5O)_2\ P\ (S)\ O\ (C_6H_4)NO_2$	0.00

[a]NADPH used, nmols/min per nmol enzyme. From Hajjar and Hodgson (1982a).

least one C-P bond is essential for oxygenation to occur; thus phosphorodithioates, such as parathion, are not FMO substrates. In contrast to the mechanism of P450 desulfuration, which involves an attack on the sulfur atom, it is thought that the oxidation of fonofos by the FMO is an attack on the phosphorus atom. Trivalent phosphorus-containing compounds are also excellent substrates for the enzyme (Smyser and Hodgson, 1985), with diethylphenylphosphine having a K_m value lower than 2.5 μM. Diethylphenylphosphonite appears also to be an excellent substrate, but its rapid nonenzymatic hydrolysis and/or oxidation precludes accurate K_m determinations. The major metabolite of both diethylphenylphosphine and its sulfide is the corresponding phosphine oxide.

IV. Sulfoxidation of Organophosphorus Compounds

Incubation of thioether-containing pesticides, such as phorate, disulfoton, or sulprofos, with microsomes and NADPH results primarily in the formation of sulfoxides and, to a lesser extent, oxons and sulfones (Fig. 3). The FMO, as well as P450, is extensively involved in catalyzing sulfoxidation reactions.

Initial experiments in our laboratory (Hajjar and Hodgson, 1982a) showed that some thioether-containing OP compounds, such as disufoton and phorate, are rapidly oxidized by purified FMO to yield the sulfoxide as the only detectable metabolite, and that this metabolite is optically active. Recently, we utilized both purified FMO and P450 isozymes to examine in detail the oxidative pathways of phorate metabolism (Levi and Hodgson, 1988b). Both P450 and FMO catalyzed the initial sulfoxidation of phorate. Subsequent oxidation reactions, however, such as formation of the sulfone and oxidative desulfuration to the corresponding oxons are catalyzed only by P450. Although both the FMO and P450 catalyze the initial sulfoxidation reaction, the products are stereochemically different, with FMO forming (−)

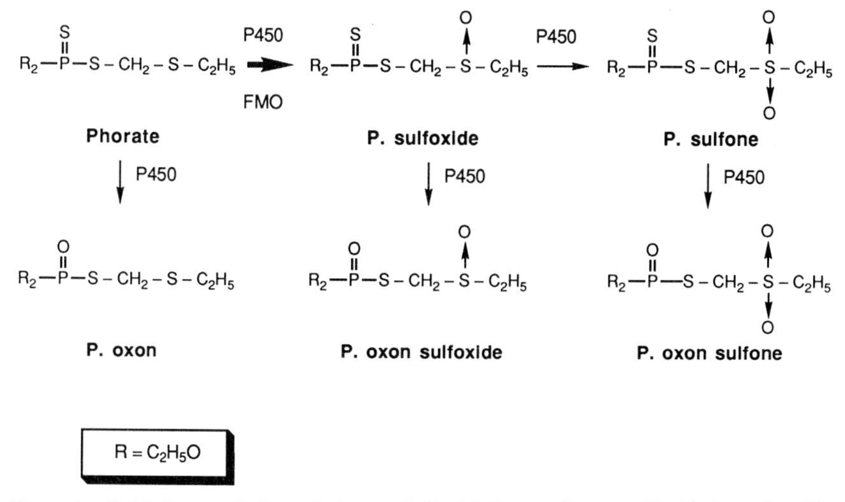

Figure 3 Oxidative metabolism of phorate. Sulfoxidation to phorate sulfoxide is catalyzed by both P450 and FMO. Other oxidations are P450 catalyzed.

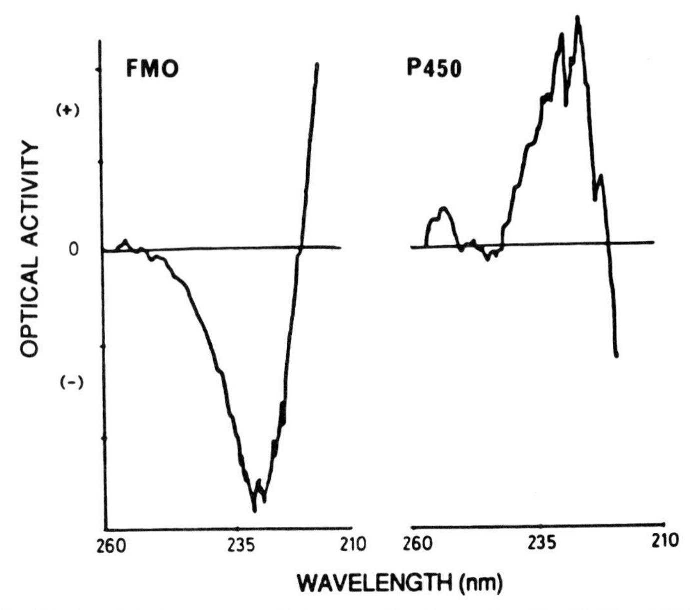

Figure 4 Circular dichroism spectra of phorate sulfoxide produced by P450 and FMO. Sulfoxides formed enzymatically are optically active, with FMO producing (−) phorate sulfoxide and P450 producing (+) phorate sulfoxide.

phorate sulfoxide and two of the P450 isozymes (P450 II B1, the principal form induced by phenobarbital, and P450 B2, a constitutive form) yielding (+) phorate sulfoxide (Fig. 4). The other three P450 isozymes examined formed racemic mixtures. Both (+) and (−) phorate sulfoxide were found to be substrates for subsequent oxidation reactions by P450, although the (+) isomer was the preferred substrate. In addition the ratios of final oxidation products differed when either the (+) or (−) sulfoxides were used as substrates, with (+) phorate sulfoxide yielding a higher percentage of phorate oxon sulfoxide (an activation reaction) than the (−) isomer. Since both FMO and P450 are involved in this complex metabolic pathway, *in vivo* the net optical activity as well as the ratio of metabolites would be a function of factors such as presence of activators, inhibitors, and inducers as well as sex and organ.

V. Relative Contributions of Flavin-Containing and Cytochrome P450 Monooxygenases in Microsomal Oxidations

Often the same substrate is metabolized by both P450 and FMO; this situation is especially prevalent with many N- and S-containing pesticides, e.g., phorate. To study the relative contributions of these two enzymes with common substrates, we developed methods to measure each separately in microsomal preparations. The most useful of these techniques is the inhibition of P450 activity by using an antibody to NADPH cytochrome P450 reductase, thus permitting measurement of FMO activity alone. A second procedure is heat treatment of microsomal preparations (50°C for 1 min), which inactivates the FMO, thus allowing determination of P450 activity, unchanged by heat treatment (Tynes and Hodgson, 1985a). Thermal inactivation, however, is ineffective with lung microsomes, since the lung FMO is more heat stable than is the liver FMO, and this necessitates the use of antireductase with lung microsomal preparations to define the relative roles of P450 and FMO.

The contribution of the two enzyme systems to the sulfoxidation of phorate in different tissues has been examined using these procedures (Table V). While the oxidation of phorate is relatively similar between males and females, there is a striking difference in the proportion of sulfoxidase activity resulting from FMO in the different tissues. In liver microsomes, P450 activity is more important (76–78% by P450 under standard assay conditions at pH 7.6). By contrast, in the kidney and lung, although overall activity is low compared to that of the liver, the relative contribution by FMO is significantly higher. This is especially dramatic in kidney microsomes from female mice, where about 90% of the activity is associated with FMO and only 10% with P450. It is important to note that the FMO rate for phorate sulfoxidation per mg of microsomal protein is similar in kidney and lung to the liver rate,

TABLE V

Relative Contributions of FMO and P450 to Microsomal Oxidation of Phorate in Mouse[a]

Tissue	Sex	Control[b]	+AR[b,c]	% FMO	% P450
Liver	M	12.7	2.8	21.7	78.3
Liver	F	14.4	3.7	24.0	76.1
Lung	M	3.3	1.9	59.1	41.3
Lung	F	5.7	3.1	54.0	46.0
Kidney	M	1.6	1.2	72.0	28.1
Kidney	F	1.9	1.7	89.5	10.5
Liver-Pb[d]	M	69.7	10.1	14.3	85.5

[a]From Kinsler *et al.* (1988, 1990).
[b]Phorate sulfoxide formed, nmols/min per mg protein.
[c]Antibody to P450 reductase.
[d]Phenobarbital-treated mice.

suggesting a crucial role for the FMO in extrahepatic oxidations of xenobiotics. Such FMO reactions may assume added significance since P450 activity is low in these tissues.

While the levels of FMO are not readily altered by classic chemical inducing and inhibiting agents, as is the case with P450, the balance of enzyme activity between P450 and FMO is easily disturbed, especially in the liver, by compounds that alter the concentration of P450 isozymes. Of special interest is a change in the balance of activity after *in vivo* exposure of animals to either inducers or inhibitors of P450 activity (Kinsler *et al.*, 1990). For example pretreatment of mice with phenobarbital increased not only the total rate of phorate oxidation (Table V), but also the proportion metabolized by P450. As a result, the percentage of products resulting from FMO activity is decreased.

An interesting situation is observed in mice pretreated with the insecticide synergist piperonyl butoxide (PBO). Piperonyl butoxide functions as a synergist by inhibiting P450 activity *in vivo* and preventing detoxication of certain insecticides, such as pyrethrins, which are detoxified by P450. Thus the toxicity of these insecticides is increased. The effect of PBO on P450 levels is biphasic; the initial inhibition of P450 is followed by an increase in P450 activity due to induction of certain P450 isozymes. This same biphasic pattern can be observed with phorate sulfoxidation. As shown in Table VI, initial inhibition of P450 activity at 2 hr results in a decrease in the total rate and the rate due to P450. As the percentage of activity due to P450 decreases, the relative contribution due to FMO increases. At a later time point, however, when P450 activity is increased, the percentage of activity due to FMO declines.

The effects of xenobiotics on the relative contributions of FMO and P450 appear to be mediated primarily by P450, since the FMO does not appear to be inducible by xenobiotics. FMO levels may, however, vary with

TABLE VI

Oxidation of Phorate in Female Mouse Liver Microsomes following *in Vivo* Treatment with Piperonyl Butoxide[a]

	Total rate[b]	Heat treatment[c]	% P450	% FMO
Control	13.8	10.49	76.0	24.0
2 hr	11.1	6.47	58.1	41.9
4 hr	12.1	9.29	77.0	23.0
8 hr	13.3	10.90	82.2	17.8
12 hr	15.4	13.77	89.2	10.8
36 hr	19.4	16.27	83.9	16.1

[a]From Kinsler *et al.* (1990).
[b]Phorate sulfoxide formed, nmol/min per microsomal protein. Values are within 10% of the mean for three determinations.

nutrition, diurnal rhythms, sex, pregnancy, and corticosteroids, although the effects appear to be both species and tissue dependent (Ziegler, 1988, Williams *et al.*, 1985). The effects of hydrocortisone treatment on the metabolism of phorate and thiobenzamide by the FMO of mouse liver and lung have also been studied (Kinsler *et al.*, 1990). The FMO activity in the liver was increased for both substrates (+82% for phorate, +52% for thiobenzamide) with only minor changes in the lung (−15% for phorate, +20% for thiobenzamide).

Such alterations in the relative contributions of the two enzyme systems may assume toxicological importance when the products from the two enzymes differ, and particularly when one metabolite is more toxic than the others. Thus prior exposure of animals to environmental agents can have a significant effect on activation–detoxication pathways and the toxicity of other xenobiotics.

VI. Species Differences

The FMO appears to be widely distributed across mammalian species, and one or more forms are present in most tissues (Dannon and Guengerich, 1982). To date, however, we have been unable to detect in insects FMO forms comparable to the mammalian FMOs (Venkatesh and Rose, 1990 unpublished results). This is in contrast to the insect P450 system, which possesses multiple isozymes, broad substrate specificities, and inductive capabilities (Hodgson, 1983; Riskallah *et al.*, 1986).

Comparison of the liver form(s) of the FMO from rat, mouse, rabbit, and pig in terms of substrate specificities show marked similarities (Tynes and Hodgson, 1985a,b; Sabourin and Hodgson, 1984; Sabourin *et al.*, 1984; Hodgson and Levi, 1989). Recently, a comparison of the derived amino acid sequences for FMO from pig and rabbit livers showed an 87% identity between these two enzymes (Lawton *et al.*, 1990).

VII. Organ Differences

While it appears unlikely that there are as many isozymic forms of the FMO as there of P450, it is now known that multiple forms of this enzyme are present in different tissues. Catalytic activities in microsomes from different tissues have been studied, and as was observed with phorate (Table V), FMO activities in lung and kidney are often as high as those in liver. One such comparison using pesticide substrates is shown in Table VII. The most extensive comparison between FMOs is between those of the liver and lung. The first indication that these FMOs might differ from one another was provided by Devereaux *et al.* (1977), who showed marked differences in the effects of Hg^{2+} on partially purified FMO preparations. Subsequently, differences in lung and liver FMOs were suggested by the studies of Ohmiya and Mehendale (1982, 1984) on chlorpromazine and imipramine metabolism in the lung and rat and rabbit. These compounds are substrates for the liver, but not the lung FMO.

FMO purified from rabbit lung (Tynes *et al.*, 1985; Williams *et al.*, 1984) has been shown to be catalytically and immunologically distinct from the liver enzyme. The mouse and rabbit lung FMOs have a unique ability for N-oxidation of the primary aliphatic amine, *n*-octylamine, a chemical commonly included in microsomal incubations to inhibit P450. In the mouse lung this compound serves not only as a substrate, but is also a positive effector of metabolism. The mouse and rabbit lung enzymes have a higher pH optimum, near 9.8, than the liver enzyme, which is around 8.8.

TABLE VII

Rates of Metabolism in Selected Pesticides by the FMO in Mouse Liver, Lung, and Kidney Microsomes[a]

Substrate	Liver	Lung	Kidney
Phorate	7.4	10.7	4.4
Disulfoton	9.0	11.1	5.9
Fenthion	6.5	4.1	2.7
Aldicarb	2.1	2.1	1.3
Croneton	5.1	4.4	2.4
Nicotine	3.6	3.2	2.3

[a]Activity expressed is nmol NADPH/min per mg microsomal protein. Incubations included 5 mg of antireductase IgG/mg microsomal protein. Mean values for three assays were within 10%. From Tynes and Hodgson (1985a).

Recently, Lawton *et al.* (1990) have derived the primary structures of the rabbit and pig liver and rabbit lung FMO enzymes from the nucleotide sequences of cloned cDNAs. The amino acid sequences between the lung and liver are 56% identical, compared to 87% identity between the liver forms of pig and rabbit. Despite the considerable differences between the liver and lung sequences, the structures of these enzymes were found to have a number of common features: their pyrophosphate-binding domains were nearly identical, and they had five putative membrane-associated regions in common. These common properties suggest that the overall structural similarity between the liver and lung enzymes is greater than would be predicted from the relative identity of their primary structures.

In addition to differences between the liver and lung enzymes, the purification of a second FMO from rabbit liver has been reported (Ozols, 1989). Thus it is now evident that there are multiple FMO isozymes with overlapping substrate specificities; moreover, it is likely that the relative proportions of these isozymes vary in different tissues within and between species. These differences will undoubtedly play an important role in organ-specific metabolism, thus influencing routes of activation and detoxication of many compounds including pesticides.

VIII. Cellular Distribution

Several recent studies on the cellular localization of FMO have been carried out. An immunohistochemical method utilizing peroxidase-labeled antibodies and diaminobenzidine revealed that in the rabbit lung, the FMO is highly localized in the nonciliated bronchiolar epithelial (Clara) cells (Overly, Lawton, Philpot, and Hodgson, 1990 unpublished results). Similar immunohistochemical studies of FMO distribution in the skin of mice and pigs (Venkatesh *et al.*, 1991) reveal significant staining in epidermis, sebaceous gland cells, and hair follicles. Since the lung and skin are often major routes of entry for environmental chemicals, such as pesticides, the presence of the FMO in these tissues is of considerable importance.

Thus an understanding of these enzymes, their distribution, and their metabolic capabilities is essential in understanding the fate of xenobiotics in the body and the mechanisms underlying the toxicities of many chemicals, including pesticides.

References

Beaty, N. B., and Ballou, D. P. (1980). Transient kinetic study of liver microsomal FAD-containing monooxygenase. *J. Biol. Chem.* 255, 3817–3819.

Beaty, N. B., and Ballou, D. P. (1981a). The reductive half-reaction of liver microsomal FAD-containing monooxygenase. *J. Biol. Chem.* **256**, 4611–4618.

Beaty, N. B., and Ballou, D. P. (1981b). The oxidative half-reaction of liver microsomal FAD-containing monooxygenase. *J. Biol. Chem.* **256**, 4619–4625.

Dannon, G. A., and Guengerich, F. P. (1982). Immunochemical comparison and quantification of microsomal flavin-containing monooxygenases in various hog, mouse, rat, rabbit, dog and human tissues. *Mol. Pharmacol.* **22**, 787–794.

Devereux, T. R., Philpot, R. M., and Fouts, J. R. (1977). The effects of Hg^{2+} on rabbit hepatic and pulmonary solubilized, partially purified *N,N*-dimethylaniline *N*-oxidases. *Chem. Biol. Interact.* **19**, 277–297.

Doerge, D. R., and Corbett, M. D. (1984). Hydroperoxyflavin-mediated oxidation of organosulfur compounds. *Mol. Pharmacol.* **26**, 348–352.

Hajjar, N. P., and Hodgson, E. (1980). Flavin adenine dinucleotide–dependent monooxygenase: Its role in the sulfoxidation of pesticides in mammals. *Science* **209**, 1134–1136.

Hajjar, N. P., and Hodgson, E. (1982a). Flavin adenine dinucleotide–dependent monooxygenase as a activation enzyme. *In* "Biological Reative Intermediates—II, Part B" (R. Snyder, D. V. Parke, J. J. Kocsis, F. J. Jollow, G. G. Gibson, and C. M. Witmer, eds.), Plenum Press, New York.

Hajjar, N. P., and Hodgson, E. (1982b). Sulfoxidation of thioether-containing pesticides by the flavin-containing dinucleotide dependent monooxygenase of pig liver microsomes. *Biochem. Pharmacol.* **31**, 745–752.

Hodgson, E. (1983). The significance of cytochrome P-450 in insects. *Insect Biochem.* **13**, 237–246.

Hodgson, E., and Levi, P. E. (1988). The flavin-containing monooxygenase as a sulfure oxidase. *In* "Metabolism of Xenobiotics" (J. W. Gorrod, H. Oelschlanger, and J. Caldwell, eds.), pp. 81–88. Taylor and Francis, London.

Hodgson, E., and Levi, P. E. (1989). Species, organ, and cellular variation in the flavin-containing monooxygenase. *Drug Metab. Drug Interact.* **6**, 219–233.

Kinsler, S., Levi, P. E., and Hodgson, E. (1988). Hepatic and extrahepatic microsomal oxidation of phorate by the cytochrome P-450 and FAD-containing monooxygenase systems in the mouse. *Pestic. Biochem. Physiol.* **31**, 54–60.

Kinsler, S., Levi, P. E., and Hodgson, E. (1990). Relative contributions of the cytochrome P-450 and flavin-containing monooxygenases to the microsomal oxidation of phorate following treatment of mice with phenobarbital, hydrocortisone, acetone, and piperonyl butoxide. *Pestic. Biochem. Physiol.* **37**, 174–181.

Lawton, M. P., Gasser, R., Tynes, R. E., Hodgson, E., and Philpot, R. M. (1990). The flavin-containing monooxygenase enzymes expressed in rabbit liver and lung are products of related but distinctly different genes. *J. Biol. Chem.* **265**, 5855–5861.

Levi, P. E., and Hodgson, E. (1988a). Metabolites resulting from oxidative and reducing processes. *In* "Intermediary Xenobiotic Metabolism in Animals" (D. J. Hutson and G. D. Paulson, eds.), pp. 119–138. Taylor and Francis, London.

Levi, P. E., and Hodgson, E. (1988b). Stereospecificity in the oxidation of phorate and phorate sulphoxide by purified FAD-containing monooxygenase and cytochrome P-450 isozymes. *Xenobiotica* **18**, 29–39.

Ohmiya, Y., and Mehendale, H. M. (1982). Metabolism of chlorpromazine by pulmonary microsomal enzymes in the rat and rabbit. *Biochem. Pharmacol.* **31**, 157–162.

Ohmiya, Y., and Mehendale, H. M. (1984). Species differences in pulmonary *N*-oxidation of chlorpromazine and imipramine. *Pharmacology* **28**, 289–295.

Ozols, J. (1989). Liver microsomes contain two distinct NADPH-monooxygenases with NH_2-terminal segments homologous to the flavin containing NADPH monooxygenase of *Pseudomonas fluorescens. Biochem. Biophys. Res. Commun.* **163**, 49–55.

Poulsen, L. L. (1981). Organic sulfur substrates for the microsomal flavin-containing monooxygenase. *In* "Reviews in Biochemical Toxicology" (E. Hodgson, J. R. Bend, R. M. Philpot, eds.), Vol. 3, pp. 33–50. Elsevier, New York.

Poulsen, L. L., and Ziegler, D. M. (1979). The microsomal FAD-containing monooxygenase: Spectral characterization and kinetic studies. *J. Biol. Chem.* **254**, 6449–6455.

Riskallah, M. R., Dauterman, W. C., and Hodgson, E. (1986). Host-plant induction of microsomal monooxygenase activity in relation to diazinon metabolism and toxicity in larvae of the tobacco budworm *Heliothis virescens* (F.). *Pestic. Biochem. Physiol.* **25**, 233–247.

Sabourin, P. J., and E. Hodgson (1984). Characterization of the purified microsomal FAD-containing monooxygenase from mouse and pig liver. *Chem. Biol. Interact.* **51**, 125–139.

Sabourin, P. J., Smyser, B. P. and Hodgson, E. (1984) Purification of the flavin-containing monooxygenase from mouse and pig liver microsomes. *Int. J. Biochem.* **16**, 713–720.

Smyser, B. P., and Hodgson, E. (1985). Metabolism of phosphorus-containing compounds by pig liver microsomal FAD-containing monooxygenase. *Biochem. Pharmacol.* **34**, 1145–1150.

Smyser, B. P., Sabourin, P. J., and Hodgson, E. (1985). Oxidation of pesticides by purified microsomal FAD-containing monooxygenase from mouse and pig liver. *Pestic. Biochem. Physiol.* **24**, 368–374.

Taylor, K. L., and Ziegler, D. M. (1987). Studies on substrate specificity of the hog liver flavin-containing monooxygenase. *Biochem. Pharmacol.* **36**, 141–146.

Tynes, R. E., and Hodgson, E. (1985a). Magnitude of involvement of the mammalian flavin-containing monooxygenase in the microsomal oxidation of pesticides. *J. Agric. Food Chem.* **33**, 471–479.

Tynes, R. E., and Hodgson, E. (1985b). Catalytic activity and substrate specificity of the flavin-containing monooxygenase in microsomal systems: Characterization of the hepatic, pulmonary, and renal enzymes of the mouse, rabbit and rat. *Arch. Biochem. Biophys.* **240**, 77–93.

Tynes, R. E., Sabourin, P. J., and Hodgson, E. (1985). Identification of distinct hepatic and pulmonary forms of microsomal flavin-containing monooxygenase in the mouse and rabbit. *Biochem. Biophys. Res. Commun.* **126**, 1069–1075.

Venkatesh, K., Levi, P. E., Inman, A. C., Monteiro-Riviere, N. A., Misra, R., and Hodgson, E. (1991). Enzymatic and immunohistochemical studies on the role of cytochrome P450 and the flavin-containing monooxygenase of mouse skin in the metabolism of pesticides and other xenobiotics. *Pestic. Biochem. Physiol.* (submitted).

Williams, D. E., Ziegler, D. M., Nordin, D. J., Hale, S. E., and Masters, B. S. S. (1984). Rabbit lung flavin-containing monooxygenase is immunochemically and catalytically distinct from the liver enzyme. *Biochem. Biophys. Res. Commun.* **125**, 116–122.

Williams, D. E., Hale, S. E., Muerhoff, A. S., and Masters, B. S. S. (1985). Rabbit lung flavin-containing monooxygenase. Purification characterization, and induction during pregnancy. *Mol. Pharmacol.* **28**, 381–390.

Ziegler, D. M. (1980). Microsomal flavin-containing monooxygenase: Oxygenation of nucleophilic nitrogen and sulfur compounds. *In* "Enzymatic Basis of Detoxification" (W.B. Jakoby, ed.), Vol. 1, pp. 201–227. Academic Press, New York.

Ziegler, D. M. (1984). Metabolic oxygenation of organic nitrogen and sulfur compounds. *In* "Drug Metabolism and Drug Toxicity" (J. R. Mitchell and M. G. Horning, eds.), pp. 33–52. Raven Press, New York.

Ziegler, D. M. (1988). Flavin-containing monooxygenases: Catalytic mechanism and substrate specificities. *Drug Metab. Rev.* **9**, 1–32.

Ziegler, D. M., and Poulsen, L. L. (1978). Hepatic microsomal mixed-function amine oxidase. *Methods Enzymol.* **52**, 155–157.

7

Role of Glutathione in the Mammalian Detoxication of Organophosphorus Insecticides

Lester G. Sultatos
Department of Pharmacology and Toxicology
The University of Medicine and Dentistry of New Jersey
Newark, New Jersey

I. Introduction

Glutathione is a tripeptide (L-γ-glutamyl-L-cysteinylglycine) involved in many cellular functions, including certain cellular transport mechanisms, cellular protection, and the metabolism of numerous xenobiotics (Meister and Anderson, 1983). Although reduced glutathione can combine spontaneously with highly reactive chemicals, resulting in their detoxication, many conjugation reactions involving glutathione are promoted by a family of enzymes known as glutathione S-transferases (EC 2.5.1.18). Numerous excellent reviews of the functions of glutathione and glutathione S-transferases have been published, and interested readers are referred to these references (Kosower and Kosower, 1978; Meister and Anderson, 1983; Reed and Fariss, 1984; Reed, 1986; Ketterer, 1988; Boyer, 1989; and Reed, 1990).

Organophosphates: Chemistry, Fate, and Effects

Since the 1960s, considerable evidence has accumulated suggesting a significant role for glutathione and glutathione S-transferases in the mammalian detoxication of certain (OP) insecticides. At least 25 such insecticides have been reported to undergo significant biotransformation as a result of conjugation with glutathione (Motoyama and Dauterman, 1980). More recently, however, several reports have raised serious doubts about the participation of glutathione in the murine biotransformation of certain OP insecticides previously thought to undergo extensive glutathione-dependent detoxication. The present report will summarize briefly the evidence that implicates glutathione in the metabolism of many OP insecticides, since previous reviews have been published concerning this topic (Motoyama and Dauterman, 1980; Fukami, 1980). This report will focus primarily on the evidence casting doubt on the hypothesis that glutathione plays a significant role in the detoxication of certain OP insecticides in the mouse.

II. Evidence Supporting a Role for Glutathione in Metabolism of Certain Organophosphorus Insecticides

The evidence that has accumulated suggesting a significant role for glutathione in the detoxication of certain OP insecticides can be classified into three major categories.

A. Identification of Metabolic Pathways *in Vitro*

Incubation of certain OP insecticides *in vitro* with hepatic 100,000 × g supernatant or partially purified glutathione transferases, fortified with glutathione, has been reported to result in the glutathione-dependent metabolism of these insecticides (Motoyama and Dauterman, 1980, and Fukami, 1980). The best substrates *in vitro* have been the dimethyl-substituted OP insecticides such as methyl parathion and azinphos-methyl, and the most frequently reported reaction is demethylation (Fig. 1) (Motoyama and Dauterman, 1980; Fukami, 1980). However, glutathione-dependent biotransformation *in vitro* of other types of OP insecticides has been documented, and reactions other than demethylation have been reported. For example, Hollingworth *et al.* (1973) demonstrated that paraoxon can undergo limited deethylation, and that both paraoxon and methyl paraoxon can undergo dearylation (Fig. 1).

Several reports have indicated that more than one glutathione S-transferase is capable of metabolizing certain OP insecticides. Hollingworth *et al.* (1973) demonstrated that alkyl and aryl transferase activities from rat liver supernatant could be separated by ammonium sulfate fractionation, while Motoyama and Dauterman (1978) separated multiple forms of glutathione S-transferases and documented that several had either alkyl or aryl transferase

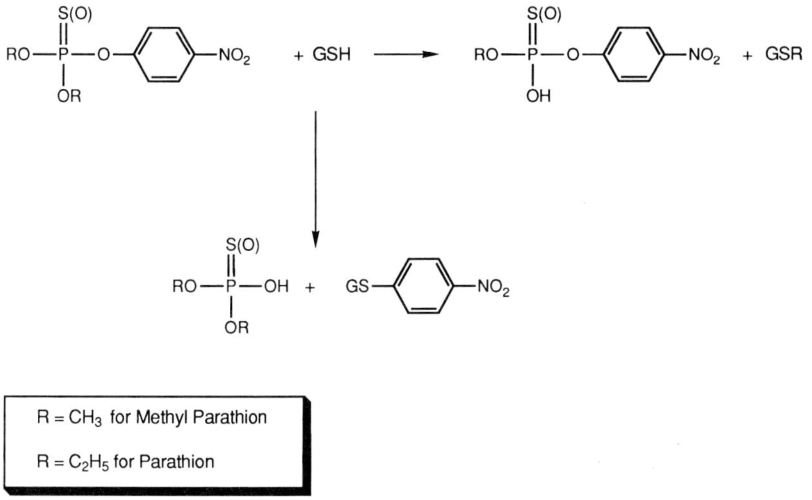

R = CH_3 for Methyl Parathion

R = C_2H_5 for Parathion

Figure 1 Glutathione-dependent pathways reported for the detoxication of methyl parathion, parathion, and their oxygen analogs. From Motoyama and Dauterman (1980); Fukami (1980).

activities. As pointed out by Motoyama and Dauterman (1980), the multiple forms separated could be divided into two distinct groups based primarily upon differences in substrate preference for either alkyl or aryl conjugations.

The characterization of metabolite profiles in blood and urine following administration of OP insecticides like methyl parathion might be expected to reveal the extent of glutathione-dependent metabolism *in vivo*. Interestingly, neither S-(methyl)glutathione nor subsequent metabolites (such as methyl mercapturic acid) could be detected in urine of mice administered methyl paraoxon, even though methyl paraoxon is known to undergo glutathione-dependent demethylation *in vitro* (Hollingworth, 1969). Hollingworth (1969) proposed that either dealkylation of methyl paraoxon *in vivo* occurs by a mechanism not producing S-(methyl)glutathione, or that S-(methyl)glutathione is rapidly degraded to metabolites not excreted in the urine. In addition, complications arise when considering the significance of metabolite profiles of OP insecticides *in vivo*, since different enzyme systems *in vitro* can attack the same chemical bonds to yield identical metabolites (Motoyama and Dauterman, 1980). For example, identification of dimethyl phosphoric acid in the blood or urine of mice following administration of methyl parathion cannot be taken as evidence for glutathione-dependent dearylation, since this metabolite has been shown to result also from cytochrome P450-catalyzed oxidations, as well as by hydrolysis of methyl paraoxon. Likewise dealkylation of certain insecticides has been reported to result from glutathione-dependent metabolism as well as P450-dependent oxidation (Appleton and Nakatsugawa, 1977).

B. Depression of Glutathione Levels by Organophosphorus Insecticides

Significantly reduced hepatic glutathione levels have been reported following administration of large doses of OP insecticides like methyl chlorpyrifos or sumithion (Fig. 2) (Hollingworth, 1969; Sultatos *et al.*, 1982). Such reductions have been thought to occur as a result of the conjugation of glutathione with insecticide molecules, much as reactive metabolites from acetaminophen can deplete hepatic glutathione levels (Mitchell *et al.*, 1973). Interestingly, reduced hepatic glutathione levels in mice occurred following administration of methyl parathion, but only at a dose so high (200 mg/kg) that mice had to be pretreated with atropine and 2-PAM to prevent rapid death (Levine and Murphy, 1977). In contrast, administration of methyl chlorpyrifos, or its diethyl-substituted analog chlorpyrifos, to mice at doses lower than their reported LD_{50} significantly reduced hepatic glutathione levels (Fig. 2). (Sultatos *et al.*, 1982).

C. Potentiation of Organophosphorus Toxicity by Diethyl Maleate and Methyl Iodide

Administration of any compound that forms an adduct with glutathione will result in the depletion of glutathione, provided the dose of chemical is high

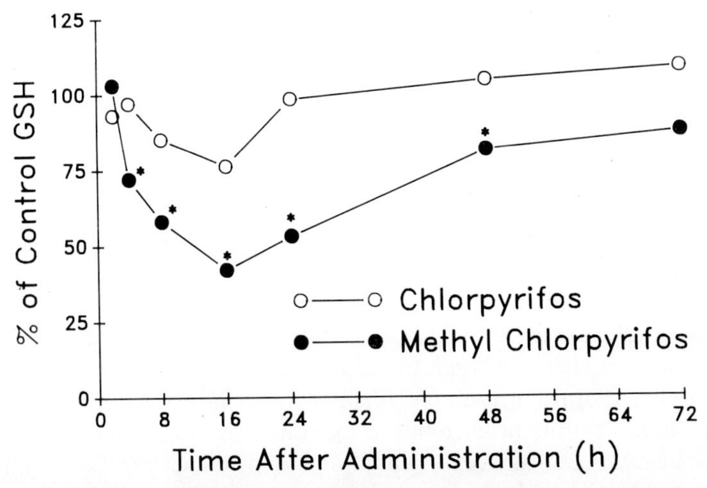

Figure 2 Hepatic glutathione (GSH) levels after intraperitoneal administration of methyl chlorpyrifos (1000 mg/kg) or chlorpyrifos (70 mg/kg) to mice. Each treated group ($n = 4$) was compared to its concurrently terminated control group. Control glutathione levels ranged from 975 to 1660 nmol/100 mg liver. *Significantly different ($p < 0.05$) from the corresponding control group by analysis of variance followed by Newman-Keul's test. From Sultatos *et al.* (1982).

enough (Plummer *et al.*, 1981). Depletion of glutathione by such agents, followed by a challenge dose of a second chemical, such as an OP insecticide, should give information regarding the role of glutathione in the metabolism of the second chemical. This approach has been applied to the metabolism *in vivo* of OP insecticides by utilizing the chemicals diethyl maleate and methyl iodide to deplete glutathione. Pretreatment of animals with either diethyl maleate or methyl iodide potentiated the acute toxicity of numerous OP insecticides like methyl chlorpyrifos and chlorpyrifos (Table I) (Fukami, 1980; Hollingworth, 1969; Sultatos *et al.*, 1982; Mirer *et al.*, 1977). The potentiation has been attributed to the loss of glutathione, and therefore the loss of a putative pathway for the detoxication of these insecticides (Fukami, 1980; Hollingworth, 1969; Sultatos *et al.*, 1982; Mirer *et al.*, 1977). Such results support the hypothesis that glutathione plays a significant role in the detoxification *in vivo* of these OP insecticides.

III. Evidence Against a Role for Glutathione in Metabolism of Certain Organophosphorus Insecticides in the Mouse

In sharp contrast to the studies outlined above, not all reports have supported the hypothesis that glutathione plays a significant role in the biotransformation of many OP insecticides *in vivo*. For example, despite the evidence that azinphos-methyl undergoes glutathione-dependent detoxification *in vitro* (Fukami, 1980; Rao and McKinley, 1969), Levine and Murphy (1977) reported that metabolism of this insecticide by such a pathway(s) does not occur to any significant extent *in vivo* in the mouse unless competing oxidative pathways are inhibited. This conclusion was based on the observation that

TABLE I

Acute Intraperitoneal Toxicities of Chlorpyrifos and Methyl Chlorpyrifos in Untreated and Diethyl Maleate-Pretreated Male Mice[a]

Insecticide	7-Day LD_{50} [mg/kg body wt (95% confidence limits)][b] Untreated	DEM pretreated[c,d]	Ratio Untreated: Pretreated (95% confidence limits)
Chlorpyrifos	192 (150–246)	96 (67–137)	2.0 (1.5–2.6)
Methyl chlorpyrifos	2325 (1626–3325)	272 (202–367)	8.5 (5.4–13.5)

[a]From Sultatos, L. G. *et al.* (1982).
[b]At least four doses and 24 to 40 animals were used for each LD_{50} determination.
[c]DEM, diethyl maleate (1 ml/kg), was administered i.p. 1 hr before challenge insecticides.
[d]DEM, pretreatment resulted in GSH levels of 399 ± 74 nmol/100mg liver, whereas control values were 1268 ± 69 nmol/100 mg liver.

upon addition of azinphos-methyl to mouse liver homogenate, glutathione levels dropped only when oxidative metabolism was inhibited.

The first suggestion that glutathione may not play a role in the *in vivo* biotransformation of an OP insecticide, previously thought to undergo extensive glutathione-dependent demethylation, was made by Dorough (1983). While evaluating the potential for a pharmacokinetic interaction between the drug acetaminophen and the insecticide sumithion, Dorough (1983) observed that depletion of mouse hepatic glutathione by acetaminophen pretreatment had no effect on the acute toxicity of sumithion, even though this insecticide clearly undergoes glutathione-dependent detoxification *in vitro* (Table II) (Motoyama and Dauterman, 1980; Fukami, 1980; Hollingworth, 1969; Hollingworth *et al.*, 1967). These results suggest that detoxification of sumithion *in vivo* in the mouse can proceed even though hepatic glutathione levels are markedly reduced. Conversely, in the same study, pretreatment of mice with methyl iodide depleted hepatic glutathione, but markedly synergized the acute toxicity of sumithion (Table II), as had been reported previously (Table II) (Hollingworth, 1969). Dorough (1983) concluded that glutathione is not significantly involved in the detoxification of sumithion *in vivo* in the mouse. Furthermore Dorough (1983) concluded that methyl iodide probably exerted its synergistic effect on the acute toxicity of sumithion by some mechanism other than depletion of glutathione. This conclusion is not unreasonable, since methyl iodide is known to be highly toxic, and can nonselectively alkylate many proteins (Barksdale and Rosenberg, 1978).

Although this report (Dorough, 1983) is highly suggestive that glutathione does not participate in the detoxication of sumithion *in vivo* in the mouse, it cannot be considered conclusive evidence. Since acetaminophen depletes hepatic glutathione by binding to reactive intermediates formed by reaction of acetaminophen with P450, these same reactive intermediates could have inhibited those forms of P450 that metabolically activate su-

TABLE II

Effect of Hepatic Glutathione–Depleting Agents on Sumithion Toxicity in Mice[a]

I.P. dose (mg/kg)	Hours between treatments	Mortality
Methyl iodide, 135	—	2/12
Sumithion, 750	—	0/9
Acetaminophen, 300	—	0/9
Methyl iodide + S	1	9/9
Acetaminophen + S	1	0/9

[a]From Dorough, J. (1983).

mithion. Similarly, metabolic activation of acetaminophen might compete with metabolic activation of sumithion. Therefore mice in the acetaminophen–sumithion group might not have been able to produce sumioxon from sumithion, thereby negating the putative potentiation of this insecticide resulting from reduced glutathione levels. Costa and Murphy (1984), however, addressed this problem by repeating Dorough's study with the insecticide dichlorvos, an OP insecticide that does not require metabolic activation. Other insecticides included in their study were methyl parathion and methyl chlorpyrifos. Depletion of hepatic glutathione by acetaminophen failed to affect the acute toxicity of any of these insecticides (Costa and Murphy, 1984). Since dichlorvos does not require metabolic activation, the failure of glutathione depletion by acetaminophen to alter the acute toxicity of this insecticide cannot be attributed to inhibition of cytochromes P450 by metabolic activation of acetaminophen. Therefore the studies of Dorough (1983) and Costa and Murphy (1984) raise serious questions regarding the role of glutathione-dependent detoxication of sumithion, dichlorvos, methyl parathion, and methyl chlorpyrifos *in vivo* in the mouse, even though each insecticide has been shown to undergo glutathione-dependent biotransformation *in vitro* (Motoyama and Dauterman, 1980; Fukami, 1980; Hollingworth, 1969; Sultatos *et al.*, 1982).

As a result of Dorough's report, Sultatos and co-workers undertook studies designed to observe the effects of manipulation of glutathione levels in mice on the acute toxicities of several OP compounds known to undergo glutathione-dependent metabolism *in vitro* (Sultatos and Woods, 1988). Glutathione levels were altered by pretreatment of mice with diethyl maleate, buthionine sulfoximine, or glutathione monoethyl ester plus diethyl maleate (Fig. 3). Buthionine sulfoximine has been shown to reduce glutathione levels as a result of inhibition of the enzyme γ-glutamylcysteine synthetase (Griffith and Meister, 1979a,b), the first step in the synthesis of glutathione. Buthionine sulfoximine has been reported to be much more selective in its action than either diethyl maleate or methyl iodide (Griffith and Meister, 1979a,b). Glutathione monoethyl ester was developed by Anderson *et al.* (1985) to increase levels of glutathione under certain circumstances. Glutathione monoethyl ester is transported into cells of the liver, kidney, spleen, pancreas, heart, and lungs in the mouse, and is subsequently hydrolyzed to form glutathione and ethanol (Anderson *et al.*, 1985; Puri and Meister, 1983). Consequently, administration of glutathione monoethyl ester to mice can afford protection against certain toxic chemicals detoxified by glutathione (Singhal *et al.*, 1987).

Administration of buthionine sulfoximine to mice, as previously indicated (Griffith and Meister, 1979a), significantly reduced hepatic glutathione levels, as did administration of diethyl maleate (Fig. 3) (Richardson and Murphy, 1975). However, administration of glutathione monoethyl ester (20 mmol/kg body weight), immediately following administration of diethyl mal-

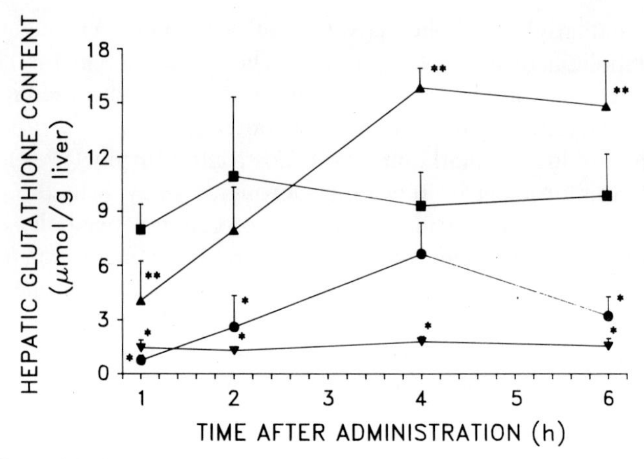

Figure 3 Effects of diethyl maleate (●), buthionine sulfoximine (▼), and glutathione monoethyl ester + buthionine sulfoximine (▲), on hepatic glutathione levels in the mouse. Controls (■) received no treatment. Each point represents the mean ± SD from four mice. Diethyl maleate was administered i.p. at a dose of 0.75 ml/kg, while buthionine sulfoximine was administered as previously outlined (Sultatos and Woods, 1988). Glutathione monoethyl ester was administered by stomach tube at a dose of 20 mmol/kg (Anderson *et al.*, 1985), immediately after diethyl maleate. Glutathione was determined as previously described (Anderson, 1985). An asterisk (*) indicates a significant difference ($p < 0.05$) from the corresponding control, whereas a double asterisk (**) indicates a significant difference ($p < 0.05$) from the corresponding control and the corresponding DEM-treated group. Statistical analyses were performed by a two-way analysis of variance, followed by the Newman-Keul's test.

eate, resulted in substantial increases of hepatic glutathione, compared to levels in mice receiving only diethyl maleate. One hour following both pretreatments, glutathione levels were lower than those of control mice, but were significantly greater than those of mice receiving only diethyl maleate. Two hours after the pretreatments, glutathione levels were equivalent to control levels, while by 4 and 6 hr, levels were greater than those of control mice.

Pretreatment of mice with diethyl maleate potentiated the acute toxicities of methyl parathion, methyl paraoxon, sumithion, azinphos-methyl chlorpyrifos (Fig. 4). Similar results have been reported by numerous investigators (Motoyama and Dauterman, 1980; Fukami, 1980; Hollingworth, 1969; Sultatos *et al.*, 1982). However, depletion of hepatic glutathione by pretreatment with buthionine sulfoximine did not potentiate the toxicities of these OP compounds (Fig. 4), indicating that the detoxication of these chemicals can proceed normally even when glutathione levels are markedly reduced. Therefore, it can be concluded that glutathione is not significantly involved in the detoxication of these insecticides in the mouse. Moreover, it can be hypothesized that diethyl maleate potentiates the acute toxicities of these chemicals by a mechanism unrelated to the ability of diethyl maleate to deplete glutathione. A comparison of the pharmacological effects of buthionine sulfoximine and diethyl maleate support this hypothesis. Buthionine sulfoximine has been reported to be extremely selective in its action (Griffith and Meister, 1979a,b). Furthermore buthionine sulfoximine has been administered to mice repeatedly, at high doses, without apparent effect other than inhibition of glutathione biosynthesis (Arrick *et al.*, 1981). Similarly Drew and Minors (1984) have reported concentrations of up to 500 m*M* buthionine sulfoximine *in vitro* had no effect on certain glutathione transferases, *p*-nitrophenyl glucuronyltransferase, phenolsulphotransferase, aniline hydroxylase, or aminopyrine demethylase. In contrast, diethyl maleate has been reported to exert many effects unrelated to depletion of glutathione (Plummer *et al.*, 1981), some of which could be mediated by impurities often present in commercially available diethyl maleate (Meister, 1985). Furthermore, diethyl maleate is hydrolyzed *in vivo* to form maleate, which can exert numerous biological effects (Thompson and Meister, 1979). Perhaps most importantly, Anders (1978) has demonstrated that the presence of diethyl maleate enhanced aniline and acetanilide hydroxylase activities in rat liver microsomes, but inhibited *N*-demethylation of benzphetamine and *O*-dealkylation of *p*-ethoxyacetanilide. Similarly Sultatos and Woods (1988) reported enhanced

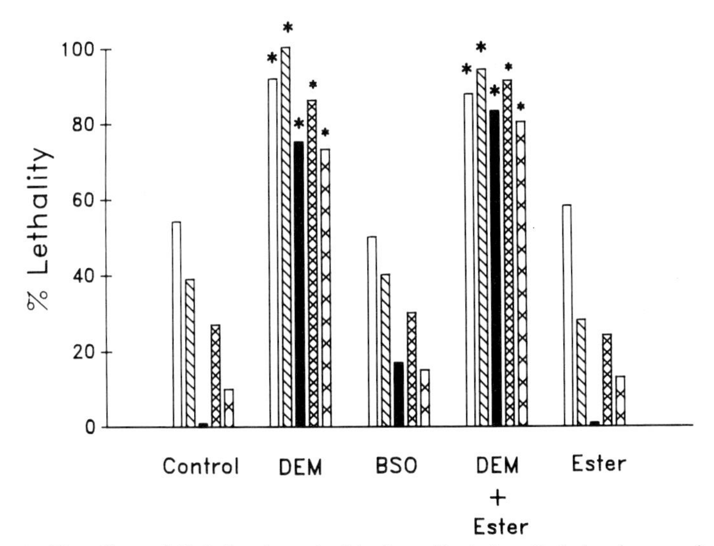

Figure 4 The effects of diethyl maleate, buthionine sulfoximine, diethyl maleate + glutathione monoethyl ester, and glutathione monoethyl ester alone, on the lethality of methyl parathion (▭) (15 mg/kg, i.p.), methyl paraoxon (▧) (5 mg/kg, i.p.), fenitrothion (■) (800 mg/kg, i.p.), methyl chlorpyrifos (⊠⊠) (950 mg/kg, i.p.), and azinphos-methyl (⊠) (8 mg/kg, i.p.). Pretreatments were the same as those in Fig. 3. An asterisk (*) indicates a significant difference from the corresponding control group by the Friedman's Block–Treatment test followed by the Two Sample Proportion test.

and decreased metabolic activation of methyl parathion and azinphos-methyl, respectively, by mouse hepatic microsomes in the presence of diethyl maleate.

Although pretreatment of mice with diethyl maleate and glutathione monoethyl ester resulted in substantial levels of hepatic glutathione (Fig. 3), the acute toxicities of methyl parathion, methyl paraoxon, sumithion, azinphos-methyl, and methyl chlorpyrifos were potentiated in these mice (Fig. 4). Consequently, just as Dorough (1983) demonstrated that the nonspecific chemical methyl iodide potentiated the acute toxicity of sumithion by a mechanism other than glutathione depletion, the present study suggests that depletion of hepatic glutathione is not the mechanism of diethyl maleate-induced potentiation of the acute toxicities of these OP chemicals in the mouse. Additionally, the fact that the acute toxicities of methyl parathion, methyl paraoxon, sumithion, azinphos-methyl, and methyl chlorpyrifos in the presence of elevated hepatic glutathione levels are unchanged (Figs. 3 and 4) lends support to the conclusion that glutathione-dependent detoxication following lethal doses of these insecticides does not occur to any significant extent *in vivo* in the mouse. Although these data suggest no explanation for the potentiation of the acute toxicity of these chemicals by diethyl maleate, previous studies have documented numerous biological effects of diethyl maleate in addition to the depletion of glutathione (Sultatos and Woods, 1988; Plummer *et al.*, 1981; Anders 1978). As a result it is not unreasonable to assume that diethyl maleate could exert other as yet undetermined effects, which could potentiate the toxicity of these OP chemicals.

Finally, Levine and Murphy (1977) suggested that detoxication of azinphos-methyl by glutathione-dependent pathways probably becomes important only after inhibition of oxidative metabolism with compounds such as piperonyl butoxide. Pretreatment of mice with piperonyl butoxide has been shown to markedly antagonize the acute toxicities of methyl parathion and azinphos-methyl (Levine and Murphy, 1977; Kamienski and Murphy, 1971). Kamienski and Murphy (1971) speculated that piperonyl butoxide could inhibit activation of these insecticides, but not affect their demethylation by glutathione S-transferases. Thus the net effect would be to reduce the amount of oxon produced, leading to decreased toxicity. Indirect support of this hypothesis was the observation that depletion of glutathione by diethyl maleate reduced the protective effect of piperonyl butoxide against methyl parathion toxicity by at least 75% (Mirer *et al.*, 1977). However, Sultatos and Woods (1988) have demonstrated that depletion of glutathione by buthionine sulfoximine did not abolish the protection against the toxicities of azinphos-methyl or methyl parathion afforded by piperonyl butoxide. They concluded, therefore, that glutathione-mediated detoxication of methyl parathion and azinphos-methyl probably does not occur to a significant extent, even in mice in which oxidative metabolism has been inhibited by piperonyl butoxide.

IV. Unanswered Questions

The studies summarized above cast serious doubts on the participation of glutathione in the detoxication of certain OP insecticides in mice. However, if glutathione does not participate in the metabolism of these chemicals *in vivo*, two very important questions remain:

1. Why does administration of certain OP insecticides like methyl chlorpyrifos reduce hepatic glutathione levels? Since many intrahepatic and extrahepatic factors regulate glutathione homeostasis (Kaplowitz, 1981), chemicals that lower hepatic glutathione levels could do so by many mechanisms other than through its utilization in their metabolism. Chemicals such as aspirin and aminopyrine lower hepatic glutathione levels *in vivo* by mechanisms unknown, although not by a direct conjugation with glutathione (Kaplowitz *et al.*, 1980; Jones *et al.*, 1978). Therefore a drop in hepatic glutathione levels following administration of any chemical is not conclusive evidence that glutathione is involved in the biotransformation of that chemical.
2. Why are OP insecticides like methyl parathion metabolized by glutathione transferases *in vitro* but not *in vivo*? Because of the high lipid solubility of these insecticides, it is possible that they do not readily gain access to the soluble forms of glutathione transferases, and instead partition into the internal membranes of hepatocytes *in vivo*. Even though it has been suggested that soluble glutathione transferases actually bind to and increase the rate of movement of nonpolar molecules from plasma membranes to intracellular sites, Boyer *et al.* (1983) pointed out that direct experimental proof of this hypothesis is lacking. These same authors explained that such a phenomenon could occur only if rates of release of nonpolar chemicals from membranes are rapid. Slower rates of release (minutes to hours) from hepatocyte membranes can limit the ability of glutathione transferases to facilitate intracellular diffusion (Boyer *et al.*, 1983). Furthermore, Boyer *et al.* (1983) reported that soluble glutathione transferases were unable to remove nonpolar electrophiles from artificial membranes, and that the rates of release of substrate from these artificial membranes were critical in determining rates of catalysis. Therefore, highly lipid-soluble chemicals, such as OP insecticides, could partition avidly into hepatocyte membranes, thereby preventing their access to the soluble glutathione transferases.

V. Summary

In conclusion, evidence from three different laboratories suggests that glutathione-dependent detoxication of several OP insecticides does not occur to a significant extent *in vivo* in the mouse, even though such reactions can be observed *in vitro*. Caution must be observed, however, when attempting to extrapolate these results to other chemicals and other species. The disposition of an insecticide within hepatocytes is a function of the structure of the insecticide as well as composition of the hepatocyte. Therefore each insecticide and species must be considered on an individual basis.

Acknowledgments

The work from the author's laboratory was supported by NIEHS Grant ES04335.

References

Anders, M. W. (1978). Inhibition and enhancement of microsomal drug metabolism by diethyl maleate. *Biochem. Pharmacol.* **27**, 1098–1101.

Anderson, M. E. (1985). Determination of glutathione and glutathione disulfide in biological samples. *Methods Enzymol.* **113**, 548–555.

Anderson, M. E., Powrie, F., Puri, R. N., and Meister, A. (1985). Glutathione monoethyl ester: Preparation, uptake by tissues, and conversion to glutathione. *Arch. Biochem. Biophys.* **239**, 538–548.

Appleton, H. T., and Nakatsugawa, T. (1977). The toxicological significance of paraoxon deethylation. *Pestic. Biochem. Physiol.* **1**, 451–465.

Arrick, B. A., Griffith, O. W., and Cerami, A. (1981). Inhibition of glutathione synthesis as a chemotherapeutic strategy for trypanosmiasis. *J. Exp. Med.* **153**, 720–725.

Barksdale, A.D. , and Rosenberg, A. (1978). Measurement of protein dissociation constants by tritium exchange. *Methods Enzymol.* **48**, 321–346.

Boyer, T. D. (1989). The glutathione S-transferases: An update. *Hepatology* **9**, 486–496.

Boyer, T. D., Zakim, D., and Vessey, D. A. (1983). Do the soluble glutathione S-transferases have direct access to membrane-bound substrates? *Biochem. Pharmacol.* **32**, 29–35.

Costa, L. G., and Murphy, S. D. (1984). Interaction between acetaminophen and organophosphates in mice. *Res. Commun. Chem. Pathol. Pharmacol.* **44**, 389–400.

Dorough, H. W. (1983). Toxicological significance of pesticide conjugates. *J. Toxicol. Clin. Toxicol.* **19**, 637–659.

Drew, R., and Miners, J. D. (1984). The effects of buthionine sulfoximine (BSO) on glutathione depletion and xenobiotic biotransformation. *Biochem. Pharmacol.* **33**, 2989–2994.

Fukami, J. (1980). Metabolism of several insecticides by glutathione S-transferase. *Pharmacol. Ther.* **10**, 473–514.

Griffith, O. W., and Meister, A. (1979a). Potent and specific inhibition of glutathione synthesis by buthionine sulfoximine (S-*n*-butyl homocysteine sulfoximine). *J. Biol. Chem.* **254**, 7558–7560.

Griffith, O. W., and Meister, A. (1979b). Glutathione, interorgan translocation, turnover, and metabolism. *Proc. Natl. Acad. Sci. U.S.A.* **76**, 5606–5610.

Hollingworth, R. M., Fukuto, T. R., and Metcalf, R. L. (1967). Selectivity of sumithione compared with methyl parathion: Influence of structure on anticholinesterase activity. *J. Agric. Food Chem.* **15**, 235–241.

Hollingworth, R. M. (1969). Dealkylation of organophosphorus esters by mouse liver enzymes *in vitro* and *in vivo*. *J. Agric. Food Chem.* **17**, 987–996.

Hollingworth, R. M., Alstott, R. L., and Litzenberg, R. D. (1973). Glutathione S-aryl transferase in the metabolism of parathion and its analogs. *Life Sci.* **13**, 191–199.

Jones, D. P., Thor, H., Anderson, B., and Orrenius, S. (1978). Detoxification reactions in isolated hepatocytes. Role of glutathione peroxidase, catalase, and formaldehyde dehydrogenase in reactions relating to N-demethylation by the cytochrome P450 system. *J. Biol. Chem.* **253**, 6031–6037.

Kamienski, F. X., and Murphy, S. D. (1971). Biphasic effects of methylenedioxyphenyl synergists on the action of hexobarbital and organophosphate insecticides in mice. *Toxicol. Appl. Pharmacol.* **18**, 883–894.

Kaplowitz, N., Kuhlenkamp, J., Goldstein, L., and Reeve, J. (1980). Effect of salicylates and phenobarbital on hepatic glutathione in the rat. *J. Pharmacol. Exp. Ther.* **212**, 240–245.

Kaplowitz, N. (1981). The importance and regulation of hepatic glutathione. *Yale J. Biol. Med.* **54**, 497–502.

Ketterer, B. (1988). Protective role of glutathione and glutathione transferases in mutagenesis and carconogenesis. *Mutat. Res.* **202**, 343–361.

Kosower, N. S., and Kosower, E. M. (1978). The glutathione status of the cell. *Int. Rev. Cytol.* **54**, 109–160.

Levine, B. S., and Murphy, S. D. (1977). Effect of piperonyl butoxide on the metabolism of dimethyl and diethyl phosphorothionate insecticides. *Toxicol. Appl. Pharmacol.* **40**, 393–406.

Meister, A. (1985). Methods for the selective modification of glutathione metabolism and study of glutathione transport. *Meth. Enzymol.* **113**, 571–583.

Meister, A., and Anderson, M.E. (1983). Glutathione. *Annu. Rev. Biochem.* **52**, 722–750.

Mirer, F. E., Levine, B. S., and Murphy, S. D. (1977). Parathion and methyl parathion toxicity and metabolism in piperonyl butoxide and diethyl maleate pretreated mice. *Chem. Biol. Interact.* **17**, 99–112.

Mitchell, J. R., Jollow, D. J., Potter, W. Z., Gillette, J. R., and Brodie, B.B. (1973). Acetaminophen-induced hepatic necrosis IV. Protective role of glutathione. *J. Pharmacol. Exp. Ther.* **187**, 211–217.

Motoyama, N., and Dauterman, W. C. (1978). Multiple forms of rat liver glutathione S-transferases: Specificity for conjugation of O-alkyl and O-aryl groups of organophosphorus insecticides. *J. Agric. Food Chem.* **26**, 1296–1301.

Motoyama, N., and Dauterman, W. C. (1980). Glutathione S-transferases: Their role in the metabolism of organophosphorus insecticides. *In* "Review of Biochemistry and Toxicology" (E. Hodgson, J. R. Bend, and R. M. Philpot, eds.), Vol. 2, pp. 49–69. Elsevier/North Holland, Amsterdam/New York.

Plummer, J. L., Smith, B. R., Sies, H., and Bend, J. R. (1981). Chemical depletion of glutathione *in vivo*. *Methods Enzymol.* **77**, 51–59.

Puri, R. N., and Meister, A. (1983). Transport of glutathione, as γ-glutamylcysteinylglycyl ester, into liver and kidney. *Proc. Natl. Acad. Sci. U.S.A.* **80**, 5258–5260.

Rao, L. N., and McKinley, W. P. (1969). Metabolism of organophosphorus insecticides by liver homogenates from different species. *Can. J. Biochem.* **47**, 1155–1159.

Reed, D. J., and Fariss, M. W. (1984). Glutathione depletion and susceptibility. *Pharmacol. Rev.* **36**, 25s–33s.

Reed, D. J. (1986). Regulation of reductive processes by glutathione. *Biochem. Pharmacol.* **35**, 7–13.

Reed, D. J. (1990). Glutathione: Toxicological implications. *Annu. Rev. Pharmacol. Toxicol.* **30**, 603–631.

Richardson, R. J., and Murphy, S. D. (1975). Effect of glutathione depletion on tissue deposition of methylmercury in rats. *Toxicol. Appl. Pharmacol.* **31**, 505–519.

Singhal, R. K., Anderson, M. E., and Meister, A. (1987). Glutathione, a first line of defense against cadmium toxicity. *FASEB J.* **1**, 220–223.

Sultatos, L. G., Costa, L. G., and Murphy, S. D. (1982). Factors involved in the differential acute toxicities of the insecticides chlorpyrifos and methyl chlorpyrifos in mice. *Toxicol. Appl. Pharmacol.* **65**, 144–152.

Sultatos, L. G., and Woods, L. (1988). The role of glutathione in the detoxification of the insecticides methyl parathion and azinphos-methyl in the mouse. *Toxicol. Appl. Pharmacol.* **96**, 168–174.

Thompson, G. A., and Meister, A. (1979). Modulation of the hydrolysis, transfer, and glutaminase activities of gamma-glutamyl transpeptidase by maleate bound at the cysteinylglycine binding site of the enzyme. *J. Biol. Chem.* **254**, 2956–2960.

8

Role of Phosphorotriester Hydrolases in the Detoxication of Organophosphorus Insecticides

Yutaka Kasai
Kao Corporation
Tochigi Research Labs
2606 Akabane Ihikaimachi
Haga, Tochigi 321-34 Japan

Takamichi Konno
Biological Research Center
Nihon Nohyaku Co. Ltd
4-31 Hondacho, Kawachi-Nagano
Osaka, Japan T586

Walter C. Dauterman
Department of Toxicology
North Carolina State University
Raleigh, North Carolina

I. Introduction

The degradation of OP insecticides by a variety of hydrolases is an important detoxication route and has been documented in a number of reviews (Dauterman, 1976; 1983a,b). Hydrolytic cleavage of C—O, C—N and other bonds, including the phosphoric anhydride bond of an OP ester, results in the formation of an anionic metabolite at neutral pH, which is a poor cholinesterase inhibitor, and thus, as a result of hydrolysis, is detoxified (Eto, 1974). Of the Phase I reactions, hydrolysis is the only reaction that does not require an expenditure of energy by the organism.

The hydrolysis of OP insecticides is mediated by a number of enzymes responsible for the cleavage of the phosphorus ester or anhydride bond (Dauterman, 1976; Ahmad and Forgash, 1976). Phosphorotriester hydrolases may attack the intact OP insecticide molecule at two sites:

Organophosphates: Chemistry, Fate, and Effects

$$(RO)_2P(O)X + H_2O \begin{cases} \rightarrow (RO)_2P(O)OH + HX \\ \rightarrow RO(OH)P(O)X + ROH \end{cases}$$

One reaction leads to the formation of a dialkyl phosphoric acid and HX and the other results in the formation of a desalkyl derivative and alcohol, Eq. (1). In both reactions the products are the constituent acids and alcohols of a *classical* ester hydrolysis. The importance of phosphorotriester hydrolases in the detoxication of certain OP compounds has generally been assumed with only limited evidence. The cleavage of the phosphate ester has often been determined in complex systems: in homogenates, in subcellular fractions, or *in vivo*. Frequently such degradation has been attributed to aryl hydrolases, largely or entirely on the evidence of chemical change. The enzyme(s) responsible have not been isolated, and the mechanism of metabolism has not been tested. Both the microsomal monooxygenase system (Yang *et al.*, 1971; Donniger *et al.*, 1972) and the glutathione S-transferases (Motoyama and Dauterman, 1980) can remove alkyl and leaving groups from many OP compounds. Thus, much of the apparent hydrolysis of these compounds in a complex system may be the result of nonhydrolyzing enzyme systems. It is quite obvious that interpretation of findings based on this approach may result in erroneous conclusions.

A variety of names have been used to identify and describe the enzymes that catalyze these reactions, such as difluorophosphoric acid (DFP)-ase, paraoxonase, A-esterase, phosphatase, phosphorylphosphatase, arylester hydrolase, aryl esterase, etc., but for the present discussion, the term phosphorotriester hydrolase will be utilized.

II. Phosphohydrolases

The fluorohydrolases (E.C. 3.8.2.1) i.e., (DFP-ase, phosphohydrolase) cleave the P—F bond of DFP and soman, tabun, and the phosphoric anhydride bond of tetraethyl pyrophosphate (TEPP), Eq. (2).

$$(iPrO)_2P(O)F + H_2O \rightarrow (iPrO)_2P(O)OH + HF \qquad (2)$$

In this case, the question of whether this is an ester hydrolysis is debatable since the hydrolysis products are two acids. The enzymes responsible for this reaction appear to be present in almost every tissue and organism examined (Mazur, 1946; Mounter *et al.*, 1955a) as well as in microorganisms (Mounter *et al.*, 1955b).

Two general types of DFPases have been described. They are the Mazur-type DFPase and the squid-nerve DFPase (Garden *et al.*, 1975). The mammalian or kidney DFPase is stimulated by $4 \times 10^{-4} M$ Mn^{2+}. The Mazur-type DFPase hydrolyzes soman faster than DFP, and the reverse occurs with the squid-nerve DFPase (Hoskin *et al.*, 1984). Human plasma and serum contains a phosphohydrolase that stereoselectively hydrolyzes soman (De Bisschop *et al.*, 1987). The mammalian enzyme has a molecular weight of 60,000 and is unstable, while the squid-nerve DFPase has a molecular weight of 26,000 and is very stable (Hoskin, 1971; Hoskin and Prusch, 1983). Miapafax, a phosphorodiamidofluoridate, inhibits hog kidney and *Escherichia coli* DFPase reversibly, but does not inhibit the squid nerve DFPase (Hoskin, 1985).

A number of DFPases or phosphohydrolases have been purified from a variety of sources such as *E. coli* (Zech and Wigand, 1975), hog kidney (Storkebaum and Witzel, 1975), protozoan *Tetrahymena thermophila* (Landis *et al.*, 1987), clam *Rangia cuneata* (Anderson *et al.*, 1988), and a thermophilic bacterium (Chettur *et al.*, 1988).

Since none of the phosphohydrolases appears to be involved in the metabolism of insecticidal OP compounds, no further discussion will be devoted to this group of enzymes.

III. Phosphorotriester Hydrolases

Aldridge (1953a) introduced the terms A-esterases and B-esterases, and subsequently demonstrated the presence of an enzyme in mammalian sera that was responsible for hydrolysis of paraoxon. The major distinction was that A-esterases (EC 3.1.1.2) were not inhibited by OP compounds, hydrolyzed *p*-nitrophenyl acetate faster than *p*-nitrophenyl butyrate and were inhibited by *p*-chloromercuribenzoic acid (PCMB) (Aldridge 1953b). Although this enzyme hydrolyzed paraoxon, it did not hydrolyze parathion, the sulfur precursor of paraoxon. At present it is assumed that arylester hydrolases hydrolyze phosphate triesters but not thiono analogs, Eq. (3). Since paraoxonase was unable to hydrolyze the monoesters of *p*-nitrophenyl phosphate (Aldridge and Reiner, 1972) the enzyme may be considered to be a phosphorotriester hydrolase.

$$(C_2H_5O)_2P(O)O-(C_6H_4)-NO_2 + H_2O \rightarrow (C_2H_5O)_2P(O)OH + HO-(C_6H_4)-NO_2 \quad (3)$$

The enzyme (paraoxonase) was purified from sheep serum, had a molecular weight of 35,000 to 50,000, a K_m 4.2 m*M* to paraoxon, and was activated by Ca^{2+} ions (Main 1960). Mackness and Walker (1983) published a partial purification from the same source. Their method was based on the preparation of a lipoprotein fraction by ultracentrifugation (Mackness *et al.*, 1985) followed by preparative polyacrylamide gel electrophoresis. The molecular weight was greater than 200,000, and it appeared that the A-esterase activity toward paraoxon was present in one or more forms of high-density lipoproteins (HDL). Eighty percent of the paroxonase activity was found in

HDL, and there was strong evidence for multiple forms of HDL "A" esterases. Levels of paraoxonase activity in plasma and liver of birds were much lower than those in mammals (Walker and Mackness, 1987). Further studies in sheep serum showed five lipoproteins that expressed differing hydrolase activity toward four OP substrates. The proteins had molecular weights of approximately 360,000 and appeared to be different species of HDL2 particles (Mackness and Walker, 1988).

Zech and Zurcher (1974) studied phosphorotriester hydrolase activity in serum from eight different mammalian species. Different K_m and V_{max} values for paraoxon were obtained depending upon the species. Rabbit serum had the highest esterase activity, and mouse serum had the lowest. These findings demonstrate that different animal species detoxify OP compounds at different rates. Chemnitius *et al.* (1983) reported that high titers of paraoxonase activity were present in vertebrate liver, the level of activity depending upon the species, while high titers of DFPase were associated with the kidney in eight animal species.

An arylesterase hydrolase was partially purified from rabbit serum and had a molecular weight of 180,000 to 200,000 (Zimmerman and Brown, 1986). Calcium ion was an absolute requirement for activity as reported by Erdos *et al.* (1960). The widely reported loss of paraoxonase activity could be overcome by combining the Ca^{2+} requirement and the presence of 0.02% sodium azide. A number of disubstituted 4-nitrophenyl phosphinates as well as methyl and ethyl paraoxon were found to be substrates for the partially purified rabbit serum paraoxonase. The enzyme also hydrolyzed acetate esters faster than butyrate esters (Grothusen *et al.*, 1986). Inhibition of phosphinate hydrolysis by PCMB is further evidence that phosphinates were hydrolyzed by rabbit serum paraoxonase, an A-esterase as classified by Aldridge (1953a).

Becker and Barbaro (1964) studied the hydrolysis of *p*-nitrophenyl ethylphosphate by rabbit plasma and found that the pH optimum was more alkaline than that reported for paraoxonase, and that the enzyme was less readily inhibited by ethylenediaminetetraacetic acid (EDTA) and Ba^{2+}. Using human serum as the enzyme source, Skrinjaric-Spoljar and Reiner (1968) concluded that paraoxon and its phosphonate analog were hydrolyzed by different enzymes. Studies on the enzyme that hydrolyzed paraoxon in rabbit serum indicated that the reaction was subject to neither substrate nor product inhibition (Lenz *et al.*, 1973). Diethyl *p*-aminophenyl phosphate and *p*-aminophenyl pinacolyl methylphosphonate were not substrates for the enzyme but were competitive inhibitors. It was concluded that paraoxonase was specific for phosphate esters and was not able to hydrolyze phosphonate esters.

Other phosphate insecticides that have been reported to be enzymatically hydrolyzed include diazoxon by rat and cockroach homogenates (Shishido and Fukami, 1972), fonofosoxon enantiomers by rat and mouse sera

(Lee *et al.*, 1978), and O,O-dimethyl-2,2-dichlorovinyl phosphate (DDVP) by mammalian plasma and liver preparations (Reiner *et al.*, 1980).

In a study by Costa and co-workers (1990), paraoxonase was purified from rabbit serum and injected into tail veins of rats. Thirty minutes later the rats were challenged with acute doses of either paraoxon or chlorpyrifos-oxon by various routes. Cholinesterase activities were measured in plasma, red blood cells, brain, and diaphragm. Rats pretreated with paraoxonase exhibited less inhibition of cholinesterase than did controls following identical doses of paraoxon or chlorpyrifos-oxon. These findings indicate that levels of serum paraoxonase can affect the *in vivo* toxicity of paraoxon and chlorpyrifos-oxon. This is similar to the findings of Main (1956) in which the acute toxicity of paraoxon in rats was reduced by injection of a partially purified preparation of rabbit serum A-esterase.

In 1959, van Asperen and Oppenoorth suggested that a relationship existed between hydrolase activity and OP insecticide resistance, which was associated with low levels of aliesterase activity found in several resistant housefly strains. Their findings led to the so-called *mutant aliesterase theory* which suggested that an increase in phosphatase activity in OP-resistant insects was due to a mutant form of the aliesterase normally found in the susceptible strain and was responsible for the degradation of the insecticide and the development of resistance (Oppenoorth and van Asperen, 1960, 1961). The evidence for this theory was based on the measurement of the disappearance of oxygen analogs as the result of a decrease in anticholinesterase activity using Warburg manometry in which a carbon dioxide–bicarbonate buffer is normally used. Attempts at the direct measurement of the metabolites formed, using radiolabeled oxygen analogs and insect preparations, were, however, inconclusive. Attempts at purification of the mutant aliesterase generally resulted in rapid loss of activity after one or two manipulations. Almost all of these studies utilized phosphate buffers and, as described below, would have decreased or inhibited phosphorotriester hydrolase activity and thus complicated the interpretation of data concerning the amount of activity and the role in resistance.

Subsequently, it was reported by McIlvain *et al.* (1984) and Zimmerman and Brown (1986) that the phosphate ion used as a buffer in the preparation and assay inhibited the mammalian arylester hydrolase activity. With this in mind, the question of the importance of insect phosphorotriester hydrolases resurfaced, since so much of the previous work had been conducted utilizing phosphate buffers in their preparation and assay. (Table I).

A resistant strain (NC-86) of the tobacco budworm, *Heliothis virescens F.* is 56-fold resistant to methyl parathion but is susceptible to carbamates and pyrethroids (Table II). This strain is also 33-fold resistant to methyl paraoxon but only twofold resistant to ethyl paraoxon (Konno *et al.*, 1989).

Biochemical studies showed that resistance was primarily the result of

TABLE I

Studies Utilizing Phosphate Buffers in the Preparation and/or Assay of Phosphorotriester Hydrolase Activity

Kojima and O'Brien (1968)	0.067*M* Phosphate buffer
Lauwerys and Murphy (1969)	0.4*M* Phosphate buffer
Welling *et al.* (1971)	0.04*M* Phosphate buffer
Miyata and Matsumura (1971)	0.02*M* Phosphate buffer
Whitehouse and Ecobichon (1975)	0.067*M* Phosphate buffer
Oppenoorth and Voerman (1975)	—[a] Phosphate buffer
Oppenoorth *et al.* (1977)	0.1*M* Phosphate buffer
Brealey *et al.* (1980)	0.1*M* Phosphate buffer
Kao *et al.* (1985)	0.1*M* Phosphate buffer

[a]Molarity not specified.

two biochemical mechanisms. The first is a decrease in cytochrome P450-dependent monooxygenase activity, which is responsible for oxidative desulfuration of methyl parathion to methyl paraoxon (Table III). The P450 content was similar, but the amount of metabolic conversion of methyl parathion to methyl paraoxon, the actual cholinesterase inhibitor, was much greater in the susceptible strain. This may be explained by P450 isozyme(s) being present in the resistant strain that did not readily metabolize methyl parathion to methyl paraoxon.

The second mechanism was an increase in phosphorotriester hydrolase activity responsible for the hydrolysis of methyl paraoxon *in vivo* (Table IV). The largest difference between the two strains occurred 1 hr after treatment and resulted in a significantly higher amount of degradation of methyl paraoxon.

The phosphorotriester hydrolase was found in 100,000 × g supernatant of whole body homogenates (Fig. 1). When soluble hydrolases were

TABLE II

Toxicities of Insecticides to Fifth Instar Larvae of *Heliothis virescens*[a]

Insecticide	LD_{50} value (μg body wt) Control	NC-86	Resistance factor NC-86/control
Methyl parathion	11.30	630.0	55.8
Methyl paraoxon	5.47	180.0	32.9
Ethyl paraoxon	9.81	8.0	1.8
Methomyl	1.51	1.47	1.0
Fenvalerate	0.44	0.58	1.3

[a]Konno *et al.* (1989).

TABLE III

Monooxygenase Activity in Fifth Instar Larvae of *Heliothis virescens*[a]

Strain	Product formed[b]		Total metabolites	P450 content (nmol/gm protein)
	Methyl paraoxon	*p*-Nitrophenol		
Control	1.69 ± 0.29	0.89 ± 0.22	2.58 ± 0.50	0.23 ± 0.02
NC-86	0.66 ± 0.05	0.50 ± 0.02	1.16 ± 0.03	0.21 ± 0.01

[a]Methyl parathion was the substrate in the presence of NADPH. From Konno *et al.* (1989).
[b]Nmol/hr/mg protein ± S.D.

assayed against methyl paraoxon, the amount of *p*-nitrophenol formed increased with time, showing a two-phase reaction in both strains. In the first 5 min, *p*-nitrophenol formation occurred rapidly, after which it gradually increased. The activity was much greater in the NC-86 strain. When the microsomal fraction was assayed, the formation of *p*-nitrophenol also occurred within the first 5 min in both strains, but no additional hydrolysis of methyl paraoxon occurred. The initial reaction is probably owing to phosphorylation of nonessential esterases, proteins, etc., which results in the release of *p*-nitrophenol.

A study of the distribution of phosphorotriester hydrolase activity in 5th instar larvae showed that the majority of the activity was associated with the cuticle and muscle, (Table V) while hemolymph, the silk gland, and ovaries had no detectible activity.

The enzyme was partially purified from the 100,000 × g supernatant of

TABLE IV

In Vivo Metabolism of Methyl Paraoxon in Fifth Instar Larvae of *Heliothis virescens*[a]

Minutes	Strain	Percentange of dose recovered			Unknown
		Methyl paraoxon	*p*NP[b]	*p*NP conj.	
5	Control	74.7	7.8	15.6	1.9
	NC-86	74.3	8.6	15.4	1.6
20	Control	61.1	8.0	29.4	1.6
	NC-86	58.4	6.4	32.7	2.6
60	Control	48.6	11.0	38.5	2.0
	NC-86	29.7	10.6	59.3	3.2
180	Control	25.3	10.6	62.1	2.0
	NC-86	12.3	10.7	71.5	5.4

[a]From Konno *et al.* (1989).
[b]*p*-Nitrophenol.

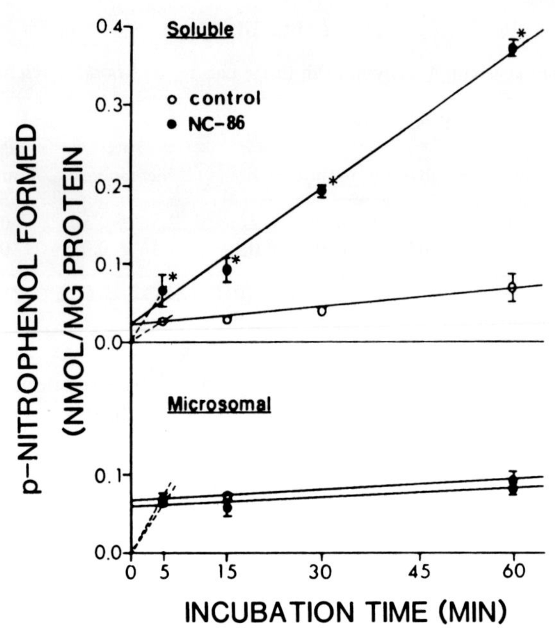

Figure 1 Soluble and microsomal hydrolase activity of *H. virescens*. Mean ± standard deviation. Asterisk indicates significant differences from control ($P < 0.05$). From Konno *et al.* (1990).

whole body homogenates by ammonium sulfate precipitation, ion-exchange chromatography, and gel filtration, resulting in a specific activity of 0.480 μmol methyl paraoxon hyd/min per mg protein with a 267-fold purification (Konno *et al.*, 1990). The optimal pH was between 8 and 9, and the molecular weight was estimated to be 120,000 by gel filtration.

Studies were conducted on the effect of metal ions on phosphorotriester

TABLE V

Tissue Distribution of Phosphorotriester Hydrolase in Fifth Instar Larvae of *Heliothis virescens*[a]

	Total activity	
Tissue	(nmol/hr/larva)	(%)
Cuticle and muscle	7.26	50.5
Fat body	2.68	18.6
Intestine	2.50	17.4
Malpighian tubules	1.24	8.6
Head	0.72	5.0
Hemolymph	0.00	0.0
Silk gland and ovaries	0.00	0.0

[a]From Konno *et al.* (1990).

TABLE VI

Effect of Metal Ions and Chelating Agent on Methyl Paraoxon Hydrolase Activity in *Heliothis virescens*[a]

Ion[b]	Specific activity (μmol/min/mg protein ± SD)	Percentage[c]
Control	0.788 ± 0.025	100
Co^{2+}	1.373 ± 0.044	176
Mn^{2+}	0.940 ± 0.040	121
Mg^{2+}	0.818 ± 0.018	105
Ca^{2+}	0.831 ± 0.024	107
Fe^{2+}	0.792 ± 0.050	102
Hg^{2+}	0.100 ± 0.003	13
Ag^{2+}	0.156 ± 0.008	20
EDTA	0.813 ± 0.012	104
Phosphate	0.660 ± 0.002	85

[a]From Konno *et al.* (1990).
[b]$1 \times 10^{-3} M$ concentration.
[c]Percentage activity compared to the control.

hydrolase activity in *Heliothis virescens* (Table VI). Cobaltous and manganous ions increased hydrolase activity 76 and 21% respectively, calcium and magnesium ions had little or no effect, and phosphate ions inhibited the reaction.

The effect of various concentrations of phosphate ions on enzyme activity is presented in Table VII. An increase in the concentration of phosphate ions decreased the enzyme activity, and all the evidence would indicate that this effect is not reversible. Therefore, many of the early studies conducted on phosphorotriester hydrolases probably had a portion of the total hydrolase activity inhibited by phosphate ions. Also in some of these studies, no Ca^{2+} was added for the mammalian enzyme or no Co^{2+} for the insect enzyme.

The substrate specificity of the insect phosphorotriester hydrolase was

TABLE VII

Effect of Phosphate Ions on Phosphorotriester Hydrolase Activity

Phosphate[a] (m*M*)	% inhibition
0	0
1	15
5	28
10	35
50	50
100	61

[a]Phosphate ions added to the enzyme solution 30 min before assay.

investigated (Table VIII). A series of O,O-dialkyl O-substituted nitrophenyl phosphates was prepared and assayed. Methyl paraoxon was the best substrate, followed by diethyl paraoxon. No hydrolysis of the di-*n*-propyl analog (compound 5) was detected. It appears that the enzyme is able to recognize the length of the alkoxy group, although the hydrolysis is at the *p*-nitrophenyl group.

The position of the nitro group on the phenyl ring appears to be critical since only substrates with the nitro group in the para-position were reactive except for compound 15. Compound 15 had the shortest half-life of all the phosphate analogs evaluated. In the methylnitrophenyl series (compounds 12–16) only compound 15 and fenitrooxon were hydrolyzed by the enzyme. No hydrolysis occurred with EPN-oxon (compound 6), which has a phos-

TABLE VIII

Substrate Specificity of the Phosphorotriester Hydrolase from *Heliothis virescens*

	K_m (m*M*)	V_{max} (μmol/min per mg protein)	Half-life (0.695 K_m/ V_{max}
O,O-dimethyl O-(2-nitrophenyl) phosphate	ND[a]	ND	
O,O-dimethyl O-(3-nitrophenyl) phosphate	ND	ND	
O,O-dimethyl O-(4-nitrophenyl) phosphate (methyl paraoxon)	0.683	0.242	1.96
O,O-diethyl O-(4-nitrophenyl) phosphate (ethyl paraoxon)	0.738	0.010	51.3
O,O-di-*n*-propyl O-(4-nitrophenyl) phosphate	ND	ND	
O-ethyl, O-(4-nitrophenyl) phenylphosphonate (EPN-oxon)	ND	ND	
4-nitrophenyl acetate	2.19	2.36	0.645
2-nitrophenyl butyrate	trace of activity		
4-nitrophenyl butyrate	0.073	2.29	0.022
2-nitrophenyl *N,N*-dimethyl carbamate	ND	ND	
3-nitrophenyl *N,N*-dimethyl carbamate	ND	ND	
O,O-dimethyl O-(3-methyl-2-nitrophenyl) phosphate	ND	ND	
O,O-dimethyl O-(4-methyl-2-nitrophenyl) phosphate	ND	ND	
O,O-dimethyl O-(5-methyl-2-nitrophenyl) phosphate	ND	ND	
O,O-dimethyl O-(2-methyl-3-nitrophenyl) phosphate	0.145	1.199	0.084
O,O-dimethyl O-(3-methyl-4-nitrophenyl) phosphate (fenitroxon)	0.661	0.168	2.73
O,O-diethyl O-(2-chloro-6-nitrophenyl) phosphate	ND	ND	

[a]Enzymatic hydrolysis was below detection.

phonate structure. This is in contrast to the findings of Galebskaya and Scherbak (1975) who reported that O-hepthyl-O (*o*-nitrophenyl) methylphosphonate was hydrolyzed by a paraoxonase present in rat blood, liver, kidney, and brain.

Compound 9, *p*-nitrophenyl butyrate, was the most reactive of all the substrates evaluated. Its enzymatic half-life was 29 times greater than that of *p*-nitrophenyl acetate. This is in contrast to the definition of A-esterases by Aldridge (1953a), in which A-esterases hydrolyze acetate esters faster than butyrate esters. It seems that insect phosphorotriester hydrolases have the substrate specificity of B-esterases and the cofactor requirements of phosphohydrolases.

IV. Conclusions

It appears that there are two types of phosphohydrolases; the kidney DFPase and the squid DFPase. The mammalian enzyme appears to require Co^{2+}/Mn^{2+} as a cofactor, while this is not required for the squid DFPase. Similarly there are two types or classes of phosphorotriester hydrolases. The enzyme from mammalian sources appears to require Ca^{2+} as a cofactor, while the phosphorotriester hydrolase from insects requires Co^{2+}. It is not clear whether this is associated with evolution in the animal kingdom or with the availability of Ca^{2+} in vertebrates.

With the insect studies, one can demonstrate the presence of phosphorotriester hydrolases only in certain insect strains that are resistant to OP insecticides. The properties of the enzyme are similar to those of the mammalian enzyme except for subcellular distribution and the cofactor requirements. The substrate specificity of the insect enzyme appears to be more narrow than that of the mammalian enzyme.

Acknowledgments

Work supported in part by PHS Grant ES-00044 from the National Institute of Environmental Health Services, U.S. Public Health Sciences. Y. Kasai of Kao Corp. and T. Konno of Nihon Nohyaku, Co. Ltd., are grateful to their respective companies for financial support during this study.

References

Ahmad, S., and Forgash, A. J. (1976). Nonoxidative enzymes in the metabolism of insecticides. *Drug. Metab. Rev.* 5, 141–164.

Aldridge, W. N. (1953a). Serum esterases. I. Two types of esterases (A and B) hydrolyzing *p*-nitrophenyl acetate, propionate and butyrate, and a method for their determination. *Biochem. J.* 53, 110–117.

Aldridge, W. N. (1953b). Serum esterases. 2. An enzyme hydrolysing diethyl *p*-nitrophenyl phosphate (E600) and its identity with the A-esterase of mammalian sera. *Biochem. J.* 53, 117–124.

Aldridge, W. N., and Reiner, E. (1972). "Enzyme Inhibitors as Substrates." North Holland, Amsterdam.

Anderson, R. S., Durst, H. D., and Landis, W. G. (1988). Organofluorophosphate-hydrolyzing activity in an estuarine clam, *Rangia cuneata. Comp. Biochem. Physiol.* **91C**, 575–578.

Becker, E. L., and Barbaro, J. F. (1964). The enzymatic hydrolysis of *p*-nitrophenyl ethyl phosphonates by mammalian plasma. *Biochem. Pharmacol.* **13**, 1219–1227.

Brealey, C. J., Walker, C. H., and Baldwin, B. C. (1980). A-esterase activities in relation to differential toxicity of pirimiphos-methyl to birds and mammals. *Pestic. Sci.* **11**, 546–554.

Chemnitius, J. M., Losch, H., Losch, K., and Zech, R. (1983). Organophosphate-detoxicating hydrolases in different vertebrate species. *Comp. Biochem. Physiol.* **76C**, 85–93.

Chettur, G., DeFrank, J. J., Gallo, B.J., Hoskin, F. C. G., Mainer, S., Robbins, F. M., Steinmann, K. E., and Walker, J. E. (1988). Soman-hydrolyzing and -detoxifying properties of an enzyme from a thermophilic bacterium. *Fundam. Appl. Toxicol.* **11**, 373–380.

Costa, L. G., McDonald, B. E., Murphy, S. D., Omenn, G. S., Richter, R. J., Motulsky, A. G., and Furlong, C. E. (1990). Serum paraoxonase and its influence on paraoxon and chlorpyrifos-oxon toxicity in rats. *Toxicol. Appl. Pharmacol.* **103**, 66–76.

Dauterman, W. C. (1976). Extramicrosomal metabolism of insecticides. *In* "Insecticide Biochemistry and Physiology" (C. F. Wilkinson, ed.), pp. 147–176. Plenum Press, New York.

Dauterman, W. C. (1983a). The role of hydrolysis in insecticide metabolism and the toxicological significance of the metabolites. *J. Clin. Toxicol.* **19**, 623–635.

Dauterman, W. C. (1983b). Role of hydrolases and glutathione S-transferases in insecticide resistance. *In* "Pesticide Resistance to Pesticides: Challenges and Prospects" (G.P. Georghiou and T. Saito, eds.), pp. 229–248. Plenum Press, New York.

De Bisschop, H. C. J. V., de Meerleer, W. A. P., van Hecke, P. R. J., and Willems, J. L. (1987). Stereoselective hydrolysis of soman in human plasma and serum. *Biochem. Pharmacol.* **36**, 3579–2586.

Donniger, C., Hutson, D. H., and Pickering, B. (1972). The oxidative dealkylation of insecticidal phosphoric acid triesters by mammalian liver enzymes. *Biochem. J.* **125**, 701–707.

Erdos, E. G., Debay, C. R., and Westman, M. P. (1960). Arylesterase in blood: Effect of calcium and inhibitors. *Biochem. Pharmacol.* **5**, 173–186.

Eto, M. (1974). "Organophosphorus Pesticides: Organic and Biological Chemistry." CRC Press, Cleveland, Ohio.

Galebskaya, L. V., and Scherbak, I. G. (1975). Enzymatic transformation of O-heptyl-O-ortho nitrophenylmethyl phosphonate, a blocking agent of cholinesterases in rat tissue homogenates. *Ukr. Biokim. Zh.* **47**, 469–473.

Garden, J. M., Hause, S.K., Hoskin, F. C. G., and Roush, A. H. (1975). Comparison of DFP-hydrolyzing enzyme purified from head ganglion and hepatopancreas of squid (*Loligo pealei*) by means of isoelectric focusing. *Comp. Biochem. Physiol.* **52C**, 95–98.

Grothusen, J. R., Bryson, P. K., Zimmerman, J. K., and Brown, T. M. (1986). Hydrolysis of 4-nitrophenyl organophosphinates by arylester hydrolase from rabbit serum. *J. Agric. Food Chem.* **34**, 513–515.

Hoskin, F. C. G. (1971). Diisopropylphosphorofluoridate and tabun: Enzymatic hydrolysis and nerve function. *Science* **172**, 1243–1245.

Hoskin, F. C. G. (1985). Inhibition of a soman- and diisopropyl phosphorofluoridate (DFP)-detoxifying enzyme by mipafax. *Biochem. Pharmacol.* **34**, 2069–2072.

Hoskin, F. C. G., and Prusch, R. D. (1983). Characterization of a DFP-hydrolyzing enzyme in squid posterior salivary gland by use of soman, DFP, and manganous ion. *Comp. Biochem. Physiol.* **75C**, 17–20.

Hoskin, F. C. G., Kirkish, M. A., and Steinmann, K. E. (1984). Two enzymes for the detoxication of organophosphorus compounds—sources, similarities, and significance. *Fundam. Appl. Toxicol.* **4**, 165–172.

Kao, L. R., Motoyama, N., and Dauterman, W. C. (1985). The purification and characterization of esterases from insecticide-resistant and susceptible houseflies. *Pestic. Biochem. Physiol.* **23**, 228–239.

Kojima, K., and O'Brien, R. D. (1968). Paraoxon-hydrolyzing enzymes in rat liver. *J. Agric. Food Chem.* **16**, 575–584.

Konno, T., Hodgson, E., and Dauterman, W. C. (1989). Studies on methyl parathion resistance in *Heliothis virescens. Pestic. Biochem. Physiol.* **33**, 189–199.

Konno, T., Kasai, Y., Rose, R. L., Hodgson, E., and Dauterman, W. C. (1990). Purification and characterization of a phosphorotriester hydrolase from methyl parathion-resistant *Heliothis virescens. Pestic. Biochem. Physiol.* **36**, 1–13.

Landis, W. G., Haley, D. M., Haley, M. V., Johnson, D. W., Durst, H. D., and Savage, R. E., Jr. (1987). Discovery of multiple organofluorophosphate hydrolyzing activities in the protozoan, *Tetrahymena thermophila. J. Appl. Toxicol.* **7**, 35–41.

Lauwerys, R. R., and Murphy, S. D. (1969). Comparison of assay methods for studying *O,O*-diethyl *O*-*p*-nitrophenyl phosphate (Paraoxon) detoxication *in vitro. Biochem. Pharmacol.* **18**, 789–800.

Lee, P. W., Allahyari, R., and Fukuto, T. R. (1978). Studies on the chiral isomers of fonofos and fonofos oxon. II. *In vitro* metabolism. *Pestic. Biochem.* **8**, 158–169.

Lenz, D. E., Deguehery, L. E., and Holton, J. S. (1973). On the nature of the serum enzyme catalyzing paraoxon hydrolysis. *Biochim. Biophys. Acta* **321**, 189–196.

Mackness, M. I., and Walker, C. H. (1983). Partial purification and properties of sheep serum 'A'-esterases. *Biochem. Pharmacol.* **32**, 2291–2296.

Mackness, M. I., and Walker, C. H. (1988). Multiple forms of sheep serum A-esterase activity associated with the high-density lipoprotein. *Biochem. J.* **250**, 539–545.

Mackness, M. I., Hallam, S. D., Peard, T., Warner, S., and Walker, C. H. (1985). The separation of sheep and human serum A-esterase activity into lipoprotein fraction by ultracentrifugation. *Comp. Biochem. Physiol.* **82B**, 675–677.

Main, A. R. (1956). The role of A-esterases in the acute toxicity of paraoxon, TEPP, and parathion. *Can. J. Biochem. Physiol.* **75**, 188–195.

Main, A. R. (1960). The purification of the enzyme hydrolyzing diethyl *p*-nitrophenyl phosphate (paraoxon) in sheep serum. *Biochem. J.* **74**, 10–20.

Mazur, A. (1946). An enzyme in animal tissue capable of hydrolyzing the phosphorus fluorine bond of acyl fluorophosphates. *J. Biol. Chem.* **164**, 271–289.

McIlvain, J. E., Timoszyk, J., and Nakatsugawa, T. (1984). Rat liver paraoxonase (Paraoxon arylesterase). *Pestic. Biochem. Physiol.* **21**, 162–169.

Miyata, N., and Matsumura, F. (1971). Partial purification of American cockroach enzymes degrading certain organophosphate insecticides. *Pestic. Biochem. Physiol.* **1**, 267–274.

Motoyama, N., and Dauterman, W. C. (1980). Glutathione S-transferases: Their role in the metabolism of organophosphorus insecticides. *Rev. Biochem. Toxicol.* **2**, 49–69.

Mounter, L. A., Dien, L. T., and Chanutin, A. (1955a). The distribution of dialkylfluorophosphatases in the tissue of various species. *J. Biol. Chem.* **215**, 691–697.

Mounter, L. A., Baxter, R. F., and Chanutin, A. (1955b). Dialkylfluorophosphatase of microorganisms. *J. Biol. Chem.* **215**, 699–704.

Oppenoorth, F. J., and van Asperen, K. (1960). Allelic genes in the housefly producing modified enzymes that cause organophosphate resistance. *Science* **132**, 298–299.

Oppenoorth, F. J., and van Asperen, K. (1961). The detoxication enzymes causing organophosphate resistance in the housefly; properties, inhibition and the action of inhibitors as synergists. *Entomol. Exp. Appl.* **4**, 311–333.

Oppenoorth, F. J., and Voerman, S. (1975). Hydrolysis of paraoxon and malaoxon in three strains of *Myzus persicae* with different degrees of parathion resistance. *Pestic. Biochem. Physiol.* **5**, 431–443.

Oppenoorth, F. J., Smissaert, H. R., Welling, W., van der Pas, L. J. T., and Hitman, K. J. (1977). Insensitive acetylcholinesterase, high glutathione S-transferase and hydrolytic activity as resistance factors in a tetrachlorvinphos-resistant strain of housefly. *Pestic. Biochem. Physiol.* **7**, 34–47.

Reiner, E., Simeon, V., and Skrinjaric-Spoljar, M. (1980). Hydrolysis of O,O-dimethyl-2,2-dichlorovinyl phosphate (DDVP) by esterases in parasitic helminths, and in vertebrate plasma and erythrocytes. *Comp. Biochem. Physiol.* **66C**, 149–152.

Shishido, T., and Fukami, J. (1972). Enzymatic hydrolysis of diazoxon by rat tissue homogenates. *Pestic. Biochem. Physiol.* **2**, 39–50.

Skrinjaric-Spoljar, M., and Reiner, E. (1968). Hydrolysis of diethyl-*p*-nitrophenyl phosphate and ethyl-*p*-nitrophenyl-ethyl phosphonate by human sera. *Biochim. Biophys. Acta* **165**, 289–292.

Storkebaum, W., and Witzel, H. (1975). Study on the enzyme-catalyzed splitting of triphosphates. *Forschungsban Landes Nordrhein-Westfalen* **2523**, 22.

van Asperen, K., and Oppenoorth, F. J. (1959). Organophosphate resistance and esterase activity in houseflies. *Entomol. Exp. Appl.* **2**, 48–57.

Walker, C. H., and Mackness, M. I. (1987). "A" esterases and their role in regulating the toxicity of organophosphates. *Arch. Toxicol.* **60**, 30–33.

Welling, W., Blaakmeer, P., Vinck, G. J., and Voerman, S. (1971). *In vitro* hydrolysis of paraoxon by parathion-resistant houseflies. *Pestic. Biochem. Physiol.* **1**, 61–70.

Whitehouse, L. W., and Ecobichon, D. J. (1975). Paraoxon formation and hydrolysis by mammalian liver. *Pestic. Biochem. Physiol.* **5**, 314–322.

Yang, R. S. H., Hodgson, E., and Dauterman, W. C. (1971). Metabolism *in vitro* of diazinon and diazoxon in rat liver. *J. Agric. Food Chem.* **19**, 10–13.

Zech, R., and Wigand, K. D. (1975). Organophosphate-detoxicating enzymes in *Escherichia coli.* Gel filtration and isoelectric focusing of DFPase (diisopropylfluorophosphatase), paraoxonase, and unspecific phosphohydrolases. *Experientia* **31**, 157–158.

Zech, R., and Zurcher, K. (1974). Organophosphate splitting serum enzymes in different mammals. *Comp. Biochem. Physiol.* **48B**, 427–433.

Zimmerman, J. K., and Brown, T. M. (1986). Partial purification of rabbit serum arylester hydrolase. *J. Agric. Food Chem.* **34**, 516–520.

9

Detoxication of Organophosphorus Compounds by Carboxylesterase

Donald M. Maxwell
U.S. Army Medical Research Institute of Chemical Defense
Aberdeen Proving Ground, Maryland

I. Introduction

Organophosphorus (OP) compounds can be detoxified by a variety of enzymatic reactions such as oxidation, reduction, hydrolysis, isomerization, dealkylation, dehalogenation, and conjugation (Eto, 1974; Mikhaylov and Scherbak, 1983; Matsumura, 1985). The importance of each of these detoxication reactions is dependent on the structure of the particular OP compound and the levels of each enzymatic activity in an organism (Miyamoto *et al.*, 1988). For highly toxic OP compounds such as sarin, soman, tabun, and

Organophosphates: Chemistry, Fate, and Effects

paraoxon, it has been found that carboxylesterase (CaE) is particularly important for detoxication (Lauwerys and Murphy, 1969; Boskovic, 1979). Several excellent reviews have described the properties of CaE and its role in detoxication of naturally occurring substrates and drugs that are carboxylesters (Krisch, 1971; Junge and Krisch, 1975; Heymann, 1980; Satoh, 1987). This chapter describes the role of CaE in the detoxication of OP compounds.

II. General Properties of Carboxylesterase

Carboxylesterase is a 60,000-dalton esterase found in many mammalian tissues—lung, liver, kidney, brain, intestine, muscle, gonads—usually as a microsomal enzyme (Satoh, 1987). Plasma CaE is found in a soluble state. CaE is a serine esterase that catalyzes the hydrolysis of carboxylesters by a two-step process in which the carboxylester acylates the active-site serine of CaE, which subsequently deacylates by the addition of water (Augustinsson 1958). CaE (EC 3.1.1.1) can be distinguished from acetylcholinesterase (AChE, EC 3.1.1.7) and butyrylcholinesterase (BuChE, EC 3.1.1.8) by the fact that AChE and BuChE react with positively charged carboxylesters, such as acetylcholine and butyrylcholine, and are readily inhibited by carbamates, while CaE does not react with positively charged substrates and is inhibited by carbamates only at high concentrations (Augustinsson, 1958). Aldridge (1953) demonstrated that all three types of esterases—AChE, BuChE and CaE—are irreversibly inhibited by OP compounds (i.e., phosphorylated). Inasmuch as the dephosphorylation of the active-site phosphorylated serine was a very slow process compared to its deacylation (Aldridge and Reiner, 1972), the phosphorylation of the active-site serine of these enzymes with radiolabeled OP compounds became a convenient method to identify their active sites (Oosterbaan and Cohen, 1964).

The amino acid sequences from peptic digest fragments of CaE containing the phosphorylated active-site serine are identical for liver CaE from pig, sheep, ox, horse, chicken, rat, and rabbit (Augusteyn *et al.*, 1969; Ozols, 1987; Long *et al.*, 1988). The entire amino acid sequence of liver CaE has been determined for rat (Long *et al.*, 1988) and rabbit (Korza and Ozols, 1988). The overall homology between the sequence for rat CaE (531 amino acids) and rabbit CaE (539 amino acids) is 68%. However, there are five regions, which constitute a total of 300 amino acids, that exhibit 87% homology. The four cysteines that form internal disulfide bridges are in the same locations in both rabbit and rat CaE. Thus, CaE enzymes from different sources are quite similar (Long *et al.*, 1988).

Although the spatial orientation of CaE in the endoplasmic reticulum is controversial, the most recent evidence suggests that CaE is located on the luminal side of the microsomal membrane (Mentlein *et al.*, 1988). Liver CaE contains no large hydrophobic regions (i.e., potential membrane-spanning domains) and can be solubilized from microsomes with concentrations of detergent too low to solubilize most microsomal proteins (Harano *et al.*, 1988), which suggests that CaE is not an integral membrane protein (Long *et al.*, 1988). CaE is synthesized by membrane-bound ribosomes, translocated into the lumen of the endoplasmic reticulum, and processed to mature size by cleavage of an extra peptide and glycosylation (Robbi and Beaufay, 1986; Harano *et al.*, 1988). However, CaE is slowly secreted ($t_{1/2}$ = 17.1 hr) into plasma (Boskovic *et al.*, 1984a) in contrast to most secretory proteins, which have $t_{1/2} < 1$ hr (Strous and Berger, 1982). This could result from a reduced ability to be transported from the endoplasmic reticulum to the Golgi apparatus (Harano *et al.*, 1988) or to prolonged retention in the Golgi (Strous and Berger, 1982), but the mechanism for the segregation of CaE from the normal traffic of secretory proteins is unknown (Robbi and Beaufay, 1986). Elucidating the mechanism of transport of slowly secreted proteins, such as CaE, has great significance for understanding CaE detoxication of OP compounds, inasmuch as plasma CaE is an important determinant of individual and species variation of OP toxicity (Section VII,A,B).

III. Problems of Nomenclature

One of the major problems in studying CaE is that the enzyme has broad substrate specificity and exists as several isoenzymes (Mentlein *et al.*, 1980). This creates difficulties in comparing observations obtained in different laboratories (Mentlein *et al.*, 1987). For example, the current IUB classification differentiates CaE (EC 3.1.1.1) from arylesterase (EC 3.1.1.2) on the basis that CaE hydrolyzes aliphatic esters while arylesterase hydrolyzes aromatic esters. However, this simple distinction is misleading, because CaE can hydrolyze aromatic esters, and arylesterase can hydrolyze some aliphatic esters as a consequence of the broad and overlapping specificity of both enzymes. Because of the difficulty in classifying these enzymes based on their substrate specificity for aromatic or aliphatic carboxylesters, it has been suggested that these enzymes be classified on the basis of their interactions with OP compounds (Walker, 1989).

CaE has been differentiated from arylesterase by its irreversible inhibition by OP compounds whereas arylesterase hydrolyzes OP compounds (Aldridge, 1953). However, this classification scheme also has difficulties, because some arylesterases hydrolyze only specific OP compounds, and another enzyme category already exists (diisopropylflurophosphatase, EC 3.8.2.1) for enzymes that hydrolyze organophosphofluoridates. These problems have led to the conclusion (Walker and Mackness, 1983; Heymann, 1989; Walker, 1989) that the classification of esterases needs significant revision. This problem is not just of heuristic interest, since at least one example exists in the literature of incorrect analysis of the mechanism of potentiation of OP toxicity because of confusion in enzyme nomenclature (Cohen, 1981).

IV. Potentiation of Organophosphorus Toxicity

The discovery that administration of O-ethyl O-*p*-nitrophenyl phenylphosphonothioate (EPN) potentiated the toxicity of malathion in animals (Frawley *et al.*, 1957) was the initial observation that led to the suggestion that CaE was important in detoxication of OP compounds. The potentiation of malathion toxicity by EPN was attributed to the inhibition of the enzymatic hydrolysis of the carboxylester bonds in the leaving group of malathion (Murphy and DuBois, 1957). Myers (1959) broadened these initial observations by demonstrating that the inhibition of CaE with tri(O-cresyl)phosphate (TOCP) potentiated the toxicity of sarin, an OP compound lacking carboxylester bonds. Fleisher *et al.* (1963) and Polak and Cohen (1969) confirmed Myers' observations with biochemical studies of the influence of EPN and TOCP, respectively, on the distribution of radiolabeled sarin. EPN and TOCP were found to reduce the amount of radiolabeled sarin bound in plasma and lung and increase the amount of sarin bound in brain, muscle, and kidney. Polak and Cohen concluded that the ^{32}P-sarin bound to plasma was bound to CaE because the *in vivo* recovery of CaE activity paralleled the disappearance of ^{32}P from plasma, and treatment with TOCP, which caused a nearly complete inactivation of plasma CaE, also produced a nearly complete reduction of the ability of plasma to bind ^{32}P-sarin.

The biochemical mechanism of the potentiation of the toxicity of paraoxon (Lauwreys and Murphy, 1969) and methylparaoxon (Benke and Murphy, 1974) by TOCP has also been examined. The potentiation of the toxicity of these OP compounds was found to be primarily the result of TOCP inhibition of tissue binding of OP compounds by esterases and not due to the inhibition of their enzymatic hydrolysis. The mechanism of TOCP inhibition of CaE was further elucidated by Eto *et al.* (1962) who discovered that the active metabolite of TOCP was 2-(O-cresyl)-4H-1:3:2-benzodioxaphosphorin oxide (CBDP). The discovery by Casida *et al.* (1963) that CBDP potentiated the toxicity of malathion stimulated investigations of the effect of CBDP on the toxicity of other OP compounds. In these studies CBDP was found to potentiate the toxicity of soman, sarin, and tabun, but not O-ethyl *S*[2-diisopropylamino)ethyl]methyl phosphonothionate (VX) (McKay *et al.*, 1971; Boskovic, 1979).

Using an OP compound such as CBDP as a pharmacological tool to demonstrate the importance of CaE as a detoxication route for other OP compounds requires careful examination of the specificity of CBDP for inhibition of CaE in comparison to AChE, whose inhibition is directly involved in the toxicity of OP compounds. A dose of CBDP must be used that inhibits CaE without inhibiting AChE, thus avoiding the difficulty of analyzing experiments in which two possible mechanisms, inhibition of AChE and inhibition of CaE, may contribute to the potentiation of OP toxicity. It has been found that 1–2 mg/kg doses of CBDP inhibit CaE without inhibiting AChE (Clement, 1984; Maxwell *et al.*, 1987a; Jimmerson *et al.*, 1989a,b). The potentiation of toxicity by OP compounds in CaE-inhibited animals pretreated with these doses of CBDP accurately reflects the influence of CaE detoxication on toxicity of OP compounds whereas other studies that used 35–50 mg/kg doses of CBDP (McKay *et al.*, 1971; Boskovic, 1979; Clement, 1984) inhibited AChE as well as CaE, and overestimated the effect of CaE detoxication.

Another complicating factor is that doses of CBDP that inhibited CaE without inhibiting AChE did not inhibit CaE in all tissues. While plasma and lung CaE were inhibited >95% by doses of 2 mg/kg CBDP, liver and kidney CaE were much less inhibited (Maxwell *et al.*, 1987a). Increasing the dose of CBDP 10-fold increased the degree of CaE inhibition in kidney and liver to nearly 100%, but this resulted in only a minor increase in toxicity by OP compounds (Clement, 1984; Maxwell *et al.*, 1987a), suggesting that plasma and lung were the important sites of detoxication of OP compounds by CaE. This observation was reflected also in experiments in which animals receiving other highly toxic OP compounds (Section V,C) instead of CBDP had significant inhibition of CaE in plasma and lung and no inhibition of CaE in other tissues.

V. Specificity of Carboxylesterase

A. Structural Specificity

A variety of OP compounds have been found to inhibit CaE at concentrations of 1 to 100 n*M* (Chow and Ecobichon, 1973; Ecobichon and Comeau, 1973; Chambers *et al.*, 1990). Kinetic measurements of the reaction of OP compounds and CaE (Ooms and Breebart-Hansen, 1965) have shown that CaE reacts with a surprising lack of structural specificity in contrast to the high degree of specificity of AChE and BuChE with OP compounds (De Jong and Benschop, 1988). The reactivity (k_i) of nearly all OP compounds for CaE has exceeded 10^5 M^{-1} min^{-1} except for those compounds with positively charged leaving groups, which are much less reactive. Recent investigations of the kinetics of inhibition of reindeer liver CaE by OP compounds have revealed $k_i > 10^7$ M^{-1} min^{-1} (Brestkin *et al.*, 1986), which has suggested that the CaE preparation used by Ooms and Breebart-Hansen was altered by their strenuous purification procedures. The milder purification procedures of subsequent investigators limited these enzyme alterations and increased the observed k_i for CaE inhibition by OP compounds (Clement, *et al.*, 1987; Maxwell, 1989).

B. Stereospecificity

The stereospecificity of CaE has been investigated with four OP compounds. For all four compounds the P(−) stereoisomers are more reactive than the P(+) stereoisomers, which is consistent with the pattern of inhibition of OP stereoisomers with AChE and BuChE (De Jong and Benschop, 1988). For O,O-diethyl malaoxon (Hassan and Dauterman, 1968), fonofos oxon (Lee *et al.*, 1978), and ethylphenylnitrophenylphosphinate (Brown *et al.*, 1986) the P(−) stereoisomers were 8.3-fold, 6.2-fold, and 3.6-fold, respectively, more reactive than the P(+) stereoisomers. Soman contains a chiral carbon as well as a chiral phosphorus. The P(−) stereoisomers of soman are 67–71 times more reactive than the corresponding P(+) stereoisomers, while the C(−) stereoisomers are approximately twice as reactive as the C(+) stereoisomers (Clement *et al.*, 1987). Inasmuch as the P(−) stereoisomers are the toxic stereoisomers for OP compounds, CaE provides an effective stereospecific detoxication of OP compounds in contrast to some other detoxication processes, such as OP hydrolases, which hydrolyze the nontoxic P(+) stereoisomers preferentially (Maxwell *et al.*, 1988b).

C. Tissue Specificity

In vivo CaE inhibition in tissues has been measured following treatment with a variety of OP compounds such as CBDP (Clement, 1984; Maxwell *et al.*, 1987a), *iso*-OMPA (Clement, 1984; Gupta and Dettbarn, 1987), soman (Maxwell *et al.*, 1988a), sarin (Boskovic *et al.*, 1984a), paraoxon (Chambers and Chambers, 1990), and tabun (Boskovic *et al.*, 1984a; Gupta *et al.*, 1987). Although there is some variation depending on the compound, route of administration, and dose, the pattern of CaE inhibition in tissues is fairly consistent. The tissues most sensitive to CaE inhibition are plasma and lung, while other tissues (kidney, liver, brain, muscle) exhibit significant inhibition only with the less-toxic OP compounds such as *iso*-OMPA or CBDP, which can be administered at high doses. There is some evidence that distinct types of CaE are found in different tissues (Sterri *et al.*, 1985b; Sterri and Fonnum, 1987). Susceptibility to CaE inhibition, however, appears to be primarily the result of the kinetics of distribution and detoxication of the OP compound. The tissue CaE that first encounters the OP compound is the CaE that is preferentially inhibited. This phenomenon has been observed also with the reaction of other tissue esterases with OP compounds (Maxwell *et al.*, 1987b; Maxwell *et al.*, 1988b).

VI. Recovery of Organophsophorus-Inhibited Carboxylesterase Activity

A. Single Dosing with Organophosphorus Compounds

The *in vivo* recovery of CaE activity, particularly in plasma, has considerable pharmacological significance. Plasma or brain CaE activity recovers to normal levels within 24 to 48 hr after inhibition by a single dose of sarin, soman, or tabun (Boskovic *et al.*, 1984a; Clement, 1989). In survivors of soman poisoning, Clement (1989) found that the return of the soman median lethal dose (LD_{50}) to control values occurred in the same time frame as the recovery of serum CaE activity. This correlation between the recovery of enzyme activity for CaE and soman LD_{50} values contrasted with the lack of correlation observed with recovery of AChE activity, which remained extensively inhibited in brain, diaphragm, and erythrocytes. This suggests that CaE activity is a better criterion of recovery from OP toxicity than is AChE activity, even though the mechanism of toxicity of OP compounds is thought to be mediated by the inhibition of AChE.

The recovery of CaE activity is also a major determinant of the ability of mammals to tolerate repetitive dosing with OP compounds. Sterri (1981) calculated the detoxication rate of soman administered sc to rats to be 0.041 mg/kg/min from data generated by repetitive dosing with 75 μg/kg at intervals ranging from 5 min to 24 hr (Sterri *et al.*, 1980). From the half-life ($t_{1/2}$) for recovery of plasma CaE after a single dose of soman (17.1 hr; 1026 min) and the dose of soman (52 μg/kg) necessary to inhibit virtually all of the plasma CaE in rats (Boskovic *et al.*, 1984a), the detoxication rate due to the recovery of plasma CaE can be predicted to be 0.035 μg/kg/min (52 μg/kg × 0.000673 min^{-1}) where $t_{1/2}$ has been converted to a first-order rate constant by the well-known relationship $k = 0.693/t_{1/2}$. Thus, the detoxication rate predicted by plasma recovery after a single dose of soman (0.035 μg/kg/min) agrees closely with the detoxication rate calculated from repetitive dosing (0.041 μg/kg/min).

In contrast to the irreversible reaction of CaE with OP compounds observed by other investigators, Clement (1982) proposed that CaE was inhibited by soman via a reversible enzyme inhibitor complex that could recover activity by release of free soman. No confirmation of this phenomenon by other laboratories has been reported, although De Jong and Van Dijk (1984) demonstrated that high fluoride concentrations could cause the formation of free soman from soman-inhibited CaE by a nucleophilic attack on the phosphophorylated active-site serine. However, this fluoride effect occurred at fluoride levels that exceed those achieved *in vivo* after soman.

B. Repetitive Dosing with Organophosphorus Compounds

Tolerance has been observed after repetitive administration with many OP compounds. To analyze the tolerance resulting from repetitive dosing with DFP, Gutpa *et al.* (1985) examined a variety of biochemical mechanisms for adaptation, including reduced levels of cholinergic receptors, reduced choline uptake, increased AChE synthesis, availability of other serine esterases and OP-hydrolyzing enzymes. Their results indicate that decreased numbers of cholinergic receptors and recovery of CaE are the major factors in the development of tolerance to OP compounds. The importance of CaE was established by the administration of CaE inhibitors (i.e., *iso*-OMPA or mipafox), which completely abolished tolerance development to DFP. Their conclusions with DFP are in close agreement with the observations previously mentioned (Section VI,A) that the tolerance to repetitive dosing with soman is also correlated with the recovery of CaE.

Studies of cross-tolerance between different OP compounds and between OP compounds and carbamates also implicate CaE as a major biochemical mechanism in the development of tolerance to OP compounds. Animals that are tolerant to one OP compound are cross-tolerant to other OP compounds. For example, mice tolerant to disulfoton were cross-tolerant to chlorpyrifos (Costa and Murphy, 1983) and OMPA (McPhillips, 1969). However, animals that are tolerant to OP compounds are not cross-tolerant to carbamates. Mice tolerant to disulfoton were not cross-tolerant to propoxur (Costa and Murphy, 1983); rats tolerant to OMPA were not cross-tolerant to physostigmine (Hagan *et al.*, 1971); and rats tolerant to DFP were not cross-tolerant to physostigmine and neostigmine (Russell *et al.*, 1975). Since carbamates are much less reactive with CaE than are OP compounds, the absence of tolerance for carbamates in OP-tolerant animals is consistent with a tolerance mechanism for OP compounds that is strongly dependent on CaE. In contrast, the development of tolerance for carbamates may be predominantly dependent on cholinergic receptor down-regulation or changes in other detoxication processes.

VII. Variation in Organophosphorus Toxicity

A. Individual Variation

The toxicity of some OP compounds is known to vary as animals develop or age (Freedman and Himwich, 1948; Brodeur and Dubois, 1963; Sterri *et al.*,

1985a; Maxwell *et al.*, 1988b; Shih *et al.*, 1990). The developmental variation in the toxicity of malathion and soman have been the most extensively studied for OP compounds. LD_{50} values for malathion correlate closely with the activity of liver CaE, which increases during the first 30 days of life in rats (Brodeur and DuBois, 1967). The increase in the LD_{50} of soman also parallels the development of CaE in liver and plasma as rats develop up to 30 days (Sterri *et al.*, 1985a) or age from 60 to 120 days (Maxwell *et al.*, 1988b). In contrast, CaE activity in lung and cholinesterase activity in lung and plasma remain unchanged during development in these animals (Sterri *et al.*, 1985a).

Changes in the cholinergic system as animals develop may influence the toxicity of OP compounds. The number of cholinergic receptors (Strong *et al.*, 1980) and the levels of acetylcholine (Ladinsky *et al.*, 1972) and cholinesterase (Kaur and Kanugo, 1970) have been reported to change with age. However, the magnitude of the changes in the cholinergic system, particularly changes in cholinesterase, do not exhibit a close relationship with OP toxicity (Shih *et al.*, 1990). Therefore, the close correlation of OP toxicity with plasma and liver CaE suggests that age-related variation in CaE is the most important parameter determining the changes in OP toxicity in animals as they age. Inasmuch as liver CaE does not exhibit inhibition after *in vivo* administration of highly toxic OP compounds while plasma CaE is usually completely inhibited (Sterri *et al.*, 1985a), plasma CaE is probably more important than liver CaE as a determinant of OP toxicity. The close correlation of liver and plasma CaE to OP toxicity may also suggest that liver is a source of plasma CaE.

B. Species Variation

If the toxicity of an OP compound is measured in a variety of species, a wide range of susceptibility is observed. Although variation in OP toxicity with species in different classes of animals (i.e., mammals, fish, amphibians, birds) may be attributable to differences in the reactivity of an OP compound with AChE (Wang and Murphy 1982a,b; Kemp and Wallace, 1990), the differences in OP toxicity observed within a class, such as mammals, have been generally attributed to differences in detoxication of OP compounds. For example, the LD_{50} of soman varies eight-fold among mice, rats, guinea pigs, rabbits, dogs, and rhesus monkeys (Maxwell *et al.*, 1987a). However, elimination of the species differences in the detoxication of soman by CaE by pretreatment of animals with the CaE inhibitor, CBDP, resulted in soman LD_{50} values in the CaE-inhibited species that were not significantly different (Maxwell *et al.*, 1987a).

Although individual variation in OP toxicity has a linear correlation to plasma CaE in developing animals (Sterri *et al.*, 1985a) or aging animals (Maxwell *et al.*, 1988b), species variation in OP toxicity has a more complex correlation to plasma CaE. Plasma CaE has a smaller influence on OP toxicity in small animals (i.e., mice) than in large animals (i.e., rabbits). This effect appears to result from the fact that CaE detoxication in plasma is a bimolecular reaction between an OP compound and plasma CaE whose effectiveness is dependent on the time available for it to react with the OP compound. This reaction time is dependent on the circulation time, which varies with animal size. If the plasma concentration of CaE for each species is multiplied by the circulation time of that species, the product (plasma CaE $\times$ t_{circ}) has a linear correlation to OP toxicity (Maxwell *et al.*, 1990), suggesting that circulation time has a major influence on the effectiveness of OP detoxication by plasma CaE.

C. Variation in Organophosphorus Compounds

The importance of CaE as a detoxication process for a variety of OP compounds has been demonstrated by the potentiation of OP toxicity in CaE-inhibited animals. The toxicities of paraoxon (Lauwerys and Murphy, 1969), methyl paraoxon (Benke and Murphy, 1974), soman, sarin, and tabun (Boskovic, 1979) were all potentiated in CaE-inhibited animals. The toxicity of VX was not increased by CaE inhibition in animals (Boskovic, 1979). The absence of potentiation of VX toxicity in CaE-inhibited animals is easily explained by the poor reactivity of VX for CaE (Maxwell, 1989), because VX is partially protonated at physiological pH, and CaE has poor reactivity for cationic compounds. However, the wide variation in the degree of *in vivo* potentiation of OP compounds does not correlate with the *in vitro* reactivities of OP compounds with CaE, inasmuch as the reactivities of many neutral or anionic OP compounds with CaE are quite similar. For example, the potentiation of paraoxon toxicity in CaE-inhibited rats is twofold (Lauwerys and Murphy, 1969), while the potentiation of soman is sixfold (Maxwell *et al.*, 1987a), but the reactivities of paraoxon and soman for CaE are nearly the same.

The explanation for this variation in OP potentiation for OP compounds with similar reactivities for CaE is found in the relationship of the reactivity of an OP compound with its pharmacological target (AChE) and its detoxication enzyme (CaE). While reactivities of OP compounds for CaE are quite similar, their reactivities for AChE can vary tremendously (Maxwell, 1989), and OP compounds with high reactivities for AChE are more toxic than are compounds with low AChE reactivities (Heath, 1961). If the effect of CaE detoxication is expressed as the difference in the LD_{50} values of OP compounds in control animals and animals whose endogenous CaE has been inhibited, the maximal effect of CaE detoxication is to increase the LD_{50} by about 1 μmol/kg (Maxwell, 1989). This effect is important for OP compounds with LD_{50} values < 2 μmol/kg such as soman, sarin, tabun, and paraoxon. On the other hand, this effect represents < 10% of the LD_{50} values of OP compounds such as DFP or dichlorvos, which have LD_{50} of 9.75 and 98.4 μmol/kg, respectively. Thus, CaE detoxication is important only for highly toxic OP compounds.

Within the group of highly toxic OP compounds in which CaE detoxication is important, the variation in the effect of CaE potentiation is not a reflection of variation in CaE reactivity, but of the dose of OP inhibition necessary to inhibit AChE sufficiently to result in death. For compounds with high reactivity for AChE, only a small concentration of an OP compound is needed to inhibit AChE, and only small amounts of the OP compound are available for CaE detoxication. For compounds with low reactivity for AChE, a higher concentration is necessary to inhibit AChE, and a larger amount of the OP compound is detoxified by CaE.

VIII. Role of Carboxylesterase in Treatment of Organophosphorus Toxicity

A. Oxime Therapy

The traditional treatment for OP toxicity is administration of oximes to reactivate OP-inhibited AChE to allow normal cholinergic neurotransmission (see discussion of oxime reactivation in Chapter 5 by Wilson *et al.*, this volume). Although effective oximes usually produce protection against OP compounds in all mammalian species, considerable variation exists in the degree of protection achieved in each species. For example, the *bis*-pyridinium oxime HI-6 in combination with atropine provided protection against 1.9 LD_{50} of soman in mice, 2.1 LD_{50} in rats, 3.5 LD_{50} in guinea pigs, 9.0 LD_{50} in dogs, and 5 LD_{50} in rhesus monkeys (Boskovic *et al.*, 1984b; Hamilton and Lundy, 1989). This variation has created uncertainty concerning the probable protection of HI-6 against soman in humans.

In recent analyses of the problem of species variation in oxime protection by HI-6 and pralidoxime chloride, it was found that equal protection against soman could be achieved with an oxime in multiple species if CaE-inhibited animals were used (Maxwell and Koplovitz, 1990; Maxwell and Brecht, 1991). The achievement of equal protection correlated with the ability of each oxime to produce equal levels of reactivation of soman-inhibited AChE in multiple species. In species whose carboxylesterase levels were high (i.e., mice), higher levels of oxime protection were achieved than in species whose carboxylesterase levels were low (i.e., guinea pigs). It is possible that oximes may also reactivate the OP-inhibited CaE to produce additional protection. Thus, reactivation could recycle OP-inhibited CaE for further covalent binding of the OP compound, thereby increasing detoxication. The detoxication potential of this process is suggested by the observation that *in vivo* oxime reactivation of soman-inhibited CaE by diacetylmonooxime in the rat increased soman detoxication enough to produce a twofold increase in the soman LD_{50} (Sterri and Fonnum, 1987; Johnsen and Fonnum, 1989).

B. Carbamate Pretreatment

The inability of oximes and atropine to provide adequate protection against some refractory OP compounds has led to the development of carbamate pretreatment, in which carbamylation of AChE effectively protects it against inhibition by OP compounds (Leadbeater *et al.*, 1985). Spontaneous decarbamylation of AChE after the OP compound is detoxified generates enough active AChE to allow normal cholinergic neurotransmission (Harris *et al.*, 1984). Carbamate pretreatment, like oxime therapy, has produced considerable variation in the degree of protection observed in various mammalian species. For example, pyridostigmine and atropine provided protection against 1.7 LD_{50} of soman in rats, 2.7 LD_{50} in rabbits, 5.3 LD_{50} in guinea pigs, and 15 LD_{50} in marmosets (Gordon *et al.*, 1978; Dirnhuber *et al.*, 1979). When carbamate pretreatment was tested in CaE-inhibited rodents, the degree of protection was found to be similar among rats, guinea pigs, and rabbits (Maxwell *et al.*, 1988a), and was also comparable to the protection achieved in nonhuman primates lacking endogenous plasma carboxylesterase. Therefore, the level of CaE detoxication of an OP compound is an important parameter in designing an animal model to test drug protection against OP toxicity.

IX. Conclusions

Among the diverse biochemical reactions involved in detoxication of OP compounds, CaE performs the role of a high affinity–low capacity detoxication process. Physiological concentrations of most OP compounds (i.e., 1–100 n*M*) react rapidly with CaE, but they react by an irreversible 1:1 stoichiometry with the active site of CaE. Therefore, the capacity of CaE to detoxify OP compounds is quantitatively limited by the number of available CaE molecules. CaE detoxication contrasts with other detoxication enzymes, such as OP hydrolases, that are high capacity–low affinity enzymes that can detoxify 10^3–10^6 OP molecules/min/active site, but have K_m for OP compounds in the m*M* concentration range. Consequently, CaE is important for detoxication of highly toxic OP compounds, in which affinity for the detoxication enzyme is more important than detoxication capacity.

CaE detoxication is a major determinant of (1) the toxic response to OP compounds during development and aging; (2) species differences in the toxic responses to OP compounds; (3) species differences in the efficacy of drug treatments against OP compounds; and (4) the development of tolerance to OP compounds. The increasingly sophisticated characterization of CaE at the subcellular and molecular level should help clarify the role of CaE in these diverse biological phenomena.

References

Aldridge, W. N. (1953). Serum esterases. 1. Two types of esterase (A and B) hydrolyzing *p*-nitrophenyl acetate, propionate, and butyrate, and a method for their determination. *Biochem. J.* 53, 110–117.

Aldridge, W. N., and Reiner, E. (1972). "Enzyme Inhibitors as Substrates," pp. 53–90. North Holland, Amsterdam.

Augsteyn, R. C., De Jersey, J., Webb, E. C., and Zerner, B. (1969). On the homology of the active-site peptides of liver carboxylesterase. *Biochim. Biophys. Acta* **171**, 128–137.

Augustinsson, K. B. (1958). Electrophoretic separation and classification of blood plasma esterases. *Nature* **131**, 1786–1789.

Benke, G. M., and Murphy, S. D. (1974). Effect of TOTP pretreatment on paraoxon and methylparaoxon detoxification in rats. *Res. Commun. Chem. Path. Pharmacol.* **8**, 665–672.

Boskovic, B. (1979). The influence of 2-(*o*-cresyl)-4H-1:3:2-benzodioxaphosphorin-2-oxide (CBDP) on organophosphate poisoning and its therapy. *Arch. Toxicol.* **42**, 207–216.

Boskovic, B., Jakanovic, M., and Maksimovic, M. (1984a). Effects of sarin, soman, and tabun on plasma and brain aliesterase activity in the rat. *In* "Cholinesterases: Fundamental and Applied Aspects" (M. Brzin, E.A. Barnard, and D. Sket, eds.), pp. 365–374. Walter de Gruyter, Berlin.

Boskovic, B., Kovacevic, V., and Jovanovic, D. (1984b). PAM-2 Cl, HI-6 and HGG-12 in soman and tabun poisoning. *Fundam. Appl. Toxicol.* **4**, S106–S115.

Brestkin, A. P., Nikolskaya, E. B., and Efimtseva, E. A. (1986). Comparative sensitivity of two carboxylesterases from the reindeer liver to various inhibitors. *Biokhimiia* **51**, 1141–1149.

Brodeur, J., and Dubois, K. P. (1963). Comparison of acute toxicity of anticholinesterase insecticides to weanling and adult male rats. *Proc. Soc. Biol. Med.* **114**, 509–511.

Brodeur, J., and DuBois, K. P. (1967). Studies on factors influencing the acute toxicity of malathion and malaoxon in rats. *Can. J. Physiol. Pharmacol.* **45**, 621–631.

Brown, T. M., Bryson, P. K., Grothusen, J. R., Joly, J. M., and Payne, G. T. (1986). "Inhibition of Xenobiotic-Degrading Hydrolysis by Organophosphinates." DTIC Report AD-A202378/6/XAB Clemson University, Clemson, South Carolina.

Casida, J. E., Baron, R. L., Eto, M., and Engel, J. L. (1963). Potentiation and neurotoxicity induced by certain organophosphates. *Biochem. Pharmacol.* **12**, 73–83.

Chambers, J. E., and Chambers, H. W. (1990). Time course of inhibition of acetylcholinesterase and aliesterases following parathion and paraoxon exposures in rats. *Toxicol. Appl. Pharmacol.* **103**, 420–429.

Chambers, H., Brown, B., and Chambers, J. E. (1990). Noncatalytic detoxication of six organophosphorus compounds by rat liver homogenate. *Pestic. Biochem. Physiol.* **36**, 308–315.

Chow, A. Y. K., and Ecobichon, D. J. (1973). Characterization of the esterases of guinea pig liver and kidney. *Biochem. Pharmacol.* **22**, 689–701.

Clement, J. G. (1982). Plasma aliesterase: A possible depot for soman (pinacolylmethylphosphonofluoridate) in the mouse. *Biochem. Pharmacol.* **31**, 4085–4088.

Clement, J. G. (1984). Role of aliesterase in organophosphate poisoning. *Fundam. Appl. Toxicol.* **4**, S96–S105.

Clement, J. G. (1989). Survivors of soman poisoning: Recovery of the soman LD_{50} to control value in the presence of extensive acetylcholinesterase inhibition. *Arch. Toxicol.* **63**, 150–154.

Clement, J. G., Benschop, H. P., De Jong, L. P. A., and Wolthuis, O. L. (1987). Stereoisomers of soman (pinacolylmethylphosphonofluoridate): Inhibition of serum carboxylic ester hydrolase and potentiation of their toxicity by CBDP [2-(2-methylphenoxy)-4H-1,3,2-benzodioxaphosphorin-2-oxide] in mice. *Toxicol. Appl. Pharmacol.* **89**, 141–143.

Cohen, S. D. (1981). Carboxylesterase inhibition and potentiation of soman toxicity. *Biochem. Pharmacol.* **49**, 105–106.

Costa, L. G., and Murphy, S. D. (1983). Unidirectional cross-tolerance between the carbamate insecticide propoxur and the organophosphate disulfoton in mice. *Fundam. Appl. Toxicol.* **3**, 483–488.

De Jong, L. P. A., and Benschop, H. P. (1988). Biochemical and toxicological implications of chirality in anticholinesterase agents. *In* "Stereoselectivity of Pesticides; Biological and Chemical Problems" (E.J. Ariens, J.J.S. Van Rensen, and W. Welling, eds.), pp. 109–149. Elsevier, Amsterdam.

De Jong, L. P. A., and Van Dijk, C. (1984). Formation of soman (1,2,2-trimethylpropylmethylphosphonofluoridate) via fluoride-induced reactivation of soman-inhibited aliesterase in rat plasma. *Biochem. Pharmacol.* **33**, 663–669.

Dirnhuber, P., French, M. C., Green, D. M., Leadbeater, L., and Stratton, J. A. (1979). The protection of primates against soman poisoning by pretreatment with pyridostigmine. *J. Pharm. Pharmacol.* **31**, 295–299.

Ecobichon, D. J., and Comeau, A. M. (1973). Hepatic aliesterase sensitivity to dichlorvos and diisopropylfluorophosphate. *Toxicol. Appl. Pharmacol.* **26**, 260–263.

Eto, M. (1974). "Organophosphorus Pesticides: Organic and Biological Chemistry," pp. 158–192. CRC Press, Cleveland, Ohio.

Eto, M., Casida, J. E., and Eto, T. (1962). Hydroxylation and cyclization reactions involved in the metabolism of tri-O-cresyl phosphate. *Biochem. Pharmacol.* **11**, 337–352.

Fleisher, J. H., Harris, L.W., Prudhomme, C., and Bursel, J. (1963). Effects of ethyl *p*-nitrophenylthiobenzene phosphonate (EPN) on the toxicity of isopropylmethylphosphonofluoridate (GB). *J. Pharmacol. Exp. Ther.* **139**, 390–396.

Frawley, J. P., Fuyat, H. N., Hagan, E. C., Blake, J. R., and Fitzhugh, O. G. (1957). Marked potentiation in mammalian toxicity from simultaneous administration of two anticholinesterase compounds. *J. Pharmacol. Exp. Ther.* **121**, 96–106.

Freedman, A. M., and Himwich, H. E. (1948). The effect of age on lethality of di-isopropyl fluorophosphate. *Am. J. Physiol.* **153**, 121–126.

Gordon, J. J., Leadbeater, L., and Maidment, M. P. (1978). The protection of animals against organophosphate poisoning by pretreatment with a carbamate. *Toxicol. Appl. Pharmacol.* **43**, 207–216.

Gupta, R. C., and Dettbarn, W.-D. (1987). *iso*-OMPA-induced potentiation of soman toxicity in rat. *Arch. Toxicol.* **61**, 58–62.

Gupta, R. C., Patterson, G. T., and Dettbarn, W.-D. (1985). Mechanisms involved in the development of tolerance to DFP toxicity. *Fundam. Appl. Toxicol.* **5**, S17–S28.

Gupta, R. C., Patterson, G. T., and Dettbarn, W.-D. (1987). Acute tabun toxicity: Biochemical and histochemical consequences in brain and skeletal muscles of rat. *Toxicology* **46**, 329–341.

Hagan, E. C., Jenner, P. M., and Jones, W. I. (1971). Increased lethal effects of acutely administered anticholinesterase in female rats prefed with similar agents. *Toxicol. Appl. Pharmacol.* **18**, 235–237.

Hamilton, M. G., and Lundy, P. M. (1989). HI-6 therapy of soman and tabun poisoning in primates and rodents. *Arch. Toxicol.* **63**, 144–149.

Harano, T., Miyata, T., Lee, S., Aoyagi, H., and Omura, T. (1988). Biosynthesis and localization of rat liver microsomal carboxylesterase E1. *J. Biochem.* **103**, 149–155.

Harris, L. W., McDonough, J. H., Stitcher, D. L., and Lennox, W. J. (1984). Protection against both lethal and behavioral effects of soman. *Drug Chem. Toxicol.* 7, 605–624.

Hassan, A., and Dauterman, W. C. (1968). Studies on the optically active isomers of *o,o*-diethyl malathion and *o,o*-diethyl malaoxon. *Biochem. Pharmacol.* **17**, 1431–1439.

Heath, D. F. (1961). "Organophosphorus Poisons." pp 177–215, Pergamon Press, New York.

Heymann, E. (1980). Carboxylesterases and amidases. *In* "Enzymatic Basis of Detoxication" (W.B. Jakoby, ed.) Vol. 2, pp. 291–323. Academic Press, New York.

Heymann, E. (1989). A proposal to overcome some general problems of the nomenclature of esterases. *In* "Enzymes Hydrolyzing Organophosphorus Compounds" (E. Reiner, W. N. Aldridge, and F. C. G. Hoskins, eds.), pp. 226–235. Ellis Horwood, Chichester, England.

Jimmerson, V. R., Shih, T.-M., Maxwell, D. M., Kaminskis, A., and Mailman, R.B. (1989a). The effect of 2-(*o*-cresyl)-4H-1:3:2-benzodioxa-phosphorin-2-oxide on tissue cholinesterase and carboxylesterase of the rat. *Fundam. Appl. Toxicol.* **13**, 568–575.

Jimmerson, V. R., Shih, T.-M., Maxwell, D. M., Kaminskis, A., and Mailman, R. B. (1989b). Cresylbenzodioxaphosphorin oxide pretreatment alters soman-induced toxicity and inhibition of tissue cholinesterase activity of the rat. *Toxicol. Lett.* **48**, 93–103.

Johnsen, H., and Fonnum, F. (1989). Detoxification of soman stereoisomers in liver and plasma. *In* "Enzymes Hydrolyzing Organophosphorus Compounds" (E. Reiner, W.N. Aldridge, and F.C.G. Hoskins, eds.), pp. 90–97. Ellis Horwood, Chichester, England.

Junge, W., and Krisch, K. (1975). The carboxylesterases/amidases of mammalian liver and their possible significance. *Crit. Rev. Toxicol.* **3**, 371–434.

Kaur, G., and Kanugo, M. S. (1970). Alterations in the activity and regulation of cholinesterase of the nervous tissue of rats of various ages. *Indian J. Biochem.* 7, 122–125.

Kemp, J. R., and Wallace, K. B. (1990). Molecular determinants of species-selective inhibition of brain acetylcholinesterase. *Toxicol. Appl. Pharmacol.* **104**, 246–258.

Korza, G., and Ozols, J. (1988). Complete covalent structure of 60-kDa esterase from 2, 3, 7, 8-tetrachlorodibenzo-*p*-dioxin-induced rabbit liver microsomes. *J. Biol. Chem.* **263**, 3486–3495.

Krisch, K. (1971). Carboxylic ester hydrolases. *In* "The Enzymes" 3rd Ed. (P. D. Boyer, ed.), Vol. 5, pp. 43–69. Academic Press, New York.

Ladinsky, H., Consolo, S., Peri, G., and Garattini, S. (1972). Acetylcholine, choline and acetyltransferase activity in the developing brain of normal and hyperthyroid rats. *J. Neurochem.* **19**, 1947–1952.

Lauwerys, R. R., and Murphy, S. D. (1969). Interaction between paraoxon and tri-*o*-tolyl phosphate in rats. *Toxicol. Appl. Pharmacol.* **14**, 348–357.

Leadbeater, L., Inns, R. H., and Rylands, J. M. (1985). Treatment of poisoning by soman. *Fundam. Appl. Toxicol.* 5, S225–S231.

Lee, P. W., Allahyari, R., and Fukuto, T. R. (1978). Studies of the chiral isomers of fonofos and fonofos oxon: 1. Toxicity and antiesterase activities. *Pestic. Biochem. Physiol.* **8**, 146–157.

Long, R. M., Satoh, H., Martin, B. M., Kimura, S., Gonzalez, F. J., and Pohl, L. R. (1988). Rat liver carboxylesterase: cDNA cloning, sequencing, and evidence for a multigene family. *Biochem. Biophys. Res. Commun.* **156**, 866–873.

Matsumura, F. (1985). "Toxicology of Insecticides" 2nd Ed. pp. 203–298. Plenum Press, New York.

Maxwell, D. M. (1989). Nerve agent specificity of scavenger protection by carboxylesterase. *In* "Supplement to the Proceeding of the Third International Symposium on Protection

Against Chemical Warfare Agents" FOA Report C40269-4.6,4.7, pp. 175–182. National Defence Research Establishment, Umea, Sweden.

Maxwell, D. M., and Koplovitz, I. (1990). Effect of endogenous carboxylesterase on HI-6 protection against soman toxicity. *J. Pharmacol. Exp. Therap.* **254**, 440–444.

Maxwell, D. M., and Brecht, K. M. (1991). The role of carboxylesterase in species variation of oxime protection against soman. *Neurosci. Biobehav. Rev.* **15**, 135–139.

Maxwell, D. M., Brecht, K. M., and O'Neill, B. L. (1987a). The effect of carboxylesterase on interspecies differences in soman toxicity. *Toxicol. Lett.* **39**, 35–42.

Maxwell, D. M., Lenz, D. E., Groff, W. A., Kaminskis, A., and Froehlich, H .L. (1987b). The effect of blood flow and detoxication on *in vivo* cholinesterase inhibition in rats. *Toxicol. Appl. Pharmacol.* **88**, 66–76.

Maxwell, D. M., Brecht, K. M., and O'Neill, B. L. (1988a). Effect of carboxylesterase inhibition on carbamate protection against soman toxicity *J. Pharmacol. Exp. Ther.* **246**, 986–991.

Maxwell, D. M., Vlahacos, C. P., and Lenz, D.E. (1988b). A pharmacodynamic model for soman in the rat. *Toxicol. Lett.* **43**, 175–188.

Maxwell, D. M., Wolfe, A. D., Ashani, Y., and Doctor, B. P. (1990). Cholinesterase and carboxylesterase as scavengers for organophosphorus agents. *In* "Cholinesterases: Structure, Function, Mechanism, Genetics, and Cell Biology" (J. Massoulie, F. Bacou, E. Barnard, A. Chatonnet, B. P. Doctor, and D. M. Quinn, eds.), pp. 206–209. American Chemical Society, Washington, D.C.

McKay, D. H., Jardine, R. V., and Adie, P. A. (1971). The synergistic action of 2-(*o*-cresyl)-4H-1:3:2-benzodioxaphosphorin-2-oxide with soman and physostigmine. *Toxicol. Appl. Pharmacol.* **20**, 474–479.

McPhillips, J. J. (1969). Altered sensitivity to drugs following repeated injections of a cholinesterase inhibitor to rats. *Toxicol. Appl. Pharmacol.* **14**, 67–73.

Mentlein, R., Heiland, S., and Heymann, E. (1980). Simultaneous purification and comparative characterization of six serine hydrolases from rat liver microsomes. *Arch. Biochem. Biophys.* **200**, 547–559.

Mentlein, R., Ronai, A., Robbi, M., Heymann, E., and Deimling, O. V. (1987). Genetic identification of rat liver carboxylesterases isolated in different laboratories. *Biochim. Biophys. Acta* **913**, 27–38.

Mentlein, R., Rix-Matzen, H., and Heymann, E. (1988). Subcellular localization of non-specific carboxylesterases, acylcarnitine hydrolase, monoacylglycerol lipase and palmitoyl-CoA hydrolase in rat liver. *Biochim. Biophys. Acta* **964**, 319–328.

Mikhaylov, S. S., and Scherbak (1983). "The Metabolism of Organophosphorus Poisons." Meditsina, Moscow.

Miyamoto, J., Kaneko, H., Hutson, D. H., Esser, H. O., Gorbach, S., and Dorn, E. (1988). "Pesticide Metabolism: Extrapolation from Animals to Man." pp. 1–37. Blackwell, Oxford, England.

Murphy, S. D., and DuBois, K. P. (1957). Quantitative measurements of inhibition of the enzymatic detoxification of malathion by EPN (ethyl *p*-nitrophenyl thionobenzenephosphonate). *Proc. Soc. Exp. Biol. Med.* **96**, 813–818.

Myers, D. K. (1959). Mechanism of the prophylactic action of diacetylmonoxime against sarin poisoning. *Biochim. Biophys. Acta* **34**, 555–557.

Ooms, A. J. J., and Breebart-Hansen, J. C. A. E. (1965). The reaction of organophosphorus compounds with hydrolytic enzymes. The inhibition of horse serum aliesterase. *Biochem. Pharmacol.* **14**, 1727–1738.

Oosterbaan, R. A., and Cohen, J. A. (1964). *In* "Structure and Activity of Enzymes" (T. W. Goodwin, J. I. Harris, and B. S. Hartley, eds.), pp. 87–95. Academic Press, New York.

Ozols, J. (1987). Isolation and purification of 60-kilodalton glycoprotein esterase from liver microsomal membranes. *J. Biol. Chem.* **262**, 15316–15321.

Polak, R. L., and Cohen, E. M. (1969). The influence of triorthocresylphosphate on the distribution of ^{32}P in the body of the rat after injection of ^{32}P-sarin. *Biochem. Pharmacol.* **18**, 813–820.

Robbi, M., and Beaufay, J. (1986). Biosynthesis of rat-liver pI-5.0 esterases in cell-free systems and in cultured hepatocytes. *Eur. J. Biochem.* **158**, 187–194.

Russell, R. W., Overstreet, D. H., Cotman, C. W., Carson, V. G., Churchill, L., Dalglish, F. W., and Vasquez, B. J. (1975). Experimental tests of hypotheses about neurochemical mechanisms underlying behavioral tolerance to the anticholinesterase diisopropylfluorophosphate. *J. Pharmacol. Exp. Ther.* **192**, 73–85.

Satoh, T. (1987). Role of carboxylesterases in xenobiotic metabolism. *In* "Reviews in Biochemical Toxicology" (E. Hodgsen, J. R. Bend, and R. M. Philpot, eds.), Vol. 8, pp. 155–181. Elsevier, New York.

Shih, T.-M., Penetar, D. M., McDonough, J. H., Romano, J. A., and King, J. M. (1990). Age-related differences in soman toxicity and in blood and brain regional cholinesterase activity. *Brain Res. Bull.* **24**, 429–436.

Sterri, S. H. (1981). Factors modifying the toxicity of organophosphorus compounds including dichlorvos. *Acta Pharmacol. Toxicol.* **49**, 67–71.

Sterri, S. H., and Fonnum, F. (1987). Carboxylesterase in guinea pig plasma and liver. Tissue-specific reactivation by diacetylmonoxime after soman inhibition *in vitro*. *Biochem. Pharmacol.* **36**, 3937–3942.

Sterri, S. H., Lyngaas, S., and Fonnum, F. (1980). Toxicity of soman after repetitive injection of sublethal doses in rat. *Acta Pharmacol. Toxicol.* **46**, 1–7.

Sterri, S. H., Berge, G., and Fonnum, F. (1985a). Esterase activity and soman toxicity in developing rat. *Acta Pharmacol. Toxicol.* **57**, 136–140.

Sterri, S. H., Johnsen, B. A., and Fonnum, F. (1985b). A radiochemical assay method for carboxylesterase, and comparisons of enzyme activity towards the substrates methyl[1-^{14}C]butyrate and 4-nitrophenyl butyrate. *Biochem. Pharmacol.* **34**, 2770–2785.

Strong, R., Hicks, P., Hsu, L., Bartus, R. T., and Enna, S. J. (1980). Age-related alterations in rodent brain cholinergic system and behavior. *Neurobiol. Aging* **1**, 59–63.

Strous, G. J. A. M., and Berger, E. G. (1982). Biosynthesis, intracellular transport, and release of the Golgi enzyme galactosyltransferase (lactose synthetase A protein) in HeLa cells. *J. Biol. Chem.* **257**, 7623–7628.

Walker, C. H. (1989). The development of an improved system of nomenclature and classification of esterases. *In* "Enzymes Hydrolyzing Organophosphorus Compounds" (E. Reiner, W. N. Aldridge and F. C. G. Hoskins, eds.), pp. 236–245. Ellis Horwood, Chichester, England.

Walker, C. H., and Mackness, M. I. (1983). Esterases: Problems of identification and classification. *Biochem. Pharmacol.* **32**, 3265–3269.

Wang, C., and Murphy, S. D. (1982a). Kinetic analysis of species differences in acetylcholinesterase sensitivity to organophosphate insecticides. *Toxicol. Appl. Pharmacol.* **66**, 409–419.

Wang, C., and Murphy, S. D. (1982b). The role of non-critical binding proteins in the sensitivity of acetylcholinesterase from different species to diisopropylfluorophosphate (DFP) *in vitro*. *Life Sci.* **31**, 139–149.

10

Hepatic Disposition of Organophosphorus Insecticides: A Synthesis of *in Vitro, in Situ* and *in Vivo* Data

Tsutomu Nakatsugawa
College of Environmental Science and Forestry
State University of New York
Syracuse, New York

I. Introduction

Many organophosphorus (OP) insecticides are latent poisons,[1] for which bioactivation is an obligatory step for toxic action. Because the parent compound does not directly cause toxicity, readily visible signs of acute poisoning are an unambiguous indicator of the active metabolite at the target macromolecule. For this reason, these chemicals have served as an excellent research tool for analyzing the complex interplay of metabolic reactions that affects toxicity *in vivo*. Our current understanding has been gained through a variety

[1]Typical anticholinesterase assays would indicate that these chemicals are two to three orders less active than their active metabolites, i.e., practically inactive. It may be pointed out, however, that the presence of a highly active impurity such as paraoxon or the S-alkyl isomer at 0.1% of a phosphorothioate, for example, would make the phosphorothioate appear 0.001 times as active as the impurity itself, even if the major constituent is totally inactive. While no systematic study exists, highly purified phosphorothioates appear to be devoid of antiesterase activity (Nakatsugawa and Dahm, 1965b; Murphy, 1966).

Organophosphates: Chemistry, Fate, and Effects

of approaches, including biochemical studies of biotransformation processes and analyses of behavior of toxicants. Inevitably, earlier work emphasized defining individual enzymes and other elements that control the level of toxic metabolite, and more recent studies have begun integrating pieces of information to gain the total image. These studies have made it clear that much of the antiesterase metabolite generated in a tissue, especially the liver, is merely a short-lived intermediate in the overall detoxication. Toxic consequences, then, depend on mechanisms that let sufficient active metabolites evade further biotransformation in the organ of origin and reach the target macromolecules. The latter molecules may be at a distant site, or in the same organ when the target organ mediates activation (see Chapter 11, this volume). In this perspective, analysis of toxicant disposition in relation to detailed functional morphology of each critical organ holds a key to successful understanding of the total toxic events. Such information has been emerging for the liver, and more recently for the skin. (See Chapter 12, this volume, for discussion of penetration and metabolism of OP compounds by the skin.)

Of the many OP insecticides, parathion has the largest database, having served as a prototype in studies of OP insecticides for nearly four decades. This chapter, therefore, considers the classical phosphorothioate, with emphasis on tracing the development of our understanding of how the liver affects the toxicity of OP insecticides. It is important to note that insight gained through these studies goes beyond parathion or OP compounds and raises a number of important questions also applicable to many other xenobiotics. Major features of biotransformation enzymes will be briefly reviewed here, with emphasis on parathion, to provide a background before their action in the liver is discussed.

II. Biotransformation Enzymes

Biotransformation of various OP esters, including parathion, produces nonionic, usually more active esters and nontoxic acidic phosphorus esters. Enzymes yielding the former received early attention for obvious reasons, predating the discovery of cytochrome P450. While P450 continues to be a major subject of research, recent studies have highlighted other oxidative enzymes, e.g., the flavin-containing monooxygenase (Chapter 6). Although structures of the acidic metabolites suggest involvement of hydrolytic enzymes, some of these compounds are products of other enzymes such as P450 and glutathione S-transferases. Although glutathione S-transferases have been most intensively studied as a major mechanism of dimethyl phosphorus ester metabolism (Chapter 7 by Sultatos, this volume), their involvement with other groups of compounds is also possible. As later discussions will illustrate, the *in vivo* significance of each enzyme cannot be assigned without carefully relating in

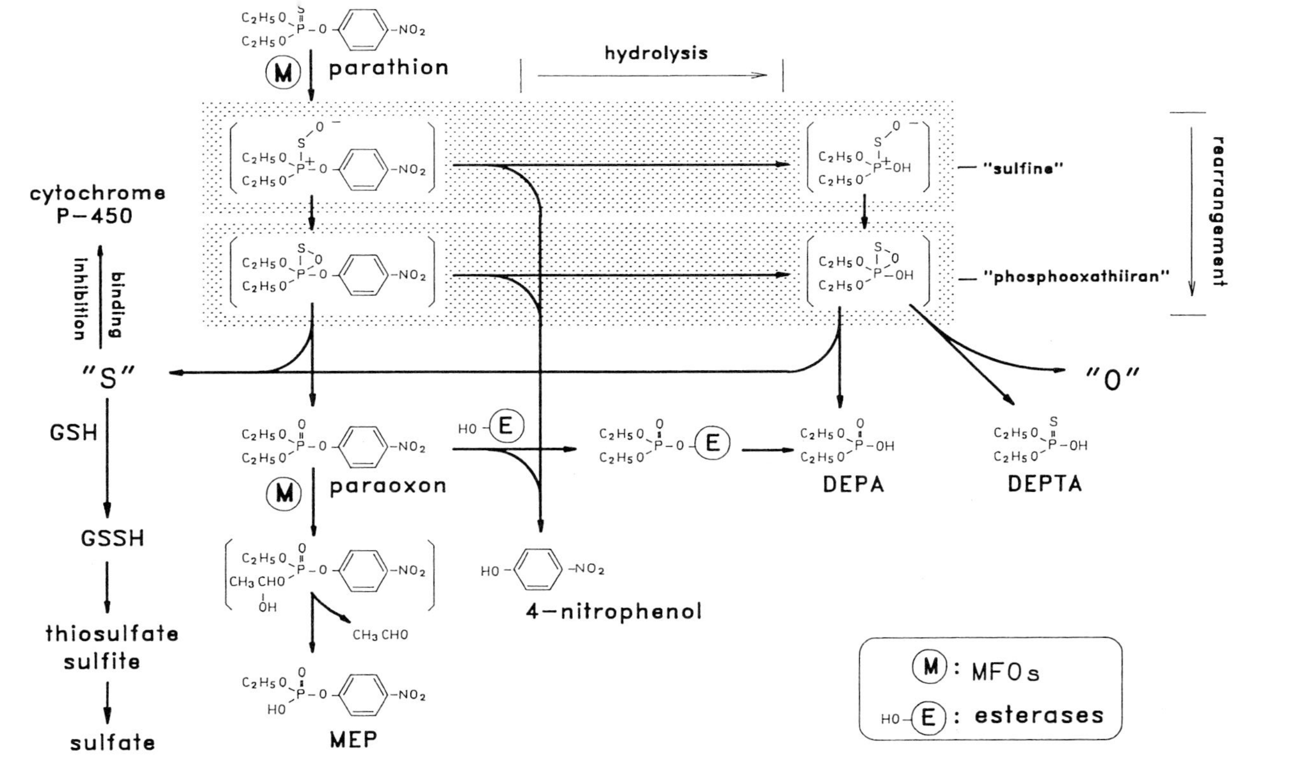

Figure 1 Major pathways of parathion biotransformation. MFO, mixed-function oxidase; MEP, monoethyl paraoxon; DEPA, diethylphosphoric acid; DEPTA, diethyl phosphorothioic acid. Structures in brackets are theoretical entities that have not been isolated. Two types of enzymes are important, i.e., MFOs involving cytochrome P450 and esterases presumably containing serine at the active site. Note that the diethyl phosphorylated esterase is a typical transient intermediate of a double-displacement reaction promptly yielding DEPA when an A-esterase type enzyme, e.g., paraoxonase, is involved. In contrast, a B-type esterase such as carboxylesterase yields a relatively stable intermediate (i.e., inhibited enzyme) and serves as a paraoxon trap both in hepatocytes and in the blood. The latter is believed to be the radioactive entity seen in autoradiographs as in Fig. 5.

vitro data to *in vivo* observations. Figure 1 presents a summary of parathion biotransformation.

Activation of parathion in liver slices was readily accomplished (Diggle and Gage, 1951) even before identification of paraoxon as the actual toxicant in parathion poisoning (Gage, 1953). Attempts to achieve cell-free activation, however, encountered considerable difficulties. Davison's demonstration (Davison, 1955) of nicotinamide adenine dinucleotide (NAD) as a cofactor for parathion activation by rat liver homogenates overcame the major hurdle, followed by a report of the superiority of nicotinamide adenine dinucleotide phosphate (NADPH) as the cofactor (O'Brien, 1959). Meanwhile, Murphy and DuBois showed microsomal nature of this enzyme with azinphosmethyl (Murphy and DuBois, 1957). A few year years later, the photochemical action spectrum of rat liver mixed-function oxidase systems established the central role of what is now known as cytochrome P450 in a variety of liver microsomal oxidations, including phosphorothioate activation (Cooper *et al.*, 1965).

Degradation of the active metabolite, paraoxon, was initially studied by Aldridge, using serum from various species of vertebrates (Aldridge, 1953a, b). He classified serum esterases hydrolyzing p-nitrophenyl esters into two classes, A-esterase and B-esterase, on the basis of their sensitivity to paraoxon (O'Brien, 1960). A-esterase, which generally hydrolyzes the acetate ester faster than it does the butyrate ester, is not inhibited by paraoxon, but rather hydrolyzes it. Paraoxon inhibits B-esterase, which has a propensity to hydrolyze the butyrate faster than the acetate. A variety of other tissues including the liver have A-esterase-like activities (Aldridge, 1953b; Neal, 1967a). The rat liver paraoxonase (paraoxon arylesterase) is a microsomal esterase, and its active form appears to be a protein calcium complex (McIlvain *et al.*, 1984). Another major acidic metabolite of paraoxon in the rat, monoethyl paraoxon (Nakatsugawa *et al.*, 1969a), is not produced by an esterase, but by a microsomal oxidase presumably cleaving the ethyl group as acetaldehyde (Ku and Dahm, 1973; Nakatsugawa and Morelli, 1976; Appleton and Nakatsugawa, 1977). A third mechanism for paraoxon disposition involves binding of paraoxon, presumably by phosphorylation, to B-esterase-type esterases (carboxylesterases, pseudocholinesterases) (Lauwerys and Murphy, 1969). GSH S-transferases also cleave parathion or paraoxon (Nakatsugawa *et al.*, 1969a; Hollingworth *et al.*, 1973), but so far appear to be of minor *in vivo* significance.

Neither the serum A-esterase (Mounter, 1954) nor liver paraoxonase (Nakatsugawa and Dahm, 1967) (microsomes without NADPH) hydrolyzes parathion. Yet diethyl phosphorothioic acid is a major urinary metabolite of

[2]The term suicidal is used here loosely, referring to the enzyme inhibition resulting from the reaction the enzyme catalyzes. More mechanism-oriented terminology limits suicidal inhibitors to those that inhibit by attacking the active site of the enzyme as reaction intermediates, not as products as is the case with parathion (Guengerich and Liebler, 1985).

parathion in the rat (Plapp and Casida, 1958). Two independent studies, one using ^{35}S-parathion (Nakatsugawa and Dahm, 1965a, 1967) and the other, ^{32}P-parathion (Neal, 1967a), defined the enzyme responsible for diethyl phosphorothioic acid as a typical microsomal mixed-function oxidase, indistinguishable from the parathion desulfuration enzyme in its requirement for NADPH and oxygen. The ^{35}S-parathion study also uncovered concurrent macromolecular binding of the sulfur metabolite of parathion desulfuration in the microsomes from cockroach fat body and rat and rabbit livers. This confirmed the possibility suggested by an earlier discovery of suicidal[2], inactivation of parathion activation enzyme in the cockroach microsomes (Nakatsugawa and Dahm, 1965b) (cf. next paragraph).

The oxidative mechanism for parathion hydrolysis is consistent with a metabolic scheme in which action of microsomal mixed-function oxidase on parathion leads to desulfuration (activation) on the one hand and to dearylation (hydrolytic degradation) on the other (Nakatsugawa and Dahm, 1967; Neal, 1967a). Whether both pathways resulted from the same enzymes was unclear, but early data suggested the latter possibility because of differential sensitivities of the two pathways to divalent cations, sulfhydryl compounds, inhibitors, and enzyme inducers, and the difference in Michaelis constants (Neal, 1967b). The subsequent proposal of a common intermediate, however, favors dual pathways from a single enzyme (Kamataki *et al.*, 1976). A study with reconstituted systems utilizing purified liver P450 isozymes showed that all six rat preparations and six rabbit preparations examined catalyzed both reactions, although the ratio of the two pathways varied widely (Guengerich, 1977). Similarly, four constitutive and two induced P450 isozymes from mouse liver yielded varying ratios of desulfuration and dearylation products from fenitrothion (Levi *et al.*, 1988). The oxidative metabolism involving concurrent activation and a degradation appears to be a general scheme for P=S esters (Nakatsugawa *et al.*, 1968; 1969b; Wolcott *et al.*, 1972; Wolcott and Neal, 1972; Yang *et al.*, 1971).

A common intermediate of liver microsomal phosphorothioate oxidation has been proposed to be a sulfur oxide or its phosphooxathiiran form (Kamataki *et al.*, 1976) based on ^{18}O studies with parathion (rabbit) (Ptashne *et al.*, 1971) and dyfonate (rat) (McBain *et al.*, 1971a). Small quantities of diethyl phosphoric acid plus 4-nitrophenol are also produced by parathion oxidation in a reconstituted system using purified rabbit liver P450 (Kamataki *et al.*, 1976). Aerobic oxygen is incorporated into the P=O moiety of the metabolites, while the oxygen of water is found in the acidic OH (Kamataki *et al.*, 1976). The postulated intermediate, however, has eluded isolation attempts with a model system (McBain *et al.*, 1971b; Wustner *et al.*, 1972).

The reactive sulfur released in the desulfuration is covalently bound to macromolecules in vitro, with a concurrent rapid loss of enzyme activity (Nakatsugawa and Dahm, 1965b, 1967; Poore and Neal, 1972; Kamataki and Neal, 1976; Morelli and Nakatsugawa, 1978). The identity of the bound sulfur as the desulfurated metabolite is supported by the quantitative equivalency of the bound sulfur to paraoxon plus the total balance sheet of ^{35}S-parathion metabolism in a rabbit liver microsomal system (Nakatsugawa and Dahm, 1967), the lack of comparable binding of ^{32}P (Poore and Neal, 1972) (rabbit liver microsomes) or ^{14}C (Kamataki and Neal, 1976; Yoshihara and Neal, 1977) (reconstituted rat liver P450 system) from labeled parathion, and chemical reactivity of bound sulfur (Neal *et al.*, 1977). Considerable progress has been made in characterizing the sulfur binding, mainly by Neal and co-workers. The sulfur binding appears to involve insertion of sulfur predominantly to sulfhydryl groups of P450 to form hydrodisulfide, but at least three other amino acids have been suggested also as target (Kamataki and Neal, 1976; Halpert *et al.*, 1980). In reconstituted systems, sulfhydryl agents slow down both sulfur binding and enzyme inactivation, but the lost activity cannot be restored (Neal *et al.*, 1977; Halpert *et al.*, 1980). Sulfhydryl compounds including glutathione can however partially restore the inactivation in rat liver microsomes (Morelli and Nakatsugawa, 1978, 1979). The exact molecular mechanism of enzyme inactivation remains unknown, although structural alteration of P450 has been suggested (Halpert *et al.*, 1980). In the presence of glutathione, detached sulfur appears to be mostly trapped as glutathione persulfide and undergoes a series of oxidations through sulfur oxyacids to become sulfate (Morelli and Nakatsugawa, 1979), which is the urinary metabolite (Nakatsugawa *et al.*, 1969a).

III. Hepatic Biotransformation *in Vivo* as a Determinant of Toxicity

The fact that a fourth of the cardiac output passes through the liver (Cahalan and Mangano, 1982), combined with the high activities of hepatic biotransformation enzymes, makes the liver the major site of biotransformation and major controller of the systemic level of toxicants. In oral exposures, essentially all the dose passes through the liver before entering the systemic circulation, making this organ metabolically dominant. Thus, studies of biotransformation have relied heavily on the liver as the source of enzymes. It must be noted, however, that biotransformation observed in vitro does not always mirror disposition *in vivo* because conditions employed in *in vitro* experiments are often far removed from the environment *in vivo* (Nakatsugawa *et al.*, 1989). As a result, difficulties have been encountered in attempts to explain the consequences of hepatic biotransformation in the toxicity *in vivo*. Such problems examined in the following paragraphs, however, have provided valuable clues to the underlying mechanisms *in vivo*.

When a toxicant is active in its original form, and its biotransformation leads to less toxic metabolites, the statement "the greater the biotransformation, the less the toxicity" is a truism. With latent poisons such as parathion or fenitrothion, however, the parallel statement the "greater the activation, the higher the toxicity" is often false. While in the former case the metabolic change in toxicity is always downward, results of activation are difficult to predict because of concurrent reactions that decrease toxicity (Menzer and Best, 1968; Levi *et al.*, 1988). In fact, many paradoxical data have been well known for a long time. For instance, male rat liver appears to activate parathion about twice as fast as the female liver (Neal, 1967a), but the male is less susceptible to parathion toxicity (DuBois, 1971). There is also a report that hepatectomy causes little loss of toxicity (Diggle and Gage, 1951) or even a potentiation (Selye and Mecs, 1974ab; Jacobsen *et al.*, 1973). Similarly, the well-known protective effect of chlorinated hydrocarbon insecticides and other chemicals against parathion toxicity, originally reported in the 1950s (Ball *et al.*, 1954), involves an increase (induction) rather than a decrease of the activation enzyme activity (Triolo and Coon, 1966; Alary and Brodeur, 1969; Vukovich *et al.*, 1971; Bass *et al.*, 1972). These data indicate that the activation in the liver does not directly translate to a toxic outcome. Numerous variations on this old theme have been presented in the literature. (See discussion by Chambers in Chapter 11 on correlation of OP toxicity with activation in nervous tissue rather than with activation by liver).

If the paraoxon degradation rate is far in excess of the parathion activation velocity, i.e., the latter is the rate-limiting step, then greater activation should result in faster overall degradation. This has led to the suggestion that the hepatic oxidation (activation and hydrolysis) hastens the decrease in the circulating levels of parathion when the detoxicative reactions in the liver and serum overcome activation (Neal, 1972). In other words, hepatic activation and subsequent degradation may be tightly coupled so that the liver as a whole serves as the detoxication organ, releasing no paraoxon (Nakatsugawa and Morelli, 1976). Excess paraoxon degradation capability over the activation potential indeed appears to be the case. The capacity (maximal possible rate) of paraoxon hydrolysis in the liver (55.5 μmol/min), lung (3.3–4.4 μmmol/min), and serum (800 μmol/min) of normal rabbits far exceeds the activation rates of the liver (1.44 μmol/min) and lung (0.015 μmol/min) (Neal, 1972). As Neal pointed out, however, lack of data on substrate levels *in vivo* makes it difficult to estimate relative contribution of various enzymes in the live animal.

These interpretations have often led to the suggestion that toxicity of parathion is owing to extrahepatic activation, and indeed activation occurs in many other tissues including the lung and brain (Poore and Neal, 1972; Forsyth and Chambers, 1989; Chambers, Chapter 11, this volume). Other possibilities, however, do exist. In a recent series of studies, Sultatos and co-workers analyzed the effluent in a mouse liver perfusion system using Krebs-Henseleit buffer containing 0 to 4% bovine serum albumin (BSA) as the perfusate (Sultatos *et al.*, 1985; Sultatos and Minor, 1986; Sultatos, 1987). An input concentration of 5 μM parathion simulated the portal concentration of the mouse given a near median lethal dose (LD_{50}) dose of parathion (Sultatos *et al.*, 1985). When parathion was infused in the perfusate containing 2% BSA, paraoxon was eluted from the liver. In contrast, a similar experiment with chlorpyrifos detected no active metabolite in the effluent. Since paraoxon was reasonably stable in the blood with an elimination half-life of 8.6 min, the authors concluded that lethal dose of parathion caused toxicity by hepatic activation, whereas chlorpyrifos was toxic due to extrahepatic activation. The liver was also shown to release methyl paraoxon from methyl parathion infused at 10 to 90 μM using perfusate containing 4% BSA (Sultatos, 1987). The effluent contained methyl paraoxon at about 1% of methyl parathion input level, together with the parent compound at 50 to 60% of the input. Pretreatment of mice with phenobarbital, however, antagonized acute toxicity and abolished paraoxon elution. Analysis of perfusion data and dialysis equilibrium experiments yielded estimates of the liver:perfusate (4% BSA) distribution coefficients: 16.4 ± 7.5 and 9.5 ± 2.7 for methyl parathion and 15.6 ± 7.5 and 19.5 ± 5.5 for parathion, respectively (Sultatos *et al.*, 1990). While these studies took clear advantage of perfusion analysis to determine hepatic activation at toxic doses, they also revealed that artificial perfusates could affect hepatic biotransformation considerably. The parathion extraction ratio at steady state was 0.19 with a perfusate containing 4% BSA, which binds 96% of parathion, but was 0.49 with 1% BSA (Sultatos and Minor, 1986). In the absence of BSA, however, all parathion partitioned into the liver and remained intact during 45 min of perfusion. These data remind us that before we understand how hepatic enzymes operate *in vivo*, we must know how the toxicant is delivered to the enzyme.

Inadequacy of our knowledge to assess enzyme action *in vivo* is also evident in the question involving another hepatic biotransformation mechanism, paraoxon deethylation. Urinary excretion of monoethyl paraoxon in the parathion-dosed rat is dose dependent, being negligible at low subtoxic doses, but nearly comparable to diethyl phosphoric acid at near-toxic doses (Appleton and Nakatsugawa, 1972; Nakatsugawa *et al.*, 1969a). Typical enzyme assays with liver homogenates showed that paraoxonase is far more active than deethylase over a range of pH (6.4–8.8), paraoxon concentration (10^{-6}–$10^{-4}M$) and ionic strength (changed by 0.08 to 0.3M KCl) (Appleton and Nakatsugawa, 1977). Induction, which antagonizes parathion toxicity, dramatically increased deethylase activity with little change in the paraoxonase activity (Ku and Dahm, 1973; Appleton and Nakatsugawa, 1977). Surprisingly, however, the relative contributions of the two enzymes *in vivo* as seen in urinary metabolites are little changed by enzyme induction. As a possible explanation, we suggested that deethylase activity is not uniformly distributed within the hepatic lobule, and induction is skewed in the non-functional region of the lobule (Appleton and Nakatsugawa, 1977). Although this prediction is still difficult to examine critically because of the deficiency of our knowledge, (cf. last section of this chapter), it has served to focus our attention on the behavior of parathion molecules within the hepatic lobule.

Phosphorothioates can affect the toxicity of other xenobiotics. Subtoxic doses of fenitrothion (25–100 mg/kg, i.p.) noncompetitively suppress aniline hydroxylase and aminopyrine demethylase activities and prolong the hexobarbital sleeping time in mice within a few hours of dosing (Uchiyama *et al.*, 1975). Immediate decline in hepatic P450 activity following doses of parathion or fenitrothion suggests sulfur binding as a possible cause of inhibition (Yoshida *et al.*, 1978). Another series of experiments shows that continuous feeding of mice (1000 ppm, 1 week) or a single dosing (100 mg/kg, p.o.) of fenitrothion strikingly synergizes toxicity of 2-sec-butylphenyl methylcarbamate by inhibiting first-pass metabolism (Takahashi *et al.*, 1984; Tsuda *et al.*, 1984). The effect seems specific to P=S compounds, but appears to involve an additional factor. For yet unknown reasons, mice receiving equitoxic carbamate doses (LD_5) after P=S treatment show drastically lower plasma concentrations of the carbamate than do the nontreatment controls (Takahashi *et al.*, 1987). The highest hepatic fenitrothion concentration is reached promptly on oral dosing, whereas the interaction peaks within a few hours. This eliminates competitive inhibition of biotransformation by fenitrothion as the major cause of the synergism (Takahashi *et al.*, 1984). While the observed effect is obviously complex, these data are also consistent with suicidal inhibition of mixed function oxidases by fenitrothion *in vivo*. Binding of sulfur does occur *in vivo* in various organs including the liver following oral administration of ^{35}S-parathion (Poore and Neal, 1972). Fenitrothion also appears to inhibit its own metabolism *in vivo* (Levi *et al.*, 1988). On the other hand, high toxicity of parathion may not normally permit sufficient uptake at a sublethal dose to achieve such inhibition *in vivo*. In fact, at a near-toxic dose of 3 mg/kg, i.p., parathion causes little enzyme inactivation or sulfur binding (0.7% of the administered dose) in the rat liver (Morelli and Nakatsugawa, 1979). Apparently hepatocytes have enough capacity to protect enzymes, possibly involving glutathione, at low doses of parathion. Addition of glutathione, however, could not prevent eventual enzyme inactivation in vitro (Morelli and Nakatsugawa, 1978). These results again illustrate the difficulties in extrapolating in vitro results to *in vivo* observations.

IV. Disposition within the Hepatic Lobule

As these examples illustrate, a sizable gap exists between the accumulated wealth of enzymological data and our understanding of how hepatic biotransformation affects the toxic outcome. We have recently directed our efforts toward filling this gap, and have succeeded in "looking into" the liver to gain a refreshing perspective, as discussed in the following section. Two observations compelled us to investigate how toxicants reach enzymes of biotransformation in the liver. First, discordance of paraoxon deethylase induction with the *in vivo* metabolic pattern suggested skewed biotransformation of paraoxon within the hepatic lobule (Appleton and Nakatsugawa, 1977). Second, experiments with hepatocytes in a Waymounth's medium suggested extremely rapid partitioning of parathion molecules between phases (Nakatsugawa *et al.*, 1980). This launched our inquiry into the translobular uptake pattern of parathion. The concept of hepatic breakthrough threshold then emerged as a useful means of analyzing hepatic disposition. Throughout these studies, our reasoning centered around the behavior of chemicals dictated by the functional morphology of the liver, which will be described briefly as background information.

A. Functional Morphology of the Liver and Hepatocellular Heterogeneity

The functions of the liver are intimately related to the serial morphology along the unidirectional flow of blood. The hepatic portal vein, which drains the entire digestive tract, supplies the major portion of the blood to the liver, with the hepatic artery supplying the rest. The venous and arterial blood vessels branch in parallel into several major lobes of the liver. The smallest functional unit of the liver, i.e., the lobule (or acinus—cf. Lamers *et al.*, 1989) comprises a mass of hepatocytes, one cell thick, arranged around capillary vessels (sinusoids), which branch out from distributing portal veins as they are joined by the arterial blood supply and converge into several central veins. The overlapping thin layers of fenestrated endothelial cells separate the vascular face of the hepatocytes from blood cells. The space between the endothelial cells and hepatocytes (space of Disse) is where hepatocytes are exposed to the incoming plasma. There the exchange of solutes takes place with great efficiency, thanks to the numerous microvilli on the surface of the hepatocytes. Figure 2 illustrates these features.

Of some 14 cell types found in the liver, the parenchymal cells (hepatocytes) are the most abundant (about 60% in rats and 85% in humans), constituting over 90% of the mass, and are responsible for most of the biotransformation. Littoral cells such as Kupffer cells and undifferentiated lining cells contribute most of the remainder (Fry and Bridges, 1979). Hepatocytes are not uniform throughout the lobule, however, and often show morphological, biochemical, and functional heterogeneity. This most likely reflects differential expression of the genome under the influence of the gradients of oxygen, hormones, and various substrates for intermediary metabolism and possibly sympathetic and parasympathetic innervations (Junger-

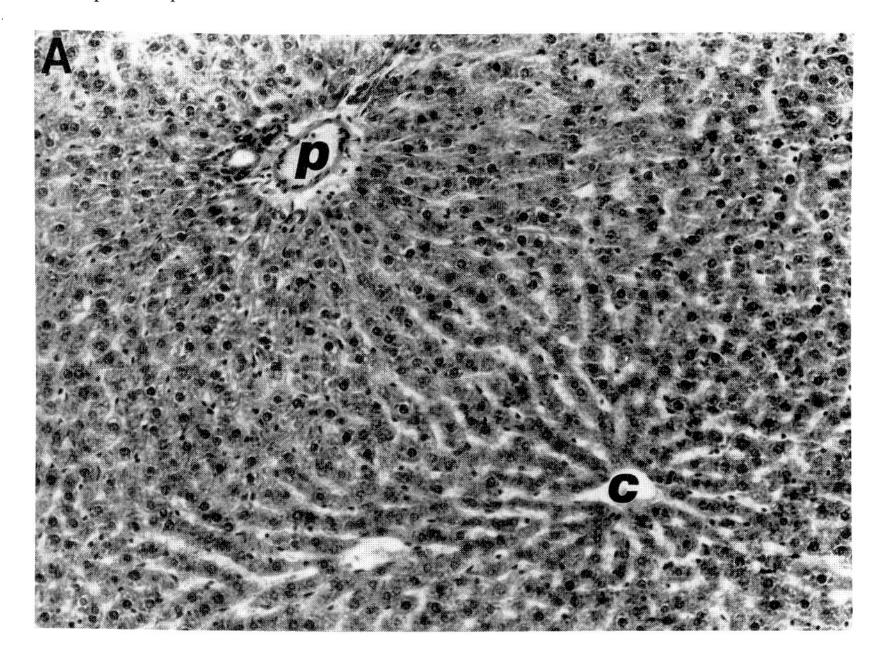

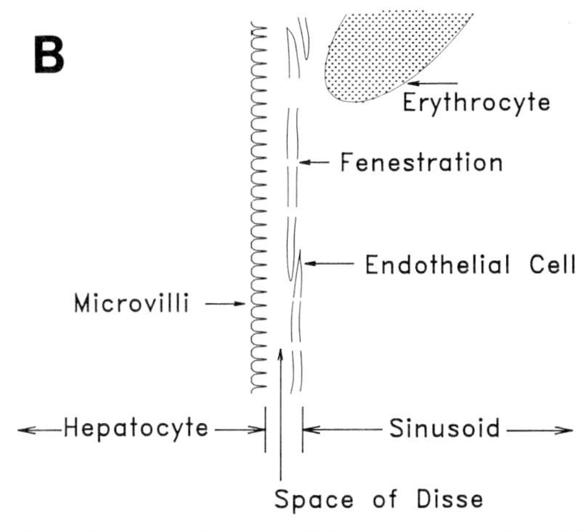

Figure 2 Functional morphology of the liver. (A) Micrograph showing a lobule of a male rat liver. P, portal vein branch; C, central venule. Dark strings of cells are hepatocytes alternating with light sinusoid passages, through which blood flows from P to C. Scale of the photo is approximately 0.7 mm across. (B) Diagrammatic view of sinusoid-hepatocyte interface, showing the vascular face of the hepatocyte rich in microvilli and covered by an overlapping layer of fenestrated endothelial cells.

mann and Katz, 1982; Gumucio and Chianale, 1988). Generally, decreasing portocentral gradients are seen for oxidative intermediary metabolism of carbohydrates and lipids, gluconeogenesis, and oxidation protection [including glutathione peroxidase and possibly glutathione (GSH)], and increasing gradients for glycolysis, liponeogenesis, and xenobiotic biotransformation (including P450). Detailed examination of intralobular distribution of P450 isozymes reveals, however, that no clear generalization can be made (Baron et al., 1984). In addition, enzyme levels are in a state of flux and fluctuate with diurnal rhythms and other physiological factors.

B. Chromatographic Translobular Migration of Parathion

Parathion, like many other uncharged lipophilic xenobiotics, is expected to enter and exit cells by reversible, passive diffusion. Our observations hinted that uptake of parathion inside the hepatic lobule is extremely rapid. Uptake of parathion by isolated hepatocytes was fast, and reached equilibrium within about 30 sec, but the actual rate of uptake at the cell surface is probably much greater. Since medium:cell volume ratio is very large (over 100) in the cell suspension, mixing becomes the rate-limiting step (Nakatsugawa *et al.*, 1980). In the liver where the blood:cell ratio is smaller than 1, transfer of diffusible molecules can occur "instantaneously" in comparison to the flow rate. We realized that narrow sinusoidal paths, unidirectional flow, and fast, reversible uptake aided by the large surface area afforded by microvilli of hepatocytes satisfy all the conditions of chromatography so that the hepatic lobule could mimic a chromatographic column. Therefore, we pulse-infused ^{35}S-labeled parathion into a rat liver being perfused with Waymouth's medium *in situ* (Nakatsugawa *et al.*, 1980). A parathion peak was indeed eluted about 35 min after infusion. Paraoxon perfused similarly with a much shorter transit time of 3 min (Bradford and Nakatsugawa, 1982). As in chromatography, addition of serum proteins to the medium increases the solvent power of the perfusate by decreasing the cell:medium partitioning and shortens the *retention time.*

To determine whether this chromatographic behavior also occurs *in vivo,* we developed a recirculating autologous blood perfusion system (Tsuda *et al.*, 1987). Perfusion with blood having a mean packed-cell volume of 30% clearly showed the expected elution of parathion with a retention time of 4.8 $\pm$ 1.7 min, which should be close to the value *in vivo*. Compared with the normal hepatic transit time for perfusate of ca. 5 sec, this still represents a significant delay of elution. The following scenario would describe what happens in the autologous blood perfusion system as well as *in vivo*. Such behavior of parathion is rather analogous to, for example, the behavior of ions in ion-pair chromatography.

Parathion molecules are mostly bound (ca. 95%) to macromolecules in the blood (Sultatos and Minor, 1986); it is the free, unbound molecules that

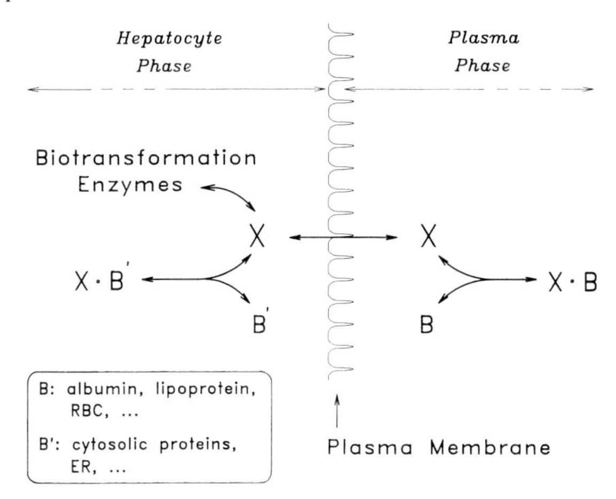

Figure 3 Uptake of parathion in the hepatic lobule. Only free unbound parathion (X) is available for rapid, passive diffusion across plasma membrane. Parathion is subject to binding to a variety of macromolecules on both sides of plasma membrane of the hepatocyte.

cross plasma membranes. As a pulse of blood carrying parathion enters the lobule, the free molecules of parathion rapidly diffuse into hepatocytes, while equilibrium between the free and bound molecules is maintained in the blood by fast binding and release (Fig. 3). Inside hepatocytes, similar equilibrium involving intracellular macromolecules is quickly established. Because of the rapidity of interactions, this will create an overall equilibrium of parathion involving intra- and extracellular macromolecules. Parathion molecules originally in the input pulse of blood are now distributed across the band comprising the pulse volume and the section of hepatocytes laterally surrounding the pulse segment. Comparatively slow flow (ca. 5 sec from the periportal to centrilobular pole) of the blood shifts the band, moving parathion molecules gradually downstream. As parathion migrates slowly down the sinusoid through a series of hepatocytes, it is gradually consumed by biotransformation. Long after the original blood pulse has exited the lobule, the fraction of parathion input that has survived metabolism emerges from the liver. We have dubbed this behavior *chromatographic translobular migration* (Nakatsugawa *et al.*, 1980), which is illustrated in Fig. 4. In exposures *in vivo*, of course, continuous input would occur in a way analogous to a series of pulses one after another.

Since the migration allows parathion nearly 4 min of interaction with a series of hepatocytes over the sinusoidal length of less than 1 mm, a dose can be totally consumed through biotransformation before it reaches the centrilobular end. The scenario predicts dose-dependent exposure of hepatocytes to the chemical within the lobule. In other words, a very low concentration

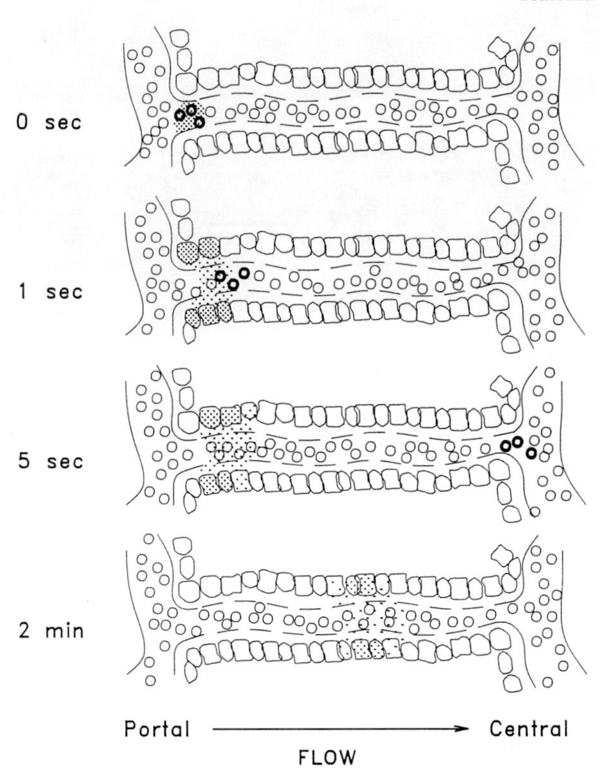

Figure 4 Chromatographic translobular migration. Parathion molecules entering the hepatic lobule are follwed in a time sequence. Simplified diagrams of the sinusoid beginning at a portal vein brancch and ending in a central vein are shown with surrounding hepatocytes in the lobule. Joining and branching of sinusoids are omitted for simplicity. A short pulse of blood containing parathion (denoted by small dots) and three red blood cells (shown as thick circles) enters the lobule at time zero. Parathion is rapidly equilibrated across the space encompassing the pulse, owing to diffusion of unbound parathion molecules (1 sec). The majority of parathion molecules is bound to blood proteins, but a significant fraction is always in the unbound form ready for diffusion, causing all the parathion molecules to move freely. The equilibrium space of rapidly diffusing parathion is slowly shifted downstream with the blood flow, while the red blood cells that originally accompanied the parathion input quickly reach the central vein (5 sec). Parathion molecules are still half way along the porto-central journey nearly 2 min after the red blood cells have left the lobule.

Figure 5 Autoradiography of 5-μm liver sections from male rats 4 hr after i.p. injection of [Ethyl-1-H^3]parathion (c.f. Fig. 1) at doses of (A) 0.1 mg/kg, (B) 0.8 mg/kg, and (C) 2.0 mg/kg. Scale of the photos is approximately 1.4 mm across. Covalent binding of paraoxon indicated by silver grains (dark area) is limited to the periportal region (P) at the low dose, extends to mid-zonal area as the medium dose and essentially reaches the central vein (C) at the high dose. Note that the darker labeling toward the central vein probably reflects higher levels of carboxylesterases. See Tsuda *et al.*, 1987 for similar autoradiographs presented at varying magnifications and for technical details.

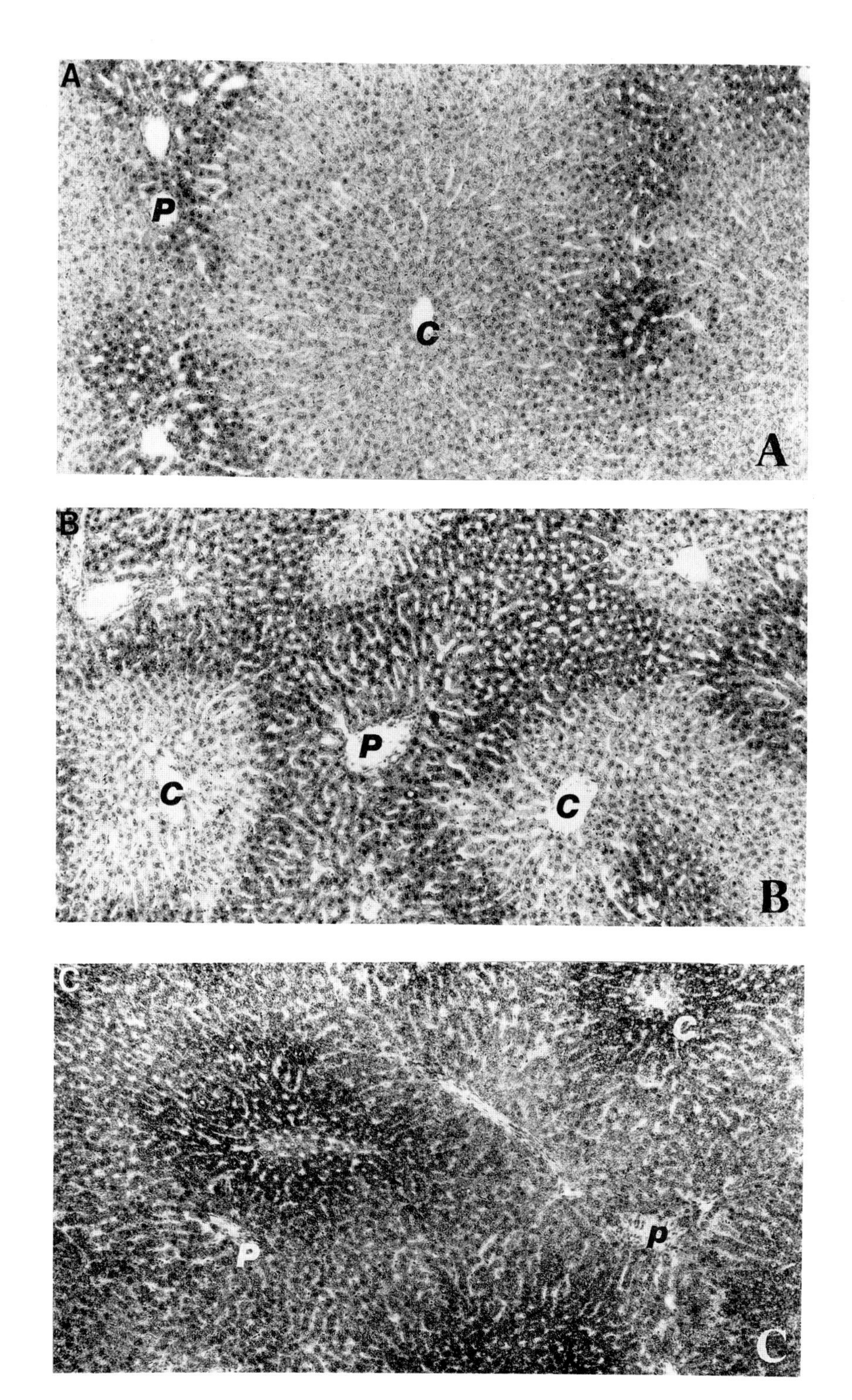

A
P
C
A
B
P
C
C
B
C
C
P
P
C

will be able to migrate only a short distance from the point of entry before total degradation, and the distance will increase with higher dosage. We took advantage of this to visualize the translobular disposition of parathion *in vivo*. Since paraoxon very rapidly binds covalently with esterases in the hepatocyte to form diethyl phosphoryl esterase, migrating [ethyl-1-^{14}C] parathion should leave a "footprint" that can be visualized by autoradiography (cf. Fig. 5). As predicted, autoradiography revealed narrow periportal labeling at a dose of 0.1 mg/kg, i.p. and broadening label with increasing doses up to 2 mg/kg (Fig. 5). We also showed that the clear areas downstream of the obvious label are in fact untouched by parathion–paraoxon; the centrilobular area remained silver grain–free after a 3-year autoradiographic exposure. In addition, autoradiography involving an *in situ* experiment established that shortly after input essentially all parathion molecules are indeed taken up by a narrow periportal band of hepatocytes, and the molecules then move down the lobule as the blood flows (Tsuda *et al.*, 1987). Oral administration of parathion to rats (Becker and Nakatsugawa, 1990a) and fenitrothion to mice (Becker and Nakatsugawa, 1990b) also gave similar results. Furthermore, we could simulate *in vivo* dosing quite well by continuously infusing these chemicals *in situ* in autologous blood perfusion (Becker and Nakatsugawa, 1990a). These results have firmly established that chromatographic translobular migration is not a perfusion artifact, but accurately characterizes the hepatic disposition of these toxicants *in vivo*.

Such behavioral features are apparently not new to hepatology; it is most likely described by the delayed-wave, flow-limited distribution model of Goresky and co-workers, known for cations since the early sixties (Goresky, 1963; Goresky et al, 1982). Lobular concentration gradients have been autoradiographically demonstrated for some solutes as a consequences of such uptake patterns (Goresky *et al.*, 1982). Parathion-like chromatographic behavior, however, has not been visualized before, nor have its implications been examined in toxicological context. Since this behavior must be strictly governed by physicochemical characteristics of a chemical, OP esters should not be unique. In fact, 1,3- and 1,2-dichlorobenzene and 4-nitroanisole have been shown to behave similarly in the autologous blood perfusion system (Tsuda *et al.*, 1988). Migration of paraoxon with a much shorter transit time (less than 1 min) has also been shown (Becker and Nakatsugawa, 1990a). Available data suggest, however, that chemicals having a lipophilicity much greater than that of parathion are not likely to behave chromatographically. Benzo(a)pyrene (water solubility: 2.0×10^{-9} *M* at 20°C) (Eisenbrand and Baumann, 1969) and 2,3,7,8-tetrachlorodibenzo-*p*-dioxin (TCDD, water solubility: 6.2×10^{-10} *M*) (Crummett and Stehl, 1973) appear to be bound to blood macromolecules so tightly that release from binding occurs gradually. As a result, a substantial fraction (30 to 50%) of the input fails to be freed from binding within a single passage through the liver (Tsuda et al., 1988). Under these conditions, hepatocytes throughout the entire lobule would be exposed to the toxicant regardless of the dose. This is in sharp contrast to the partial exposure at low doses of parathion. Solubility characteristics of OP esters vary widely as indicated by log P (i.e., octanol–water partition coefficient), e.g., dimethoate (2.71), fenitrothion (3.38), parathion (3.81), chlorpyrifos (5.11) and leptophos (6.31) (Verschueren, 1983). Therefore, it is not expected that all OP esters will behave like parathion.

C. Hepatic Breakthrough Threshold

At issue regarding the inverse relationship between parathion activation and toxicity is the question "How is hepatically formed paraoxon totally consumed within the liver or released to systemic circulation?" Armed with knowledge on the hepatic behavior of parathion and paraoxon and the technique of recirculating autologous blood liver perfusion, we were now ready to tackle this long-standing question from a new perspective. One of the corollaries of the concept of chromatographic translobular migration is that, for a biodegradable chemical, a hepatic release threshold dose (Tsuda *et al.*, 1987) should exist below which the chemical does not survive hepatic metabolism to exit the liver. In fact, periportally limited radiolabeling of liver lobules (see Fig. 5) implies such a threshold; i.e. since parathion or paraoxon did not reach centrilobular hepatocytes at the dosages employed, neither chemical could have entered the systemic circulation.

The hepatic breakthrough threshold for parathion may be defined as the portal vein concentration of parathion that will just result in leakage of parathion or paraoxon into the systemic blood (Becker and Nakatsugawa, 1990a). As the first autoradiographs obtained with parathion (i.p.) showed, the labeling spreads up to 2 mg/kg, at which the label covers all but the sporadic areas adjoining the central vein. Interestingly, the 2 mg/kg dose is very nearly the threshold dose of acute toxicity. The data, therefore, suggest that appearance of parathion or paraoxon in the systemic circulation coincides with the lowest toxic dose. Although autoradiography cannot differentiate between the parent compound and the active metabolite, separate thresholds are predicted *a priori* for elution of parathion and paraoxon. For brevity, we shall refer to the former as T(P=S) and the latter as T(P=O).

To determine which is eluted at the toxicity threshold dose, we directly sampled blood exiting the rat liver under anesthesia 1 hr (usually the time of maximal toxic signs) after an oral parathion dose that was rarely lethal (Becker and Nakatsugawa, 1990a). We also collected samples from both portal and jugular veins. Because toxic signs exhibited by individual animals given the same dose varied considerably, individual data were kept for both toxic signs and blood analysis. After a parathion dose of 10 mg/kg, paraoxon, but no parathion, was eluted from the liver. Apparently enough paraoxon had to be eluted to survive in the blood; Toxicity was best associated with the appearance of paraoxon in the jugular vein. Additional data suggested that breakthrough of paraoxon occurred below 10 mg/kg parathion, but above 5 mg/kg. Breakthrough of parathion occurred when 15 mg/kg was lethal, but not when the same dose was only moderately toxic. Experiments involving continuous parathion infusion *in situ* indicated T(P=S) of about 3 μM parathion and T(P=O) of slightly less than 1 μM parathion.

Autoradiography was consistent with these results for both oral dosing and continuous infusion. Therefore, a higher T(P=S) than T(P=O) characterizes hepatic disposition of parathion in normal male rats. The close link between the toxic signs and elution indicates that variation in the thresholds is a major basis for different tolerances among individual animals. Since no parathion enters the systemic circulation at sublethal doses, toxicity is clearly attributable to paraoxon originating in the liver. Conceivably, a much higher lethal dose could cause both paraoxon and parathion to enter systemic circulation; then both the liver and extrahepatic tissue would contribute to toxicity.

In dichlorodiphenyldichloroethylene (DDE)-pretreated rats, nearly 10 times more tolerant than the control, the order of the two thresholds is reversed. Parathion elution was detected even at a nontoxic dose of 10 mg/kg, whereas paraoxon was absent in the hepatic vein even when a toxic dose of 100 mg/kg was given. Paraoxon was detected, however, in the jugular vein, undoubtedly due to extrahepatic activation of systemic parathion. Since autoradiography with a 30 mg/kg dose revealed exposure of the entire lobule, T(P=S) was most likely to be not much greater than 30 mg/kg. Paraoxon breakthrough did not occur even at nonlethal but toxic 150 mg/kg. We could not demonstrate elution of either compound in continuous infusion of up to 7.7 μM, a practical limit of infusion concentration.

These data suggest that DDE pretreatment increased T(P=S) by a factor of perhaps not more than 3, but increased T(P=O) by 15 or more. In DDE pretreated rats, the absence of paraoxon after parathion breakthrough suggests centrilobular hepatocytes do not allow paraoxon to escape. Since toxicity is absent at doses exceeding T(P=S), more than a minimal breakthrough of parathion appears necessary for extrahepatic activation to cause toxicity. In contrast, in normal rats, parathion is not likely to have reached the centrilobular end when the paraoxon breakthrough occurs. Apparently, paraoxon is not totally destroyed within the cell of origin, but is gradually metabolized as it migrates through other cells downstream.

Fenitrothion also behaves similarly and exhibits breakthrough thresholds like parathion in both male rats and female mice. Blood analysis similar to those with parathion showed that in mice T(P=S) is much lower than T(P=O) (Becker and Nakatsugawa, 1990b). In mice dosed with 20 to 1000 mg/kg fenitrothion, the parent compound was detected even at the lowest dose, and clear elution occurred above 100 mg/kg. No fenitrooxon was detected at any dose even when toxic signs were evident. Therefore, toxicity is unquestionably owing to extrahepatic activation, just as in the case of parathion toxicity in DDE pretreated rats.

In rats dosed with 10 to 100 mg/kg of fenitrothion, breakthrough of fenitrothion and fenitrooxon occurred nearly coincident with the appearance of toxic effects (20 to 25 mg/kg), with fenitrooxon appearing at a slightly lower dose. Both hepatic and extrahepatic activation may contribute to toxicity. This case is not far from that of parathion in normal rats described in a preceding paragraph, with the exception of the more distinct difference between T(P=S) and T(P=O) for parathion.

Paradoxical observations such as the inverse relationship between parathion activation and toxicity as well as increased toxicity after partial hepatectomy have been considered evidence for extrahepatic activation as the cause of poisoning. Once the chromatographic behavior and its inherent thresholds are taken into account, however, *the detoxicative liver* and *the liver as the critical source of paraoxon* are not mutually exclusive. As long as the input is below T(P=O), the liver clearly serves as an organ of detoxication since no *oxon* is released. An increase in activation would raise T(P=S) directly and T(P=O) indirectly, thereby increasing the tolerance. If paraoxon degradation is also enhanced, it will further raise T(P=O), and hence the tolerance. Partial hepatectomy can conceivably increase the toxicity either by decreasing T(P=S) and T(P=O) due to increased blood flow (same blood volume per less liver mass) or by increased portal concentration of parathion due to the decreased splanchnic and portal flow rates [but same flow per liver mass, hence no change in T(P=S) or T(P=O)]. Generally, toxicity develops either when coupling breaks down with the elution of *oxon* and/or when the systemic *thion* level becomes sufficient for extrahepatic activation.

As pointed out in a preceding section, OP esters that are much more lipophilic than parathion or fenitrothion may not behave in the same manner; however, principles gained through these studies do permit certain predictions of behavior and facilitate an understanding of their toxic action. For instance, if leptophos or perhaps chlorpyrifos behaves more like benzo(a)pyrene cited earlier (Tsuda *et al.*, 1988), a significant fraction of the parent chemical should always be expected to appear in the systemic blood. Whether oxon has a breakthrough threshold would depend on its behavior, a function of lipophilicity.

V. Unanswered Questions, Current Reasoning, and Future Issues

While analyses of the behavior of these chemicals have greatly facilitated our understanding of toxicity, many questions remain. Why does no paraoxon elute when a high parathion level is undoubtedly achieved in the liver? Does *suicidal inactivation* occur in DDE pretreated rats? Which is the key to increased tolerance of DDE-pretreated rats, increased parathion metabolism or induction of paraoxon degradation enzymes? How much change in T(P=O) or T(P=S) is caused by, for example, a 2× induction of various enzymes? Before these quantitative interpretations are attempted, a few critical somewhat interrelated items must be considered. The following may be offered as only partial answers or hints.

A 2× increase in threshold, for example, may not reflect a 2× induction of biotransformation enzymes for parathion-like compounds because of the chromatographic behavior. Since a fraction of the incoming dose is biotransformated as molecules migrate through each hepatocyte, the metabolic rate may be *compounded* as migration proceeds, especially at low enzyme saturation. This scenario is further complicated by hepatocellular heterogeneity (see preceding section), which could introduce *variable interest rate* into the serial processing. Not only is heterogeneity a major issue in kinetic modeling, but it is also of obvious importance in low-dose biotransformation involving only a part of the hepatic lobule. Heterogeneity would also be critical in estimation of enzyme induction and inactivation. If, for instance, induction occurs mainly in the centrilobular region, whole liver homogenates will provide only an average figure. Similarly, if enzymes are inactivated only in the periportal hepatocytes composing 10% of the total cells, a routine assay will reveal 1% inactivation, essentially a negative result. A number of methods have been devised to analyze partial populations of the hepatocytes, as reviewed previously (Nakatsugawa and Tsuda, 1983; Nakatsugawa and Timoszyk, 1988). For instance, comparing metabolism during antero- and retrograde perfusion can reveal certain differences between periportal and centrilobular subpopulations of hepatocytes (Bradford and Nakatsugawa, 1982). We also have devised a method of fluorescence-labeling a partial population of hepatocytes by adapting the chromatographic translobular migration of fluorescein diacetate. Labeled cells were isolated by the usual collagenase procedure and separated individually by a fluorescence-activated cell sorter (Nakatsugawa and Timoszyk, 1988). The centrilobular hepatocytes contained both paraoxon deethylase and paraoxonase at twice the level as that of the periportal cells. Anticipated differential enzyme induction, however, was not demonstrated by the enzyme assays employed or by perfusion experiments. Validity of such assays, however, now appears questionable for reasons discussed in the following paragraphs.

Although many isozymes for various biotransformation enzymes have been recognized (Guengerich and Liebler, 1985), many of our enzyme data for *in vivo* observations are derived from classical enzyme assays in cell-free preparations. As one attempts to estimate biotransformation *in vivo*, it becomes clear that both the identity of the enzyme and the substrate concentration are uncertain. Most assays of biotransformation enzymes employ *optimal* conditions that provide the maximal rate following the routine practice for enzymes of intermediary metabolism. Infrequently, both V_{max} and K_m are also defined, but using the range of substrate concentrations that are practical, not the range pertinent to *in vivo* situations. For our purpose, however, this is far from satisfactory. Near-LD_{50} doses appear to give a portal parathion concentration in the 10^{-6} to 10^{-5} *M* range, although anesthesia could complicate the measurement (Schoenig, 1975; Sultatos *et al.*, 1985; Becker and Nakatsugawa, 1990a). Since 95% would be bound in the blood, actual free concentration would be in the 10^{-7} to 10^{-6} *M* range both in the blood and in hepatocytes in the periportal region and probably less further downstream. With this background, a hypothetical case will illustrate the point. Assume for parathion oxidase only two isozymes with identical optima for pH, etc. Further assume that

$$V_1 \text{ (}V_{max}\text{ for isozyme 1)} = 100 \text{ nmol/min}; V_2 = 1{,}000 \text{ nmol/min}$$

for a given amount of hepatocytes and

$$K_{m1} \text{ (Michaelis constant for isozyme 1)} = 1 \times 10^{-7} M, K_{m2} = 1 \times 10^{-5} M.$$

In other words, V_{max} for isozyme 2 is 10× that for isozyme 1 and K_m for isozyme 2 is 100× that for isozyme 1. The total activity is the sum of the rates for the two isozymes as given by the Michaelis–Menten equation:

$$v = V_1 (S) / [K_{m1} + (S)] + V_2 (S) / [K_{m2} + (S)]$$

Then if parathion available near the enzyme in hepatocytes is 1×10^{-7} M, the rate will be about 60 nmol/min, of which 50 is from isozyme 1 and 10 is from isozyme 2. *In* other words, isozyme 1 is the major enzyme under these conditions, working at half-maximal rate, while isozyme 2 is a minor component, participating at 1% of maximum. In our typical enzyme assay with a cell-free system, substrate concentration may be selected to give a high rate convenient for assay, e.g., 1×10^{-5} *M*. Then the rate will be 600 mnol/min, of which 100 is by isozyme 1 and 500 by isozyme 2. Thus, these assay conditions saturate isozyme 1, but bring into action isozyme 2, which nearly masks the isozyme 1.

In reality, of course, there are more isozymes, each of which will respond differently to enzyme inhibitors and inducers. Whether such a combination of V_{max} and K_m occurs *in vivo* is not certain, but our data suggest that it may not be unrealistic. Prompted by the need to attain nonsuicidal parathion activation *in vitro* as it occurs *in vivo*, we realized that simulation of *substrate buffering* by extracellular or *extramicrosomal* proteins would restore *in vivo*–like conditions (Nakatsugawa and Becker, 1987; Nakatsugawa *et al.*, 1989). Indeed, use of 5% bovine serum albumin permitted 20-min linear reactions in vitro appropriate for kinetic experiments. K_ms for parathion oxidation in a cell-free system were 2×10^{-6} to 4×10^{-6} *M* (Nakatsugawa *et al.*, 1989). Since bovine serum albumin binds 95% of parathion (Sultatos and

Minor, 1986), actual K_m would be 1×10^{-7} to 2×10^{-7} *M*. This contrasts with literature values of approximately 2×10^{-5} to 5×10^{-5} *M* (Norman *et al.*, 1974; Neal 1967b), which were obtained without substrate buffering. Clearly we need to stop using *optimal* conditions by habit, and aim our enzyme assays at biotransformation *in vivo*. Only then will we be able to integrate *in vitro* and *in situ* data to learn how biotransformation works *in vivo*.

Clues provided by various toxic phenomena have indeed yielded useful knowledge, including the chromatographic translobular migration, hepatic breakthrough threshold, and substrate buffering, but much remains to be discovered. Undoubtedly, efforts will continue to integrate the hepatic functions into the complex system of the whole body. While the impact of serial biotransformation is most direct in oral exposure to these chemicals, awareness of the mechanisms should greatly facilitate analysis of parenteral toxicity as well. For example, when dermal exposure produces the circulating concentration below the hepatic breakthrough threshold, any mathematical model would have to incorporate the liver as the terminal sink. We may also ask basic biological questions. How important are hepatic thresholds for a variety of xenobiotics in our foods? If biotransformation represents the animal's response to natural xenobiotics, could these features of hepatic disposition give any advantage for survival? There are undoubtedly a number of toxicological implications yet to be explored. Would a change in hepatic blood flow under stress alter the toxicity? Parathion toxicity to quail increases at extreme temperatures (Rattner *et al.*, 1987). Also knowledge of these hepatic mechanisms may eventually contribute to assessment of health risks. The basic significance of this new knowledge, however, may lie in the conceptual framework it offers in studies of the hepatic disposition of toxicants, for which OP insecticides have been an excellent model.

Acknowledgments

Research at SUNY ESF described in this article was supported by National Institute of Environmental Health Sciences Grant ES01019, DHHS, and a grant from Sankyo Co., Japan.

References

Alary, J. G., and Brodeur, J. (1969). Studies on the mechanism of phenobarbital-induced protection against parathion in adult female rats. *J. Pharmacol. Exp. Ther.* **169**, 159–167.

Aldridge, W. N. (1953a). Serum esterases. 1. Two types of esterase (A and B) hydrolysing *p*-nitrophenyl acetate, propionate and butyrate, and a method for their determination. *Biochem. J.* 53, 110–117.

Aldridge, W. N. (1953b). Serum esterases. 2. An enzyme hydrolysing diethyl *p*-nitrophenyl phosphate (E600) and its identity with the A-esterase of mammalian sera. *Biochem. J.* 53, 117–124.

Appleton, H. T., and Nakatsugawa, T. (1972). Paraoxon deethylation in the metabolism of parathion. *Pestic. Biochem. Physiol.* **2,** 286–294

Appleton, H. T., and Nakatsugawa T. (1977). The toxicological significance of paraoxon deethylation. *Pestic. Biochem. Physiol.* 7, 451–465.

Ball, W. L., Sinclair, J. W., Crevier, M., and Kay, K. (1954). Modification of parathion's toxicity for rats by pretreatment with chlorinated hydrocarbon insecticides. *Can. J. Biochem. Physiol.* **32,** 440–445.

Baron, J., Kawabata, T. T., Knapp, S. A., Voigt, J. M., Redick, J. A., Jacoby, W. B., and Geungerich, F. P. (1984). Intrahepatic distribution of xenobiotic-metabolizing enzymes. *In* "Foreign Compound Metabolism" (J. Caldwell and G.D. Paulson, eds.), pp. 17–36. Taylor and Francis, London.

Bass, S. W., Triolo, A. J., and Coon, J. M. (1972). Effect of DDT on the toxicity and metabolism of parathion in mice. *Toxicol. Appl. Pharmacol.* **22,** 684–693.

Becker, J. M., and Nakatsugawa, T. (1990a). Hepatic breakthrough thresholds for parathion and paraoxon and their implications to toxicity in normal and DDE-pretreated rats. *Pestic. Biochem. Physiol.* **36,** 83–98.

Becker, J. M., and Nakatsugawa, T. (1990b). Significance of hepatic breakthrough threshold in fenitrothion toxicity in male rats and femal mice. *Pestic. Biochem. Physiol.* **38,** 34–40.

Bradford, W.L., and Nakatsugawa, T. (1982). Perfusion analysis of periportal and centrilobular metabolism of paraoxon in the rat liver. *Pestic. Biochem. Physiol.* **18,** 298–303.

Cahalan, M. K., and Mangano, D. T. (1982). Liver function and dysfunction with anesthesia and surgery. *In* "Hepatology, Textbook of Liver Diseases" (D. Zakim and T. D. Boyer, eds.), W.B. Saunders, Philadelphia, Pennsylvania. pp. 1250–1261.

Cooper, D. Y., Levin, S., Narasimhulu, S., Rosenthal, O., and Estabrook, R.W. (1965). Photochemical action spectrum of the terminal oxidase of mixed function oxidase systems. *Science* **147,** 400–403.

Crummett, W. B., and Stehl, R. H. (1973). Determination of chlorinated dibenzo-*p*-dioxins and dibenzofurans in various materials. *Environ. Health Perspect.* **5,** 15–25.

Davison, A. N. (1955). The conversion of shradan (OMPA) and parathion into inhibitors of cholinesterase by mammalian liver. *Biochem. J.* **61,** 203–209.

Diggle, W. M., and Gage J. C. (1951). Cholinesterase inhibition by parathion *in vivo. Nature* **168,** 998.

DuBois, K. P. (1971). The toxicity of organophosphorus compounds to mammals. *Bull. W.H.O.* **44,** 233–240.

Eisenbrand, J., and Baumann, K. (1969). Water-solubility of 3,4-benzopyrene and formation of a water-soluble complex with caffeine. *Z. Lebensm. Unters. Forsch.* **140,** 158–163.

Forsyth, C. S., and Chambers, J. E. (1989). Activation and degradation of the phosphorothionate insecticides parathion and EPN by rat brain. *Biochem. Pharmacol.* **38,** 1597–1603.

Fry, J. R., and Bridges, J. W. (1979). Use of primary hepatocyte cultures in biochemical toxicology. *Rev. Biochem. Toxicol.* **1,** 201–247.

Gage, J. C. (1953). A cholinesterase inhibitor derived from O,O-diethyl O-*p*-nitrophenyl thiophosphate *in vivo. Biochem. J.* **54,** 426–430.

Goresky, C. A. (1963). A linear method for determining liver sinusoidal and extravascular volumes. *Am. J. Physiol.* **204,** 626–640.

Goresky, C. A., Huet, P. M., and Villeneuve, J. P. (1982). Blood-tissue exchange and blood flow in the liver. *In* "Hepatology, a Textbook of Liver Disease," (D. Zakim and T. D. Boyer, eds.), pp. 32–63. W.B. Saunders, Philadelphia, Pennsylvania.

Guengerich, F. P. (1977). Separation and purification of multiple forms of microsomal cytochrome P450. Activities of different forms of cytochrome P450 towards several compounds of environmental interest. *J. Biol. Chem.* **252,** 3970–3979.

Guengerich, F. P., and Liebler, D. C. (1985). Enzymatic activation of chemicals to toxic metabolites. *Crit. Rev. Toxicol.* **14,** 259–307.

Gumucio, J. J., and Chianale, J. (1988). Liver cell heterogeneity and liver function. *In* "The Liver: Biology and Pathology" 2nd Ed. (I. M. Arias, W. B. Jakoby, H. Popper, D. Schachter, and D. A. Shafritz, eds.), pp. 931–947. Raven Press, New York.

Halpert, J., Hammond, D., and Neal, R. A. (1980). Inactivation of purified rat liver cytochrome P450 during the metabolism of parathion (diethyl *p*-nitrophenyl phosphorothionate). *J. Biol. Chem.* **255**, 1080–1089.

Hollingworth, R. M., Alstott, R. L., and Litzenberg, R. D. (1973). Glutathionne S-aryl transferase in the metabolism of parathion and its analogs. *Life Sci.* **13**, 191–199.

Jacobsen, P. L., Spear, R. C., and Wei, E. (1973). Parathion and diisopropylfluorophosphate (DFP) toxicity in partially hepatectomized rats. *Toxicol. Appl. Pharmacol.* **26**, 314–317.

Jungermann, K., and Katz, N. (1982). Functional hepatocellular heterogeneity. *Hepatology* **2**, 385–395.

Kamataki, T., and Neal, R. A. (1976). Metabolism of diethyl *p*-nitrophenyl phosphorothionate (parathion) by a reconstituted mixed-function oxidase enzyme system: Studies of the covalent binding of the sulfur atom. *Mol. Pharmacol.* **12**, 933–944.

Kamataki, T., Lin, M. C. M. L., Belcher, D. H., and Neal, R. A. (1976). Studies of the metabolism of parathion with an apparently homogeneous preparation of rabbit liver cytochrome P450. *Drug Metab. Dispos.* **4**, 180–189.

Ku, T.-Y., and Dahm, P. A. (1973). Effect of liver enzyme induction on paraoxon metabolism in the rat. *Pestic. Biochem. Physiol.* **3**, 175–188.

Lamers, W. H., Hilberts, A., Furt, E., Smith, J., Jones, G. N., van Noorden, C. J. F., Janzen, J. W. G., Charles, R., and Moorman, A. F. M. (1989). Hepatic enzyme zonation: A reevaluation of the concept of the liver acinus. *Hepatology* **10**, 72–76.

Lauwerys, R. R., and Murphy, S. D. (1969). Interaction between paraoxon and tri-o-tolyl phosphate in rats. *Toxicol. Appl. Pharmacol.* **14**, 348–357.

Levi, P. E., Holllingworth, R. M., and Hodgson, E. (1988). Differences in oxidative dearylation and desulfuration of fenitrothion by cytochrome P450 isozymes and in the subsequent inhibition of monooxygenase activity. *Pestic. Biochem. Physiol,* **32**, 224–231.

McBain, J. B., Yamamoto, I., and Casida, J. E. (1971a). Mechanism of activation and deactivation of Dyfonate (O-ethyl *S*-phenyl ethylphosphonodithioate) by rat liver microsomes. *Life Sci.* **10**, 947–954.

McBain, J. B., Yamamoto, I., and Casida, J. E. (1971b). Oxygenated intermediate in peracid and microsomal oxidations of the organophosphonothionate insecticide Dyfonate. *Life Sci.* **10**, 1311–1319.

McIlvain, J. E., Timoszyk, J., and Nakatsugawa, T. (1984). Rat liver paraoxonase (paraoxon arylesterase). *Pestic. Biochem. Physiol.* **21**, 162–169.

Menzer, R. E., and Best, N. H. (1968). Effect of phenobarbital on the toxicity of several organophosphorus insecticides. *Toxicol. Appl. Pharmacol.* **13**, 37–42.

Morelli, M. A., and Nakatsugawa, T. (1978). Inactivation in vitro of microsomal oxidases during parathion metabolism. *Biochem. Pharmacol.* **27**, 293–299.

Morelli, M. A., and Nakatsugawa, T. (1979). Sulfur oxyacid production as a consequence of parathion desulfuration. *Pestic. Biochem. Physiol.* **10**, 243–250.

Mounter, L. A. (1954). Some studies of enzymatic effects of rabbit serum. *J. Biol. Chem.* **209**, 813–817.

Murphy, S. D. (1966). Liver metabolism and toxicity of thiophosphate insecticides in mammalian, avian, and piscine species. *Proc. Soc. Exp. Biol. Med.* **123**, 392–398.

Murphy S. D., and DuBois, K. P. (1957). Enzymatic conversion of the dimethoxy ester of benzotriazine dithiophosphoric acid to an anticholinesterase agent. *J. Pharmacol. Exp. Ther.* **119**, 572–583.

Nakatsugawa, T., and Becker, J. M. (1987). *In vitro* systems in the study of mechanisms of pesticide metabolism in animals. *In* "Pesticides Science and Biotechnology" (R. Greenhalgh and T. R. Roberts, eds.), pp. 523–526. Blackwell Scientific Publications, Boston, Massachusetts.

Nakatsugawa, T., and Dahm, P. A. (1965a) Parathion metabolism by liver microsomes, abstract #57. *Bull. Entomol. Soc. Am.* **11**, 157.

Nakatsugawa, T., and Dahm, P. A. (1965b). Parathion activation enzymes in the fat body microsomes of the American cockroach. *J. Econ. Entomol.* **58**, 500–509.

Nakatsugawa, T., and Dahm, P. A. (1967). Microsomal metabolism of parathion. *Biochem, Pharmacol.* **16**, 25–38.

Nakatsugawa, T., and Morelli, M. A. (1976). Microsomal oxidation and insecticide metabolism. *In* "Insecticide Biochemistry and Physiology" (C. F. Wilkinson, ed.), pp. 61–114. Plenum Press, New York.

Nakatsugawa, T., and Timoszyk, J. (1988). Fluorescence labeling and sorting of hepatocyte subpopulations to determine the intralobular heterogeneity of paraoxon-metabolizing enzymes in DDE-treated and control rats. *Pestic. Biochem. Physiol.* **30**, 113–124.

Nakatsugawa, T., and Tsuda, S. (1983). Metabolism studies with liver homogenate, hepatocyte suspension, and perfused liver. *In* "IUPAC Pesticide Chemistry, Human Welfare, and the Environment," (J. Miyamoto, ed.), Vol. 3, pp. 395–400. Pergamon Press, New York.

Nakatsugawa, T., Tolman, N. M., and Dahm, P. A. (1968). Degradation and activation of parathion analogs by microsomal enzymes. *Biochem. Pharmacol.* **17**, 1517–1528.

Nakatsugawa, T., Tolman, N. M., and Dahm, P. A. (1969a). Degradation of parathion in the rat. *Biochem, Pharmacol.* **18**, 1103–1114.

Nakatsugawa, T., Tolman, N. M., and Dahm, P. A. (1969b). Oxidative degradation of diazinon by rat liver microsomes. *Biochem. Pharmacol.* **18**, 685–688.

Nakatsugawa, T., Bradford, W. L., and Usui, K. (1980). Hepatic disposition of parathion: Uptake by isolated hepatocytes and chromatographic translobular migration. *Pestic. Biochem. Physiol.* **14**, 13–25.

Nakatsugawa, T., Timoszyk, J., and Becker, J. M. (1989). Substrate delivery as a critical element in the study of intermediary metabolites of lipophilic xenobiotics in vitro. *In* "Intermediary Xenobiotic Metabolism in Animals: Methodology, Mechanisms and Significance" (D. H. Hutson, J. Caldwell, and G. D. Paulson, eds.), pp. 335-353. Taylor and Francis, London.

Neal, R. A. (1967a). Studies on the metabolism of diethyl 4-nitrophenyl phosphorothionate (parathion) *in vitro. Biochem. J.* **103**, 183–191.

Neal, R. A. (1967b). Studies of the enzymic mechanism of the metabolism of diethyl 4-nitrophenyl phosphorothionate (parathion) by rat liver microsomes. *Biochem. J.* **105**, 289–297.

Neal, R. A., (1972). A comparison of the in vitro metabolism of parathion in the lung and liver of the rabbit. *Toxicol. Appl. Pharmacol.* **23**, 123–130.

Neal, R. A., Kamataki, T., Lin, M., Ptashne, K. A., Dalvi, R. R., and Poore, R. E. (1977) Studies of the formation of reactive intermediates of parathion. *In* "Biological Reactive Intermediates" (D. J. Jollow, J. J. Kocsis, R. Snyder, and H. Vainio, eds.), pp. 320–334. Plenum Press, New York.

Norman, B. J., Poore, R. E., and Neal, R. A. (1974). Studies of the binding of sulfur released in the mixed-function oxidase-catalyzed metabolism of diethyl *p*-nitrophenyl phosphorothionate (parathion) to diethyl *p*-nitrophenyl phosphate. *Biochem. Pharmacol.* **23**, 1733–1744.

O'Brien, R. D. (1959). Activation of thionophosphates by liver microsomes. *Nature* **183**, 121–122.

O'Brien, R. D. (1960). "Toxic Phosphorus Esters, Chemistry, Metabolism, and Biological Effects." Academic Press, New York.

Plapp, F. W., and Casida, J. E. (1958). Hydrolysis of the alkyl-phosphate bond in certain dialkyl aryl phosphorothioate insecticides by rats, cockroaches, and alkali. *J. Econ. Entomol.* **51**, 800–803.

Poore, R. E., and Neal, R. A. (1972). Evidence for extrahepatic metabolism of parathion. *Toxicol. Appl. Pharmacol.* **23**, 759–768.

Ptashne, K. A., Wolcott, R. M., and Neal, R. A. (1971). Oxygen-18 studies on the chemical mechanisms of the mixed function oxidase catalyzed desulfuration and dearylation reactions of parathion. *J. Pharmacol. Exp. Ther.* **179**, 380–385.

Rattner, B. A., Becker, J. M., and Nakatsugawa, T. (1987). Enhancement of parathion toxicity to quail by heat and cold exposure. *Pestic. Biochem. Physiol.* **27**, 330–339.

Schoenig, G. P. (1975). "Studies on the Hepatic Metabolism, Absorption, Binding and Blood Concentration of Parathion in the Albino Rat." Ph.D. Thesis, University of California, Davis.

Selye, H., and Mecs, I. (1974a). Effect upon drug toxicity of surgical interference with hepatic or renal function—Part 1. *Acta Hepato-Gastroenterol.* **21**, 191–202.

Selye, H., and Mecs, I. (1974b). Effect upon drug toxicity of surgical interference with hepatic or renal function. *Acta Hepato-Gastroenterol.* **21**, 266–273.

Sultatos, L. G. (1987). The role of the liver in mediating the acute toxicity of the pesticide methyl parathion in the mouse. *Drug Metab. Dispos.* **15**, 613–617.

Sultatos, L. G., and Minor, L. D. (1986). Factors affecting the biotransformation of the pesticide parathion by the isolated perfused mouse liver. *Drug Metab. Dispos.* **14**, 214–220.

Sultatos, L. G., Minor, L. D., and Murphy, S. D. (1985). Metabolic activation of phosphorothioatte pesticides: Role of the liver. *J. Pharmacol. Exp. Ther.* **232**, 624–628.

Sultatos, L. G., Kim, B., and Woods, L. (1990). Evaluation of estimations in vitro of tissue/blood distribution coefficients for organophosphate insecticides. *Toxicol. Appl. Pharmacol.* **103**, 52–55.

Takahashi, J., Miyaoka, T., Tsuda, S., and Shirasu, Y. (1984). Potentiated toxicity of 2-*sec*-butylphenyl methylcarbamate (BPMC) by *O,O*-dimethyl *O*-(3-methyl-4-nitrophenyl-phosphorothioate (fenitrothion) in mice; Relationship between acute toxicity and metabolism of BPMC. *Fundam. Appl. Toxicol.* **4**, 718–723.

Takahashi, H., Kato, A., Yamashita, E., Naito, Y., Tsuda, S., and Shirasu, Y. (1987). Potentiations of *N*-methylcarbamate toxicities by organophosphorus insecticides in male mice. *Fundam. Appl. Toxicol.* **8**, 139–146.

Triolo, A. J., and Coon, J. M. (1966). Toxicologic interactions of chlorinated hydrocarbon and organophosphate insecticides. *J. Agric. Food Chem.* **14**, 549–555.

Tsuda, S., Miyaoka, T., Iwasaki, M., and Shirasu, Y. (1984). Pharmacokinetic analysis of increased toxicity of 2-*sec*-butylphenyl methylcarbamate (BPMC) by fenitrothion pretreatment in mice. *Fundam. Appl. Toxicol.* **4**, 724–730.

Tsuda, S., Sherman, W., Rosenberg, A., Timoszyk, J., Becker, J. M., Keadtisuke, S., and Nakatsugawa, T. (1987). Rapid periportal uptake and translobular migration of parathion with concurrent metabolism in the rat liver in vivo. *Pestic. Biochem. Physiol.* **28**, 201–215.

Tsuda, S., Rosenberg, A., and Nakatsugawa, T. (1988). Translobular uptake patterns of environmental toxicants in the rat liver. *Bull. Environ. Contam. Toxicol.* **40**, 410–417.

Uchiyama, M., Yoshida, T., Homma, K., and Hongo, T. (1975). Inhibition of hepatic drug-metabolizing enzymes by thiophosphate insecticides and its drug toxicological implications. Biochem. Pharmacol. 24, 1221–1225.

Verschueren, K. (1983). "Handbook of Environmental Data on Organic Chemicals." 2nd Ed. Van Nostrand Reinhold, New York.

Vukovich, R. A., Triolo, A. J., and Coon, J. M. (1971). The effect of chlorpromazine on the toxicity and biotransformation of parathion in mice. *J. Pharmacol. Exp. Ther.* **178**, 395–401.

Wolcott, R. M., and Neal, R. A. (1972). Effect of structure on the rate of the mixed-function oxidase-catalyzed metabolism of a series of parathion analogs. *Toxicol. Appl. Pharmacol.* **22**, 676–683.

Wolcott, R. M., Vaughn, W. K., and Neal, R. A. (1972). Comparison of the mixed-function oxidase-catalyzed metabolism of a series of dialkyl *p*-nitrophenyl phosphorothionates. *Toxicol. Appl. Pharmacol.* **22**, 213–220.

Wustner, D. A., Desmarchelier, J., and Fukuto, T. R. (1972). Structure for the oxygenated product of peracid oxidation of Dyfonate insecticide (O-ethyl *S*-phenyl ethylphosphonodithioate). *Life Sci.* **11**, 583–588.

Yang, R. S. H., Hodgson, E., and Dauterman, W. C. (1971). Metabolism in vitro of diazinon and diazoxon in rat liver. *J. Agric. Food Chem.* **19**, 10–13.

Yoshida, T., Homma, K., Suzuki, Y., and Uchiyama, M. (1978). Effect of organophosphorus insecticides on hepatic microsomal cytochrome P450 in mice. *J. Pestic. Sci.* **3**, 21–26.

Yoshihara, S., and Neal, R. A. (1977). Comparison of the metabolism of parathion by a rat liver reconstituted mixed-function oxidase enzyme system and by a system containing cumene hydroperoxide and purified rat liver cytochrome P450. *Drug Metab. Dispos.* **5**, 191–197.

11

The Role of Target Site Activation of Phosphorothionates in Acute Toxicity

Janice E. Chambers
College of Veterinary Medicine
Mississippi State University
Mississippi State, Mississippi

I. Introduction

Organophosphorus (OP) insecticides display a wide range of acute mammalian toxicities, which have been well documented (Gaines, 1960, 1969; Gaines and Linder, 1986; Worthing and Walker, 1987; Meister, 1989). The oral median lethal dose (LD_{50}) for rats ranges from a few mg/kg for such compounds as demeton, parathion, mevinphos, and phorate to g/kg for compounds such as malathion, temophos, and ronnel. The potential danger of serious or lethal accidental poisonings to humans or other vertebrate nontarget organisms will, therefore, vary greatly among compounds.

The majority of OP insecticides used are phosphorothionates, which must be bioactivated primarily by cytochrome P450–dependent monooxygenases (P450) to the oxon or phosphate metabolites, as discussed in Chapters 1, 6, and 10, this volume. This bioactivation, the desulfuration reaction, is essential for converting the parent insecticide into a potent anticholinesterase. We have observed an increase in anticholinesterase potency of three orders of

Organophosphates: Chemistry, Fate, and Effects

magnitude between the phosphorothionates parathion, methyl parathion, and O-ethyl O-*p*-nitrophenyl phenylphosphonothioate (EPN) and their corresponding oxons (Chambers *et al.*, 1989a; Forsyth and Chambers, 1989).

Our studies elucidating biochemical factors responsible for acute toxicity differences of phosphorothionate insecticides have concentrated on six insecticides that display a wide range of acute toxicities. The rat oral LD_{50}s for these compounds from literature sources are given in Table I, as are the intraperitoneal 90-min lethal doses determined in our laboratories. It should be noted that our investigations established 500 mg/kg as the maximal tested dosage. However, based upon overt signs of poisoning as well as on brain acetylcholinesterase (AChE) inhibition observed in animals studied at this dosage (Chambers *et al.*, 1990), there is a difference in the toxicities of the three compounds tested at the 500 mg/kg dosage, with leptophos the most toxic, chlorpyrifos intermediate, and chlorpyrifos-methyl the least toxic.

II. Anticholinesterase Potency

The most obvious fact that could influence acute toxicity levels is the sensitivity of the target enzyme, nervous system AChE, to inhibition by the active metabolite, the oxon. Target enzyme sensitivity was assessed by 15 min I_{50} values to AChE of whole rat brains (Chambers *et al.*, 1990). These I_{50} values are also given in Table I. As can be seen by the values presented, there is no correlation between anticholinesterase potency and acute toxicity levels. In fact, some reversals were observed, notably with methyl parathion, which is extremely toxic and has the least potent oxon, and chlorpyrifos and chlorpyrifos-methyl, which are moderately or weakly toxic and have the most potent oxons. Thus, the ability of the oxon to inhibit the target enzyme appears to have only a minor role in determining the overall acute toxicity level.

III. Hepatic Bioactivation Activity

Clearly, the rate of oxon generation could also play a very important role in determining the acute toxicity level. An animal might not be able to detoxify or adapt to oxon generated rapidly, and therefore the parent insecticide would be more toxic. The liver is typically assumed to be responsible for most, if not all, of the generation of reactive metabolites in vertebrates, because of both its large size and its high specific activities of xenobiotic metabolizing enzymes, most importantly P450. Therefore, if hepatic oxon generation rate or capacity were a major determinant of acute toxicity level, then a correlation

TABLE I

Acute Toxicity Levels and Biochemical Parameters in Rats Related to Six Phosphorothionate Insecticides

Insecticides	LD_{50}(mg/kg)[a]	90-min LD (mg/kg)[b]	Brain AChE I_{50}(nM)[b]	Liver desulfuration[c]	Liver AliE I_{50} (nM)[b]	Brain desulfuration[c]
Parathion	4–13	8	22.5	5.1–12.0	1.3	640–790
Methyl parathion	14–24	12.5	89.3	19.0–25.8	290.0	860–2770
EPN	8–36	60	19.3	7.1–10.6	1.7	810–1000
Leptophos	19–20	500+	29.9	7.2–27.5	1.1	45.6–51.1
Chlorpyrifos	82–163	500+	4.0	7.1–26.7	0.75	12.7–13.0
Chlorpyrifos-methyl	1630–2140	500+	1.8	3.6–8.7	0.79	7.2–7.6

[a]Rat oral LD_{50}, data from Gaines (1960) for parathion, methyl parathion, and EPN; Gaines and Linder (1986) for leptophos; Gaines (1969) for chlorpyrifos; and Worthing and Walker (1987) for chlorpyrifos-methyl.

[b]Minimum lethal dose at 90 min, compounds administered intraperitoneally. Information from Chambers *et al.*, (1990). AChE, acetylcholinesterase; AliE, aliesterase.

[c]Hepatic microsomal desulfuration, expressed as nmol/min/g; composite microsomal plus mitochondrial brain desulfuration, expressed as pmol/min/g. Data from Chambers *et al.* (1989a) for methyl parathion, Forsyth and Chambers (1989) for parathion and EPN, and Chambers and Chambers (1989) for leptophos, chlorpyrifos, and chlorpyrifos-methyl.

should be expected between acute toxicity level and hepatic microsomal desulfuration activity. These activities are also listed in Table I. Wide ranges are given because hepatic sex differences in xenobiotic metabolism were observed, with males showing greater activity than females in all cases. The hepatic desulfuration activities for the various compounds are all within one order of magnitude of one another, whereas the acute toxicity levels span three orders of magnitude. Also, the differences in hepatic bioactivation do not correlate with acute toxicity levels. Thus the potential of the liver to bioactivate the phosphorothionate also does not seem to play a major role in determining the acute toxicity level.

IV. Protection by Aliesterases in the Liver and the Plasma

The potential of the aliesterases (carboxylesterases) to sequester organophosphates has been discussed by Maxwell in Chapter 9 of this volume. These serine esterases are quite sensitive to oxon inhibition and would be expected to detoxify a substantial amount of oxon as it is generated. The hepatic aliesterases display high specific activities (Chambers *et al.,* 1990), and reside in close proximity to the hepatic desulfuration activity. Thus they would be expected to trap large amounts of the hepatically generated oxon before the metabolite can escape the liver and may contribute substantially to the covalent binding of parathion or its metabolites observed in the perfused rat liver (Tsuda *et al.,* 1987). Very small proportions of phosphorothionate and oxon exited the isolated liver perfused with phosphorothionate, and the phosphorothionate occurred in higher concentrations than the oxon (Sultatos and Minor, 1986; Sultatos, 1987; Nakatsugawa, Chapter 10 of this volume). These data strongly suggest that the majority of phosphorothionate is metabolized or bound in the liver, and that the majority of any reactive metabolite generated is sequestered or detoxified. Much of this latter detoxication may be by phosphorylation of the aliesterases. Oxons would be expected to be labile, based on their reactivity, and they would be expected to react with the nearby aliesterases. Five of the six oxons we have studied are more potent as hepatic aliesterase inhibitors (by 2- to 20-fold) than as brain AChE inhibitors; the exception is methyl paraoxon, which is considerably less potent. These data, also presented in Table I, suggest that for all compounds tested except methyl parathion, the hepatic aliesterases should be able to sequester substantial amounts of the hepatically generated oxon, preventing its release into the bloodstream. In fact, an excellent correlation exists between the amount of aliesterase activity present in a liver homogenate and the homogenate's ability to detoxify paraoxon (Chambers *et al.,* 1990). This noncatalytic detoxication would, of course, be limited by the saturation of the aliesterases. With a high-dosage exposure to the six phosphorothionates tested (i.e., the

90-min lethal dose or 500 mg/kg), the level of hepatic aliesterase inhibition at 90 min was similar to or greater than the level of brain AChE inhibition for all compounds except methyl parathion, as would be predicted by the *in vitro* sensitivities of AChE and the aliesterases to the oxons (Chambers *et al.*, 1990). Also, in *in vivo* exposures to parathion, hepatic aliesterases were inhibited faster than was brain AChE (Chambers and Chambers, 1990).

Aliesterases also exist in substantial levels in rat plasma (Chambers et al., 1990). Further protection against the hepatically generated oxon would be afforded by these aliesterases, which are very substantially inhibited following an exposure of rats to all six of the phosphorothionates tested. Therefore, hepatically generated oxon that did escape the liver would be detoxified readily in the plasma, although this detoxication mechanism is also saturable.

V. Target Site Phosphorothionate Bioactivation and Other Monooxygenase Activities in the Brain

Thus, with the detoxication potential present in both the plasma and the liver (the latter also including monooxygenase-mediated phosphorothionate detoxication as well as A-esterase–mediated oxon hydrolysis), we have predicted that very little of the hepatically generated oxon would be able to escape from both the liver and the bloodstream to enter the target nervous tissue. The importance of extrahepatic activation of parathion has been suggested by others (Norman and Neal, 1976; Sultatos *et al.*, 1985). We have thus hypothesized that the brain itself is responsible for generating the oxon that ultimately inhibits the AChE. Although such desulfuration activity in brain would be expected to be low, its close proximity to the target enzyme would afford limited amounts of the reactive metabolite a much greater opportunity to reach the target enzyme.

A. Xenobiotic Metabolizing Enzymes in the Brain

There is a growing body of information on xenobiotic metabolizing enzyme activities within mammalian brain; this information has been recently reviewed by Minn *et al.* (1991). A variety of these activities have been found in brain tissue, although at substantially lower levels than corresponding activities in hepatic microsomes: cytochrome P450, NADPH-cytochrome *c* reductase, 7-ethoxycoumarin *O*-deethylase, aminopyrine *N*-demethylase, morphine *N*-demethylase, *d*-amphetamine hydroxylase, parathion desulfurase, aryl hydrocarbon hydroxylase (using naphthalene as the substrate), and benzo(a)pyrene hydroxylase (Das *et al.*, 1982; Fishman *et al.*, 1976; Ghersi-Egea *et al.*, 1987a,b; Liccione and Maines, 1989; Marietta *et al.*, 1979; Mesnil *et al.*, 1985; Nabeshima *et al.*, 1981; Norman and Neal, 1976; Qato and Maines, 1985; Srivastava *et al.*, 1983; Walther *et al.*, 1986 and 1987). Activities were present in microsomes and/or mitochondria. Cytochrome P450 has also been observed in microsomes and mitochondria from brain microvessels (Ghersi-Egea *et al.*, 1988). We have reported, in rat whole brain microsomal and crude mitochondrial preparations, the presence of NADPH-cytochrome *c* reductase, aminopyrine *N*-demethylase, and the desulfuration of six phosphorothionate insecticides (Chambers and Chambers, 1989; Chambers and Forsyth, 1989; Chambers *et al.*, 1989a; Forsyth and Chambers, 1989) and have observed aniline hydroxylase activity. There were no observable gender differences in these activities.

Studies of the inducibility of brain monooxygenase activities have suggested that the brain activities are less responsive than the liver to classical inducers, and are generally more responsive to 3-methylcholanthrene (3-MC)-type inducers than to phenobarbital (PB). The following were either not induced or induced very weakly: cytochrome P450, NADPH-cytochrome *c* reductase, 7-ethoxycoumarin O-deethylase, and 7-ethoxyresorufin O-deethylase (Ghersi-Egea *et al.*, 1987b; Liccione and Maines, 1989; Nabeshima *et al.*, 1981; Srivastava *et al.*, 1983; Strobel *et al.*, 1989; Walther *et al.*, 1987). Our own studies have indicated no induction by either PB of β-naphthoflavone (BNF) of any of the monooxygenase activities investigated thus far, i.e., parathion desulfuration, aminopyrine *N*-demethylase, NADPH-cytochrome *c* reductase (Chambers and Forsyth, 1989), or aniline hydroxylase.

B. Brain Cytochrome P450 Characterization

One of the most important functions of the brain monooxygenases may be the processing of steroid hormones, which then act within the brain on reproductive system activity and reproductive behavior. Aromatase, which converts androgens to estrogens, was identified in the hypothalamus–preoptic area, a region with important endocrine and reproductive functions (Lieberburg *et al.*, 1977; Roselli *et al.*, 1984). A P450-dependent catechol estrogen-forming activity has been found in the brain (Paul *et al.*, 1977). Additionally rat brain mitochondria possess P450 with cholesterol side-chain cleavage (SCC) ability, and rat brain reacts with antibodies raised to the adrenal mitochondrial P450XIA1 possessing SCC activity (Le Goascogne *et al.*, 1987; Walther *et al.*, 1987). Since the side-chain cleavage reaction is the first step of steroidogenesis, this activity supports the idea that neurosteroids may be produced by the brain for modulation of the brain's activity.

The rat brain contains a form of P450 with a similar antigenicity to hepatic cytochrome P450IA1 (P450c in rats; 3-MC inducible) (Kapitulnik *et al.*, 1987). Studies with antibodies raised to hepatic P450IIB1 (P450b in rats; PB-inducible) and cytochrome P450IA1 have suggested that one constitutive brain microsomal P450 immunologically resembles cytochrome P450IIB1 and is not responsible for steroid hydroxylations. A form immunologically similar to cytochrome P450IA1 was not constitutive but was inducible by BNF (Naslund *et al.,* 1988). However, these researchers did not do a similar characterization of the mitochondrial cytochrome P450.

Thus the literature strongly suggests that at least three P450 forms exist in the brain, identical to or antigenically related to P450IA1, P450IIB1, and P450XIA1 ($P450_{scc}$). It is unknown at present whether these and/or other P450s catalyze desulfuration or whether desulfuration is catalyzed by P450s normally involved in steroid hormone metabolism.

C. Brain Desulfuration Activity

Because of the documented evidence for low but significant levels of P450 and its associated activities within the brain, the ability of brain to activate the six phosphorothionates of interest was investigated. Since P450 has been shown to exist in mitochondria in amounts similar to those in the microsomes, both fractions were investigated. Desulfuration activity was found for all six phosphorothionates, with similar activity displayed within both subcellular fractions for a given phosphorothionate (Chambers and Chambers, 1989; Chambers and Forsyth, 1989; Chambers *et al.,* 1989a; Forsyth and Chambers, 1989). Although this even distribution of activity between the two compartments appeared to be at variance with reports of others who found P450 in higher concentrations in purified brain and microvessel mitochondria than in corresponding microsomes (Walther *et al.,* 1986; Ghersi-Egea *et al.,* 1988), these researchers expressed their P450 concentrations in terms of protein content of the preparation. Our results, however, were expressed in terms of wet weight of tissue, because it was felt that more useful comparisons could be made between the fractions, so the two data sets may not be contradictory.

These studies on phosphorothionate desulfuration in brain microsomal and mitochondrial fractions revealed an excellent correlation between brain desulfuration activity and acute toxicity levels. The composite activities (mitochondrial plus microsomal to yield a representative brain potential activity) are also listed in Table I. It can be readily seen that the three most toxic insecticides (parathion, methyl parathion, and EPN) are activated by the brain to the greatest extent and that the three least toxic insecticides (leptophos, chlorpyrifos, and chlorpyrifos-methyl) are activated to the least extent. Additionally, within this latter group, the bioactivation by the brain correlates with acute toxicity level, i.e., leptophos > chlorpyrifos > chlorpyrifos-methyl. In comparing the brain to the liver, the liver displayed a 6- to 30-fold higher desulfuration activity than the brain per unit wet weight for the three more toxic compounds (Chambers and Chambers, 1989). Thus, the data strongly suggest that the ability of the brain to bioactivate the phosphorothionates is critical in setting the overall acute toxicity level of the insecticide, although clearly other dispositional and metabolic factors will also be of significance in determining the precise level of toxicity.

VI. Target Site Phosphorothionate Bioactivation *in Vivo*

These data are strongly supportive of the importance of brain desulfuration in phosphorothionate acute toxicity. However, an examination of the data clearly reveals the fact that brain desulfuration activity is far lower than hepatic activity on a wet-weight basis, and the difference would be even more dramatic if total organ weight were taken into account. The *in vitro* presence of phosphorothionate desulfuration activity in the brain clearly does not prove that these monooxygenases are capable of generating oxon *in vivo*. We have been left with the question as to whether such low activities would in reality contribute any active metabolite. Two types of experiments have been conducted to eliminate the liver as a source of oxon in an intact animal. Following intravenous administration of a phosphorothionate, parathion, oxon production was monitored by measuring the degree of brain AChE inhibition.

In the first of these experiments, the posterior portion of an anesthetized rat was removed from the circulation by ligation of the aorta just posterior to the diaphragm. Because of the tremendous amount of the body (estimated to be about 60%) that had been removed from circulation, a very short incubation time of 15 min was employed. Although what is considered to be a realistic dosage of parathion (2.4 mg/kg) failed to result in statistically significant brain AChE inhibition, a very high dosage (48 mg/kg) did in this short time (Table II) (Chambers *et al.,* 1989b). These data strongly suggested that the brain was capable of generating the oxon that poisoned it. Although the liver clearly could not have generated the oxon in this experiment, there were three problems with the experiment: first, that anterior nonbrain tissue could not be ruled out as a source of oxon; second, the dose required was unrealistically high; and, third, that the time frame employed was unrealistically short.

A second experiment was conducted to address these problems. The procedure employed was a partial hepatectomy in which 70% of the liver was surgically removed; this was the maximal amount that could be removed without serious disruption of the circulation (Chambers *et al.,* 1991). Since the rats were allowed to recover from anesthesia by this less disruptive procedure, a more realistic lower dosage and longer time could be employed. A parathion dosage of 1.2 mg/kg was used, which gave about 70% brain AChE inhibition 30 min after injection. This degree of inhibition is similar to that seen during lethal parathion intoxication and at a time approximating the time required for a minimally lethal dose of parathion to kill. The fact that no liver or plasma aliesterase inhibition was observed clearly indicated that neither the residual liver nor any extrahepatic tissues other than the brain were

TABLE II

Inhibition of Rat Cerebral Cortex AChE Activity[a]

Condition	Parathion dosage (mg/kg)	Time (min)	Specific activity[b]	% Inhibition
Intact	0	15	55.6 ± 2.8 (3)	
	2.4	0	58.0 ± 3.5 (4)	—
	2.4	15	3.0 ± 1.0 (5)[c]	95
Ligated	0	15	51.6 ± 2.0 (3)	
	2.4	0	50.8 ± 8.8 (4)	2
	2.4	15	48.8 ± 1.7 (5)	5
	48	0	44.2 ± 3.5 (5)	14
	48	15	14.7 ± 2.2 (4)[c]	72

[a]Information from Chambers *et al.* (1989b).
[b]Specific activity expressed as nmole product formed per min/mg protein, mean ± S.E.M. (N).
[c]Mean significantly different from control mean ($P < 0.05$).

producing oxon that could have caused the brain AChE inhibition. The experiment was unable to assess whether this bioactivation was occurring in neurons, glial cells, or microvessels, any of which are possibilities. This experiment indicated that the brain is capable of bioactivating sufficient parathion to be responsible for substantial brain AChE inhibition at a dose and time consistent with an actual parathion intoxication. Brain bioactivation may not contribute substantially to intoxication at very high doses of phosphorothionate, when liver and blood protective mechanisms are saturated, and hepatically generated oxon can escape from both the liver and the blood. However, brain bioactivation may well be of great significance in lower-dose exposures in which the protective mechanisms can function effectively, especially in dermal exposures, in which first-pass extraction of phosphorothionate by the brain from the bloodstream is likely.

VII. Summary

Our experiments to this point have indicated that, while brain AChE sensitivity to oxon inhibition and hepatic phosphorothionate desulfuration activity do not correlate with insecticide acute toxicity level, brain phosphorothionate bioactivation can occur *in vivo*, and desulfuration activity, albeit low, does correlate. The brain, despite its low bioactivation activity, is capable of generating sufficient reactive metabolite to contribute substantially, if not solely, to the signs of intoxication; this bioactivation can occur at a dosage of phosphorothionate and in a time consistent with an acute intoxication. Thus, the evidence strongly suggests that target-site activation of phosphorothionates may be an important determinant of phosphorothionate acute toxicity level.

Acknowledgments

The financial support of the National Institutes of Health from Research Grant ES04394 and from Research Career Development Award ES00190 is gratefully acknowledged. All phosphorothionates and oxons were gifts of Dr. Howard Chambers, Department of Entomology, Mississippi State University. In addition to the coauthors of the cited publications, the author also gratefully acknowledges the laboratory assistance of Marilyn Alldread, Michael Bassett, Scott Boone, Russell Carr, and Amanda Holland.

References

Chambers, J. E., and Chambers, H. W. (1989). Oxidative desulfuration of chlorpyrifos, chlorpyrifos-methyl and leptophos by rat brain and liver. *J. Biochem. Toxicol.* **4,** 201–203.

Chambers, J. E., and Chambers, H. W. (1990). Time course of inhibition of acetylcholinesterase and aliesterases following parathion and paraoxon exposures in rats. *Toxicol. Appl. Pharmacol.* **103,** 420–429.

Chambers, J. E., and Forsyth, C. S. (1989). Lack of inducibility of brain monooxygenase activities including parathion desulfuration. *J. Biochem. Toxicol.* **4,** 65–70.

Chambers, J. E., Forsyth, C. S., and Chambers, H. W. (1989a). Bioactivation and detoxication of organophosphorus insecticides in rat brain. *In* "Intermediary Xenobiotic Metabolism: Methodology, Mechanisms and Significance" (J. Caldwell, D. H. Hutson, and G. D. Paulson, eds.), pp. 99–115. Taylor and Francis, Basingstoke, U.K.

Chambers, J. E., Munson, J. R., and Chambers, H. W. (1989b). Activation of the phosphorothionate insecticide parathion by rat brain *in situ. Biochem. Biophys. Res. Commun.* **165,** 327–333.

Chambers, H. W., Brown, B., and Chambers, J. E. (1990). Non-catalytic detoxication of six organophosphorus compounds by rat liver homogenates. *Pestic. Biochem. Physiol.* **36,** 308–315.

Chambers, J. E., Chambers, H. W., and Snawder, J. E. (1991). Target site bioactivation of the neurotoxic organophosphorus insecticide parathion in partially hepatectomized rats. *Life Sci.* **48,** 1023–1029.

Das, M., Seth, P. K., Dixit, R., and Mukhtar, H. (1982). Arylhydrocarbon hydroxylase of rat brain mitochondria: Properties of, and effect of inhibitors and inducers on, enzyme activity. *Arch. Biochem. Biophys.* **217,** 205–215..

Fishman, J., Hahn, E. F., and Norton, B. I. (1976). *N*-Demethylation of morphine in rat brain is localized in sites with high opiate receptor content. *Nature* **261,** 64–5.

Forsyth, C. S., and Chambers, J. E. (1989). Activation and degradation of the phosphorothionate insecticides parathion and EPN by rat brain. *Biochem. Pharmacol.* **38,** 1597–1603.

Gaines, T. B. (1960). The acute toxicity of pesticides in rats. *Toxicol. Appl. Pharmacol.* **2,** 88–99.

Gaines, T. B. (1969). Acute toxicity of pesticides. *Toxicol. Appl. Pharmacol.* **14,** 515–534.

Gaines, T. B., and Linder, R. E. (1986). Acute toxicity of pesticides in adult and weanling rats. *Fundam. Appl. Toxicol.* **7,** 299–308.

Ghersi-Egea, J. F., Walther, B., Minn, A., and Siest, G. (1987a). Quantitative measurement of cerebral cytochrome P450 by second derivative spectrophotometry. *J. Neurosci. Methods* **20,** 261–269.

Ghersi-Egea, J. F., Walther, B., Perrin, R., Minn, A., and Siest, G. (1987b). Inducibility of rat brain drug-metabolizing enzymes. *Eur. J. Drug Metab. Pharmacokin.* **12,** 263–265.

Ghersi-Egea, J. F., Minn, A., and Siest, G. (1988). A new aspect of the protective functions of the blood–brain barrier: Activities of four drug-metabolizing enzymes in isolated rat brain microvessels. *Life Sci.* **42,** 2414–2423.

Kapitulnik, J., Gelboin, H. V., Guengerich, F. P., and Jacobowitz, D. M. (1987). Immunohistochemical localization of cytochrome P450 in rat brain. *Neuroscience* **20**, 829–833.

Le Goascogne, C., Robel, P., Gouezou, M., Sananes, N., Baulieu, E.-E., and Waterman, M. (1987). Neurosteroids: Cytochrome $P450_{scc}$ in rat brain. *Science* **237**, 1212–1215.

Liccione, J. J., and Maines, M. D. (1989). Manganese-mediated increase in the rat brain mitochondrial cytochrome P450 and drug metabolism activity: Susceptibility of the striatum. *J. Pharmacol. Exp. Ther.* **248**, 222–228.

Lieberburg, I., Wallach, G., and McEwen, B. S. (1977). The effects of an inhibitor of aromatization (1,4,6-androstatriene-3,17-dione) and an anti-estrogen (Cl-628) on *in vivo* formed testosterone metabolites recovered from neonatal rat brain tissues and purified cell nuclei. Implication for sexual differentiation of the rat brain. *Brain Res.* **128**, 176–181.

Marietta, M. P., Vessell, E. S., Hartman, R. D., Weiz, J., and Dvorchik, B. H. (1979). Characterization of cytochrome P450-dependent aminopyrine *N*-demethylation. *J. Pharmac. Exp. Ther.* **208**, 271–9.

Meister, R. T. (1989). "Farm Chemicals Handbook." Meister Publishing, Willoughby, Ohio.

Mesnil, M., Testa, B., and Jenner, P. (1985). Aryl hydrocarbon hydroxylase in rat brain microsomes. *Biochem. Pharmacol.* **34**, 435–436.

Minn, A., Ghersi-Egea, J.-F., Perrin, R., Leininger, B., and Siest, G. (1991). Drug metabolizing enzymes in the brain and cerebral microvessels. *Brain Res. Reviews* **16**, 65–82.

Nabeshima, T., Fontenot, J., and Ho, I. K. (1981). Effects of chronic administration of pentobarbital or morphine on the brain microsomal cytochrome P450 system. *Biochem. Pharmacol.* **30**, 1142–1145.

Naslund, B., Glaumann, H., Warner, M., Gustafsson, J.-A., and Hansson, T. (1988). Cytochrome P450b and c in the rat brain and pituitary gland. *Mol. Pharmacol.* **31**, 31–37.

Norman, B. J., and Neal, R. A. (1976). Examination of the metabolism *in vitro* of parathion (diethyl *p*-nitrophenyl phosphorothionate) by rat lung and brain. *Biochem. Pharmacol.* **25**, 37–45.

Paul, S. M., Axelrod, J., and Diliberto, E. J. (1977). Catechol estrogen-forming enzyme of brain: Demonstration of a cytochrome P450 monooxygenase. *Endocrinology* **101**, 1604–1610.

Qato, M. K., and Maines, M. D. (1985). Regulation of heme and drug metabolism activities in the brain by manganese. *Biochem. Biophys. Res. Commun.* **128**, 18–24.

Roselli, C. E., Ellinwood, W. E., and Resko, J. A. (1984). Regulation of brain aromatase activity in rats. *Endocrinology 144*, 192–200.

Srivastava, S. P., Seth, P. K., and Mukhtar, H. (1983). 7-Ethoxycoumarin *O*-deethylase activity in rat brain microsomes. *Biochem. Pharmacol.* **32**, 3657–3660.

Strobel, H. W., Cattaneo, E., Adesnik, M., and Maggi, A. (1989). Brain cytochromes P450 are responsive to phenobarbital and tricyclic amines. *Pharmacol. Res.* **21**, 169–175.

Sultatos, L. G. (1987). The role of the liver in mediating the acute toxicity of the pesticide methyl parathion in the mouse. *Drug Metab. Dispos.* **15**, 613–617.

Sultatos, L. G., and Minor, L. D. (1986). Factors affecting the biotransformation of the pesticide parathion by the isolated perfused mouse liver. *Drug Metab. Dispos.* **14**, 214–220.

Sultatos, L. G., Minor, L. D., and Murphy, S. D. (1985). Metabolic activation of phosphorothionate pesticides: Role of the liver. *J. Pharmacol. Exp. Ther.* **232**, 624–628.

Tsuda, S., Sherman, W., Rosenberg, A., Timoszyk, J., Becker, J. M. Keadtisuke, S., and Nakatsugawa, T. (1987). Rapid periportal uptake and translobular migration of parathion with concurrent metabolism in the rat liver *in vivo*. *Pestic. Biochem. Physiol.* **28**, 201–215.

Walther, B., Ghersi-Egea, J. F., Minn, A., and Siest, G. (1986). Subcellular distribution of cytochrome P450 in the brain. *Brain Res.* **375**, 338–344.

Walther, B., Ghersi-Egea, J.-F., Minn, A., and Siest, G. (1987). Brain mitochondrial cytochrome $P450_{scc}$: Spectral and catalytic properties. *Arch. Biochem. Biophys.* **254**, 592–596.

Worthing, C. R., and Walker, S. B. (eds.) (1987). "The Pesticide Manual," 8th Ed. British Crop Protection Council, Thornton Heath, United Kingdom.

12

Transdermal Penetration and Metabolism of Organophosphate Insecticides

J. Edmond Riviere
Shao-Kuang Chang
Cutaneous Pharmacology and Toxicology Center
North Carolina State University
Raleigh, North Carolina

I. Overview

Penetration with subsequent absorption of topically applied pesticides through the skin is a major route for systemic effect. Although this has been appreciated for many years (Wang *et al.*, 1989; Honeycutt *et al.*, 1985), techniques have recently become available that should significantly increase our understanding of the mechanism of the percutaneous absorption and cutaneous metabolism of pesticides and provide a sounder base upon which data collected in animal and *in vitro* systems can be used to estimate human risk. When a pesticide is applied to the skin, it has a number of potential fates, only a few of which result in penetration or absorption through the skin. As can be seen in Fig. 1, the chemical must reach the stratum corneum (the dead outermost layer of skin) to begin the process of penetration.

One important variable is the formulation of the pesticide or the nature of the application vehicle. Many laboratory studies use simple monocom-

Organophosphates: Chemistry, Fate, and Effects

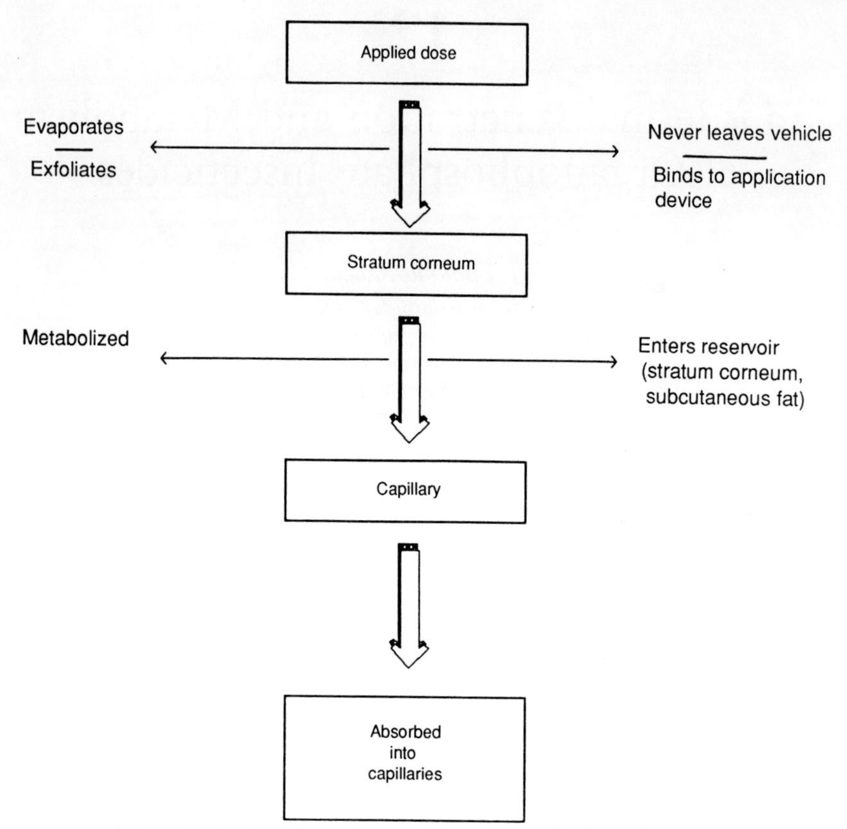

Figure 1 Possible fates of a pesticide applied topically to the skin.

ponent organic solvents (e.g., ethanol, acetone) to solubilize the pesticide for application to the skin. However, many field situations involve exposure to pesticides in complex formulations, granular preparations, or mixed with soil, which make extrapolations between the two settings problematic. The vehicle may have many effects on the process of percutaneous absorption. The rate and extent of pesticide release from the vehicle may be a primary determinant of topical bioavailability. For example, chemicals exposed in soil may not be completely available for absorption because of binding to soil constituents (Skowronski *et al.*, 1988). Recent work suggests that the vehicle itself will penetrate the stratum corneum, changing the solubility of the intercellular lipid matrix (primary route for passive absorption of pesticides) and thus changing the partitioning coefficient of the pesticide in the skin. The concentration of pesticide in this matrix is the primary determinant of extent of absorption. A substantial fraction of the dose of a volatile pesticide may evaporate before absorption occurs. Similarly, pesticide binding to exfoliated corneacytes or to application devices (laboratory setting) confounds estimating the extent of absorption. In order to be absorbed through skin, the pesticide must solubilize in the stratum corneum lipid environment for further penetration to occur by diffusion down its concentration gradient. Some compound may also penetrate by so-called shunt pathways; however, this is considered a minor route except for hydrophilic chemicals, which exclude most pesticides. These topics are well covered in the text of Bronaugh and Maibach (1989).

Another important variable is whether the chemical that has penetrated the stratum corneum is absorbed intact into the dermal microcirculation or is metabolized by the viable epidermis or cutaneous appendages (hair follicles, sweat glands). Although such first-pass cutaneous biotransformation has been studied for steriods, certain organic compounds and pro-drug pharmaceuticals (Kao, 1989), very little attention has been focused on skin metabolism of pesticides. For compounds not metabolized by the skin, a significant fraction of the more lipid-soluble pesticides may partition into cutaneous depots and be released only over a prolonged period. This may result in a biphasic absorption pattern that may modulate the toxicologic risk.

An area that is often ignored is the interaction of environmental factors with penetration, absorption, and metabolism. For example, what is the effect of temperature or humidity on pesticide percutaneous absorption? These factors could potentially have significant impact on the risk-assessment process.

Ideally, the importance of these factors would be addressed with *in vivo* human data, as often occurs in the pharmaceutical sector. However, this is generally not possible with agricultural chemicals, and thus, *in vitro* human and *in vivo* animal models must be employed to assess pesticide absorption. The first major obstacle facing these studies is in defining the limitations inherent in *in vitro* systems and in defining the factors that limit interspecies extrapolations.

II. Experimental Models

A. Overview

Because of the necessity to rely on experimental model data, it is instructive to review the assumptions inherent in using models. Skin is a complex organ that in addition to its barrier properties, is also involved in thermoregulation, nutrient metabolism, and endocrine regulation. Because of these additional functions, a significant number of evolutionary adaptations have occurred in different species, resulting in significant anatomical and physiological differences between species. The most obvious differences that affect assessing pesticide percutaneous absorption are the species differences in hair/fur density and skin thickness. In most comparative pesticide absorption studies reported in the literature, small laboratory animals (mice, rats, rabbits) generally show much greater rates of absorption than those reported for humans. This is probably owing to a thinner skin and extensive pelage. In contrast, pig and monkey studies generally are closer to human values (Wester and Maibach, 1977). Recent development of hairless rodents and human xenograft–rodent models has shown much promise (Reifenrath *et al.,* 1984). However, *in vivo* studies have methodological limitations that restrict their utility in studying mechanisms of percutaneous absorption and metabolism.

B. *In Vivo* Models

The primary strategy used to assess the percutaneous absorption of topcially applied pesticides in intact animals involves monitoring concentrations of absorbed chemical in blood or excreta. Because of the low concentrations often encountered in pesticide absorption studies, determining topical bioavailability by assaying blood concentrations is difficult. The most common technique employed in animals and humans is measuring compound appearance in urine and feces (Carver and Riviere, 1989; Wester and Maibach, 1985). An intravenous injection study is often independently conducted to determine the fraction of chemical excreted in urine. Results are then used as a correction factor so that only urine excretion is monitored to assess bioavailability. This technique is the primary method used to generate the *in vivo* human database on pesticide absorption, against which most animal data are compared (Maibach and Feldman, 1975). More sophisticated pharmacokinetic analyses (deconvolution linear system analysis) have also been performed on pesticide urinary excretion data (Fisher *et al.*, 1985). Finally, when small rodents are used, total mass balance studies may be conducted in which pesticide concentrations in urine, feces, exhaled air, and the carcass are measured (Shah *et al.*, 1981).

The major limitation to *in vivo* studies, in addition to the problem of direct toxicity to the host, is the confounding factor of systemic disposition and metabolism. Because absorbed pesticide concentrations are assayed only in blood or excreta, systemic metabolic processes (e.g., liver) confound interpretation of the data. The specific contribution of the skin to compound metabolism cannot be discerned, although its importance can be inferred when the fraction of parent compound excreted in urine is different after topical versus parenteral administration.

C. *In Vitro* Models

The most widely used method for assessing pesticide percutaneous absorption involves the use of diffusion cells, in which the skin is a membrane separating donor and receptor chambers. In many cases, the donor chamber facing the epidermis is left open to the air. In static systems, the receptor chamber is a fixed volume of perfusate, while in flow-through systems, the receptor constantly flows under the dermis to simulate blood perfusion. The skin sample employed may be full thickness or of a defined thickness produced by a dermatome. In some cases, epidermal membranes alone may be studied. Skin may be harvested from any species including humans (Bronaugh, 1989). Recent developments include the use of cell culture-derived, human skin-equivalent membranes.

The major problem encountered with these systems is that the anatomical pathway of penetration is not identical to that seen *in vivo* because an intact capillary network is not present. Compounds must traverse the dermis before entering the receptor chamber, a distance greater than that normally traveled *in vivo* to reach blood perfusing a capillary. For lipid-soluble pesticides, a reservoir may form, and a compound may never enter the receptor solution, especially if an aqueous buffer is employed. This problem may partially be avoided by assaying the compound in both the receptor fluid and dermis, although time-concentration profiles are difficult to generate using this technique. Additionally, the formation of the reservoir *in vitro* could directly affect penetration kinetics.

III. Assessing Environmental Effects

A. *In Vitro* Penetration Studies

As alluded to earlier, environmental variables could have a major impact on the rate and extent of pesticide percutaneous absorption. In order to illustrate this, results from our laboratory on the percutaneous absorption of parathion in flow-through diffusion cells will be presented. The environmental variables studied were temperature and humidity. The importance of these variables to assessing percutaneous absorption in the field is obvious. Three levels of relative humidity were studied: 20, 60, and 90%. These conditions were produced by enclosing the flow-through diffusion cells in a chamber through which air of a defined humidity and temperature is circulated. Pig skin was utilized in all of these studies. As can be seen in Fig. 2, high relative humidity significantly increased parathion absorption (n = 4 replicates).

Two temperatures were independently controlled in these studies; temperature of the perfusate bathing the dermal receptor chambers and the temperature of the air in the donor chamber. As can be seen in Fig. 3, only when the temperature of both the perfusate and air were increased to 42°C did a significant increase in penetration occur at the dose studied.

These *in vitro* studies suggest that varying temperature and humidity may have significant effects on the rate of parathion absorption. These could

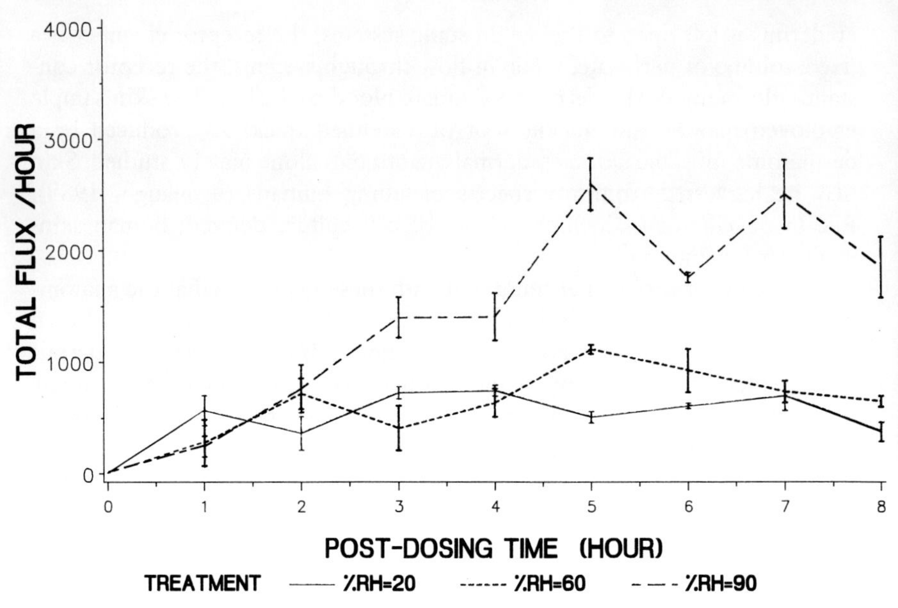

Figure 2. Effect of humidity on the *in vitro* percutaneous absorption of parathion in pig skin (4 μg/cm^2).

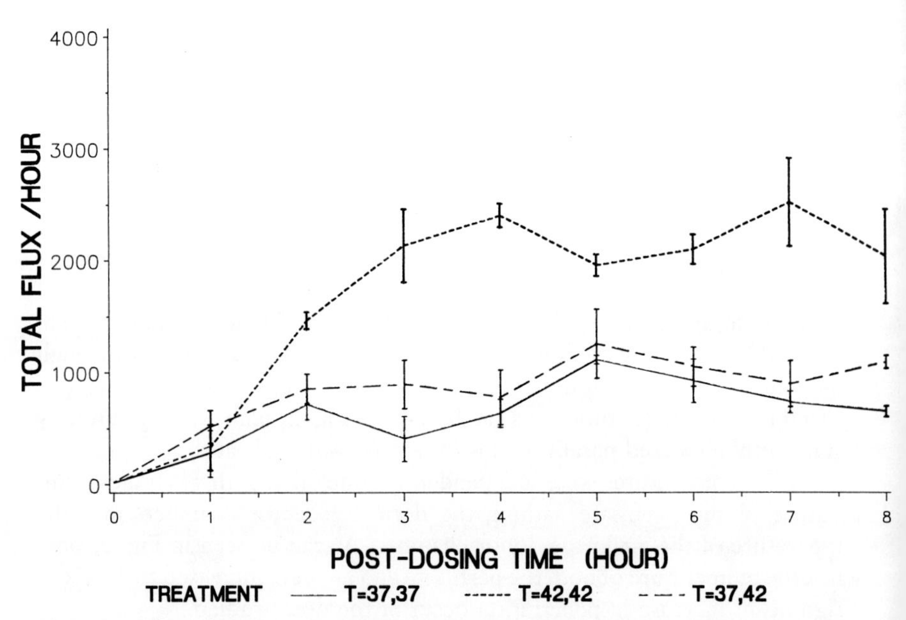

Figure 3. Effect of temperature on the *in vivo* percutaneous absorption of parathion in pig skin (4 μg/cm^2).

relate to changes in the state of stratum corneum hydration or to the fluidity of intercellular epidermal lipids. Since these changes were detected *in vitro* , changes in blood flow, which might be expected *in vivo* , were not involved. However, cutaneous blood flow would be expected to change with increased temperature and may affect the results seen *in vivo* .

B. *In Vitro* Cutaneous Metabolism

Another potential effect of these environmental conditions on parathion absorption involves their modulation of cutaneous metabolism. Parathion is metabolized to *p*-nitrophenol and paraoxon (see discussion of metabolism in Chapters 1 and 10; also Fig. 10, Chapter 1). Previous studies in an isolated perfused skin preparation (later discussion) demonstrated P450-mediated conversion of parathion to paraoxon in porcine skin. The effect of temperature and humidity on cutaneous metabolism of parathion was also studied in *in vitro* diffusion cells. The conditions varied were the same as described in Section III, A above.

Increasing relative humidity to 90% increased the production of *p*-nitrophenol (Fig. 4). Increasing temperature of perfusate or ambient air tended to increase the production of paraoxon (Fig. 5) compared to normal temperatures. These effects may have been mediated through changes in

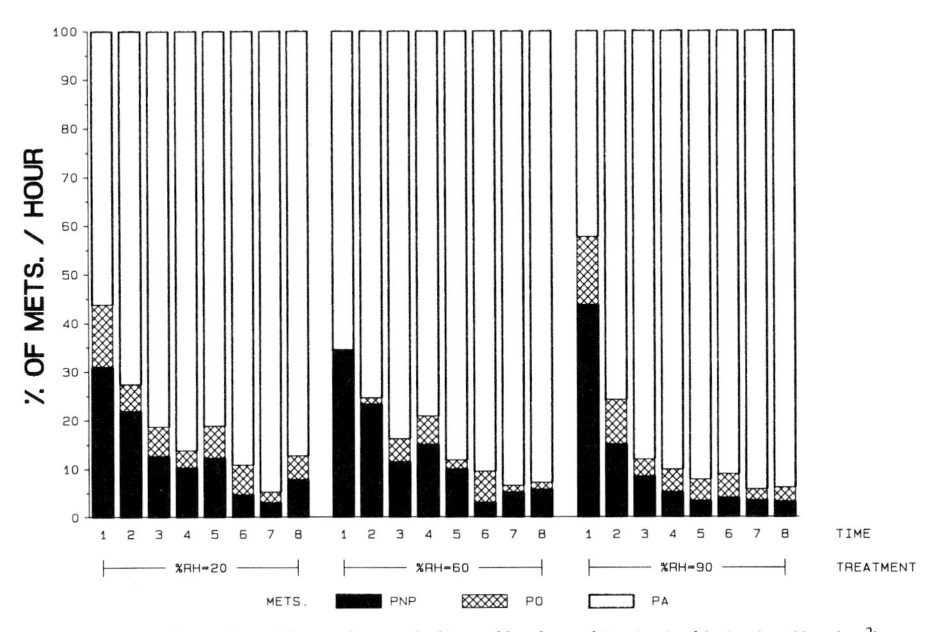

Figure 4. Effect of humidity on the metabolic profile of parathion in pig skin *in vitro* (4 μg/cm^2).

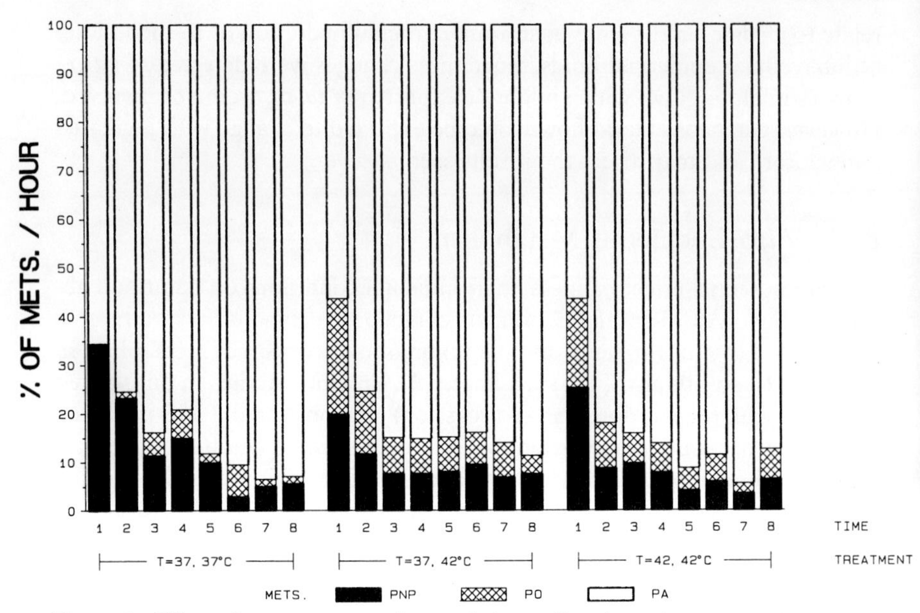

Figure 5. Effect of temperature on the metabolic profile of parathion in pig skin *in vitro* (4 $\mu g/cm^2$).

enzyme activity (e.g., temperature-induced increased in cytochrome activity for paraoxon production) or changes in the absolute or relative penetration of parent compound or metabolite.

C. Overall Effects of Environmental Variables

When these *in vitro* studies are considered together, one can begin to appreciate the potential impact of altered temperature and/or humidity on parathion absorption and cutaneous biotransformation. In order to investigate the importance of these variables in physiologically normal skin, *in vivo* studies should be conducted. However, for the reason discussed above, the resolution of *in vivo* studies for discerning these effects, especially cutaneous metabolism, is limited. A promising approach is the use of isolated perfused skin preparations.

IV. Isolated Perfused Porcine Skin Flap

In order to illustrate the advantages of perfused skin preparations, we shall discuss the isolated perfused porcine skin flap (IPPSF) developed in our laboratory for assessing xenobiotic percutaneous absorption in a biologically relevant model. This model was developed as an *in vitro* system that possesses a viable epidermis and dermis and a functional microcirculation, most of the attributes normally associated with *in vivo* studies. The creation of this model, the maintenance of a viable preparation *in vitro* and for use in percutaneous absorption and cutaneous toxicology studies, has been adequately reported elsewhere (Riviere, *et al.*, 1986, 1989/1990; Monteiro-Riviere *et al.*, 1987; Carver *et al.*, 1989; Williams *et al.*, 1990; King and Monteiro-Riviere, 1990).

Briefly, the IPPSF is created in a two-stage surgical procedure. A single pedicle, axial-patterned, tubed skin flap is created in weanling Yorkshire swine, based on the superficial epigastric artery and vein. Two days later, the artery is cannulated, and the skin flap, transferred to the specially designed organ-perfusion chamber depicted in Fig. 6. The skin flap is perfused with a Krebb's bicarbonate-based albumin buffer solution containing glucose as an energy source. Viability is assessed by monitoring arterial and venous perfusate glucose concentrations and calculating glucose utilization over the course of an experiment (Fig. 7) (Riviere *et al.*, 1986; Monteiro-Riviere, 1990).

Percutaneous absorption studies are conducted by placing a compound on the surface of the IPPSF and assaying venous perfusate for compound

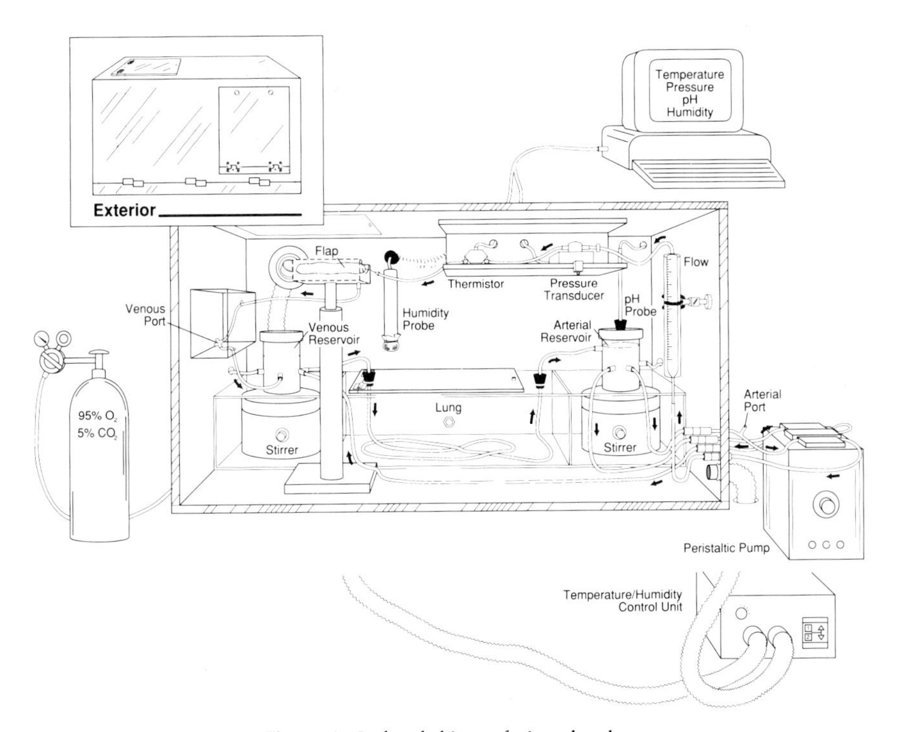

Figure 6 Isolated skin-perfusion chamber.

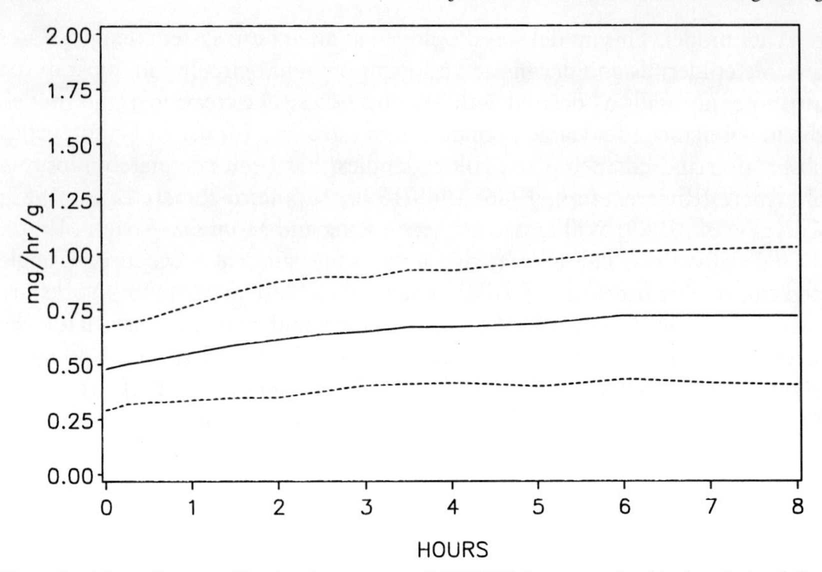

Figure 7 Mean glucose utilization for a group of 49 IPPSF demonstrating biochemical stability of the time course of a typical absorption experiment (mean ± SE).

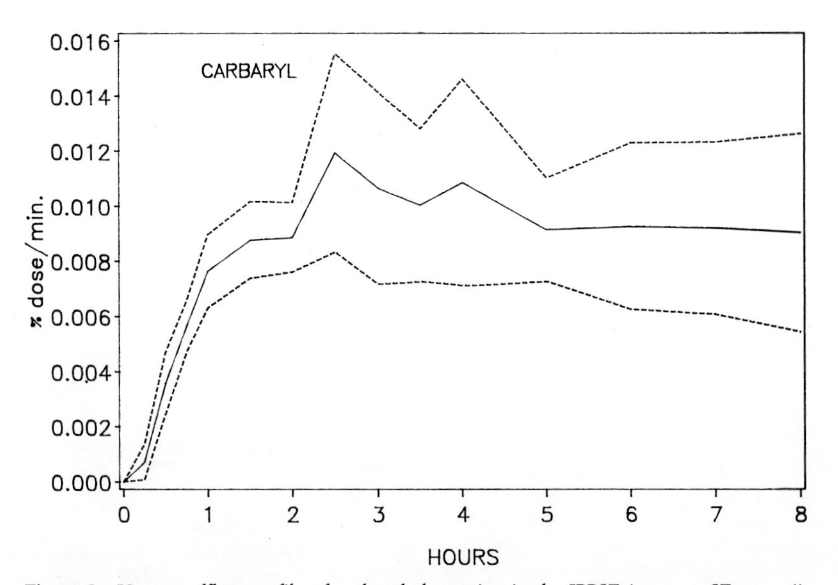

Figure 8 Venous efflux profile of carbaryl absorption in the IPPSF (mean ± SE; n = 4).

concentration. The main advantage of this approach is that in such an isolated preparation, the venous effluent collected is free from systemic influence, and thus truly reflects the skin's contribution to absorption and subsequent metabolism. Typical mean (± SE) percutaneous absorption profiles for carbaryl (Fig. 8) and lindane (Fig. 9) demonstrate the chemical specificity of the profiles observed. For a series of seven compounds, the correlation of IPPSF predicted to reported *in vivo* total absorption over 6 days in pigs was excellent ($R = 0.973$) (Williams *et al.*, 1990). To calculate a 6-day absorption from an 8-hour IPPSF experiment, the venous profiles were fitted to a pharmacokinetic model and these parameters then used to extrapolate to the later time frame (Williams and Riviere, 1989; Williams *et al.*, 1990).

Recent studies employing the IPPSF have also demonstrated that topically applied parathion is metabolized to paraoxon and *p*-nitrophenol in the IPPSF (Carver *et al.*, 1990). This reaction may be modulated by occlusion of the application site. Furthermore, metabolism may be completely blocked if cutaneous P450 activity is inhibited. These studies demonstrate the utility of a model such as the IPPSF in predicting percutaneous absorption *in vivo* and in studying the mechanism of cutaneous biotransformation.

V. Conclusions

The process of pesticide percutaneous absorption is a complex phenomenon involving the interaction of physiochemical and metabolic processes. A num-

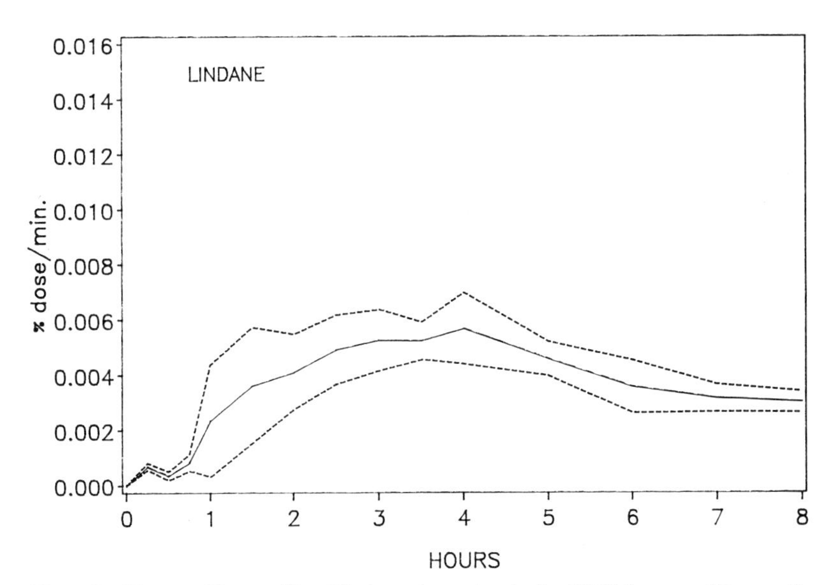

Figure 9 Venous efflux profile of lindane absorption in the IPPSF (mean ± SE; n = 4).

ber of approaches are currently being used to study this problem, including *in vitro* and animal models. In order for these to shed light on the *in vivo* absorption of pesticides in humans, the experimental model selected should first be validated by comparing absorption kinetics in human volunteers. Because of species-specific anatomical and metabolic handling of different chemical entities, extrapolation of an "unknown" pesticide from any model system to humans is difficult. In such screening studies, often only the relative extent of penetration may be estimated, especially since site-to-site and interindividual variability within humans may be greater than the interspecies variability for some chemicals. This would preclude precise quantitative estimates being made.

In vitro models designed to study the mechanisms of pesticide penetration, and metabolism should be capable of responding as close to the *in vivo* situation as possible. Flow-through diffusion cells and isolated perfused skin preparations are useful tools to probe mechanisms of percutaneous absorption. The mechanisms behind the interaction of environmental factors on pesticide penetration and cutaneous metabolism deserve further study. Finally, the nature of variability in percutaneous absorption within an individual, between individuals, and between species should be defined and correlated to measurable parameters if meaningful risk-assessment predictions are to be made.

References

Bronaugh, R. L. (1989). Determination of percutaneous absorption by *in vitro* techniques. *In* "Percutaneous Absorption" 2nd Ed. (R. L. Bronaugh and H. I. Maibach, eds.), pp. 239–258. Marcel Dekker, New York.

Bronaugh, R. L., and Maibach, H. I. (1989). "Percutaneous Absorption" 2nd Ed. Marcel Dekker, New York.

Carver, M. P., and Riviere, J. E. (1989). Percutaneous absorption and excretion of xenobiotics after topical and intravenous administration to pigs. *Fundam. Appl. Toxicol.* **13,** 714–722.

Carver M. P., Williams, P. L., and Riviere, J. E. (1989). The isolated perfused porcine skin flap (IPPSF). III. Percutaneous absorption pharmacokinetis of organophosphates, steroids, benzoic acid, and caffeine. *Toxicol. Appl. Pharmacol.* **97,** 324–337.

Carver, M. P., Levi, P. E., and Riviere, J. E. (1990). Parathion metabolism during percutaneous absorption in perfused porcine skin. *Pestic. Biochem. Physiol.,* **38,** 245–254.

Fisher, H. L., Most, B., and Hall, L. L. (1985). Dermal absorption of pesticides calculated by deconvolution. *J. Appl. Toxicol.* **5,** 163–177.

Honeycutt, R. C., Zweig, G., and Ragsdale, N. N. (1985). "Dermal Exposure Related to Pesticide Use." American Chemical Society, Washington, D.C.

Kao, J. (1989). The influence of metabolism on percutaneous absorption. In "Percutaneous Absorption" 2nd Ed. (R. L. Bronaugh and H. I. Maibach, eds.) pp. 259–282. Marcel Dekker, New York.

King, J. R., and Monteiro-Riviere, N. A. (1990). Cutaneous toxicity of 2-chloroethyl methyl sulfide in isolated perfused porcine skin. *Toxicol. Appl. Pharmacol.* **104,** 167–179.

Maibach, H. I., and Feldman, R. J. (1975). Systemic absorption of pesticides through the skin of man. *In* "Occupational Exposure to Pesticides: Report to the Federal Working Group on Pest Management from the Task Group on Occupational Exposure to Pesticides." Appendix B, pp. 120–127. U.S. Government Printing Office, 0-551-026, Washington, D.C.

Monteiro-Riviere, N. A. (1990). Specialized technique: Isolated perfused porcine skin flap. *In* "Methods for Skin Absorption" (B. W. Kemppainen and W. G. Reifenrath, eds.), pp. 175–189, CRC Press, Boca Raton, Florida.

Monteiro-Riviere, N. A., Bowman, K. F., Scheidt, V. J., and Riviere, J. E. (1987). The isolated perfused porcine skin flap (IPPSF). II. Ultrastructural and histological characterization of epidermal viability. *In Vitro Toxicol.* **1**, 241–252.

Reifenrath, W. G., Chellquist, E. M., Shipwash, E. A., Jederberg, W. W., and Krueger, G. G. (1984). Percutaneous penetration in the hairless dog, weanling pig, and grafted athymic nude mouse: Evaluation of models for predicting skin penetration in man. *Br. J. Dermatol.* 3 (Suppl. 27), 123–135.

Riviere, J. E., Bowman, K. F., Monteiro-Riviere, N. A., Carver, M. P., and Dix, L. P. (1986). The isolated perfused procine skin flap (IPPSF). I. A novel *in vitro* model for percutaneous absorption and cutaneous toxicology studies. *Fundam Appl. Toxicol.* **7**, 444–453.

Riviere, J. E., Sage, B., and Monteiro-Riviere, N. A. (1989/1990). Transdermal lidocaine iontophoresis in isolated perfused porcine skin. *Cutan. Ocular Toxicol.* **9**, 493–504.

Shah, P. V., Monroe, R. J., and Guthrie, F. E. (1981). Comparative rates of dermal penetration of insecticides in mice. *Toxicol. Appl. Pharmacol.* **59**, 414–423.

Skowronski, G. A., Turkall, R. M., and Abdel-Rahman, M. S. (1988). Soil absorption alters bioavailability of benzene in dermally exposed male rats. *Am. Ind. Hyg. J.* **49**, 506–511.

Wang, R. G. M., Franklin, C. A., Honeycutt, R. C., and Reinert, J. C. (1989). "Biological Monitoring for Pesticide Exposure." American Chemical Society, Washington, D.C.

Wester, R. C., and Maibach, H. I. (1977). Percutaneous absorption in man and animals: A perspective, *In* "Cutaneous Toxicity" (V. A. Drill and P. Lazar, eds.), pp. 111–126 Academic Press, New York.

Wester R. C., and Maibach, H. I. (1985). *In vivo* percutaneous absorption and decontamination of pesticides in humans. *J. Toxicol. Environ. Health* **16**, 25–37.

Williams, P. L., and Riviere, J. E. (1989). Definition of a physiological pharmacokinetic model of cutaneous drug distribution using the isolated perfused porcine skin flap (IPPSF). *J. Pharm. Sci.* **78**, 550–555.

Williams, P. L., Carver, M. P., and Riviere, J. E. (1990). A physiologically relevant pharmacokinetic model of xenobiotic percutaneous absorption utilizing the isolated perfused porcine skin flap (IPPSF). *J. Pharm. Sci.* **79**, 305–311.

III

Toxic Effects—Noncholinergic Biochemical

13

Direct Actions of Organophosphorus Anticholinesterases on Muscarinic Receptors

Amira T. Eldefrawi
Department of Pharmacology and
Experimental Therapeutics
School of Medicine
University of Maryland at Baltimore
Baltimore, Maryland

David Jett
Department of Pharmacology and
Experimental Therapeutics
School of Medicine
University of Maryland at Baltimore and
U.S. Fish and Wildlife Service
Patuxent Wildlife Research Center
Laurel, Maryland

Mohyee E. Eldefrawi
Department of Pharmacology and
Experimental Therapeutics
School of Medicine
University of Maryland at Baltimore
Baltimore, Maryland

Organophosphates: Chemistry, Fate, and Effects

I. Introduction

A. Historical Evidence for Effects of Organophosphorus Compounds on Acetylcholine Receptors

It is easy to distinguish between acetylcholinesterase (AChE) and acetylcholine (ACh) receptors on the basis of their different functions, drug specificities and structures. In the late 1960s, some thought that the nicotinic acetylcholine receptor (nAChR) and AChE were part of the same protein for both bound ACh and the classic receptor-inhibitors D tubocurarine (Belleau, *et al.*, 1970) and α-bungarotoxin (Stalc and Zupancic, 1972), as well as ambenonium derivatives (Webb, 1965). With the similar locations of AChE and nAChR at synapses in the neuromuscular and brain nicotinic junctions, and the consequence of AChE inhibition on nAChR function, it was extremely difficult to distinguish a direct effect of an anti-AChE on the nAChR before the two proteins were purified in the 1970s. An early report by Bartels and Nachmansohn (1969) revealed that, as expected, the organophosphorus (OP) compounds paraoxon, diisopropylfluorophosphate (DFP), and echothiophate enhanced ACh-induced depolarization of the postsynaptic membrane in electric eel electroplax. However, at higher concentrations of the OP compounds, the membrane hyperpolarized, suggesting that the OP compounds were directly inhibiting the nAChR. This was followed by the demonstration of inhibition of [^{3}H]ACh binding to the *Torpedo* electric organ after all AChE was inhibited (Eldefrawi *et al.*, 1971). In frog muscle, the nAChR conductances were inhibited directly by DFP (Kuba *et al.*, 1973). The nAChR channel was also inhibited by the potent nerve gas VX (*o*-ethyl *S*[2-(diisopropylamino)ethyl)]methyl phosphonothionate (Aracava *et al.*, 1987; Bakry *et al.*, 1988). Based on effects of the OP compounds on binding of [^{3}H]ACh and [^{3}H]phencyclidine or [^{3}H]perhydrohistrionicotoxins to the nAChR ionic channel, it was found that soman and echothiophate in micromolar concentrations acted as partial agonists and induced receptor desensitization, while VX was a potent open-channel blocker and enhanced receptor desensitization (Bakry *et al.*, 1988). On the other hand, DFP was suggested to induce receptor desensitization by binding to a third site on the nAChR (Eldefrawi *et al.*, 1988).

There were numerous reports on decreased muscarinic acetylcholine receptor (mAChR) numbers based on binding of [^{3}H]quinuclidinyl benzilate (QNB) in brains of animals treated chronically with low levels of OP anti-AChEs (Gazit *et al.*, 1979; Ehlert *et al.*, 1980). This was not necessarily owing to a direct action of the OP compound on mAChR, since inhibition of AChE increases ACh concentration in the synaptic gap, and continued activation of mAChRs would result in such down-regulation. In addition, however, soman was found to depress ganglionic synaptic transmission after AChE was inhibited, and mAChR antagonists protected against it (Yarowsky *et al.*, 1984). The first evidence for a direct action of OPs on mAChRs was based on inhibition by OPs of [^{3}H] QNB binding to bovine brain receptors (Volpe *et al.*, 1985). The direct action of OP compounds on mAChRs is the focus of this presentation. Unlike the evidence on nAChRs, which is based on the easily accessible nAChRs, studies on mAChRs are mostly from mammalian brain, where several mAChR subtypes are found, thereby complicating analysis of the data.

B. Muscarinic Receptor Subtypes

Characterization of different mAChR subtypes has generated much interest in the field of pharmacology and experimental therapeutics because of the potential for development of selective therapeutics without side effects (Bonner, 1989; Goyal, 1989; Hulme *et al.*, 1990). The first evidence for division of mAChRs into distinct subtypes was the observation that pirenzepine (PZ) bound to neuronal mAChR with much higher affinity than to sites in the heart and smooth muscle (Hammer *et al.*, 1980). This discovery led to the classification of two subtypes of mAChRs: a high-affinity M_1 subtype found primarily in the CNS, and a low-affinity M_2 subtype found in heart and smooth muscle.

It had been known for quite some time that the nicotinic antagonist, gallamine had potent cardioselective muscarinic effects, and as a result of studies with PZ and cardioselective muscarinic antagonists, there soon emerged a third subtype, M_3, that was non-M_1 and not found in the heart but rather in smooth muscle and exocrine glands. It is important to note that no agonists or antagonists are selective for a single subtype, and pharmacological classification is based solely on relative affinities of the drugs for the receptor subtype.

The advent of molecular cloning techniques suitable for studying the mAChR receptor protein provided a second important tool for subtype classification. The primary amino acid sequences of five distinct human mAChRs have been determined by cloning and sequence analysis of cDNAs or genomic DNAs; hybridization studies indicate there may be several others (Bonner, 1989a). Transfection experiments with frog *Xenopus* oocytes lacking endogenous receptors have produced cell lines that express a single subtype. Pharmacological profiles of subtypes using this technique parallel those determined by conventional methods such that the rat gene (designated by a lower case m, e.g., m1, m2, etc.) products correspond to pharamcologically defined subtypes (designated by an upper case M and numerical subscript, e.g., M_1, M_2, etc.). On the other hand, the five human genes are designated as HM1, HM2, etc. (Table I). Thus, to avoid confusion, the gene product designation of receptor subtypes will be used in the remainder of this discus-

TABLE I

Subtypes of Muscarinic Receptors

Rat gene[a]	m1	m2	m3	m4	m5
Human gene[a]	HM1	HM2	HM4	HM3	HM5
Pharmaceutically defined subtype[b]	M_1	M_2	M_3	—	—
Primary tissue distribution[c]	Neuronal; ganglionic	Cardiac; smooth muscle	Glandular; smooth muscle	Brain striatum	hippocampus, brainstem[d]
Effectors and responses[e]	PI hydrolysis; cGMP formation	cAMP inhibition; opening of K^+ channels in heart and CNS	PI hydrolysis; cGMP formation	cAMP inhibition	PI hydrolysis; cGMP formation
Selective antagonists[f]	Pirenzepine; (+)–telenzepine	AF-DX-116; AF-DX 384; himbacine; methoctramine; gallamine	HHSiD; *p*-fluoro-HHSiD 4-DAMP	—	—

[a]Gene product designations are used.
[b]Nomenclature follows that recommended by the Fourth International Symposium Subtype of Muscarinic Receptors, From Levine and Birdsall, (1989).
[c]Subtypes are widely distributed; only primary areas are given for clarity and comparative purposes. Arease with highest concentrations are underscored. From van Delft *et al.* (1989) and Hume *et al.* (1990).
[d]From Weiner and Brann (1989).
[e]From Ashkenazi *et al.* (1989); Lechleiter *et al.* (1989); Peralta *et al.* (1988a); McKinney and Richelson (1984).
[f]From Mutschler *et al.* (1989) and Hulme *et al.* (1990).

sion. The general tissue distribution of mRNAs encoding mAChR is congruent with what is known about the distribution of pharmacologically defined M_{1-3}. Sequence analysis of cloned mAChR subtypes has revealed that the receptor protein is composed of seven transmembrane α-helical segments that are highly conserved among the subtypes (Peralta *et al.*, 1988b). The last large cytoplasmic loop is much more divergent than the transmembrane segments among the subtypes and is believed to be involved in subtype-selective coupling to different effectors.

Muscarinic receptors belong to a superfamily of seven-helix, G-protein–coupled receptors including adrenergic and serotonergic receptors, K-receptors, rhodopsin, and opsin. Generally, m1, m3, and m5 gene products are coupled to phospholipase C by one of two G proteins (Gp and Gp*, pertussis–insensitive and pertussisensitive, respectively). Stimulation of the receptor activates phosphoinositide (PI) hydrolysis. On the other hand, m2 and m4 gene products are coupled to adenylate cyclase by an inhibitory G protein (G_i), and inhibition of cyclic adenosine monophosphate (cAMP) formation results from receptor activation (Table I). G_i is sensitive to ADP-ribosylation by pertussis toxin. Recently, porcine atrial M_2 receptors were shown to interact with at least three kinds of G-proteins that differ in the molecular weight of their α subunits (Ikegaya *et al.*, 1990). Although certain mAChR subtypes may couple to one effector more efficiently, complete selective- coupling does not exist for any of the subtypes. Selective or *preferential* coupling to a particular signal-transduction mechanism depends on the cell type and the species (Hulme *et al.*, 1990; Tietje *et al.*, 1990). Assignment of one subtype to a given response is further complicated by the effects of one type of effector on another, even on the same cell surface (e.g., the rise in cytosolic Ca^{2+} levels due to PI hydrolysis will activate protein kinase C and lead to the activation of adenylate cyclase). Despite these caveats to the selective-coupling hypothesis, the general pattern illustrated in Table I is correct for most mammalian tissues. Coupling specificity has been correlated with the degree of amino acid sequence divergence in the cytoplasmic loop between m1, m3, m5, and m2, m4 gene products.

Muscarinic receptors also couple to other effectors and mediate other cellular responses. Ca^{2+}-dependent stimulation of m1, m3, and m5 gene products activates guanylate cyclase, which catalyzes the formation of cyclic guanosine monophosphate (cGMP). However, it is not known whether cGMP is actually involved in ionic events. Stimulation of M_1 and M_3 receptors can also lead to the release of arachidonic acid, K^+ and Cl^- channel activation, and inhibition of a special neuronal K^+ channel termed the M current (Fukuda *et al.*, 1988). The M_2 receptors in cardiac muscle are coupled to K^+ channels by a special G-protein (G_k), and stimulation leads to increases in K^+ levels. The M_2-mediated decreases in cAMP levels in cardiac tissue works in conjunction with these increases in K^+ levels to inhibit cardiac muscle activity (Goyal *et al.,* 1989). Recent advance in molecular cloning techniques and continuous screening of drugs for subtype selectivity will undoubtably lead to the discovery of new mAChR subtypes as already detected in rats.

II. Inhibition by Organophosphorus Compounds of Radioligand Binding to Muscarinic Receptors

A. Brain

Mammalian brain is the tissue of choice for study of mAChRs because it has all the receptor subtypes. The effects of OP compounds were first studied primarily on [^{3}H]QNB binding, where effects were observed only with high concentrations of OP compounds because QNB has similar affinities on all the subtypes while OP compounds have a high affinity for only certain of the subtypes. Thus, when we investigated the effects of OP compounds on rat brain mAChR subtypes (using [^{3}H]PZ to label mostly M_1 and [^{3}H]*cis*-methyldioxolane [^{3}H]CD) to label a high-affinity subpopulation of M_2 receptors, the M_2 subtype was discovered to be the most sensitive to these OP compounds (Bakry *et al.,* 1988). Inhibition of [^{3}H]CD binding by OP compounds was by n*M* to μ*M* concentrations (Fig. 1) with $K_{0.5}$ values in n*M* of 3, 10, 40,

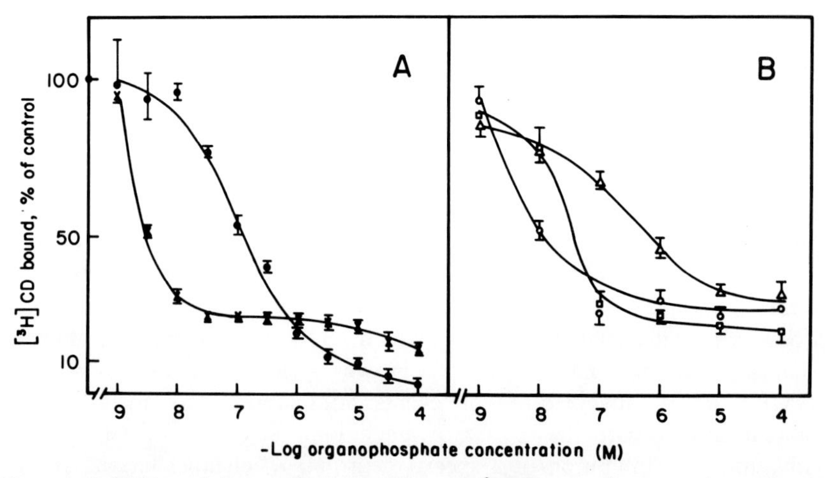

Figure 1 Inhibition of the specific binding of 5 nM [^{3}H]CD to rat brain membranes by OP anti-AChEs. (A). Echothiophate (●) and VX (Ø). (B) Soman (○), sarin (Δ) and tabun (❑). Symbols and bars are means of triplicates measurements of three experiements ± SD. From Bakry *et al.* (1988).

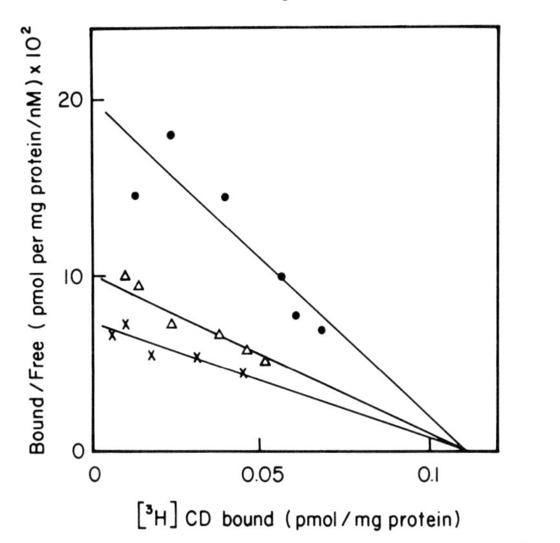

Figure 2 Scatchard plots of saturation isotherms of the specific binding of [^{3}H]CD to rat brain membranes in absence (●) and presence of 3 n*M* VX (Ø) or 100 n*M* echothiophate (Δ). The common intercept on the X-axis suggests competitive inhibition. From Bakry *et al.* (1988).

100, and 800 for VX, soman, sarin, echothiophate, and tabun, respectively. This effect was reversible and appeared to be at the ACh and CD binding site (Fig. 2). Other data confirmed that the M_1 subtype is insensitive to OP compounds, and added that M_2 and M_3 are highly sensitive to even much lower concentrations (< p*M*) of paraoxon (Katz and Marquis, 1989). Thus on calf brain caudate nuclei, paraoxon modulated [^{3}H]QNB binding at concentrations below those needed to affect AChE. Pretreatment of the membranes with a high concentration of both the M_2-selective antagonist, AF-DX116 and the M_3-selective antagonist, 4-diphenylacetoxy-*N*-methylpiperidine methiodide (4-DAMP), protected against paraoxon inhibition of [^{3}H]QNB binding. On the other hand, the M_1-selective antagonist PZ showed no protective effect.

B. Heart

Muscarinic receptors in cardiac muscle are almost all of the M_2 subtype, though a few are of the M_1 subtype (Watson *et al.*, 1986). Thus, the study of OP effects on cardiac mAChRs avoids the complications of dealing with the multiple mAChR subtypes that occur in the brain. Using a ligand that labels a high-affinity population of the M_2 receptors in rat heart, i.e., [^{3}H]CD, we found that soman, VX, sarin, and tabun inhibited the binding with $K_{0.5}$ of 0.8,

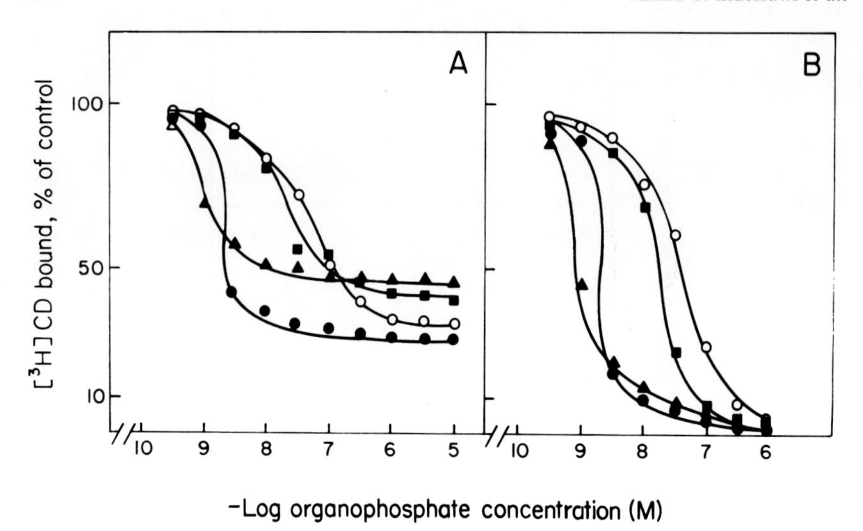

Figure 3 Inhibition of [^{3}H]CD binding at 5 n*M* to cardiac muscle membranes by VX (●), soman (▲), sarin (■), and tabun (○). (A). Observed displacement. (B) Corrected displacement after subtracting the OP-sensitive portion of [^{3}H]CD binding. Symbols are means of six measurements. From Silveira *et al.* (1990).

2, 20, and 50 n*M*, respectively (Silveira *et al.*, 1990). In general, these M_2 receptors have lower affinities for OP insecticides than do the nerve agents (Fig. 3) with $K_{0.5}$ of 200 n*M* for paraoxon and > 1 μ*M* for EPN, coumaphos, doxathion, dichlorvos, and chlorpyriphos. The M_2 receptors of the heart differ from those of the brain in being inhibited with Ni^{2+} and *N*-ethylmaleimide, while the brain receptors are stimulated by the former and inhibited mildly by the latter (Bakry *et al.*, 1988). The inhibition of [^{3}H]CD binding by as little as n*M* concentration of echothiophate is evidence for the high affinity that the M_2 receptor has for certain OP compounds. However, the resistance to inhibition by OP compounds of a portion of [^{3}H]CD binding (Fig. 3) suggests that not all the CD-binding cardiac M_2 receptors have such a high affinity for OP compounds. Like the binding to rat brain, inhibition of [^{3}H]CD binding to cardiac receptors by two OP compounds appears to be competitive (Fig. 4), suggesting their binding to the ACh and CD binding site. Most of the mAChRs in the striatum were believed to be of the M_2 subtype because of their low affinity for PZ and inhibition of cAMP synthesis, similar to that of the M_2 cardiac receptor. More recently, however, the complete drug specificity of this striatal receptor revealed that it is different from the cardiac receptors and similar to that of the m4 receptor gene product (McKinney *et al.*, 1989; Ehlert *et al.*, 1989).

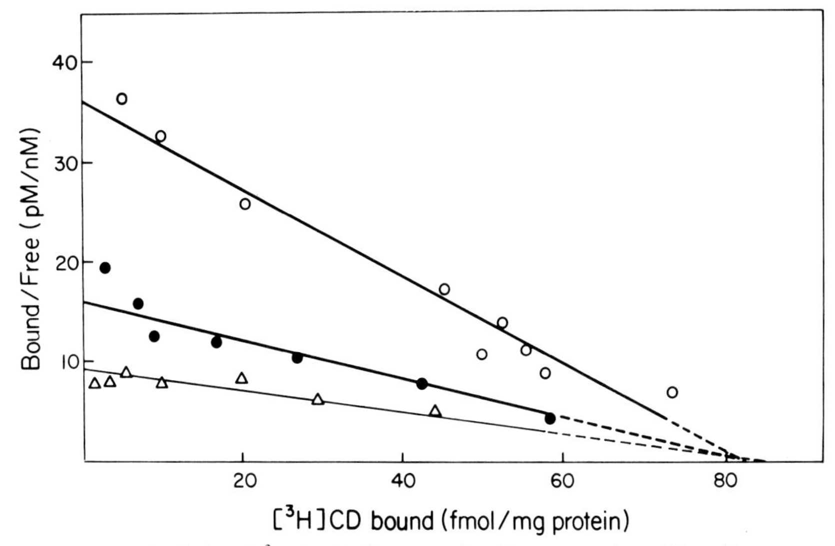

Figure 4 Scratchard plot of [^{3}H]CD binding to cardiac M_2 receptos alone (○) and in presence of 20 *nM* sarin (●) or 200 *nM* paraoxon (Δ). The common intercept on the X-axis suggests competitive inhibition. From Silveira *et al.* (1990).

III. Action of Organophosphorus Compounds on Muscarinic Receptor Function

Inhibition by a drug or toxicant of binding of a specific radioligand to a receptor does not indicate whether this chemical acts as a agonist or antagonist. Only its effect on receptor function can reveal its mechanism of action. Assays for receptor function are complicated, so very few have been attempted.

A. Inhibition of Muscarinic Receptor–Regulated Synthesis of cGMP in Neuroblastoma Cultures

The neuroblastoma N1E-115 cells, which are known to respond to mAChR activation by increased synthesis of cGMP, were used to study the effect of OP compounds on mAChR. Neither the nerve agents nor echothiophate at 1–100 μ*M* could stimulate [^{3}H]cGMP synthesis, but echothiophate and VX produced 50% inhibition (IC_{50}) with 150 and 3 μ*M*, respectively (Fig. 5). These data suggest that OP compounds at high concentrations act as antagonists of this mAChR in neuroblastoma N1E-115.

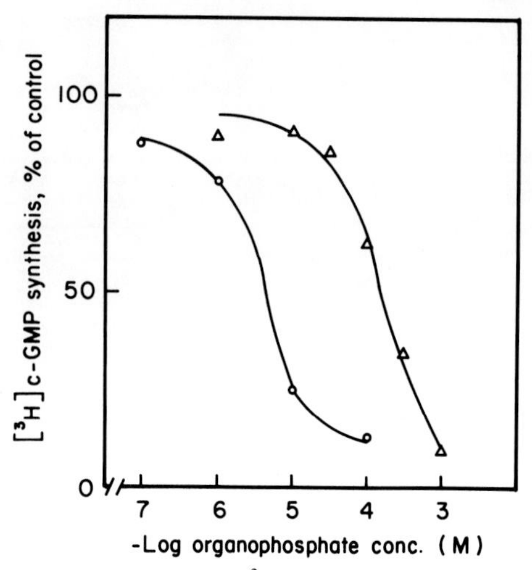

Figure 5 Inhibition of carbachol-stimulated [^{3}H]cGMP synthesis in neuroblastoma cells by VX (○) and echothiophate (Δ). Cells were pretreated with the OP compound for 30 min, then 600 μM carbachol was added for 30 sec before 50% trichloroacetic acid was added to stop the reaction. Symbols are means of triplicate experiements with S.E. < 10%. From Bakry *et al.* (1988).

B. Activation of Muscarinic Receptor–Regulated cAMP Synthesis in Rat Brain

In brain cells another functional assay was utilized. Adenylate cyclase is activated directly with forskolin or indirectly via coupled Gs proteins (e.g., β-adrenergic receptor) and is inhibited indirectly by receptors coupled to G_i-proteins, such as mAChR. Thus, mAChR function was assayed by the inhibition of the forskolin-activated [^{3}H]cAMP synthesis in rat brain striatal cells. Paraoxon inhibited it in a dose-dependent manner as did carbachol and CD (Fig. 6), and this inhibition was completely blocked by the muscarinic antagonist atropine (Jett *et al.*, 1991). When both paraoxon and carbachol were added at a concentration at which each produces maximal inhibition, there was no further increase. However, when low concentrations (0.3 μ*M)* of carbachol and paraoxon were used together, there was additive inhibition of 39 ± 4% compared to 23 ± 3% for carbachol and 19 ± 3% for paraoxon. The data suggest that carbachol and paraoxon act on the same receptor and that as little as 10 n*M* paraoxon acts like carbachol, causing activation of the mAChR in the striatum.

C. Effects on Phosphoinositide Hydrolysis

Short-term administration of DFP (18 hr) resulted in reversible significant (25%) down-regulation of mouse brain mAChRs, as indicated by the reduc-

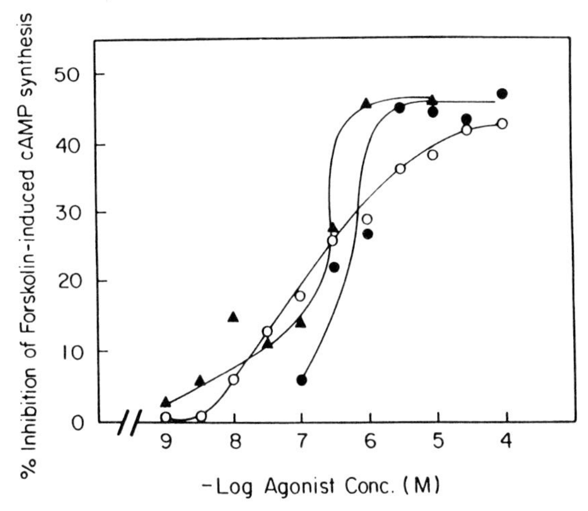

Figure 6 Inhibition of forskolin-activiated adenylate cyclase activity (measured by synthesis of [^{3}H]cAMP in dissociated rat striatal cells) by different concentrations of two known muscarinic agonists [carbachol (●) and CD (▲)] and paraoxon (○). Activity of 100% represents [^{3}H]cAMP synthesis in the presence of 25 μ*M* forskolin in absence of any OP compound. From Jett *et al.* (1991).

tion in maximal [^{3}H]*N*-methylscopolamine binding without change in affinity, which is reversed in 36 hr (Cioffi and El-Fakahany, 1988). Neither mAChR-mediated PI hydrolysis, nor a particular mAChR-binding conformation was changed. No effect was observed either on carbamylcholine-stimulated PI hydrolysis in rat brain after acute disulfoton treatment (Costa *et al.*, 1986). The lack of change in this functional response may be owing to the low sensitivity of M_1 or M_3 receptors to OP compounds compared to the high sensitivity of the M_2 receptor of the heart (Silveira *et al.*, 1990) and the putative M_4 of the striatum, both of which are coupled to adenylate cyclase.

IV. Toxicological and Therapeutic Implications

Acute toxicity of an OP compound is an expression not only of its potency on the primary target, but also of penetration, elimination, bioactivation, biodegradation, and pharmacodynamics. In comparing the antiAChE activities of parathon, methylparathion, chlorpyriphos and methylchlorpyriphos, leptophos, and EPN, there was poor correlation between rat oral toxicity and inhibition of AChE by their oxygen analogs (Chambers *et al.*, 1990). These results suggest possible contributions to toxicity of action at a secondary target.

The significance of the discovery of direct action of OP antiChEs on mAChRs is the very high affinity certain mAChRs have for particular OP compounds, in particular the M_2 subtype of the heart and the putative M_4 subtype in the striatum. The affinities of these receptors are several orders of magnitude lower than the affinity toward AChE (100 μ*M)* (Eldefrawi, 1985). Nevertheless, AChE is believed to be the primary critical target for OP compounds, because of its progressive irreversible phosphorylation by the OP compound and possible aging. In addition, there is a poor correlation between AChE inhibition by OP compounds and their affinities for the M_2 receptor; moreover, the effects of the OP compounds on the M_2 receptors are reversible.

In the heart, ACh released from the vagus nerve activates M_2 receptors on the heart muscle to reduce the force of contraction and the beating frequency, and may also stimulate presynaptic M_2 receptors on cholinergic (autoreceptors) and noradrenergic (heteroreceptors) nerve fibers (Bognar *et al.*, 1990). Thus, an important aspect of the agonist action of an OP compound on the postsynaptic cardiac muscle mAChRs is that receptor activation would exaggerate the effect of excess ACh, which accumulates in the synaptic gap when AChE is inhibited by the OP compound. This effect would have considerable impact on OP toxicity, especially at the early stages of exposure, when a small proportion of AChE is inhibited, as postsynaptic mAChRs are being activitated by the excess ACh as well as by the OP compound. Thus, the therapeutic value of the mAChR inhibitor atropine on the heart during OP intoxication is of particular significance.

On the other hand, in the brain the predominant location of M_2 and possibly M_4 receptors is presynaptic at the nerve terminals. Activation by excess ACh or OP compounds of these, and the presynaptic mAChRs in the heart, would reduce ACh release by the nerve. The effect of the OP compound would be to counteract in part the toxicity that results from AChE inhibition. Adding to the complexity of OP actions is the recent finding that paraoxon may act directly on the G-protein–adenylate cyclase system (Eldefrawi, unpublished) and that muscarinic receptor activation of protein kinase C induces c-fos and c-jun oncogenes (Gutkind and Novotny, 1991; Trejo and Brown, 1991).

Acknowledgments

The research reported from our laboratories has been supported in part by U.S. Army Grant DAAK11-84-006 and NIH Grant NIEHS ES02594.

References

Aracava, Y., Deshpande, S. S., Rickett, D. L., Brossi, A., Schonenberger, B., and Albuquerque, E.X. (1987). The molecular basis of anticholinesterase actions on nicotinic and glutamatergic synapses. *Ann. N.Y. Acad. Sci.* **505**, 2261–255.

Ashkenazi, A., Peralta, E. G., Winslow, J. W., Ramachandran, J., and Capon, D. J. (1989). Functional diversity of muscarinic receptor subtypes in cellular signal transduction and growth. Subtypes of muscarinic receptors IV. *Trends Pharmacol. Sci.* **10**,(Suppl.),16–22.

Bakry, N. M. S., El-Rashidy, A. H., Eldefrawi, A. T., and Eldefrawi, M. E. (1988). Direct actions of organophospate anticholinesterases on nicotinic and muscarinic acetylcholine receptors. *J. Biochem. Toxicol.* **3**: 23–259.

Bartels, E., and Nachmansohn, D. (1969). Organophosphate inhibitors of acetylcholine-receptor and -esterase tested on the electroplax. *Arch. Biochem. Biophys.* **133**, 1–10.

Belleau, B., Ditullio, V., and Tsai, Y.-H. (1970). Kinetic effects of leptocurares and pachycurares on the methanesulfonylation of acetylcholinesterase. *Mol. Pharmacol.* **6**: 41–45.

Bognar, I. T., Beinhauer, B., Kann, P., and Fuder, H. (1990). Different muscarinic receptors mediate autoinhibition of acetylcholine release and vagally induce vasoconstriction in the rate isolated perfused heart. *Naynyn-Schmiedeberg's Arch. Pharmacol.* **341**: 279–287.

Bonner, T. I. (1989a). New subtypes of muscarinic acetylcholine receptors. *Trends Pharmacol. Sci.* **10** Suppl. 11–15.

Bonner, T. I. (1989b). The molecular basis of muscarinic receptor diversity. Trends *Neurosci.* **12**: 148–151.

Chambers, H., Brown, B., and Chambers, J. E. (1990). Noncatalytic detoxication of six organophosphorus compounds by rat liver homogenates. *Pestic. Biochem. Physiol.* **36**, 30–31S.

Cioffi, C.L., and El-Fakahany, E. E. (1988). Lack of alterations in muscarinic receptor sybtypes and phosphoinsitide hydrolysis upon acute DFP treatment. *Eur. J. Pharmacol.* **156**, 35–45.

Costa, L. G., Kaylor, G., and Murphy, S. D. (1986). Charbachol- and norephinephrine-stimulated phosphoinositide metabolism in rat brain: Effect of chronic cholinesterase inhibition. *J. Pharmacol. Exp. Ther.* **251**, 32–37.

Ehlert, F. J., Dumont, Y., Roeske, W. R., and Yamamura, A. J. (1980). Muscarinic receptor binding in rat brain using the agonist [^{3}H]-*cis*-methyldioxolane. *Life Sci.* **26**, 961–967.

Ehlert, F.J., Delen, F. M., Yun, S. H., Friedman, D. J., and Self, D. W. (1989). Coupling subtypes of the muscarinic receptor to adenylate cyclase in the corpus straitum and heart. *J. Pharmacol. Exp. Ther.* **251**, 660--671.

Eldefrawi, A. T. (1985). Cholinesterases and anticholinesterases. *In* "Comprehensive Insect Physiology, Biochemistry and Pharmacology. Insect Control" (G.A. Kerkut and L.I. Gilbert, eds.), Vol. 12, pp. 102–124. Pergamon Press, Oxford.

Eldefrawi, M. E., Britten, A. G., and O'Brien, R. D. (1971). Action of organophosphates on binding of cholinergic ligands. *Pestic. Biochem. Physiol.* **1**, 101–108.

Eldefrawi, M. E., Schweizer, G., Bakry, N. M., and Valdes, J. J. (1988). Desensitization of the nicotinic acetylcholine receptor by diisopropylfluorophosphate. *J. Biochem. Toxicol.* **3**, 21–32.

Fukuda, K., Higashida, H., Kubo, T., Maeda, A., Akiba, I., Bujo, H., Mishina, M., and Numa, S. (1988). Selective coupling with K^+ current of muscarinic acetylcholine receptor subtypes in NG108-15 cells. *Nature* **335**, 355–358.

Gazit, H., Silman, I., and Dudai, Y. (1979). Administration of an organophosphate causes a decrease in muscarinic receptor levels in rat brain. *Brain Res.* **335**, 355–358.

Goyal, R. K. (1989). Muscarinic receptor subtypes. Physiology and clinical implcations. *N. Engl. J. Med.* **321**: 1022–1029.

Gutkind, J. S., Novotny, E. A., Braun, M. R., and Robbins, K. C. (1991). Muscarinic acetylcholine receptor subtypes as agonist-dependent oncogenes. *Proc. Natl. Acad. Sci. USA* **88**, 4703–4707.

Haga, T., Haga, K., Berstein, G., Nishiyama, T., Uchiyama, H., and Ichiyama, A. (1988). Molecular properties of muscarinic receptors. Subtypes of Muscarinic Receptors III. *Trends Pharmacol. Sci.* **9**, (Suppl.), 12–18.

Hammer, R., Berrie, C. P., Birdsall, N. J. M., Burgen, A. S. V., and Hulme, E.C. (1980). Pirenzepine distinguishes between subclasses of muscarinic receptors. *Nature* **283**, 90–92.

Hulme, E. C., Birdsall, N. J. M., and Buckley, N. J. (1990). Muscarinic receptor subtypes. *Annu. Rev. Pharmacol. Toxicol.* **30**, 633–673.

Ikegaya, T., Nishiyama, T., Haga, K., Haga, T., Ichiyama, A., Kobayashi, A., and Yamazaki, N. (1990). Interaction of atrial muscarinic receptors with three kinds of GTP-binding proteins. *J. Mol. Cell. Cardiol.* **22**, 343–351.

Jett, D. A., Abdallah, E. A. M., El-Fakahany, E. E., Eldefrawi, M. E. and Eldefrawi, A. T. (1991). High-affinity activation by paraoxon of a muscarinic receptor subtype in rat brain straitum. *Pestic. Biochem. Physiol.* **39**, 149–157.

Katz, L., and Marquis, J. K. (1989). Modulation of central muscarinic receptor binding *in vitro* by ultralow levels of the organophosphate paraoxon. *Toxicol. Appl. Pharmacol.* **101**, 114–123.

Kuba, K., Albuquerque, E. X., and Barnard, E. A. (1973). Diisopropylfluorophosphoate: Suppression of ionic conductance of the cholinergic receptor. *Science* **181**, 853–856.

Lechleiter, J., Peralta, E. and Clapham, D. (1989). Diverse functions of muscarinic acetylcholine receptor subtypes. *Trends Pharmacol. Sci.* **10** (Suppl.), 34–38.

Levine, R. R., and Birdsall, N. J. M. (eds.) (1989). Subtypes of muscarinic receptors IV. *Trends Pharmacol. Sci.* **10** (Suppl.), pp. 119 Elsevier Publ.

McKinney, M., Anderson, D., Forray, C., and El-Fakahany, E. E. (1989). Characterization of the striatal M_2 muscarinic receptor mediating inhibition of cyclic AMP using selective antagonists: A comparison with the brainstem M_2 receptor, *J. Pharmacol. Exp. Ther.* **250**, 565–572.

Mustchler, E., Moser, U., Wess, J., and Lambrecht, G. (1989). Muscarinic receptor subtypes: Agonsits and antagonists. *Prog. Pharmacol. Clinical Pharmacol.* 7, 13–31.

Peralta, E. G., Ashkenazi, A., Winslow, J. W., Ramachandran, J., and Capon, D. J. (1988a). Differential regulation of PI hydrolysis and adenylyl cyclase by muscarinic receptor subtypes. *Nature* **334**, 434–437.

Peralta, E. G., Winslow, J. W., Ashkenazi, A., Smith, D. H., Ramachandran, J., and Capon, D. J. (1988b). Structural basis of muscarinic acetylcholine receptor subtype diversity. *Trends Pharmacol.* (Suppl.) 6–11.

Silveira, C. L. P., Eldefrawi, A. T., and Elderfrawi, M. E. (1990). Putative M_2 muscarinic receptors of rat heart have high affinity for organophosphorus anticholinesterases. *Tox. Appl. Pharmacol.* **103**, 474–481.

Stalc, A., and Zupancic, A. O. (1972). Effect of α-bungarotoxin on acetylcholinesterase bound to mouse diaphragm endplates. *Nature* **239**, 91–93.

Tietje, K. M., Goldman, P. S., and Nathanson, N. M. (1990). Cloning and functional analysis of a gene encoding a novel muscarinic acetylcholine receptor expressed in chick heart and brain. *J. Biol. Chem.* **265**: 2828–2834.

Trejo, J., and Brown, J. H. (1991). *c-fos* and *c-jun* are induced by muscarinic receptor activation of protein kinase C but are differentially regulated by intracellular calcium. *J. Biol. Chem.* **266**, 7876–7882.

Watson, M., Roeske, W. R. and Yamamura, H. I. (1986). [^{3}H]Pirenzepine and (-)-[^{3}H]quinuclidinyl benzilate binding to rat cortical and cardiac muscarinic cholinergic sites. II. Characterization and regulation of antagonist binding to putative muscarinic subtypes. *J. Pharmacol. Exp. Therap.* **237**, 419–427.

Webb, G. D. (1965). Affinity of benzoquinonium and ambenonium derivatives for the acetylcholine receptor, tested on the electroplax and for acetylcholinesterase in solution. *Biochim. Biophys. Acta* **102**, 172–184.

Weiner, D. M., and Brann, M. R. (1989). Distribution of m1–m5 muscarinic receptor nRNAs in rat brain. Muscarinic receptor subtypes IV. *Trends Pharmacol. Sci.* **10** (Suppl.) 115.

van Delft, A. M. L., Hagan, J. J., and Tonnaer, J. A. D. M. (1989). Muscarinic receptors in the central nervous system. *Prog. Pharmacol. Clin. Pharmacol.* 7, 93–117.

Volpe, L. S., Biagioni, T. M. and Marquis, J. K. (1985). *In vitro* modulation of bovine caudate muscarinic receptor number by organophosphates and carbamates. *Tox. Appl. Pharmacol.* **78**, 226–234.

Yarowsky, P., Fowler, J. C., Taylor, G., and Weinreich, D. (1984). Noncholinesterase actions of an irreverisble acetylcholinesterase inhibitor on synaptic transmission and membrane properties in autonomic ganglia. *Cell. Mol. Neurobiol.* **4**, 351–366.

14

Role of Second-Messenger Systems In Response to Organophosphorus Compounds

Lucio G. Costa
Department of Environmental Health, SC-34
University of Washington
Seattle, Washington

I. Introduction

The biological activity of organophosphorus (OP) insecticides is attributed to their reaction with the enzyme acetylcholinesterase (AChE) and other cholinesterases (Murphy, 1986) (see also Chapters 1, 4, and 9, this volume).

Inhibition of AChE by an OP compound causes a rapid accumulation at cholinergic synapses of acetylcholine (ACh), a major neurotransmitter in the central and peripheral nervous systems. Unlike other neurotransmitters, such as norepinephrine or gamma-aminobutyric acid (GABA), which have more than one way to be removed from the synaptic cleft, acetylcholine depends entirely on AChE for its inactivation.

Signs and symptoms of acute poisoning with an OP insecticide may be classified into muscarinic (parasympathetic), nicotinic (sympathetic and motor), and central nervous system (CNS) manifestations, according to the site of action (Namba *et al.,* 1971). The interval between exposure and onset of symptoms varies with the route and degree of exposure and the chemical nature of the OP compound. According to a recent retrospective study on 236 patients with acute OP poisoning, more than 90% of the cases showed per-

Organophosphates: Chemistry, Fate, and Effects

ipheral muscarinic symptoms (Hirshberg and Lerman, 1984). Among these, miosis was the most prevalent specific sign. Forty percent of the patients had CNS symptoms, but only 17% presented a combination of muscarinic, nicotinic, and CNS signs.

If cholinesterase is to be considered the primary target for the action of OP compounds, cholinergic receptors represent the secondary target. The interaction of accumulated ACh with its receptors triggers a cascade of intracellular events that lead to the clinical signs of OP poisoning. This chapter will review briefly some of the biochemical steps that follow receptor activation by endogenous ACh and will focus on the muscarinic receptor, because of its major role in the central and peripheral effects of OP compounds.

II. Acetylcholine Muscarinic Receptors

Until a few years ago, all muscarinic receptors were thought to be alike, but at least three different pharmacologically identifiable types and at least five different molecular forms have now been delineated (Wolfe, 1989; Goyal, 1989; for additional discussion of receptors and binding affinities of subtypes, refer to Eldefrawi *et al.*, Chapter 13, this volume).The three pharmacologically defined muscarinic receptors, represented by a capital M, include M_1, which has a high affinity for pirenzepine and telenzepine; M_2, which has a high affinity for AFDX-116 [AFDX-116 (11[[2-(diethylamino)methyl]-1-piperidinyl] acetyl]-5,11-dihydro-6H-pyrido [2,3-b] [1,4] benzodiazepine-6-one)] and oxotremorine; and M_3which has a high affinity for 4-DAMP [4-DAMP(4-diphenylacetoxy-N-methylpiperidine methiodide)] (Wolfe, 1989; Goyal, 1989). The five molecular forms are denoted by a lower-case m, m_1 to m_5. Receptor binding and functional studies have shown that muscarinic receptor subtypes are differentially distributed. Cardiac tissue has primarily M_2 receptors, while smooth muscle has both M_2 and M_3 receptors. Muscarinic receptors in secretory and glandular cells appear to be of the M_3 subtype. All three subtypes are present in the nervous system although with different regional distribution. Thus, for example, the cerebral cortex and the hippocampus are highly rich in M_1 receptors, while in the medulla–pons and the cerebellum, the M_2 subtype is predominant.

All intracellular responses that result from stimulation of muscarinic receptors are thought to be mediated by G proteins, which bind the guanine nucleotide guanosine triphosphate (GTP) and are part of the mechanism that transduces signals across the cell membrane (Nathanson, 1987; Casey and Gilman, 1988). The muscarinic m_1, m_3, and m_5 receptors couple primarily with a G protein referred to as Gp, while m_2 and m_4 couple primarily with a G protein known as Gi (Ashkenazi *et al.*, 1989a). This implies that M_1 and M_3 receptors are coupled to the phosphoinositide pathway, while M_2 receptors are linked to the cyclic AMP pathway. However, receptor subtypes and the various G proteins with which they couple overlap substantially. So, for example, in the heart, muscarinic receptors are also coupled with Gp, and their activation leads to an increase of phosphatidylinositol turnover (Woodcock *et al.*, 1987). Moreover, there are interactions between the intracellular mediators: for example, stimulation of phosphatidylinositol turnover can affect the intracellular levels of cyclic AMP (Kendall-Harden, 1989).

The existence of different subtypes of muscarinic receptors coupled to different signal-transduction systems, their widespread distribution, and their role in controlling functions of many organs in the body have several clinical implications. The most relevant is that specific effects in different systems (e.g., nervous system, cardiovascular system, respiratory systems, gastrointestinal tract) may be better antagonized by the use of subtype-specific receptor antagonists, which might, therefore, find a role in OP poisoning.

III. The Phosphoinositide–Protein Kinase C Pathway

Interaction of ACh with the appropriate subtype of muscarinic receptor activates a phosphoinositidase (phosphoinositidase C), which hydrolyzes phosphatidylinositol 4,5-bisphosphate (PtdIns 4,5-P_2) to inositol 1,4,5-trisphosphate (Ins 1,4,5-P_3) and diacylglycerol (DG) (Berridge, 1987; Costa, 1990). A GTP-binding protein is believed to couple the receptor to phosphoinositidase C. The exact nature of such protein (possibly Gp) has not been fully elucidated, although the current view is that there might be more than one protein (Lo and Hughes, 1987).

Ins 1,4,5-P_3 binds to specific and saturable receptor sites in, or near, the endoplasmic reticulum (Worley *et al.*, 1989) and causes the mobilization of calcium ions in the cytosol (Stundermann *et al.*, 1988). Ins 1,4,5-P_3 is then dephosphorylated by phosphatases to generate Ins 1,4-P_2, Ins 1-P, and Ins. This latter reaction, catalyzed by an Ins 1-P phosphatase was the first to be shown to be inhibited by lithium ions (Drummond *et al.*, 1987). In addition to fostering hypotheses on the possible mechanism involved in the antimanic action of lithium, application of this discovery has allowed the easy measurement of InsPs accumulation in tissue slices (Berridge *et al.*, 1982). Ins 1,4,5-P_3 can also be phosphorylated by a 3-kinase to form inositol 1,3,4,5-tetrakisphosphate (Ins 1,3,4,5-P_4), which in turn is dephosphorylated to Ins 1,3,4-P_3 (Irvine *et al.*, 1988). It has been suggested that Ins 1,3,4-P_3 might be involved in long-term cellular effect (e.g., regulation of gene transcription or cell division); however, evidence for this is still lacking. Ins 1,3,4,5-P_4 might play a role in modulating the mobilization of intracellular Ca^{2+} by Ins 1,4,5-P_3 and/or in regulating Ca^{2+} entry from outside the cell (Irvine *et al.*, 1988). Ins 1,3,4,5,6-P_5 and Ins-P_6 have been suggested to serve as storage for phosphates, the same roles they have in higher plants. Both have been shown to induce dose-dependent changes in heart rate and blood pressure in the rat (Vallejo *et al.,* 1987), and specific binding sites for Ins-P_6 have been identified in mammalian brain (Hawkins *et al.,* 1990). The cyclic inositol 1-2,4,5-triphosphate is also capable of mobilizing Ca^{2+} and is formed and metabolized at a slower rate. It could, therefore, act as a long-term messenger, although it is still of doubtful physiological significance.

Thus, a major consequence of the activation of the phosphoinositide system is a change in the intracellular concentration of Ca^{2+}. Ca^{2+} plays a pivotal role in synaptic transmission and in cell functions (Llinas, 1982). Calcium-activated mechanisms are believed to be involved in the toxic effect of several chemicals (Orrenius and Nicotera, 1987; Pounds and Rosen, 1988). Increased free intracellular Ca^{2+} by toxic agents has been suggested as an index of potential neurotoxicity (Komulainen and Bondy, 1988).

The other product of phosphoinositide hydrolysis, DG, is capable of activating a novel protein kinase, PKC (Nishizuka, 1988). In most tissues, PKC is present in its inactive form in the cytosol and is translocated to membranes when cells are stimulated; PKC requires Ca^{2+} and phospholipids, particularly phosphatidylserine, for its activation. It is now evident that at least seven subspecies of PKC exist in nerve tissue, and their structures, deduced by analysis of their DNA sequences, have been elucidated (Nishizuka, 1988; Kikkawa *et al.,* 1988). These different PKCs, one of which, the gamma, is expressed only in the brain and spinal cord but not in other tissues, including peripheral nerves, differ in their specific activity in different brain areas, in their developmental profile, and in their activation requirements. In addition to DG, the gamma form can also be activated by arachidonic acid and by other eicosanoids (Kikkawa *et al.,* 1988; Rana and Hokin, 1990), suggesting that production of these compounds upon receptor stimulation might have differential intracellular effects depending on the forms of PKC present in a certain cell, and that PKC might be regulated independent of DG and Ca^{2+}, i.e., in a manner not necessarily linked to phosphoinositide turnover.

PKC is the receptor for a class of tumor promoters, the phorbol esters (Ashendel, 1985). The use of these compounds as direct activators of PKC is proving extremely useful in investigating the role of PKC in cellular functions. A large number of proteins have been shown to be substrates for PKC; these include such diverse molecules as receptors, ion channels, cytoskeletal proteins, and enzymes (Nishizuka, 1988). PKC has been shown to provide both a *positive forward action* leading to a synergistic interaction with the Ca^{2+} pathway (e.g., in control of gene expression), and a *negative feedback control* over various steps of the cell-signaling process (Berridge, 1987; Nishizuka, 1988).

It is now well established that phorbol esters enhance the release of various neurotransmitters in a number of neuronal preparations (Kikkawa and Nishizuka, 1986). This effect is believed to be owing to activation of PKC, since it is not induced by inactive phorbol esters and is inhibited by PKC antagonists, such as H-7. The involvement of PKC in neurotransmitter release might also be linked to its role in the maintenance of long-term potentiation (LTP), a rapidly induced, persistent increase in synaptic efficacy (Brown *et al.*, 1988). Since LTP is a leading candidate for a synaptic mechanism of rapid learning in mammals, a role for PKC in memory formation has also been suggested (Chiarugi *et al.*, 1989).

Additionally, there is increasing evidence that PI metabolism plays a relevant role in the control of cell proliferation (Vicentini and Villareal, 1986). For example, in the nervous system, proliferation of astrocytes has been shown to be associated with activation of the phosphoinositide–PKC system by acetylcholine (Ashkenazi *et al.*, 1989b).

IV. The Regulation of Cyclic AMP and Cyclic GMP Metabolism

The role of cyclic AMP as a mediator of several physiological processes is well established (Robison *et al.*, 1971). Activation of a number of receptors, coupled through the G protein Gs to adenylate cyclase, causes the conversion of ATP to cyclic AMP. Cyclic AMP can activate a cAMP-dependent protein kinase (PKA), which, in turn, phosphorylates a number of intracellular substrates. Cyclic AMP is then metabolized to 5′-AMP by the action of phosphodiesterases (Beavo *et al.*, 1982). These enzymes represent the target for commonly used drugs such as caffeine. The muscarinic M_2 receptor shares with other receptors (e.g., alpha$_2$-adrenergic, A_1-adenosine), the capacity to inhibit adenylate cyclase. Similar to activation of adenylate cyclase by beta adenergic agonists, receptor-mediated inhibition of this enzyme involves a G protein, which is known as Gi. Inhibition of adenylate cyclase leads to a decrease in the intracellular levels of cyclic AMP. Although a direct association of cyclic AMP with muscarinic receptor–mediated responses has been unambiguously demonstrated in only a few cases, in several tissues, particularly those rich in M_2 receptor subtype, physiological effects of acetylcholine involve alterations of cyclic AMP metabolism (Kendall-Harden, 1989). In addition to causing inhibition of adenylate cyclase through an interaction with Gi, there is limited evidence that activation of muscarinic receptors may lead to activation of phosphodiesterase. This also will lead to a decrease in intracellular levels of cyclic AMP (Kendall-Harden, 1989).

The observation that muscarinic receptors can stimulate phosphodiesterase, together with the observation that activation of alpha$_1$-adrenoceptors can produce a similar event, has led to a proposed mechanism for such stimulation. Since both alpha$_1$-adrenoceptors and muscarinic receptors are coupled to phospholipase C, this model considers the interaction between the phosphoinositide and the cyclic AMP pathway (Kendall-Harden, 1989). Elevation of cytoplasmatic Ca^{2+} levels occurs as a consequence of the increased Ins-1,4,5 P_3 levels, and an activation of a Ca^{2+} calmodulin-regulated phosphodiesterade ensues. Thus, while inhibition of cyclic AMP accumulation through interaction with Gi can be seen as a direct mechanism, the muscarinic receptor–mediated attenuation of cyclic AMP accumulation probably should be considered more as a modulatory mechanism that follows receptor-stimulated phosphoinositide hydrolysis (Kendall-Harden, 1989). Although detailed pharmacological studies on muscarinic receptor–mediated activation of phosphodiesterase have not been carried out, it would appear that all subtypes of muscarinic receptors are capable of affecting cyclic AMP levels, the M_2 subtype, directly through inhibition of adenylate cyclase, and the M_1 and M_3 subtype, indirectly, as a consequence of intracellular calcium following phosphoinositide hydrolysis. It is becoming apparent, however, that several muscarinic receptor subtypes couple to each of the second-messenger response systems with varying degrees of efficiency.

Activation of muscarinic receptors also causes the elevation in intracellular cyclic GMP (guanosine 3′,5′-monophospate) levels. This has generally been considered an indirect effect, involving an additional second messenger, since muscarinic agonists cannot activate guanylate cyclase (the enzyme that converts GTP to cyclic GMP) in broken-cell preparations (McKinney and Richelson, 1989). The two most likely candidates for this role are calcium ions or a metabolite of arachidonic acid. Evidence for a role of calcium comes from experiments showing that extracellular calcium ions are necessary for the muscarinic receptor stimulation of cyclic GMP (Schultz *et al.,* 1973), and that calcium channel antagonists block this effect (El-Fakahany and Richelson, 1983). The role of a metabolite of arachidonic acid is suggested by the finding that inhibitors of lipoxygenase (but not cycloxygenase) block the cyclic GMP response (Snider *et al.,* 1984; McKinney and Richelson, 1986).

The muscarinic receptor involved in the cyclic GMP response appears to be the M_1 subtype (McKinney and Richelson, 1989) with a pharmacological profile similar to that of stimulation of phosphoinositide metabolism. A link between these two biochemical responses is suggested by some evidence (e.g., Ohsako and Deguchi, 1981), but not by other (e.g., Kendall, 1986).

Increases in intracellular levels of cyclic GMP contribute to some of the effects of muscarinic receptor stimulations in various tissue such as smooth muscle, heart, and brain (McKinney and Richelson, 1989). Most of these effects are probably mediated by protein phosphorylation regulated by cyclic GMP–dependent protein kinases.

V. Organophosphorus Compounds and Second-Messenger Systems

The literature on the effects of OP compounds on second-messenger systems is still scarce. Most research has focused on the interaction of OP compounds with their primary target (acetylcholinesterase) and the secondary targets (cholinergic receptors), and only recently have the intracellular events that follow receptor activation been investigated. The studies summarized here should, therefore, serve as stimulus for further research, which would better elucidate the effects of OP compounds on second-messenger systems.

Measurements of second-messenger responses as indices of receptor function following OP exposure has received some attention. Repeated exposure to OP insecticides has been shown to cause a decrease in the density of muscarinic receptors in brain and peripheral tissues (Costa *et al.*, 1982a,b; Costa, 1988). This down-regulation of muscarinic receptors appears to play a relevant role in the development of tolerance to OP toxicity (see Chapter 15, Hoskins and Ho, this volume; Costa *et al.*, 1982a; Costa, 1988). Although several studies have identified an alteration in muscarinic receptor binding following prolonged OP exposure, only a few have investigated whether a corresponding alteration of receptor function was present. This is important in order to infer that development of tolerance is owing, at least in part, to a functional adaptation of muscarinic receptors (Costa *et al.*, 1982a). One study found that repeated administration of diisopropylfluorophosphate to rats caused a decrease in the carbachol-evoked inhibition of dopamine-stimulated adenylate cyclase in striatum (Olianas *et al.*, 1984). This effect, as discussed above, is believed to be mediated by activation of the M_2-subtype of muscarinic receptors (Gil and Wolfe, 1985). The decreased inhibition of adenylate cyclase paralleled a decrease of muscarinic receptor density in the same brain area.

Following a 2-week treatment with disulfoton, (*O,O*-diethyl *S*-[2-ethylthio-ethyl] phosphate), the ability of carbachol to stimulate accumulation of Ins 1-P in cerebral cortex slices (an index of activation of phosphoinositide metabolism, which in this brain area is mediated by M_1 and M_3 receptors) was also decreased (Costa *et al.*, 1986). The effect appeared to be specific for the muscarinic receptors, since the action of norepinephrine on phosphoinositide metabolism system was not affected (Costa *et al.*, 1986). The decrease and the recovery of the responsiveness to carbachol paralleled the decrease and recovery of muscarinic receptor binding, suggesting that the former was a consequence of the latter and indicating the presence of a tight receptor–effector coupling in this brain area. On the other hand, following an acute exposure to either disulfoton or diisopropylfluorophosphate, no changes in carbachol-stimulated phosphoinositide metabolism were found (Costa *et al.*, 1986; Cioffi and El-Fakahany, 1988), in agreement with the lack of alterations of ^{3}H-quinuclidinyl benzilate (QNB) binding following acute exposure (Costa *et al.*, 1982b). The loss of surface receptors, labeled by ^{3}H-*N*-methylscopolamine, observed following acute exposure to OP compounds is, therefore, not accompanied by physiological desensitization of the acetylcholine response (Cioffi and El-Fakahany, 1988).

Savolainen *et al.*, (1988a,b) have examined the role of inositol phosphates in the acute neurotoxic effects of OP compounds. Administration of 1.5 mg/kg diisopropylfluorophosphate to rats induced salivation, lacrimation, and tremors but no convulsions or increases in brain regional Ins 1-P concentrations (Savolainen *et al.*, 1988a). However, administration of LiCl increased the potential of diisopropylfluorophosphate to cause convulsions, and these convulsions were associated with a marked increase of brain regional Ins 1-P concentrations. A link between the inositol lipid signaling system and convulsions induced by soman had also been suggested (Savolainen *et al.*, 1988b). Soman-induced convulsions and increase in brain Ins 1-P were potentiated by lithium and antagonized by atropine as well as by diazepam, suggesting an involvement of both the cholinergic and GABAergic system (Savolainen *et al.*, 1988b). The exact mechanism underlying such potentiation by LiCl of the neurotoxic effects of direct or indirect cholinergic agonists is still obscure. It has also been suggested, however, that the alteration by LiCl of the inositol lipid signaling system may potentiate the disregulation of acetylcholine synthesis observed during status epilepticus (Jope *et al.*, 1987). Similar observations of lithium-induced potentiation of physostigmine or pilocarpine neurotoxicity had been reported previously (Samples *et al.*, 1977; Honchar *et al.*, 1983). An increase in Ins 1-P has also been observed after administration of malaoxon, even in the absence of lithium (Hirvonen *et al.*, 1989). This increase was modest and transient in nonconvulsing animals, while levels of Ins 1-P increased fourfold for at least 72 hr in convulsing rats. The authors suggest that Ins 1-P elevation in the nonconvulsing rats might have been caused by malaoxon-induced cholinergic stimulation, whereas in convulsing rats, the marked Ins 1-P elevation may be attributable partly to cholinergic stimulation and partly to the seizure process itself (Hirvonen *et al.*, 1989).

Recently, utilizing microwave irradiation as a method for animal sacrifice, Mobley (1990) demonstrated an increase in Ins 1,4,5-P_3 in neocortex and striatum of rats administered soman. No changes were observed in cerebellum or medulla–pons. This increase in Ins 1,4,5-P_3, which occurred shortly following administration of subconvulsive doses of soman, is probably owing to activation of muscarinic receptors by acetylcholine.

Certain OP compounds also might interfere with inositol metabolism by interacting with key enzymes in the phosphoinositide cycle. For example, parathion and diazinon, as well as their oxygen analogs, were found to stimulate (by 10 to 57%) the hydrolysis of PtdIns in rat brain microsomes (Davies and Holub, 1983). This effect, however was observed only at millimolar concentrations, greatly exceeding those required to inhibit acetylcholinesterase.

A few studies have examined the effects of OP compounds on cyclic AMP and cyclic GMP levels in brain or plasma. Soman, tabun, and sarin increased cyclic GMP levels in rat striatum 15 min. following administration; however, an inhibition of guanylate cyclase was observed at two hrs., perhaps as a compensatory mechanism brought into play by the elevated cyclic GMP levels (Liu *et al.,* 1986). Soman and sarin also increased cyclic AMP levels in striatum and in plasma (Stitcher *et al.,* 1977; Liu *et al.,* 1986) as did VX (Sevaljevic *et al.,* 1981). This finding has been interpreted as possibly being caused by the acetylcholine-induced release of catecholamines. While these effects appear to be mediated by accumulated acetylcholine, there is limited evidence that other factors may also play a role. For example, noncholinergic mechanisms have been suggested for soman-induced convulsions and increases in cerebellar cyclic GMP levels (Lundy and Shaw, 1983), on the basis of an antagonism of this effect by benzodiazepines. This latter finding is in agreement with the aforementioned results on Ins 1-P accumulation in convulsing animals and the effects of diazepam (Savolainen *et al.,* 1988b). Furthermore, Sevaljevic *et al.,* (1984) observed that the synaptosomal membrane-associated adenylate cyclase was more active *in vitro,* following *in vivo* exposure to soman.

In addition to *indirect* activation of cholinergic receptors by OP compounds, as a result of accumulated acetylcholine, recent evidence suggests that certain OP compounds can interact *directly* with muscarinic receptors (see previous chapter). Paraoxon, dichlorvos, and tetraethylpyrophosphate were found to inhibit the binding of ^{3}H-(QNB) in bovine caudate in a noncompetitive manner, with 50% inhibition (IC_{50}) of 5 to 50 n*M* (Volpe *et al.,* 1985). Further studies have shown that the effect of paraoxon appears to be selective for the M_2 and M_3 subtype (Katz and Marquis, 1989). Two other OP compounds, ecothiophate and VX (*o*-ethyl-*s*[2-(diisopropylamino)ethyl]methylphosphorothionate) were found to also inhibit the binding of ^{3}H-QNB and ^{3}H-pirenzepine (an M_1-selective muscarinic ligand) in rat brain; these two OP compounds, together with soman, tabun, and sarin, also inhibited the binding of ^{3}H-*cis*-methyldioxolane (a muscarinic agonist, possibly M_2-selective) with IC_{50} ranging from 3 to 800 n*M* (Bakry *et al.,* 1988). Binding studies, however, do not provide evidence of whether the OP compounds act as agonists or antagonists at the muscarinic receptors. Bakry *et al.* (1988) reported that in neuroblastoma cells, VX and ecotiophate competitively inhibited carbachol-stimulated accumulation of cyclic GMP, a response preferentially mediated by M_1 or M_3 receptor, suggesting that they may act as antagonists at these sites. On the other hand, paraoxon inhibited cyclic AMP accumulation (i.e., acted as a muscarinic agonist) in rat striatum by interacting with a population of M_2 receptors labeled by ^{3}H-*cis*-methyldioxolane (Jett *et al.,* 1990). In human SK-N-SH neuroblastoma cells, paraoxon appears to stimulate phosphoinositide hydrolysis by interacting with M_2 and/or M_3 receptors (Katz and Marquis, 1990; Katz, 1990, personal communication). Although more in-depth studies are needed to better evaluate the functional interaction of OP compounds with muscarinic receptors, it is already apparent that specific direct interaction with receptor subtypes might lead to different cellular responses and, therefore, differentially contribute to their overall toxic effects. A mechanistic question that should also be addressed is whether these effects of OP compounds are due to an *acetylcholine-like* action or to other mechanisms such as phosphorylation of the receptor or of other targets, as in direct activation of phospholipase C.

VI. Summary and Conclusions

Interaction of endogenous acetylcholine, which accumulates following acetylcholinesterase inhibition by OP compounds, with muscarinic receptors leads to various biochemical effects, involving the metabolism of membrane phosphoinositides, of cyclic AMP, and of cyclic GMP. Changes in the intracellular levels of free calcium, diacylglycerol, and cyclic nucleotides initiate a cascade of biochemical steps that, as a result of phosphorylation of specific substrates, lead to a final physiological or toxic response. Although activation of second-messenger systems by OP compounds is mostly a secondary effect, mediated by acetylcholine, evidence of possible direct effects of OP compounds, through an interaction with muscarinic receptors or other receptor systems, or through effects on other targets, is starting to accumulate. Measurements of changes in second messengers, in addition to being useful as indicators of changes of receptor function following acute and/or chronic OP poisoning, might also be fruitful to investigate some of the still elusive, yet relevant, aspects of their toxicity as, for example, the CNS effects often ascribed to chronic exposure.

Acknowledgments

This chapter is dedicated to the memory of Sheldon D. Murphy, whose work on organophosphorus toxicology has greatly contributed to the advancement of the field. Research by the author was supported by Grants ES-03424 and ES-04296 from the National Institute of Environmental Health Sciences and by a grant from the Fondazione Clinica del Lavoro, Pavia. The secretarial assistance of Claudia Thomas is gratefully acknowledged.

References

Ashendel, C. L. (1985). The phorbol ester receptor: A phospholipid-regulated protein kinase. *Biochim. Biophys. Acta* **822**, 219–242.

Ashkenazi, A., Peralta, E. G., Winslow, J. W., Rawachendran, J., and Copon, D.J. (1989a). Functionally distinct G proteins selectively couple different receptors to PI hydrolysis in the same cell. *Cell* **56**, 487–493.

Ashkenazi, A., Ramachandran, J., and Capon, D. J. (1989b). Acetylcholine analogue stimulated DNA synthesis in brain-derived cells via specific muscarinic receptor subtypes. *Nature* **340**, 146–150.

Bakry, N. M. S., El-Rashidy, A. M., Eldefrawi, A. T., and Eldefrawi, M. E. (1988). Direct actions of organophosphate anticholinesterases on nicotinic and muscarinic acetylcholine receptors. *J. Biochem. Toxicol.* **3**, 235–259.

Beavo, J. A., Hansen, R. S., Harrison, S. A., Hurrwitz, R. L., Martins, T. J., and Mumby, M. C. (1982). Identification and properties of cyclic nucleotide phosphodiesterases. *Mol. Cell. Endocrinol.* **28**, 387–410.

Berridge, M. J., Downes, C. P., and Hanley, M. R. (1982). Lithium amplifies agonist-dependent phosphatidylinositol responses in brain and salivary glands. *Biochem. J.* **206**, 587–595.

Berridge, M. J. (1987). Inositol triphosphate and diacylglycerol: Two interacting second messengers. *Annu. Rev. Biochem.* **56**, 159–193.

Brown, T. H., Chapman, P. F., Kairiss, E. W., and Keenan, C. L. (1988). Long-term synaptic potentiation. *Science* **242**, 724–728.

Casey, P. J., and Gilman, A. G. (1988). G Protein involvement in receptor–effector coupling. *J. Biol. Chem.* **263**, 2577–2580.

Chiarugi, V. P., Ruggiero, M., and Corradetti, R. (1989). Oncogenes, protein kinase C, neuronal differentiation and memory. *Neurochem. Int.* **14**, 1–9.

Cioffi, C. L., and El-Fakahany, E. E. (1988). Lack of alterations in muscarinic receptor subtypes and phosphoinositide hydrolysis upon acute DFP treatment. *Eur. J. Pharmacol.* **156**, 35–45.

Costa, L. G., Kaylor, G., and Murphy, S. D. (1986). Carbachol- and norepinephrine-stimulated phosphoinositide metabolism in rat brain: Effect of chronic cholinesterase inhibition. *J. Phamacol. Exp. Ther.* **239**, 32–37.

Costa, L. G., Schwab, B. W., and Murphy, S. D. (1982a). Tolerance to anticholinesterase compounds in mammals. *Toxicology* **25**, 79–87.

Costa, L. G., Schwab, B. W., and Murphy, S. D. (1982b). Differential alterations of cholinergic muscarinic receptors during chronic and acute tolerance to organophosphorus insecticides. *Biochem. Pharmacol.* **31**, 3407–3413.

Costa, L. G. (1988). Organophosphorus compounds. *In* "Recent Advances in Nervous System Toxicology" (C.L. Galli, L. Manzo, and P.S. Spencer, eds.), pp. 203–246. Plenum Press, New York.

Costa, L. G. (1990). The phosphoinositide/protein kinase C system as a potential target for neurotoxicity. *Pharmacol. Res.* **22**, 393–408.

Davies, D. B., and Holub, B. J. (1983). Comparative effects of organophosphorus insecticides on the activity of acetylcholinesterase, diacylglycerol kinase and phosphatidylinositol phosphodiesterse in rat brain microsomes. *Pestic. Biochem. Physiol.* **20**, 92–99.

Drummond, A. H., Joels, L. A. and Hughes, P. J. (1987). The interaction of lithium ions with inositol lipid signaling systems. *Biochem. Soc. Trans.* **15**, 32–35.

El-Fakahany, E., and Richelson, E. (1983). Effect of some calcium antagonists on muscarinic receptor–mediated cyclic GMP formation. *J. Neurochem.* **40,** 705–710.

Gil, D. W., and Wolfe, B. D. (1985). Pirenzepine distinguishes between muscarinic receptor–mediated phosphoinositide breakdown and inhibition of adenylate cyclase. *J. Pharmacol. Exp. Ther.* **232,** 608–616.

Goyal, R. K. (1989). Muscarinic receptor subtypes. Physiology and clinical implications. *N. Engl. J. Med.* **321,** 1022–1029.

Hawkins, P. T., Reynolds, D. J. M., Poyner, D. R., and Hanley, M.R. (1990). Identification of a novel inositol phosphate recognition site: Specific ^{3}H-inositol hexakisphosphate binding to brain regions and cerebellar membranes. *Biochem. Biophys. Res. Commun.* **167,** 819–827.

Hirshberg, A., and Lerman, Y. (1984). Clinical problems in organophosphate insecticide poisoning: The use of a computerized information system. *Fundam. Appl. Toxicol. 4,* S209–S214.

Hirvonen, M. R., Komulanainen, H., Polgarvi, L., and Savolainen, K. (1989). Time-course of malaxon-induced alterations in brain regional inositol-1-phosphate levels in convulsing and nonconvulsing rats. *Neurochem. Res.* **14,** 143–147.

Honchar, M. P., Olney, J. W., and Sherman, W. R. (1983). Systemic cholinergic agents induce seizures and brain damage in lithium-treated rats. *Science* **200,** 323–325.

Irvine, R. F., Moor, R. M., Pollock, W. K., Smith, P. M., and Wreggett, K.A. (1988). Inositol phosphates: Proliferation, metabolism, and function. *Philos. Trans. R. Soc. Lond. B* **320,** 281–298.

Jett, D., Abdallah, E., El-Fakahany, E., and Eldefrawi, A. (1990). High-affinity agonist action of paraoxon on the M_2 muscarinic receptor in rat brain. *Toxicologist* **10,** 341.

Jope, R. S., Simonato, M., and Lally, K. (1987). Acetylcholine content in rat brain is elevated by status epilepticus induced by lithium and pilocarpine. *J. Neurochem.* **49,** 944–951.

Katz, L. S., and Marquis, J. K. (1989). Modulation of central muscarinic receptor binding *in vitro* by ultralow levels of the organophosphate paraoxon. *Toxicol. Appl. Pharmacol.* **101,** 114–123.

Katz, L. S., and Marquis, J. K. (1990). Selective inhibition of muscarinic receptor binding by the organophosphate paraoxon in the human SK-N-SH cell line. *Toxicologist* **10,** 341.

Kendall, D. A. (1986). Cyclic GMP and inositol phosphate accumulation do not share common origins in rat brain slices. *J. Neurochem.* **47,** 1483–1489.

Kendall-Harden, T. (1989). Muscarinic cholinergic receptor-mediated regulation of cyclic AMP metabolism. *In* "The Muscarinic Receptors" (J.H. Brown, ed.), pp. 221–258. Humana Press, Clifton, New Jersey.

Kikkawa, U., and Nishizuka, Y. (1986). The role of protein kinase C in transmembrane signaling. *Annu. Rev. Cell Biol.* **2,** 149–178.

Kikkawa, U., Ogita, K., Shearman, M. S., Ase, K., Sekiguchi, K., Naor, Z., Ido, M., Nishizuka, Y., Saito, N., Tanaka, C., Ono, Y., Fujii, T., and Igarashi, K. (1988). The heterogeneity and differential expression of protein kinase C in nervous tissue. *Philos. Trans. R. Soc. Lond. B* **320,** 313–324.

Komulainen, H., and Bondy, S. C. (1988). Increased free intracellular Ca^{2+} by toxic agents: An index of potential neurotoxicity. *Trends Pharmacol. Sci.* **9,** 154–156.

Liu, D. D., Watanabe, H. K., Ho, I. K.,and Hoskins, B. (1986). Acute effects of soman, sarin, and tabun on cyclic nucleotide metabolism in rat striatum. *J. Toxicol. Environ. Health* **19,** 23–32.

Llinas, R. R. (1982). Calcium in synaptic transmission. *Sci. Am.* **247,** 56–65.

Lo, W. W. Y., and Hughes, J. (1987). Receptor-phosphoinositidase C coupling. Multiple G proteins. *FEBS Lett.* **244,** 1–3.

Lundy, P. M., and Shaw, R. K. (1983). Modification of cholinergically induced convulsive activity and cyclic GMP levels in the CNS. *Neuropharmacol.* **22,** 55–63.

McKinney, M., and Richelson, E. (1986). Blockade of N1E-115 murine neuroblastoma muscarinic receptor function by agents that affect the metabolism of arachidonic acid. *Biochem. Pharmacol.* **35,** 2389–2397.

McKinney, M., and Richelson, E. (1989). Muscarinic receptor regulation of cyclic GMP and eicosanoid production. "The Muscarinic Receptors" (J.H. Brown, ed.), pp. 309–340, Humana Press, Clifton, New Jersey.

Mobley, P. L. (1990). The cholinesterase inhibitor soman increases inositol triphosphate in rat brain. *Neuropharmacology* **29,** 189–191.

Murphy, S. D. (1986). Toxic effects of pesticides. *In* "Toxicology: The Basic Science of Poisons" (J. Doull, C. D. Klaassen, and M. O. Amdur, eds.), pp. 519–581. Macmillan, New York.

Namba, T., Nolte, C. T., Jacknel, J., and Grob, D. (1971). Poisoning due to organophosphate insecticides. Acute and chronic manifestations. *Am. J. Med.* **50,** 475–492.

Nathanson, N. M. (1987). Molecular properties of the muscarinic acetylcholine receptors. *Annu. Rev. Neurosci.* **10,** 195–236.

Nishizuka, Y. (1988). The molecular heterogeneity of protein kinase C and its implication for cellular regulation. *Nature* **334,** 661–665.

Ohsaka, S., and Deguchi, T. (1981). Stimulation by phosphatidic acid of calcium influx and cyclic GMP synthesis in neuroblastoma cells. *J. Biol. Chem.* **256,** 10945–10948.

Olianas, M. C., Onali, P., Schwartz, J. P., Neff, N. H., and Costa, E. (1984). The muscarinic receptor adenylate cyclase complex of rat striatum: Desensitization following chronic inhibition of acetylcholinesterase activity. *J. Neurochem.* **42,** 1439–1443.

Orrenius, S., and Nicotera, P. (1987). On the role of calcium in chemical toxicity. *Arch. Toxicol.* (Suppl) **11,** 11–19.

Pounds, J. G., and Rosen, J. F. (1988). Cellular Ca^{2+} homeostasis and Ca^{2+}-mediated cell processes as critical targets for toxicant action: Conceptual and methodological pitfalls. *Toxicol. Appl. Pharmacol.* **94,** 331–341.

Rana, R. S., and Hokin, L. E. (1990). Role of phosphoinositides in transmembrane signaling. *Physiol. Rev.* **70,** 115–164.

Robison, G. A., Butcher, R. W., and Sutherland, E. W. (1971). "Cyclic AMP", Academic Press, New York.

Samples, J. R., Janowsky, D. S., Pechnick, R., and Judd, L. L. (1977). Lethal effects of physostigmine plus lithium in rats. *Psychopharmacol.* **52,** 307–309.

Savolainen, K. M., Terry, J. B., Nelson, S. R., Samson, F. E., and Pazdernik, T. L. (1988a). Convulsions and cerebral inositol-1-phosphate levels in rats treated with diisopropylfluorophosphate. *Pharmacol. Toxicol.* **63,**137–138.

Savolainen, K. M., Nelson, S. R., Samson, F. E., Pazdernik, T. L. (1988b). Soman-induced convulsions affect the inositol lipid signaling system: Potentiation by lithium; attenuation by atropine and diazepam. *Toxicol. Appl. Pharmacol.* **96,** 305–314.

Schultz, G., Hardman, J. G., Schultz, K., Baird, C. E., and Sutherland, E. W. (1973). The importance of calcium ions for the regulation of guanosine 3′,5′-cyclic monophosphate levels. *Proc. Natl. Acad. Sci. U.S.A.* **70,** 3889–3893.

Sevaljevic, L., Krtolica, K., Poznanovic, G., Boskovic, B., and Maksimovic, M. (1981). The effect of organophosphate poisoning on plasma cyclic AMP in rats. *Biochem. Pharmacol.* **30,** 2725–2727.

Sevaljevic, L., Krtolica, K., and Boskovic, B. (1984). The effect of soman poisoning on phosphorylating capability and adenylate cyclase activity of isolated synaptosomal membranes. *Biochem. Pharmacol.* **33,** 3714–3716.

Snider, R. M., McKinney, M., Forray, C., and Richelson, E. (1984). Neurotransmitter receptors mediate cyclic GMP formation by involvement of arachidonic acid and lipoxygenase. *Proc. Natl. Acad. Sci. U.S.A.* **81,** 3905–3909.

Stitcher, D. L., Harris, L. W., Moore, R. D., and Heyl, W. C. (1977). Synthesis of cholinesterase following poisoning with irreversible anticholinesterase: Effect of theophylline and N^6, O^2-dibutyryl adenosine 3′,5′-monophosphate on synthesis and survival. *Toxicol. Appl. Pharmacol.* **41,** 79–90.

Stundermann, K. A., Harris, G. D., and Lovenberg, W. (1988). Characterization of inositol 1,4,5-triphosphate-stimulated calcium release from rat cerebellum microsomal fractions. *Biochem. J.* **255,** 677–683.

Vallejo, M., Jackson, T., Lightman, S., and Hanley, M. R. (1987). Occurrence and extracellular actions of inositol pentakis and hexakis-phosphate in mammalian brain. *Nature* **330,** 656–658.

Vicentini, L. M., and Villereal, M. L. (1986). Inositol phosphates turnover, cytosolic Ca^{2+} and pH: Putative signals for the control of cell growth. *Life Sci.* **38,** 2269–2276.

Volpe, L. S., Biagioni, T. M., and Marquis, J. K. (1985). *In vitro* modulation of bovine caudate muscarinic receptor number by organophosphates and carbamates. *Toxicol. Appl. Pharmacol.* **78,** 226–234.

Wolfe, B. B. (1989). Subtypes of muscarinic cholinergic receptors: Ligand binding, functional studies, and cloning. *In* "The Muscarinic Receptors" (J.H. Brown, ed.), pp. 125–150. Humana Press, Clinton, New Jersey.

Woodcock, E. A., Leving, E., and McLeod, J. K. (1987). A comparison of muscarinic acetylcholine receptors coupled to phosphatidylinositol turnover and to adenylate cyclase in guinea-pig atria and ventricles. *Eur. J. Pharmacol.* **133,** 283–289.

Worley, P. F., Baraban, J. M., and Snyder, S. H. (1989). Inositol 1,4,5-triphosphate receptor binding: Autoradiographic localization in rat brain. *J. Neurosci.* **9,** 339–346.

15

Tolerance to Organophosphorus Cholinesterase Inhibitors

Beth Hoskins
Ing K. Ho
Department of Pharmacology and Toxicology
University of Mississippi Medical Center
Jackson, Mississippi

I. Introduction

The toxicity of organophosphorus (OP) cholinesterase inhibitors has been extensively studied. Toxic symptoms arise because of the irreversible inhibition of cholinesterase by the OP compound, resulting in accumulation of endogenous acetylcholine (ACh) in nerve tissue and effector organs (see also Chapter 1 by Chambers, this volume). Symptoms of acute toxicity in animals and humans exposed to these compounds are typical of cholinergic overactivity and include salivation, lacrimation, sweating, involuntary defecation and urination, muscular twitching, weakness, tremors, and convulsions. Many studies have demonstrated that in the subacute or chronic situation in which sublethal doses of these compounds are injected repeatedly or fed in the diet to animals, acute symptoms of cholinergic overactivity are observed initially. In time, however, the animals will no longer respond with these obvious signs after each dose, and their general appearance, growth, and behavior will appear normal. At sacrifice, these apparently normal animals will be shown to have markedly inhibited blood and nerve-tissue cholines-

Organophosphates: Chemistry, Fate, and Effects

terase activities and their brain ACh levels will also be elevated. The following are examples of these studies.

It has been reported that rats fed 20 ppm of parathion for 1 year were apparently normal as judged by tests of behavior, rate of growth, mortality, and pathological changes (Barnes and Denz, 1951). Another study reported that daily injections of octamethyl pyrophosphoramide (OMPA), 0.5 mg/kg, permitted rats to withstand a daily dose of 1 mg/kg, which would have killed them otherwise (Rider *et al.*, 1952). Also, it was reported that rats could tolerate 50% (1 mg/kg) of the LD_{50} of disulfoton given repeatedly at daily intervals for a period of 60 days, while brain cholinesterase activity was depressed to 15% of the normal level (Bombinski and Dubois, 1958). During these studies, the rats exhibited marked symptoms and precipitous decreases in body weight during the first 10 days of treatment; after this period, they gained weight and failed to exhibit toxic symptoms when additional daily doses of the same magnitude were given. Costa and Murphy (1982) reported that when mice were administered disulfoton daily for 14 days, signs of poisoning disappeared after 5 to 8 days of treatment, while brain cholinesterase activity was inhibited 75 to 93%.

Several reports of behavioral tolerance of rats to diisopropylfluorophosphate (DFP) have come from the laboratories of Overstreet and Russell (Overstreet, 1973; Overstreet *et al.*, 1974; Russell *et al.*, 1969; 1971a, b,c; 1975). A dosing schedule of 1 mg/kg, followed at 3-day intervals for 22 days with doses of 0.5 mg/kg, maintained brain cholinesterase activity at approximately 30% of control levels. However, despite the low cholinesterase activity, tolerance became evident during measurements of consummatory behavior and in motor function tests.

A systematic assessment of tolerance development to DFP in terms of growth rates and consummatory behaviors was carried out in our laboratories (Lim *et al.*, 1983). Not only was tolerance to this compound demonstrated during continuous administration of sublethal doses, but continued exposure after the tolerance had developed resulted also in suprabaseline consummatory behaviors, while growth rates remained the same as control growth rates. In addition to these studies on body weights and consummatory behaviors, we also investigated DFP-induced tremors, chewing movements, and hind-limb abduction (Lim *et al.*, 1987a). In short, we found that although tolerance to DFP-induced tremors and hind-limb abduction occurred within 1 month, chewing still occurred even after 3 months of treatment and, in fact, became more intense. These studies therefore suggest that rats do not develop tolerance to all signs and symptoms of OP-induced toxicity.

There is also evidence of development of tolerance to the OP nerve gases. Sterri *et al.* (1980) reported that when rats were exposed daily to one half LD_{50} doses of soman, some of the animals did not show symptoms of soman poisoning and survived a total exposure of 4 to 7 times the acute LD_{50}, while brain and diaphragm cholinesterase activity declined steadily during repeated soman exposure. In another study, they reported that after injection of one half LD_{50} doses of soman to guinea pigs and mice, 70% of the guinea pigs and 36% of the mice survived a total exposure to 5.5 times the acute LD_{50} doses (Sterri *et al.*, 1981). In our own studies with the nerve gases (Dulaney *et al.*, 1985), we found that when rats were treated daily with low doses of soman (16% LD_{50}) or sarin (13–26% LD_{50}), toxic symptoms such as tremors and convulsions were very infrequent or absent, and all animals survived after 60 days of treatment. In a later study (Fernando *et al.*, 1985), we noted that increasing the doses of soman and sarin to 50 to 60% of their LD_{50} doses and administering these agents at intervals of 4 days caused variable neurotoxicity and increasing mortalities. After 10 injections, the survival rates were 31 and 54%, when brain AChE activity was 14 and 25% of control for soman and sarin, respectively. Therefore, we concluded that although tolerance does, indeed, develop to the nerve gases, it is of low degree and may be partly related to peripheral dispositional mechanisms.

From these examples, it is clear that tolerance does develop to OP toxicity. Not only is this true, but the evidence available also suggests that the chronic administration of an OP cholinesterase inhibitor is accompanied by the development of cross-tolerance to the cholinomimetic effects of other cholinergic agents. For example, rats rendered tolerant to disulfoton were found to be less sensitive to the lethal effects of carbachol (Brodeur and Dubois, 1964; Costa *et al.*, 1981; Schwab and Murphy, 1981) and less susceptible to the subacute lethal action of OMPA (McPhillips, 1969). Likewise, Costa and Murphy (1983) demonstrated that when rats developed tolerance to disulfoton (2 mg/kg for 10 days), the antinociceptive effect of nicotine was markedly reduced.

Mechanisms of tolerance development to OP cholinesterase inhibitors have been studied for the past 3 decades. The exact mechanism is still unclear, however, as the following literature review will demonstrate.

II. Metabolic Dispositional Tolerance

Although OP compounds have been reported to inhibit microsomal mixed-function oxidase activity when given acutely (Rosenberg and Coon, 1958), Stevens *et al.* (1972) showed that cytochrome P450 and microsomal enzyme activities are induced when rats are repeatedly administered organophosphorus compounds. Since pretreatment of animals with phenobarbital (Clement, 1983; Dubois, 1969; Menzer and Best, 1968), chlorinated hydrocarbons (Menzer, 1970), and hormones (Selye, 1970) reduced the toxicity of OP compounds, it has been proposed that enhancement of metabolic disposition of these compounds may be one aspect of tolerance development to them; i.e., the induction of hepatic microsomal enzymes by OP compounds increases their own metabolism and decreases their half-lives in the body (see Chapter 10 by Nakatsugawa, this volume, for more detailed discussion of induction, disposition, and toxicity). While this hypothesis seems to be plausible, numerous studies do not support the contention that the dispositional aspect plays a major role in the tolerance phenomenon. It has been consistently reported that cholinesterase activities in the brain and blood are persistently inhibited during repeated administration of these compounds (Barnes and Denz, 1954; Russell *et al.*, 1971a,b,c; Sivam *et al.*, 1983b; Sterri *et al.*, 1980, 1981; Wecker *et al.*, 1977). Furthermore, it has also been reported that sarin (Polak and Cohen, 1970) and DFP (Myers, 1952) bind to plasma and red blood cell aliesterase to a large extent (70%), and only a very small portion (18%) of the administered dose is actually available to inhibit cholinesterase. Sterri *et al.*, (1980, 1981) have even suggested that tolerance to soman may be because of its storage in adipose tissue and its binding to plasma proteins such as aliesterase and cholinesterase. They demonstrated that when guinea pigs and mice were exposed daily to half-LD_{50} doses of soman, the plasma cholinesterase activity was more than 90% inhibited 1 hour after soman administration and returned to between 40 and 50% of control levels within 24 hr, while plasma aliesterase activity was about 70% inhibited after 1 hr and was fully restored within 24 hr.

An enzyme or enzyme complex capable of hydrolyzing DFP or soman has been found in the squid and has been named DFPase or somanase (Chemnitius *et al.*, 1983; Hoskin *et al.*, 1966; Hoskin and Roush, 1982). Pai (1983) reported on studies of a phosphatase purified from a bacterium that catalyzes the hydrolysis of OP pesticides. The enzyme was termed FNT phosphatase. The discovery of these enzymes has raised the possibility that during subacute administration of OP cholinesterase inhibitors, these enzymes may be induced, and that such induction may lead to the acceleration of hydrolysis of the OP compounds in animals exposed to them.

Another enzyme system that is distinctively different from cholinesterase and can be phosphorylated by OP compounds has been termed neurotoxic esterase (NTE) (Johnson, 1975; 1976). Neurotoxic agents such as phosphates, phosphonate, and phosphoramidates have been shown to inhibit this enzyme. Therefore, with the possibility in mind that NTE may serve as a depot for these agents, we studied the inhibition and recovery of rat brain NTE, AChE, and butyrylcholinesterase (BuChE) activities after acute and subacute administration of DFP (Lim *et al.*, 1989). We found that DFP displayed different specificities in inhibiting these enzymes. Inhibition was greatest for BuChE, followed by AChE and NTE. Recovery was most rapid for BuChE, followed by NTE and AChE. The recovery rates of AChE and BuChE following acute and subacute treatment were similar. However, the recovery rate of NTE in subacutely treated rats was significantly faster than that in acutely treated rats. These results suggest that regeneration of these enzymes may be involved in tolerance to OP compounds. It is certainly clear that the roles of the various dispositional factors, such as various enzyme systems and adipose tissue storage, in the development of tolerance to OP compounds remain to be elucidated.

III. Involvement of Changes in Cholinergic Functional States

A. Synthetic and Degradative Cholinergic Enzymes

Acetylcholine synthesis is regulated by the concentration of ACh, coenzyme A (CoA), and choline as well as by the activity of choline acetyltransferase (CAT) at synaptic sites. Although parathion was shown to inhibit synaptosomal CAT activity (Muramatsu and Kuriyama, 1976), many investigators have reported that this enzyme is not affected by either acute or chronic administration of DFP (Russell *et al.*, 1975), paraoxon (Wecker *et al.*, 1977), disulfoton (Stavinoha *et al.*, 1969), or soman, sarin, and tabun (Ho *et al.*, 1983). It is unlikely that changes in the activity or availability of CAT are involved in the development of tolerance to OP compounds.

Cholinesterase is one of the most efficient enzymes known and has the capacity to hydrolyze 300,000 molecules of ACh per molecule of enzyme per minute. OP compounds phosphorylate the serine residue in the esteratic subsite of the active center. In addition, the phosphorylated enzyme is converted by dealkylation to the inactive form, which can no longer be reactivated either spontaneously or by oximes (Davies and Green, 1956; Hobbiger, 1956)(see also Chapter 5, by Wilson *et al.*, this volume on oxime reactivation). This dealkylation of the phosphorylated enzyme is known as aging. Chronic administration of OP compounds usually maintains ChE activity at between 20 and 30% of control levels (Russell *et al.*, 1975; Wecker *et al.*, 1977). We found that after chronic DFP treatment, the degree of AChE inhibition remained at a steady level despite the regression of DFP-induced overactivity (Sivam *et al.*, 1983b). Therefore, the data indicate that tolerance to OP compounds is not due to increased synthesis or activity of the enzyme. On the other hand, studies in our laboratories (Lim *et al.*, 1986) have suggested that muscarinic receptor density (to be discussed later) is intimately related to AChE activity.

B. Acetylcholine Levels

As early as the late 1940s, it was reported that tissues of rats poisoned by parathion contained markedly reduced ChE activity and that the increase in free ACh in treated brain was due to the irreversible inhibition of ChE in nerve tissues and affected organs (Dubois *et al.*, 1949). Subsequently, numerous investigators have reported that acute and subacute treatment with OP compounds markedly increases the total ACh level in synapses through their inhibition of ChE activity.

Brodeur and Dubois (1964) treated rats with disulfoton (1.2 mg/kg) for 30 days. They reported that the free ACh of the rat brain was elevated approximately the same extent by each successive dose, but the bound portion of total brain ACh did not show significant changes in the tolerant rats when compared with controls. They suggested that the increase in the total level was mainly attributable to the increase in free ACh rather than to a shift in the ratio of free to bound ACh.

Wecker *et al.* (1977) reported that the total increase in ACh in brains from rats treated chronically with paraoxon was less than half that in acutely treated rats. The free ACh levels were the same in both acutely and chronically treated rats, while levels of bound ACh were higher in the acutely treated animals. They observed no increase in total ACh levels in rats that were treated daily with paraoxon for 3 days. They therefore suggested that constant AChE inhibition might lead to an alteration in the presynaptic mobilization and storage of ACh.

After we treated rats with DFP both acutely and daily for 14 days, we monitored total, free, and bound ACh levels in striatum, hippocampus, and frontal cortex (Lim *et al.*, 1987c). We found that 30 min after daily administration of DFP, the total and free ACh levels were significantly increased and remained constant after each successive dose. In striatum and frontal cortex, the levels of bound ACh were also significantly increased; however, they were comparable to control levels after the 14th injection. Thirty min after a challenge dose of DFP was administered to saline-treated rats and to those treated subacutely with DFP, total ACh levels were significantly increased in hippocampus and frontal cortex and were not significantly different between the two treatment groups of rats. However, the level of total ACh in striatum was increased less in tolerant rats than in the acutely treated ones. The levels of free and bound ACh after acute administration of DFP were markedly increased in all three brain regions. After subacute administration of DFP, the levels of bound ACh were significantly increased in hippocampus and frontal cortex, but not in striatum. The increase in levels of bound ACh after subacute DFP treatment was less than the increase in the acutely treated rats in all three brain regions; however, the duration of the elevation of free ACh in striatum was shorter in subacutely treated rats. These results substantiate those of Wecker *et al.* (1977) and suggest that the presynaptic cholinergic storage sites for ACh might be changed during subacute administration of OP compounds and that this may play a role in the development of tolerance to these compounds.

C. Cholinergic Receptors

Brodeur and Dubois (1964) were the first to suggest that tolerance to OP compounds was due to the development of subsensitivity of cholinergic receptors to ACh. They reasoned that cholinergic receptors became less sensitive because the LD_{50} of carbachol was significantly higher in disulfoton-tolerant rats. Similar conclusions were reached by Chippendale *et al.* (1972) and Overstreet (1973) using other cholinergic and anticholinergic drugs.

1. Muscarinic Receptors

Overstreet *et al.*, (1974) measured the operant responses to muscarinic agonists and antagonists in DFP-tolerant rats and concluded that muscarinic subsensitivity was developed to muscarinic agonists when brain ChE activity fell to approximately 40% of the control activity. Since those studies, the evidence has consistently revealed that tolerance is associated with a subsensitivity to muscarinic agonists; this has led to the suggestion that decreased receptor density and/or affinity may be involved in the tolerance phenomenon.

Using the specific muscarinic antagonist [^{3}H]quinuclidinyl benzilate ([^{3}H]QNB), Ehlert and Kokka (1977) first demonstrated a decreased binding of this ligand in the ileum of rats that had been repeatedly treated with DFP. They later reported decreased [^{3}H]QNB binding and affinity of muscarinic receptors in rat brain and showed that muscarinic receptors displayed a greater decrease in affinity for muscarinic agonists than for atropine (Ehlert *et al.*, 1980). We also found that chronic administration of DFP reduced the number of muscarinic sites in rat striata without affecting the affinity of these sites for [^{3}H]QNB (Sivam *et al.*, 1983b). Similar results were obtained during studies of muscarinic receptor-binding characteristics after repeated treatment with other OP compounds, such as paraoxon (Smit *et al.*, 1980) and disulfoton (Costa and Murphy, 1982, 1983; Costa *et al.*, 1982).

Churchill *et al.* (1984) reported that the down-regulation of muscarinic receptors during chronic OP treatment is not unified throughout the brain. They showed down-regulation to 60 to 85% of control in cortex, striatum, lateral septum, hippocampus, superior colliculus, and pons. However, receptors were not altered in thalamus, hypothalamic nuclei, cerebellum, and reticular formation of the brain stem.

2. Nicotinic Receptors

Indirect evidence suggests that the nicotinic cholinergic receptors are also involved in the development of tolerance to OP compounds. McPhillips (1969) reported that disulfoton-tolerant rats were more sensitive to the toxic action of nicotinic antagonists. Overstreet *et al.* (1974) reported that supersensitivity of nicotinic receptors developed during chronic treatment with DFP when AChE activity dropped to 30% of normal, and that nicotinic receptors were more resistant to change than were muscarinic receptors. Based on their studies of cross-tolerance between DFP and the reversible anticholinesterases, Russell *et al.,* (1975) suggested that muscarinic and nicotinic receptors are involved differentially in tolerance development to OP compounds.

In contrast to the above studies, three groups of investigators (Schwartz and Kellar, 1983; Costa and Murphy, 1983; Lim *et al.,* 1987b) have reported that chronic treatment of rats with OP compounds resulted in decreased numbers of nicotinic receptors in brain tissues. Therefore, the exact involvement of nicotinic receptors in the process of tolerance development to OP compounds remains to be elucidated.

D. Choline Uptake

Although several studies have reported that chronic treatment of rats with OP compounds caused no alterations in choline uptake (Russell *et al.,* 1979, 1981; Costa and Murphy, 1983), Yamada *et al.* (1983a,b) reported that DFP administration to guinea pigs caused a significant decrease in [^{14}C] choline uptake in the striatum, hippocampus, and ileum longitudinal muscle. Studies in our laboratory (Lim *et al.,* 1987b) are in agreement with those of Yamada and colleagues. We found that after subacute treatment of rats with DFP, the maximal velocity of high-affinity choline uptake was significantly decreased in the striatum (33%) and in the hippocampus (53%) without changes in Michaelis constant (K_m) values. We therefore suggested that along with the down-regulation of postsynaptic receptors, subsensitivity of presynaptic functions of the cholinergic synapse also develops during subacute administration of OP compounds; this may play a role in the development of tolerance to these agents. We have also found that chronic stimulation of cholinergic receptors by reversible or irreversible inhibition of AChE (i.e., by administration of physostigmine or DFP, respectively) or by administration of the muscarinic receptor agonist, oxotremorine, resulted in supersensitivity to muscarinic antagonist–induced motor excitation (Fernando *et al.,* 1986).

IV. Other Mechanisms

Although down-regulation of cholinergic receptors appears to play the major role in the development of tolerance to OP compounds, other neuronal activities that interact with cholinergic pathways cannot be overlooked. An early suggestion of the involvement of other neurotransmitter systems in OP tolerance came from the studies of Russell *et al.* (1975), which demonstrated that control and DFP-tolerant animals behaved similarly when they were treated with 907-methyl-*p*-tyrosine, which depletes norepinephrine and dopamine in the brain.

There is ample evidence of interactions between neurotransmitters in the brain. For example, it has been demonstrated that an imbalance of dopaminergic and cholinergic activity in the basal ganglia is associated with motor dysfunction, particularly parkinsonism (Hornykiewicz, 1966; 1975). Furthermore, not only does acetylcholine appear to regulate gamma-amino butyric acid (GABA) synthesis (Roberts and Hammerschlag, 1972), but drugs that increase GABAergic activity in the brain also decrease dopamine turnover (Lahti and Losey, 1974).

We (Sivam *et al.,* 1983b) demonstrated that acute administration of DFP increased the number of dopamine and GABA receptors without affecting muscarinic receptor characteristics. As expected, subacute administration of DFP reduced the number of muscarinic sites without affecting their affinity; however, dopamine and GABA receptor densities increased, although not to the extent found following acute DFP treatment (suggesting a return toward control receptor characteristics). Since *in vitro* addition of DFP to striatal membranes did not affect any of these receptor types, the data indicated an involvement of GABAergic and dopaminergic systems in the development of tolerance to DFP. In another study the same year (Sivam *et al.,* 1983a), we found that acute, but not subacute, treatment of rats with DFP resulted in increased levels of glutamate and GABA, whereas subacute treatment decreased GABA uptake and release (decreased GABA turnover). Our studies of the dopaminergic system (Fernando *et al.,* 1984) revealed that striatal dopamine turnover apparently increased following acute treatment of rats with DFP, but decreased after chronic treatment. We therefore suggested that the changes in the GABAergic and dopaminergic systems arose secondary to an elevation of brain acetylcholine following cholinesterase inhibition. Acutely, it appears that these two systems may act singularly or in combination, to increase in activity perhaps in efforts to counteract the enhanced cholinergic activity induced by OP compounds. Prolonged changes in the levels and/or turnover of these neurotransmitters could be responsible for the increased postsynaptic densities of their receptors and might therefore partially mediate tolerance to DFP.

Therefore, we have proposed that GABAergic and dopaminergic neurons interact with cholinergic neurons after acute and subacute exposure to OP compounds, and because of these interactions, tolerance to OP compounds develops. Specifically, we propose that under control conditions, a balance between excitatory cholinergic and the inhibitory GABAergic and dopaminergic activities is maintained in the striatum. After acute exposure to an OP compound, inhibition of AChE results in increased cholinergic activity countered by increases in dopaminergic and GABAergic activities. After subacute exposure to an OP compound, synaptic concentrations of ACh are still high, but the muscarinic receptor characteristics have changed such that there is a decrease in number of receptors as well as in receptor sensitivity to ACh. Thus, the effect of increased concentrations of ACh in this situation results in a *normal* cholinergic (excitatory) response that, in turn, is balanced by normal inhibitory responses of the GABAergic and dopaminergic systems due to a return to normal of the inhibitory receptor populations. It is very important that further studies of the involvement of other neurotransmitter systems in the toxicity of and development of tolerance to OP compounds be carried out, because these agents are widely used as insecticides and may also be used as chemical warfare agents. A knowledge of the mechanisms involved in development of tolerance to these substances will allow manipulation of tolerance. Such manipulation could be used to hasten the development of tolerance in humans and animals exposed to them during their manufacture and agricultural (and possible chemicl warfare) use.

Acknowledgments

The studies from the authors' laboratories were supported by a contract from the United States Army Medical Research and Development Command, DAMD17-85-C-5036.

References

Barnes, J. M., and Denz, F. A. (1951). The chronic toxicity of *p*-nitrophenyl diethyl thiophosphate (E.605), a long-term feeding experiment with rats. *J. Hyg.* **49,** 430–441.

Barnes, J. M. and Denz, F. A. (1954). The reaction of rats to diets containing octamethyl pyrophosphoramide (Schradan) and *o,o*-diethyl-*s*-ethylmercaptoethanol thiophosphate ("Systox"). *Br. J. Industr. Med.* **11,** 11–19.

Bombinski, T. J., and Dubois, K. P. (1958). Toxicity and mechanism of action of Disyston. *AMA. Arch. Ind. Health,* **17,** 192–199.

Brodeur, J., and Dubois, K. P. (1964). Studies on the mechanism of acquired tolerance by *o,o*-diethyl-*s*-2-(ethylthio) ethyl phosphorodithioate (Di-syston). *Arch. Int. Pharmacodyn.* **149,** 560–570.

Chemnituis, J. M., Losch, H., Losch, K., and Zech, R. (1983). Organophosphate-detoxicating hydrolases in different vertebrate species. *Comp. Biochem. Phys.* **766,** 85–93.

Chippendale, T., Zawolkow, G. A., Russell, R. W., and Overstreet, D. H. (1972). Tolerance to low acetylcholinesterase: Modification behavior without acute behavioral change. *Psychopharmacologia,* **26,** 127–139.

Churchill, L., Pazdernik, T. L., Samson, F., and Nelson, S. R. (1984). Topographical distribution of down-regulated muscarinic receptors in rat brains after repeated exposure to diisopropylphosphonofluoridate. *Neurosci.* **11,** 463–372.

Clement, J. G. (1983). Effect of pretreatment with sodium phenobarbital on the toxicity of soman in mice. *Biochem. Pharm.* **32,** 1411–1415.

Costa, L. G., and Murphy, S. D. (1982). Passive avoidance retention in mice tolerant to the organophosphorus insecticide disulfoton. *Toxicol. Appl. Pharmacol.* **65**, 451–458.
Costa, L. G., and Murphy, S. D. (1983). [^{3}H]-Nicotine binding in rat brain: Alteration after chronic acetylcholinesterase inhibition. *J. Pharmacol. Exp. Ther.* **226**, 392–397.
Costa, L. G., Schwab, B. W., Hand, H., and Murphy, S. D. (1981). Reduced [^{3}H]quinuclidinyl benzilate binding to muscarinic receptors in disulfoton-tolerant mice. *Toxicol. Appl. Pharmacol.* **60**, 441–450.
Costa, L. G., Schwab, B. W., and Murphy, S. D. (1982). Differential alterations of cholinergic muscarinic receptors during chronic and acute tolerance to organophosphorus insecticides. *Biochem. Pharmacol.* **31**, 3407–3417.
Davies, D. R., and Green, A. L. (1956). The kinetics of reactivation of oximes of cholinesterase inhibited by organophosphorus compounds. *Biochem. J.* **63**, 529–535.
Dubois, K. P. (1969). Combined effects of pesticides. *Can. Med. Assoc. J.* **100**, 173–179.
Dubois, K. P., Doull, J., Salerno, P. R., and Coon, J. M. (1949).Studies on the toxicity and mechanisms of action of *p*-nitrophenyl diethylthionophosphate (parathion). *J. Pharmacol. Exp. Ther.* **95**, 79–91.
Dulaney, M. D., Jr., Hoskins, B., and Ho, I. K. (1985). Studies on low-dose subacute administration of soman, sarin, and tabun in the rat. *Acta Pharmacol. Toxicol.* **57**, 234–241.
Ehlert, F. J., and Kokka, N. (1977). Decrease in [^{3}H]-quinuclidinyl benzilate binding to muscarinic cholinergic receptor in the longitudinal muscle of the rat ileum following chronic administration of diisopropylfluorophosphate. *Proc. West. Pharmacol. Soc.* **20**, 1–7.
Ehlert, F. J., Kokka, N., and Fairhurst, A. S. (1980). Altered [^{3}H]-quinuclidinyl benzilate binding in the striatum of rats following chronic cholinesterase inhibition with diisopropylfluorophosphate. *Mol. Pharmacol.* **17**, 24–30.
Fernando, J. C. R., Hoskins, B., and Ho, I. K. (1984). Effect on striatal dopamine metabolism and differential motor behavior tolerance following chronic cholinesterase inhibition with diisopropylfluorophosphate. *Pharmacol. Biochem. Behav.* **20**, 951–957.
Fernando, J. C. R., Lim, D. K., Hoskins, B., and Ho, I. K. (1985). Variability of neurotoxicity of and lack of tolerance to the anticholinesterases soman and sarin in the rat. *Res. Commun. Chem. Pathol. Pharmacol.* **48**, 415–429.
Fernando, J. C. R., Hoskins, B., and Ho, I. K. (1986) The role of dopamine in behavioral supersensitivity to muscarinic antagonists following cholinesterase inhibition. *Life Sci.* **29**, 2169–2176.
Ho, I.K., Sivam, S. P., and Hoskins, B. (1983). Acute toxicity of diisopropylfluorophosphate, tabun, sarin, and soman in rats: Lethality in relation to cholinergic and GABAergic enzymes activities. *Fed. Proc.* **42**, 656.
Hornykiewicz, O. (1966). Dopamine (3-hydroxtyramine) and brain function. *Pharmacol. Rev.* **18**, 925–975.
Hornykiewicz, O. (1975). Parkinson's disease and its chemotherapy. *Biochem. Pharmacol.* **24**, 1061–1065.
Hobbiger, F. (1956). Chemical reactivation of phosphorylated human and bovine true cholinesterases. *Br. J. Pharmacol.* **11**, 295–303.
Hoskin, F. C. G., and Roush, A. H. (1982). Hydrolysis of nerve gas by squid-type diiospropylphosphorofluoridate hydrolysing enzyme on agarose resin. *Science* **215**, 1255–1257.
Hoskin, F. C. G., Rosenberg, P., and Brzin, M. (1966). Reexamination of the effect of DFP on electrical and cholinesterase activity of squid giant axon. *Proc. Natl. Acad. Sci. U.S.A.* **55**, 1231–1234.
Johnson, M. K. (1975). Organophosphorus esters causing delayed neurotoxic effects. Mechanism of action and structure/activity studies. *Arch. Toxicol.* **34**, 259–288.

Johnson, M. K. (1976). Mechanism of protection against the delayed neurotoxic effect of organophosphorus esters. *Fed. Proc.* **35**, 73–74.

Lahti, R. A., and Losey, E. G. (1974). Antagonism of the effects of chlorpromazine and morphine on dopamine metabolism by GABA. *Res. Commun. Chem. Pathol. Pharmacol.* **7**, 31–40.

Lim, D. K., Hoskins, B., and Ho, I. K. (1983). Assessment of diisopropylfluorophosphate (DFP) toxicity and tolerance in rats. *Res. Commun. Chem. Pathol. Pharmacol.* **39**, 399–418.

Lim, D. K., Hoskins, B., and Ho, I. K. (1986). Correlation of muscarinic receptor density and acetylcholinesterase activity in repeated DFP-treated rats after the termination of DFP administration. *Eur. J. Pharmacol.* **123**, 223–228.

Lim, D. K., Fernando, J. C. R., Hoskins, B., and Ho, I. K. (1987a). Quantitative assessment of tolerance development to diisopropylfluorophosphate. *Pharmacol. Biochem. Behav.* **26**, 218–286.

Lim, D. K., Hoskins, B., and Ho, I. K. (1987b). Evidence for the involvement of presynaptic cholinergic functions in tolerance to diisopropylfluorophosphate. *Toxicol. Appl. Pharmacol.* **90**, 465–476.

Lim, D. K., Porter, A. B., Hoskins, B., and Ho, I. K. (1987c). Changes in ACh levels in rat brain during subacute administration of diisopropylfluorophosphate. *Toxicol. Appl. Pharmacol.* **90**, 477–489.

Lim, D. K., Hoskins, B., and Ho, I. K. (1989). Effects of diisopropylfluorophosphate on brain acetylcholinesterase, butyrylcholinesterase, and neurotoxic esterase in rats. *Biomed. Environ. Sci.* **2**, 295–304.

McPhillips, J. J. (1969). Altered sensitivity to drugs following repeated injections of a cholinesterase inhibitor to rats. *Toxicol. Appl. Pharmacol.* **14**, 67–73.

Menzer, R. E. (1970). Effect of chlorinated hydrocarbons in the diet on the toxicity of several organophosphorus insecticides. *Toxicol. Appl. Pharmacol.* **16**, 446–452.

Menzer, R. E., and Best, N. H. (1968). Effect of phenobarbital on the toxicity of several organophosphorus insecticides. *Toxicol. Appl. Pharmacol.* **13**, 37–42.

Muramatsu, M., and Kuriyama, K. (1976). Effect of organophosphorus compounds on acetylcholine synthesis in brain. *Jpn. J. Pharmacol.* **26**, 249–254.

Myers, D. K. (1952). Competition of aliesterase in rat serum with the pseudocholinesterase for diisopropylfluorophosphate. *Science,* **115**, 568–570.

Overstreet, D. H. (1973). The effects of pilocarpine on the drinking behavior of rats following acute and chronic treatment with diisopropylfluorophosphate and during withdrawal. *Behav. Biol.* **9**, 257–263.

Overstreet, D. H., Russell, R. W., Vasquez, B. J., and Dalglish, F. W. (1974). Involvement of muscarinic and nicotinic receptors in behavioral tolerance to DFP. *Pharmacol. Biochem. Behavior* **2**, 45–54.

Pai, S. B. (1983). Purification of a bacterial organophosphate-hydrolysing phosphatase by cibacron sepharose affinity chromatography. *Biochem. Biophys. Res. Commun.* **110**, 421–416.

Polak, L. R., and Cohen, E. M. (1970). The finding of sarin in the blood plasma of the rat. *Biochem. Pharmacol.* **19**, 877–881.

Rider, J. A., Ellonwood, L. Z., and Coon, J. M. (1952). Production of tolerance in the rat to octamethyl pyrophosphoramide (OMPA). *Proc. Soc. Exp. Biol. Med.* **81**, 455–459.

Roberts, E., and Hammerschlag, R. (1972). Amino acid transmitters, *In* "Basic Neurochemistry" (R.W. Albers, G.J. Siegel, R. Katzman and B.W. Agranoff, eds.), pp. 218–245.

Rosenberg, P., and Coon, J. M. (1958). Increase of hexobarbital sleeping time by certain anticholinesterase OMPA, ZPN, malathion, chlorothion phostex. *Proc. Soc. Exp. Biol. Med.* **98**, 650–652.

Russell, R. W., Warburton, and Segal, D. S. (1969). Behavioral tolerance during chronic changes in the cholinergic system. *Commun. Behav. Biol.* **4**, 121–128.

Russell, R. W., Vasquez, B. J., Overstreet, D. H., and Dalglish, F. W. (1971a). Effects of cholinolytic agents on behavior following development of tolerance to low cholinesterase activity. *Psychopharmacologia,* **20,** 32–41.

Russell, R. W., Vasquez, B. J., Overstreet, D. H., and Dalglish, F. W. (1971b). Consummatory behavior during tolerance to and withdrawal from chronic depression of cholinesterase activity. *Physiol. Behav.* 7, 523–528.

Russell, R. W., Warbutton, D. M., Vasquez, B. J., Overstreet, D. H., and Dalglish, F. W. (1971c). Acquisition of new responses by rats during chronic depression of acetylcholinesterase activity. *J. Comp. Physiol. Psychol.* 77, 228–233.

Russell, R. W., Overstreet, D. H., Cotman, C. W., Carson, V. G., Churchill, L., Dalglish, F. W., and Vasquez, B. J. (1975). Experimental tests of hypotheses about neurochemical mechanisms underlying behavioral tolerance to the anticholinesterase diisopropylfluorophosphate. *J. Pharmacol. Exp. Ther.* **192,** 73–85.

Russell,R. W., Carson, V. G., Jope, R. S., Booth, R. A., and Macre, J. (1979). Development of behavioral tolerance: A search for subcellular mechanism. *Psychopharmacol.* **66,** 155–158.

Russell, R. W., Carson, V. G., Booth, R. A., and Jenden, D. J. (1981). Mechanisms of tolerance to the anticholinesterase, DFP: Acetylcholine levels and dynamics in the rat brain. *Neuropharmacol.* **20,** 1197–1201.

Schwab, B. W., and Murphy, D. D. (1981). Induction of anticholinesterase tolerance in rats with doses of disulfoton that produce no cholinergic signs. *J. Toxicol. Environ. Health.* **8,** 199–204.

Schwartz, R. D., and Kellar, K. J. (1983). Nicotinic cholinergic receptor binding sites in the brain: Regulation *in vivo. Science,* **220,** 214–216.

Selye, H. (1970). Resistance to various pesticides induced by catatoxic steroids. *Arch. Environ. Health* **21,** 706–710.

Sivam, S. P., Nabeshima, T., Lim, D. K., Hoskins, B., and Ho, I. K. (1983a). Diisopropylfluorophosphate and GABA synaptic function: Effect on levels, enzymes, release, and uptake in the rat striatum. *Res. Commun. Chem. Pathol. Pharmacol.* **42,** 51–60.

Sivam, S. P., Norris, J. C., Lim, D. K., Hoskins, B., and Ho, I. K. (1983b). Effects of acute and chronic cholinesterase inhibitors with diisopropylfluorophosphate on muscarinic, dopamine, and GABA receptors of the rat striatum. *J. Neurochem.* **40,** 1414–1422.

Smit, M. F., Ehlert, F. J., Yamamura, S., Roeske, W. R., and Yamamura, I. H. (1980). Differential regulation of muscarinic agonist binding sites following chronic cholinesterase inhibition. *Eur. J. Pharmacol.* **66,** 379–380.

Stavinoha, W. B., Ryan, L. C., and Smith, P. W. (1969). Biochemical effects of the organophosphorus cholinesterase inhibitor on the rat brain. *Ann. N.Y. Acad. Sci.* **160,** 378–382.

Sterri, S. H., Lyngaas, S., and Fonnum, F. (1980). Toxicity of soman after repetitive injection of sublethal doses in rat. *Acta Pharmacol. Toxicol.* **46,** 1–7.

Sterri, S. H., Lyngaas, and Fonnum, F. (1981). Toxicity of soman after repetitive injection of sublethal doses in guinea pig and mouse. *Acta Pharmacol. Toxicol.* **49,** 8–13.

Stevens, J. T., Stitzel, R. E., and McPhillips, J. J. (1972). Effects of anticholinesterase insecticides on hepatic microsomal metabolism. *J. Pharmacol. Exp. Ther.* **181,** 576–583.

Wecker, L., Mobley, P. L., and Dettbarn, W. D. (1977). Central cholinergic mechanisms underlying adaptation to reduced cholinesterase activity. *Biochem. Pharmacol.* **26,** 633–637.

Yamada, S., Isogai, M., Okudaira, H., and Hayashi, (1983a). Regional adaption of muscarinic receptors and choline uptake in brain following repeated administration of diisopropylfluorophosphate and atropine. *Brain Res.* **268,** 315–320.

Yamada, S., Isogai, M., Okudaira, H., and Hayashi, E. (1983b). Correlation between cholinesterase inhibition and reduction in muscarinic receptors and choline uptake by repeated diisopropylfluorophosphate administration: Antagonism by physostigmine and atropine. *J. Pharm. Exp. Ther.* **226,** 519–525.

16

Interactions of Organophosphorus Compounds with Neurotoxic Esterase

Rudy J. Richardson
Toxicology Program
The University of Michigan
Ann Arbor, Michigan

Organophosphates: Chemistry, Fate, and Effects

I. Introduction

A. Why Study Organophosphorus Compound Interactions with Neuropathy Target Esterase (NTE)?

This chapter focuses on the interactions of organophosphorus (OP) compounds with neurotoxic esterase (neuropathy target esterase, NTE; see definition in Section I,C). The understanding of these interactions achieved thus far has led to improved methods for predicting the neuropathic potential of OP compounds (Johnson, 1982; Davis *et al.*, 1985) and for the biomonitoring of neuropathic OP exposures (Lotti *et al.*, 1983; Richardson and Dudek, 1983). An important result of such advances is that the risk of organophosphate-induced delayed neuropathy (OPIDN; see definition in Section I,B) can be minimized (Johnson, 1980; Lotti *et al.*, 1984). However, discoveries have been made of interactions that promote or potentiate OPIDN following its initiation with subthreshold doses of neuropathic OP compounds (Pope and Padilla, 1989a, 1990; Caroldi *et al.*, 1990; Lotti *et al.*, 1991; Moretto *et al.*, 1991b; Peraica, 1991; Pope *et al.*, 1991). These findings have important implications, not only for the safety evaluation and regulation of OP and other compounds that interact with NTE, but also for the acceptance of the proposed role of NTE in the biochemical mechanism of OPIDN (Richardson, 1984; Johnson, 1990) (see Section V,D). A complete description of the pathogenesis of OPIDN including full knowledge of how NTE may be involved in this process will benefit public health and experimental neurology by enhancing the safe use of OP compounds and by providing new approaches for studying degenerative diseases of the peripheral nerves and spinal cord (Davis and Richardson, 1980; Richardson, 1983, 1984; Zech and Chemnitius, 1987; Carrington, 1989; Lotti, 1991).

B. Definition of Organophosphate-Induced Delayed Neuropathy (OPIDN)

OPIDN is a neurodegenerative disorder characterized by sensory loss and ataxia with concomitant distal degeneration of sensory and motor axons in ascending and descending spinal cord tracts as well as in peripheral nerves. Clinical signs occur approximately 2 to 3 weeks following absorption of a suprathreshold dose of a neuropathic OP compound (Davis and Richardson, 1980; Abou-Donia, 1981). The syndrome of OPIDN and its newly discovered variants (Abou-Donia and Lapadula, 1990) are described in more detail in Chapter 17, by Abou-Donia, this volume.

C. Definition of NTE

NTE is the putative target site attacked by neuropathic OP compounds to form the initial biochemical lesion that leads to OPIDN (Johnson, 1974,

1990; Richardson, 1984). NTE is operationally defined as a phenyl valerate hydrolase activity resistant to inhibition by diethyl 4-nitrophenylphosphate (paraoxon) and sensitive to inhibition by *N,N'*-di-2-propyl phosphorodiamidofluoridate (mipafox), under specified conditions of pH, ionic strength, temperature, preincubation time with inhibitors, incubation time with substrate, and inhibitor and substrate concentrations (Johnson, 1982). This enzyme may therefore be regarded as a subset of the carboxylic ester hydrolases (EC 3.1.1.1) (Chemnitius and Zech, 1983). Moreover, since NTE is inhibited by di-2-propyl phosphorofluoridate (DFP) and certain other OP compounds, it behaves like a serine esterase in the B-esterase category (Aldridge and Reiner, 1972), although an active-site serine has not been directly demonstrated for NTE. On the basis of its operational definition, an alternative name, neurotoxicant-sensitive esterase (NTSE), has been proposed for this enzyme, which has no known physiological substrate or function (Dudek and Richardson, 1982; Zech and Chemnitius, 1987). An assay for NTE was originally developed and later improved by Johnson (1975a, 1977). Recently, the assay conditions were optimized by Kayyali *et al.*, (1989, 1991).

D. Characterization of NTE

1. Species and Tissue Distribution

NTE activity has been characterized best in hen and chick brain, but the enzyme has been detected in neural and neuroendocrine tissue from a variety of species, including man (Dudek and Richardson, 1982; Johnson, 1982; Fulton and Chambers, 1985; Novak and Padilla, 1986; Moretto and Lotti, 1988). In contrast to the highly uneven distribution of brain acetylcholinesterase (AChE) activity, NTE activity varies only by about a factor of two across anatomical regions in hen, rat, and human brain. The highest relative activities of NTE are found in cerebral cortical areas of hen and human brain, and in hypothalamus of rat brain. Hen spinal cord and sciatic nerve NTE activities are about 20 and 2%, respectively, of that found in brain, and a similar distribution is found for the human nervous system (Lotti and Johnson, 1980; Dudek and Richardson, 1982; Hollingsworth *et al.*, 1984). Some nonneural tissues, including placenta, spleen, lymphocytes, and testis also possess substantial levels of NTE activity (Gurba and Richardson, 1983; Richardson and Dudek, 1983; Williams, 1983; Bertoncin *et al.*, 1985; Lotti *et al.*, 1985), but the highest activity in a given species is found in brain. The properties of NTE described below refer to hen or chick brain NTE unless otherwise indicated.

2. Subcellular Distribution

NTE is predominantly a membrane-associated enzyme that requires detergents or organic solvents for its solubilization (Davis and Richardson, 1987; Schwab *et al.*, 1985; Pope and Padilla, 1989b). Moreover, the activity of the enzyme exhibits lipid dependency, and is stimulated by certain phospholipids, most notably, phosphatidylcholine (Davis and Richardson, 1987; Pope and Padilla, 1989c). However, a small amount (about 3%) of the brain activity is found in the cytosolic fraction (Richardson *et al.*, 1979), and recent work indicates that a significantly higher portion of the activity in peripheral nerve is soluble (Vilanova *et al.,* 1990). Differential centrifugation of hen brain homogenates results in about 70% of the NTE activity pelleting in the P_3 ("microsomal") fraction, but the relative contribution made to this fraction by neurons versus glia or intracellular versus surface membranes is not known.

3. Enzymology and Protein Chemistry

Enzymological studies indicate that the activity associated with brain NTE is comprised of a single enzyme (Johnson, 1975b, 1990). The apparent identification of two isoforms by iterative elimination of exponential mipafox inhibition curves (Chemnitius *et al.,* 1983) has been shown to arise from the formation of a significant amount of Michaelis complex at high mipafox concentrations (Carrington and Abou-Donia, 1986). Likewise, the biphasic heat inactivation of NTE has been explained on the basis of the formation of a partially active intermediate during the denaturation process (Reiner *et al.,* 1987). However, the soluble NTE from peripheral nerve appears to contain two components that differ substantially in their sensitivity to mipafox (Vilanova *et al.,* 1990).

NTE was first identified by differential phosphorylation using [^{32}P]-DFP rather than by differential hydrolysis using phenyl valerate (or phenyl phenyl acetate in early experiments)(Johnson, 1969a,b). Separations of similarly prepared radiolabeled proteins by gel exclusion chromatography, gel electrophoresis, or sucrose density gradient centrifugation after prior solubilization in detergents indicate that the phosphorylated subunit of NTE has a relative molecular mass (M_r) of 148 to 178 kilodaltons (kDa)(Williams and Johnson, 1981; Carrington and Abou-Donia, 1985; Novak and Padilla, 1986; Pope and Padilla, 1989b; Thomas *et al.,* 1989). However, there are other phosphorylated proteins of lower M_r identified in these preparations that should be investigated further as possible alternative targets or components of a larger NTE complex.

II. Inhibition of Esterases by Organophosphorus Compounds

A. Types of Inhibition

The type and extent of toxicity produced by an OP compound depends on the concentration of active metabolite available for binding to target esterases, and the relative degree of net inhibition achieved with one target versus another. For example, the intended action of OP insecticides is lethality in target species due to a critical level of AChE inhibition. An untoward action of these compounds is OPIDN in nontarget species attributable, in part, to a critical level of NTE inhibition (Murphy, 1986). The net inhibition achieved at a specific time for a given esterase depends in turn on the inhibition kinetics: the rates of reactions between the OP compound and esterase that result in inhibition of esterase activity. Esterase inhibition by OP compounds is considered to be irreversible, because a covalent bond is formed between the esterase and the OP compound, and enzyme activity often returns relatively slowly by hydrolytic cleavage of this bond (reactivation). Even in cases of rapid reactivation, the acylating inhibition of esterases by OP compounds is termed irreversible to distinguish it from cases of truly reversible inhibition, in which the intact inhibitor molecule can be dissociated from the enzyme. In irreversible inhibition, part of the original OP molecule is lost during acylation of the enzyme (Aldridge and Reiner, 1972). In order to understand more fully the biochemical basis for postulating a pathogenic role for NTE in OPIDN, and to provide a quantitative framework for evaluating the relative risk of OPIDN from a given OP compound relative to other compounds, it will be necessary first to examine the kinetics of irreversible inhibition of esterases in some detail.

B. Kinetics of Irreversible Inhibition

1. Introduction

Thorough treatments of the kinetics of irreversible inhibition of esterases by OP compounds may be found in the reference sources by Aldridge and Reiner (1972) and Main (1980); also, a convenient summary is provided in the paper by Clothier *et al.,* (1981). The discussion of kinetics and the equations presented throughout Section II,B of this chapter represent a synthesis drawn from these sources; therefore, citations of these sources will not be repeated in this section.

2. Reaction Scheme and Kinetic Constants

The relevant partial reactions and associated rate constants that describe the interactions between an esterase and an OP inhibitor are shown in Fig. 1. The esterase is denoted by EOH to emphasize the activated serine hydroxyl group presumed to be in the active site of all B-esterases (carboxylesterases inhibited by OP oxons). PX represents an OP oxon of the general form, R(R′)P(O)X, where R and R′ typically may be simple or substituted alkyl or aryl groups attached to pentacovalent phosphorus through a C, N, O, or S atom, and X is an acidic moiety that functions as the primary leaving group when phosphorus undergoes nucleophilic attack by the activated serine hydroxyl group of the esterase. The oxon form of the OP is required to provide a sufficiently

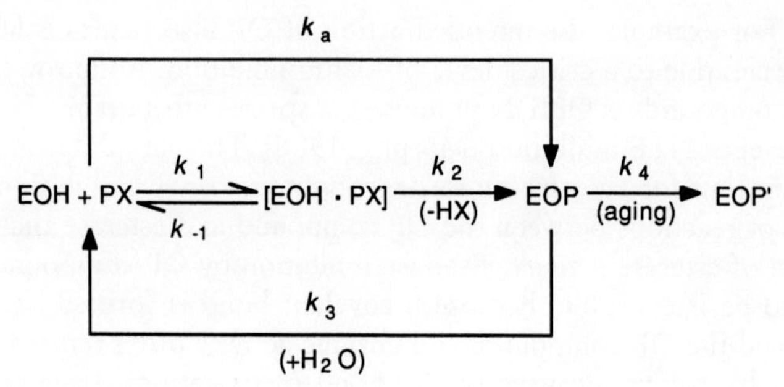

Figure 1 Partial reactions and associated rate constants for interactions between an esterase (EOH) and an organophosporus inhibitor (PX). Adapted from Clothier *et al.* (1981, Scheme I, p. 307).

electropositive center at the phosphorus atom. (See also Chapters 1 and 4 of this volume for discussion of factors affecting charges on the OP phosphorus and the serine hydroxyl group.) Such OP compounds are mechanism-based inhibitors: their tetrahedral geometry mimics the tetrahedral transition state of the carboxylic ester substrate of the enzyme. In fact, with the exception of the aging reaction (see Section II,B,6), the series of reactions between an esterase and an OP is entirely analogous to the reactions between an esterase with its substrate.

The first interaction between an esterase and an OP inhibitor is a reversible association to form a Michaelis-type complex, which can react further to form an acylated (in this case, phosphorylated) enzyme (Fig. 1). The reversible first step is characterized by rate constants for the forward (k_1) and reverse (k_{-1}) reactions. For most OP inhibitors, the concentration of the Michaelis-type complex present at any time during the reaction sequence is close to zero, and the rate constant for phosphorylation (k_2) is quite high. Furthermore, the rate constant for spontaneous reactivation due to hydrolysis of the phosphorylated enzyme (k_3) is typically much smaller than k_2, so that it usually suffices to measure the overall bimolecular rate constant of inhibition (k_a) as an index of the inhibitory power of an OP compound. The relative magnitude of these rate constants distinguishes a substrate from an inhibitor. For substrates, the acylated form of the enzyme is a transient species, because the rate of reactivation is relatively fast. For inhibitors, the active site of the enzyme is blocked at the acylation step, because spontaneous reactivation is relatively slow.

The bimolecular rate constant of inhibition, k_a, is the most straightforward kinetic constant to obtain experimentally, requiring only a measurement of the disappearance of activity of the starting esterase, EOH, as a function of time of preincubation with the inhibitor, PX. The inhibitory power of an OP compound, however, is a combination of its binding affinity for the esterase,

K_a, together with the rate of phosphorylation, k_2. As can be seen from Eq. (1) below, the binding affinity is analogous to the Michaelis constant for the reaction of an enzyme with its substrate, where K_a corresponds to the concentration of PX required to achieve a half-maximal rate of production of phosphorylated enzyme, EOP.

$$K_a = (k_{-1} + k_2)/k_1 \quad (1)$$

When the concentration of inhibitor is much less than K_a, the bimolecular rate constant of inhibition is simply the ratio of the phosphorylation rate constant and the affinity constant, as shown in Eq. (2).

$$k_a = k_2/K_a \quad (2)$$

3. Ideal First-Order Kinetics

Methods for determining k_a, k_2, or K_a begin by solving the differential equation that describes the kinetics of the reaction sequence depicted in Fig. 1. This solution is given by Eq. (3).

$$\ln(v/v_o) = -k_2[PX]t/([PX] + K_a) \quad (3)$$

In Eq. 3, (v/v_o) is the ratio of the rate at time t divided by the initial rate, and [PX] is the concentration of inhibitor. The ratio of the percentage of control activity (% activity) at time t over the initial percentage (100%) is commonly substituted for the ratio of rates. If we also let $k' = k_2[PX]/([PX]+K_a)$, then we have the following:

$$\ln(\%\ \text{activity}/100) = 2.303 \log(\%\ \text{activity}/100) = -k't \quad (4)$$

Solving Eq. (4) for log (% activity) gives Eq. (5):

$$\log(\%\ \text{activity}) = -(k't/2.303) + \log(100) = -(k't/2.303) + 2 \quad (5)$$

In Eq. (5), k' is a first-order rate constant for a given inhibitor concentration, [PX], such that [PX]>10[EOH]. If the concentration of Michaelis-type complex is small, and the rate of spontaneous reactivation (k_3) is negligible, plots of log (% activity) versus time (t) at a given inhibitor concentration will be straight lines that extrapolate through the "origin" (intercept on the ordinate when t = 0; log 100% = 2), with slopes given by $k'/2.303$. Furthermore, plots of k' versus [PX] will be linear with a slope equal to the bimolecular rate constant of inhibition, k_a:

$$k' = k_a[PX] \quad (6)$$

Substituting Eq. (6) into Eq. (5) gives Eq. (7), the rate equation that describes the kinetics of inhibition of esterases by many OP compounds when the reaction is carried out under pseudo first-order conditions. When the results of such an inhibition experiment can be described by Eq. (7), k_a can readily be obtained, but k_2 and K_a cannot be determined by this approach. Note that Eq. (7) indicates that enzyme activity is first order with respect to preincubation time at a given constant value of [PX], and also first order with respect to [PX] at a constant value of preincubation time (provided that [PX]>10[EOH]).

$$\log(\%\ \text{activity}) = -(k_a[\text{PX}]t/2.303) + 2 \quad (7)$$

4. Deviations from Ideal Kinetics

If the plots of log (% activity) versus time are linear and pass through the origin (2), but k' is not proportional to [PX], then there is an appreciable concentration of a Michaelis-type complex. In this case, the constants k_2 and K_a can be determined from Eq. (8) below, obtained by combining Eq. (3) and Eq. (4) and rearranging:

$$[\text{PX}]/k' = K_a/k_2 + [\text{PX}]/k_2 \quad (8)$$

A Wilkinson-type plot of Eq. (8) ([PX]/k' versus [PX]) will give a straight line with a slope of $1/k_2$, an intercept on the ordinate of K_a/k_2, and an intercept on the abscissa of $-K_a$. Thus, from this plot, the affinity and phosphorylation constants can be determined, and k_a may then be calculated from Eq. (2).

The primary semilog plots from an inhibition experiment may give straight lines with k' values that vary with [PX] and intercepts that do not extrapolate through the origin. This indicates that a significant concentration of a Michaelis-type or other complex exists both during the preincubation and after addition of substrate. In this situation, further experiments are necessary to examine the effect of dilution, substrate concentration, or both on the magnitude of the intercept shift on the ordinate, in order to determine k_2 and either K_a or K_i, where K_i is the equilibrium constant for the dissociation of a reversible complex that does not react further to give phosphorylated enzyme.

If the primary semilog plots are not first order, and k' decreases with time, there may be an approach to a steady state due to significant reactivation of phosphorylated enzyme (EOP) by hydrolysis to yield EOH. In this case, the initial rates can be determined and used to determine k_a and possibly K_a and k_2, as described. It may also be of interest to determine the reactivation rate constant, k_3. Deviation from first-order kinetics may be the result of [PX] being too low to achieve pseudo first-order conditions, which is easily remedied by increasing the inhibitor concentration, unless the solubility of PX is limiting. Another reason for obtaining curvilinear semilog plots is loss of inhibitor due to hydrolysis or enzymatic disposal by other enzymes in the preparation; this possibility can be checked by varying [PX] at a fixed preincubation time to see if linear plots of log (% activity) versus [PX] can be obtained at higher [PX]. Finally, when working with crude enzyme preparations from tissues, departures from first-order kinetics may be because of the presence of more than one enzyme hydrolyzing the same substrate. For hen brain AChE and NTE, this problem can be circumvented or minimized through the use of selective substrates and differential inhibition of nontarget enzymes. An advantage of kinetic methods is the power to dissect out and characterize the activity of interest from a complex mixture, often with considerable mathematical precision.

5. Spontaneous Reactivation

The rate constant for spontaneous reactivation, k_3, is treated as a first-order rate constant, owing to the fact that the concentration of the reactivator (water) is essentially constant and in great excess. This rate constant may be determined by preincubating an enzyme preparation under conditions that rapidly produce nearly complete inhibition. The inhibitor is then removed or diluted to the extent that inhibition of free enzyme will not proceed at an appreciable rate, and the return of enzyme activity is measured in aliquots at timed intervals. A plot of log (% activity) versus time will give a straight line with a slope of $-k_3/2.303$ from which k_3 may be calculated. Since reactivation is a first-order process, the half-life of reactivation is given by $0.693/k_3$. Apparent half-lives of spontaneous reactivation range from minutes to weeks, depending on pH, temperature, the identity of the B-esterase, and the structure of the acyl group of the inhibitor attached to the enzyme. Spontaneous reactivation may appear to be slow or nonexistent if aging has occurred, because the aged form of the organophosphorylated esterase is stable to hydrolysis and will not reactivate.

6. Aging

The aging reaction, characterized by a first-order rate constant, k_4, involves scission of an R-group from the OP moiety of the organophosphorylated esterase, yielding a negatively charged monosubstituted phosphoryl group still attached to the active site of the enzyme (denoted by EOP′ in Fig. 1). Aging can occur when at least one of the R-groups is attached to phosphorus through a labile bond, such as R—O—P or R—NH—P. If the R-groups are attached to phosphorus through carbon atoms, aging will not occur, because of the relative stability of the C—P bond. In an enzyme inhibition experiment, usually only active enzyme is being measured. Therefore, if the aging reaction occurs, it will not interfere with determinations of EOH activity, since both the phosphorylated (EOP) and phosphorylated-aged (EOP′) esterases are enzymatically inactive. However, aging does render the enzyme nonreactivatable, not only by water (spontaneous reactivation), but even by powerful nucleophiles such as certain oximes or fluoride ion.

The technique of measuring the rate of aging is similar to that used for measuring spontaneous reactivation. An enzyme preparation is preincubated with a high concentration of inhibitor to produce nearly complete inhibition as rapidly as possible. The inhibitor is then quickly removed or diluted to a concentration too low to cause inhibition of reactivated enzyme. At timed intervals after the inhibition period, aliquots are treated with fluoride or another reactivator and the amount of reactivated enzyme activity relative to a nonreactivated sample is measured. As the enzyme ages, less activity will be restored by the reactivator. A plot of log (% activity restored) versus time will be linear, with a slope equal to $-k_4/2.303$. Half-lives of aging vary by orders of magnitude, from seconds to days or more, depending upon pH, temperature, the identity of the B-esterase, and the structure of the phosphoryl moiety attached to the enzyme. Aging has been shown to occur with a variety of esterases, including AChE and NTE (Clothier and Johnson, 1979, 1980). Aging is rapid if AChE is inhibited with OP compounds having highly branched alkyl R-groups (favoring carbonium ion stabilization), and slow with OP compounds having straight-chain alkyl R-groups; just the opposite is true for aging of NTE. The evidence to date indicates that aging of AChE proceeds by an S_N1 mechanism, whereas aging for NTE proceeds by an S_N2 mechanism. Aging of NTE may differ in another important respect from that of AChE. For at least some compounds, NTE aging involves an intramolecular transfer of the cleaved group to an unknown site on the phosphorylated subunit of NTE (referred to as site Z), whereas AChE aging involves loss of the cleaved group from the molecule (Clothier and Johnson, 1979, 1980; Williams, 1983). Furthermore, it is now clear that the toxicological consequences of the aging of AChE are completely different from those of the aging of NTE, as discussed in Section III (Murphy, 1986; Johnson, 1987).

C. The I_{50} as a Measure of Potency of Inhibition

A commonly used measure of inhibitory power is the I_{50}, the concentration of inhibitor required to produce 50% inhibition of activity under specified conditions of preincubation time, temperature, enzyme concentration, pH, and other ionic conditions. When the inhibition kinetics are described by Eq. (7) above, substitution of I_{50} for [PX] and 50% for (% activity) yields the following relationship between the I_{50} and k_a for a given preincubation time with inhibitor:

$$I_{50} = 0.693/k_a t \qquad (9)$$

The conversion of k_a into an I_{50} is useful because concentrations may be more attractive than rate constants for visualizing relative inhibitory potency, just as half-lives may have greater intuitive appeal than rate constants for conveying information about the relative time required to complete a process.

Often, the I_{50} is determined directly by measuring the enzymatic activity remaining after preincubation for a fixed interval at various [PX], plotting (% activity) versus [PX], and reading the [PX] required for 50% inhibition from the curve. Determined in this way, the I_{50} can be thought of as the midpoint in the inhibitor titration curve of an enzyme activity. Such empirical fixed-time I_{50} values must not be used to calculate apparent k_a values (Aldridge and Reiner, 1972; Main, 1980).

If ideal first-order kinetics are obtained [see Eq. (7) above], the I_{50} may be conveniently calculated from the linear relationship between log (% activity) and [PX] at a constant time of preincubation. Because the I_{50} is a function of time, and the kinetics of inhibition may be complex, the I_{50} is of less value than a kinetic rate constant as a single measure of inhibitory potency. However, even in cases of nonideal inhibition kinetics, as long as its limitations are understood, the I_{50} can be a useful concept and a practical tool for comparing relative potencies of inhibitors or relative sensitivities of different enzymes to a given inhibitor (Aldridge and Reiner, 1972; Main, 1980) (see Section IV,B).

III. Conventional View of the Role of NTE in OPIDN

A. Aging Inhibitors of NTE Cause OPIDN

Single doses of neuropathic OP compounds administered to hens in amounts that produce a critical level of inhibition (the so-called threshold) of about 70% or more of whole-brain NTE the day after dosing result in OPIDN in pair-dosed animals 2 to 3 weeks after dosing. OPIDN does not appear to result from NTE inhibition significantly below the critical level, nor does it result from treatment with compounds that do not inhibit NTE. However, the ability to produce a critical level of NTE inhibition is not the sole prerequisite for the induction of OPIDN. It is now apparent that while all compounds capable of causing OPIDN inhibit NTE *in vivo,* some compounds inhibit NTE without causing OPIDN. Among the OP inhibitors of NTE, phosphates, phosphonates, and phosphoramidates cause OPIDN. These compounds all contain at least one —O— or —NH— bridge linking an R-group to phosphorus, and are therefore considered capable of undergoing the aging reaction. OP inhibitors of the phosphinate class inhibit NTE without producing OPIDN. Phosphinates have their R-groups linked to phosphorus through a carbon atom, and do not undergo the aging reaction, because of the greater stability of the P—C bond compared with the P—O or P—N bond (see Section II,B,6). Phenylmethylsulfonyl fluoride (PMSF) and the carbamate inhibitors of NTE are also incapable of aging and do not produce OPIDN (Davis and Richardson, 1980; Johnson, 1982; Davis *et al.,* 1985; Carrington, 1989).

B. Nonaging Inhibitors of NTE Protect against OPIDN

It would be difficult to attribute any role in the mechanism of OPIDN to NTE if the only effect of nonaging inhibitors of NTE were not to induce OPIDN.

Since NTE is inhibited by both aging and nonaging inhibitors, the implication is that there must be another target that is inhibited only by aging inhibitors, which are the compounds that produce OPIDN. In order to rule out a pathogenic role for NTE altogether, it would be necessary to show that inhibition of NTE by nonaging inhibitors not only did not produce OPIDN, but also that such inhibition would have no effect on the induction of OPIDN by OP compounds known both to inhibit NTE and to cause OPIDN.

Aging does not affect the initial stages of esterase inhibition leading to phosphorylated enzyme, but if aging does occur, spontaneous reactivation is blocked, resulting in prolonged inhibition (see Section II,B,5). Conversely, if aging does not occur, spontaneous reactivation is possible, and may be relatively rapid with some compounds, resulting in short-lived inhibition. Therefore, it is possible to postulate that OPIDN requires prolonged inhibition of NTE, and that nonaging inhibitors fail this requirement because of rapid reactivation. However, prolonged inhibition above the critical level by repeated dosing with rapidly reactivating carbamates does not result in OPIDN. Also, since PMSF does not reactivate, and some phosphinates have reactivation rates comparable to those of neuropathic OP compounds, single doses of these compounds produce prolonged inhibition of NTE, but OPIDN does not occur. Thus, it appears that nonaging inhibitors of NTE do not produce OPIDN, even if reactivation rates are taken into account (Johnson, 1982).

The first crucial experiment to test the hypothesis that NTE is irrelevant to the induction of OPIDN was carried out by Johnson and Lauwerys (1969). Carbamate inhibitors of NTE were given to hens, followed by known neuropathic doses of DFP. As long as the DFP was given while the NTE was still inhibited by the carbamate, the animals were protected from the development of OPIDN. The spontaneous reactivation rate for the carbamates used is high (70% reactivation within hours); accordingly, it was found that the protection afforded by these carbamates lasted only for a similar period. After this time, sufficient NTE had been regenerated to allow access of a critical amount of the active site for inhibition by a neuropathic OP compound (Johnson, 1982).

Later experiments with PMSF and phosphinate inhibitors of NTE showed that these compounds were also protective against OPIDN. Moreover, the protection afforded by these compounds was found to last for hours to days, in keeping with the duration of NTE inhibition (Johnson, 1970, 1974). Furthermore, if the order of presentation was reversed, with the neuropathic OP compound given before the nonaging NTE inhibitor, there was no protection against OPIDN (Johnson, 1982; Carrington, 1989). In an interesting variation, it was shown that pretreatment with a carbamate inhibitor of NTE followed by PMSF resulted in protection against subsequent neuropathic OP treatment that lasted only hours rather than days (Johnson, 1974). Thus, it is possible to protect against protection, with the length of protection being determined by the reactivation rate of the nonaging NTE inhibitor that is given. Protection has been demonstrated and correlated with the extent and duration of NTE inhibition by the nonaging inhibitor in several animal models using several different combinations of protectant and neurotoxicant (Johnson, 1982; Carrington, 1989). It appears that inhibition of at least 30% of the NTE activity by a nonaging inhibitor is required in order to achieve protection against a subsequent dose of a neuropathic OP compound.

C. OPIDN is Initiated by Concerted Inhibition and Aging of NTE

If about 30% or more inhibition of NTE by a nonaging compound is needed for protection against OPIDN, the implication is that protection works by preventing inhibition of the critical level of about 70% of the NTE by a neuropathic OP compound. Since the only consistent difference between neuropathic and protective compounds seems to be their ability to undergo the aging reaction, the conclusion that has been drawn is that OPIDN is not caused simply by inhibition of NTE, but by inhibition and aging of a critical amount of NTE (Johnson, 1982; Richardson, 1984; Davis *et al.*, 1985). Aging rates have been measured for several neuropathic OP compounds and have been found to be rapid (half-lives on the order of minutes)(Clothier and Johnson, 1980). Thus, the period of clinical quiescence between dosing and appearance of clinical signs is not a result of the time required for the aging reaction. Furthermore, by the time clinical signs have fully developed, NTE activity has returned nearly to control levels. The return of activity is attributed to resynthesis, and occurs with a half-life of approximately 5–7 days (Johnson, 1974). Taken together, these observations indicate that OPIDN is a consequence of the relatively rapid formation of a critical level of aged NTE (NTEOP′). What has come to be the conventional view of the role of NTE in OPIDN is that the event that initiates the neurodegenerative process is not the disappearance of NTE activity, but the appearance of NTEOP′ (Richardson, 1984; Johnson, 1987).

D. Consequences of Inhibition and Aging of AChE versus NTE

It is important to realize that the mechanism and target for acute cholinergic neurotoxicity of OP compounds are entirely different from the mechanism and target for OPIDN. In the case of acute cholinergic neurotoxicity, only a critical level of AChE inhibition is required, because the mechanism depends only on the loss of catalytic activity of AChE. Inhibition of AChE by either an aging or nonaging inhibitor of this enzyme results in the same toxic response, because aging does not affect inhibition *per se*. Inhibition of AChE by an aging inhibitor affects only the duration of inhibition and the therapy that can be employed: spontaneous reactivation will not occur, and therapeutic oxime reactivators will not be effective. In the case of OPIDN, both inhibition and aging of a critical level of NTE are required. Loss of catalytic activity of NTE appears to irrelevant to the production of lesions. Aging of NTE might be regarded as a xenobiotically produced posttranslational modification of the NTE protein that acts as a pathogenetic trigger to initiate a cascade of events leading ultimately to axonal degeneration (Richardson, 1984; Murphy, 1986; Johnson, 1987).

IV. Predictors of Neuropathic Potential of Organophosphorus Compounds

A. Relative Potency of Inhibition of NTE and AChE *in Vivo*

In order to assess the risk of OPIDN from exposure to an OP, it is useful first to determine the relative potency of the OP oxon for inhibition of NTE versus AChE. For an OP insecticide, it is helpful to think of AChE and lethality as the intended target and action, and to consider NTE and OPIDN as the unintended target and action of the compound. In practical terms, a compound that is a potent inhibitor of AChE but a poor inhibitor of NTE will result in morbidity or mortality from cholinergic poisoning at a dose below that required for initiating OPIDN. A good way to illustrate this point is to borrow the concept of the therapeutic index (TI) from the safety evaluation of pharmaceuticals. The TI is defined as the ratio of the TD_{50} over the ED_{50}, where the TD_{50} is the median toxic dose (the undesired effect), and the ED_{50} is the median effective dose (the desired effect) in a population, as shown in Eq. (10):

$$TI = TD_{50}/ED_{50} \tag{10}$$

For maximal safety, the TI should be as large as possible. That is, the dose–response curve for the toxic (undesired) effect should lie as far to the right of the dose–response curve for the effective (desired) effect as possible (Klaassen, 1986). For an OP insecticide, the odd wrinkle in the use of the TI concept is that both the TD_{50} and the ED_{50} are measures of toxic effects. The ED_{50} is the median dose for the desired effect of lethality due to AChE inhibition, and the TD_{50} is the median dose for the undesired effect of OPIDN. In order to maximize the TI, the TD_{50} must occur at a dose as far as possible above the median lethal dose. For AChE inhibitors, this paradoxical situation can occur only under conditions in which the lethal effects are blocked by treatment with atropine. Clearly, there would be reason for concern if an OP insecticide could produce OPIDN at a dose below that required for acute cholinergic neurotoxicity.

To convert the TI into a form more applicable to the safety evaluation of OP compounds, the neuropathic dose (NPD, dose required to produce clinical signs of OPIDN) may be substituted for the TD_{50}, and the LD_{50} substituted for the ED_{50}, to yield a new ratio that we could call the neuropathy safety index (NSI):

$$NSI = NPD/LD_{50} \tag{11}$$

The larger the NSI value, the safer the OP compound with respect to its ability to produce OPIDN.

To avoid using an LD_{50} and having to administer doses high enough to produce clinical neuropathy, an index based on the *in vivo* susceptibility of the relevant targets could be devised; it might be called the neuropathy target index (NTI):

$$NTI = ED_{50}(NTE)/ED_{50}(AChE) \tag{12}$$

In Eq. (12), the ED_{50} values refer to the median effective doses for inhibition of NTE or AChE in a given tissue (e.g., brain). Again, higher ratios would imply greater safety with respect to the relative ability of the compound to induce OPIDN.

B. Relative Potency of Inhibition of NTE and AChE *in Vitro*

Since NTE and AChE are found in the same target tissues, and it is the same oxon form of the OP compound that inhibits either NTE or AChE, it is possible to move to an *in vitro* assessment of relative potency, such as a ratio of I_{50} values for inhibition of the two target enzymes. This ratio could be called the neuropathy target ratio (NTR):

$$NTR = I_{50}(NTE)/I_{50}(AChE) \tag{13}$$

In order to compensate for deviations from ideal kinetics, or for differences in the kinetics of inhibition between the two enzymes, it would be good procedure to use an inverted ratio of k_a values, or to determine each I_{50} from its corresponding k_a (see Eq. (9)).

Lotti and Johnson (1978) have compiled data on I_{50} ratios for a variety of OP compounds and have shown that this ratio correlates well with the dose required *in vivo* to produce OPIDN. In this chapter, the NTR is the reciprocal of the I_{50} ratio calculated by Lotti and Johnson, so that the ratios presented here increase as the relative safety with respect to OPIDN increases. Overall, NTRs greater than 1 correlate with a dose greater than the LD_{50} being required to produce OPIDN. If the ratio is close to unity, the OPIDN dose is close to the LD_{50}, and if the NTR is less than one, the dose required for OPIDN is less than the LD_{50}. This relationship is most clearly seen in a homologous series of compounds, as exemplified by derivatives of dichlorvos (dimethyl 2,2-dichlorovinyl phosphate) made by substituting alkyl groups of increasing chain length for the methyl groups in the parent compound. Some indicators of the relative tendency of these compounds to produce acute cholinergic versus delayed neuropathic effects are listed in Table I.

As shown in Table I, the correlation between the NTR (I_{50} ratio) and the NSI (NPD/LD_{50}) is striking. In each case, large ratios indicate a relatively greater degree of expected safety with respect to OPIDN. To appreciate the potential use of these indicators, it is important to realize that dichlorvos has had wide application as an insecticide, with no reports of OPIDN. Note that it is possible to produce OPIDN in hens with dichlorvos, but only after administering two successive daily doses of over nine times the LD_{50}. Not only is this dose large relative to the LD_{50}, but it is also large in absolute terms. Heroic measures are required to protect the animals from the acute cholinergic effects of the compound at these doses and to keep the animals alive long enough to exhibit signs of OPIDN. These are conditions that would not be encountered with the intended use of this compound, which would indicate that the practical risk of developing OPIDN from acute exposures encountered during normal handling of dichlorvos would be nil. On the other hand, a compound like di-*n*-pentyl dichlorvos is at the opposite end of the spectrum. Here is a compound that clearly presents a hazard of OPIDN under conditions of normal use: the absolute dose required for OPIDN is quite low, and well below the LD_{50}. In fact, hens given neuropathic doses of di-*n*-pentyl dichlorvos show no signs of acute cholinergic neurotoxicity at all. Note that

TABLE 1

Comparison of *in Vivo* and *in Vitro* Indicators of Acute Cholinergic Neurotoxicity and OPIDN for Dichlorvos Homologs[a]

Compound	LD_{50} (mg/kg)	NPD[b] (mg/kg)	I_{50}(NTE)[c] (μM)	I_{50}(AChE)[c] (μM)	NSI[d]	NTR[e]
Dichlorvos[f]	11	2×100	27	0.59	18	46
Diethyl dichlorvos	3	18	2.3	0.37	6	6.2
Di-*n*-propyl dichlorvos	10	2	0.052	0.14	0.2	0.37
Di-*n*-pentyl dichlorvos	26	2	0.0016	0.052	0.08	0.03

[a]Data from Lotti and Johnson (1978).
[b]NPD, neuropathic dose, the dose required to produce clear clinical signs of OPIDN within the 3-week observation period. Compounds administered as single subcutaneous doses except for dichlorvos, which required two successive daily doses to produce OPIDN. Prophylaxis against acute cholinergic neurotoxicity was administered as required.
[c]I_{50}, concentration required to inhibit 50% of the hen brain NTE (neurotoxic esterase) or AChE (acetylcholinesterase) activity; determined from fixed-time incubations normalized to 20 min at 37°C.
[d]NSI, neuropathy safety index, NPD/LD_{50}.
[e]NTR, neuropathy target ratio, I_{50}(NTE)/I_{50}(AChE).
[f]Dichlorvos, the dimethyl parent compound, dimethyl 2,2-dichlorovinyl phosphate.

the hazard of di-*n*-pentyl dichlorvos is accurately predicted by the I_{50} ratio (see Table 1)(Lotti and Johnson, 1978; Johnson, 1982).

C. Ability of NTE Inhibitors to Age

It must be stressed again that mere inhibition of NTE is insufficient to cause OPIDN. The inhibited enzyme apparently must also undergo aging. Direct measurement of aging is not yet a routine matter and would not currently be a practical test to incorporate into screening protocols. However, some generalizations can be made from direct measurements of aging and from inferences about aging drawn from structure–activity correlations of NTE inhibitors that cause OPIDN. Overall, it appears that phosphate, phosphonate, and phosphoramidate OP classes are ageable by virtue of having R-groups attached to phosphorus by labile —O— or —NH—linkages. The rate of aging is slowed by R-group structures that favor carbonium ion stabilization (conditions that favor aging of AChE). Chirality is apparently also a determinant of aging, since the C+,P− stereoisomer of soman (pinacolyl methylphosphonofluoridate and the D(+) stereoisomer of EPN oxon (ethyl 4-nitrophenyl phenylphosphonate) inhibit NTE, but do not age, or do so at extremely slow rates. In the case of EPN and its oxon, the acute toxicity is low enough to test the ability of the nonaging stereoisomers of these compounds to protect against OPIDN. Even though the enzyme is inhibited by a phosphonate species, the chirality of the inhibitor prevents aging, and the prediction of protection on the basis of nonaging inhibition is borne out (Johnson *et al.*, 1986; Johnson and Read, 1987).

V. Challenges to the Conventional View of the Role of NTE in OPIDN

A. Lack of an Established Physiological Role for NTE

Many questions have been raised concerning the validity of positing a causal role for NTE inhibition and aging in the mechanism of OPIDN. Part of the difficulty is that comparatively little is known about NTE at the level of molecular structure, because the protein has not been isolated, and its encoding gene has not been cloned. Also, apart from its apparent role in OPIDN, nothing is known about any physiological function that NTE might have, although it would appear that its esteratic activity is not an essential component of its function. Even though there is now a conventional view of the role of NTE in OPIDN, the idea that this protein possesses enzymatic activity that may be superfluous to its function has been regarded as unconventional and difficult to accept. These conceptual stumbling blocks should become less troublesome if parallels or precedents could be found in other biological systems. However, in the final analysis, analogies and speculations must give way to direct demonstration of the role of NTE in OPIDN. Such a demonstration will be difficult before NTE is fully characterized and its physiological function is established (Richardson, 1984; Zech and Chemnitius, 1987; Carrington, 1989; Johnson, 1990).

B. Species and Age Differences in Susceptibility to OPIDN

In susceptible species, young animals possess apparent NTE activity, but inhibition and aging of NTE does not result in OPIDN following a single dose of a neuropathic OP compound until the animal reaches a critical age. Likewise, many species are known to possess apparent NTE activity in target and other tissues, but only some of these species are susceptible to the clinical manifestations of OPIDN following single doses of neuropathic OP compounds. Thus far, it appears that these differences in susceptibility are not the result of differences in the target enzyme, including its susceptibility to inhibition, or its rate of aging. The answer does not lie in differences in biotransformation, because the same differences in susceptibility are observed when direct-acting (oxon) forms of OP compounds that do not require metabolic activation in order to inhibit NTE are given in doses that produce equivalent NTE inhibitions across different ages and species. It seems plausible that the initial molecular and cellular events resulting from inhibition and aging of NTE could be the same in all species and ages of animals. Susceptible animals may either possess activating mechanisms for propagating the biochemical lesion into tissue injury, or lack effective repair mechanisms (Davis and Richardson, 1980; Richardson, 1984; Moretto *et al.,* 1991a).

C. NTE in Nontarget Tissues

NTE is found in highest concentrations in brain and lymphocytes, and in lesser concentrations in spinal cord, peripheral nerves, and many other tissues. However, the only consistently deleterious effect of aging inhibitors of NTE that has been identified so far is distal degeneration of spinal cord and peripheral nerve axons. A helpful analogy here is the presence of AChE in mammalian erythrocytes. Inhibition of red cell AChE is a useful biomarker of exposure to OP insecticides, but nothing is known of the function of AChE in red cells. Furthermore, there are no apparent ill effects in these cells due to AChE inhibition. It is possible to explain the toxic effects of AChE inhibition in target tissues, because the physiological role of the enzyme in these sites is reasonably well understood. Although the physiological role of NTE is unknown for any tissue, it could be anticipated that ill effects would occur in some sites and not others, even if the cellular role of the protein were the same in each site. Another point to consider is that deleterious effects may not have been noted in nontarget tissues because of a lack of knowledge of what endpoints to examine. Toxicity may be functional as well as structural, and without knowing the physiological function of NTE, it is not surprising that injurious changes have not yet been detected as a result of the inhibition of NTE in nontarget tissues. Indeed, toxicant-induced functional changes at the molecular and cellular level generally precede the appearance of histopathologically identifiable lesions in any tissue. It is reasonable to expect that the molecular and cellular consequences of the inhibition and aging of NTE may be compensated or repaired in some tissues but not in the axons that ultimately display visible signs of injury (Richardson, 1984; Johnson, 1990).

D. Promotion/Potentiation of OPIDN by Nonaging Inhibitors of NTE

The most recent and interesting challenge to the conventional role of NTE in OPIDN has come from observations that nonaging inhibitors of NTE can do more than protect against OPIDN from subsequently administered neuropathic OP compounds. In addition, nonaging NTE inhibitors can promote OPIDN if given after a subthreshold dose of a neuropathic OP compound, and potentiate the severity of OPIDN if given after a suprathreshold dose of a neuropathic OP compound (Pope and Padilla, 1989a, 1990; Caroldi *et al.*, 1990; Lotti *et al.*, 1991; Moretto *et al.*, 1991b; Pope *et al.*, 1991).

The term promotion is used in the setting of OPIDN in the same way as it is used in the context of the classical initiation–promotion model of carcinogenesis (Williams and Weisburger, 1986). That is, the endpoint in question (in this case, OPIDN) does not arise from pretreatment with the promoter (nonaging NTE inhibitor) alone, or from pretreatment with promoter followed by initiator (aging NTE inhibitor). OPIDN does arise from pretreatment with a sufficiently high dose of initiator, or from two or more doses of initiator that result in suprathreshold inhibition of NTE. In keeping with an initiation–promotion model, the ability of an initiator to produce OPIDN on its own could be construed as self-promotion.

The term potentiation is used to refer to the action of a nonaging NTE inhibitor that causes the emergence or exacerbation of OPIDN caused by an aging NTE inhibitor, where it is understood that the nonaging inhibitor on its own does not cause the toxic endpoint (OPIDN)(Klaassen, 1986). The problem with the term potentiation in this context is that a potentiator would generally be considered to be able to exacerbate the toxic endpoint if given before the agent whose toxicity is being potentiated. In the case of OPIDN and nonaging inhibitors of NTE, the outcome changes completely depending on the order of presentation of the aging and nonaging inhibitor. If the nonaging inhibitor is given first, the result is protection from OPIDN; if the nonaging inhibitor is given second, the result is promotion (emergence of clinical OPIDN from an otherwise subclinical case) or potentiation (intensification of signs from an otherwise marginal or mild case).

Some of the work on OPIDN promotion/potentiation has shown that young chicks normally resistant to clinical OPIDN develop clinical signs after treatment with a nonaging inhibitor of NTE (PMSF), as if this treatment had lowered the age threshold for OPIDN (Moretto *et al.,* 1991b; Peraica *et al.,* 1991; Pope *et al.,* 1991). These and other findings on the promotion/potentiation of OPIDN have many exciting implications for the risk assessment and regulation of OP compounds and other NTE inhibitors, as well as for the advancement of insight into the role of NTE in OPIDN.

VI. Conclusions

A. Toward a Modified View of the Role of NTE in OPIDN

Recent findings (Section V,D) concerning new interactions of NTE inhibitors will force a reexamination of current thinking about the role of NTE in OPIDN, ultimately yielding either a complete shift in the prevailing paradigm, or at least a refinement of the existing model.

An immediate practical consequence of the new findings is that nonaging NTE inhibitors must now be examined as possible promoters or potentiators of OPIDN. Currently used OP compounds will need to be reevaluated to assess whether latent OPIDN may be expressed following posttreatment with a nonaging NTE inhibitor. Also, appropriate sulfonate esters or carbamates should be examined for NTE inhibitor potential and, if found to be sufficiently potent, evaluated for toxicological interactions with OP inhibitors of NTE. In light of a report showing that lymphocyte NTE was markedly depressed in patients with alcoholic neuropathy (Fournier *et al.,* 1987), it may be advisable to examine some compounds not currently thought to act as NTE inhibitors (e.g., commonly employed solvents, drugs, and neurotoxicants thought to act by non-NTE mechanisms) in order to discover any effects on NTE, and to then examine possible interactions with OP compounds.

Finally, it will be important to determine whether nonaging inhibitors of NTE produce a neurological defect that represents the emergence or intensification of OPIDN rather than that of some other neurodegenerative disorder. If the effect is indeed OPIDN in each case, it will then be of interest to determine whether NTE inhibition is obligatory for promotion or potentiation. The obvious alternative hypothesis must also be scrupulously examined: promotion or potentiation could depend on the inhibition of a different esterase or protease with an inhibitor sensitivity overlapping that of NTE. The theme of a different esterase could be extended to different tissue compartments containing NTE. For example, it would be of interest to determine whether leukocyte NTE plays an active role in OPIDN.

B. Significance of NTE-Inhibitor Interactions

This chapter has presented a summary of what is known about the way OP compounds and some other compounds interact with NTE. Through the application of the kinetics of irreversible enzyme inhibition, it is possible to assess the nature and potency of inhibition of NTE and other esterases in order to compare compounds and their relative tendency to inhibit different esterase targets. Although much work remains to be done to characterize NTE and to ascertain its physiological function, a useful working model has arisen regarding the role of NTE in OPIDN. The current model indicates that concerted inhibition and aging of NTE yields a modified protein, NTEOP′, that initiates axonal degeneration. This process can be blocked by pretreatment with NTE inhibitors that do not age, presumably preventing the formation of a critical level of NTEOP′. The latest findings discussed in this chapter on promoting and potentiating interactions of NTE inhibitors may cause a modification of the conventional view of the role of NTE in OPIDN, but the significance of the existing model is at least threefold. First, in toxicology the model provides a quantifiable way to predict the potential of an OP compound to produce OPIDN, or to predict which compounds could be used as prophylactic agents against OPIDN. Second, in industrial hygiene and occupational medicine, NTE in relatively accessible sites, such as circulating lymphocytes, furnishes the potential for biomonitoring in cases of neuropathic OP exposures. Finally, in neurobiology and experimental neurology, the ability to produce a well-defined pattern of axonal degeneration and to trace the origins of this degeneration to a molecular event that can be specifically blocked yields an unusual opportunity for the study of fundamental processes of neural degeneration, regeneration, and repair.

Acknowledgments

Work reported from the author's laboratory has been supported by NIH Grants ES01611 and ES02770, NIH Training Grant ES07062, EPA Grant R805339, USAMRDC Contract DAMD1783C3187, NIAAA Center Grant AA07378, a grant from the American Diabetes Association, a postdoctoral fellowship from the Monsanto Fund, predoctoral fellowships from Stauffer Chemical Company, faculty grants and fellowships from the Horace H. Rackham School of Graduate Studies and the Phoenix Memorial Project, The University of Michigan, and gifts from Shell Development Company, FMC Corporation, Ciba-Geigy, Ltd., and The Dow Chemical Company. The figure for this chapter was produced by U.S. Kayyali.

References

Abou-Donia, M. B. (1981). Organophosphorus ester-induced delayed neurotoxicity. *Annu. Rev. Pharmacol. Toxicol.* **21,** 511–548.

Abou-Donia, M.B., and Lapadula, D.M. (1990). Mechanisms of organophosphorus ester–induced delayed neurotoxicity: Type I and type II. *Annu. Rev. Pharmacol. Toxicol.* **30,** 405–440.

Aldridge, W. N., and Reiner, E. (1972). *Enzyme Inhibitors as Substrates: Interactions of Esterases with Esters of Organophosphorus and Carbamic Acids,* Frontiers of Biology, Vol. 12 (A. Neuberger and E. L. Tatum, eds.), North-Holland Publishing, Amsterdam.

Bertoncin, D., Russolo, A., Caroldi, S., and Lotti, M. (1985). Neuropathy target esterase in human lymphocytes. *Arch. Environ. Health* **40,** 139–144.

Caroldi, S., Capodicasa, E., Moretto, A., and Lotti, M. (1990). Phenylmethylsulfonyl fluoride precipitates delayed neuropathy after single noneffective doses of diisopropylfluorophosphate in hens. *Toxicologist* **10,** 183.

Carrington, C. D. (1989). Prophylaxis and the mechanism for the initiation of organophosphorous compound–induced delayed neuropathy. *Arch. Toxicol.* **63,** 165–172.

Carrington, C. D., and Abou-Donia, M. B. (1985). Characterization of [^{3}H]diisopropylphosphorofluoridate–binding proteins in hen brain. *Biochem. J.* **228,** 537–544.

Carrington, C. D., and Abou-Donia, M. B. (1986). Kinetics of substrate hydrolysis and inhibition by mipafox of paraoxon-preinhibited hen brain esterase activity. *Biochem. J.* **236,** 503–507.

Chemnitius, J. M., and Zech, R. (1983). Inhibition of brain carboxylesterases by neurotoxic and nonneurotoxic organophosphorus compounds. *Mol. Pharmacol.* **23,** 717–723.

Chemnitius, J. M., Haselmeyer, K. H., and Zech, R. (1983). Neurotoxic esterase. Identification of two isozymes in hen brain. *Arch. Toxicol.* **53,** 235–244.

Clothier, B., and Johnson, M. K. (1979). Rapid aging of neurotoxic esterase after inhibition by di-isopropylphosphorofluoridate. *Biochem. J.* **177,** 549–558.

Clothier, B., and Johnson, M. K. (1980). Reactivation and aging of neurotoxic esterase inhibited by a variety of organophosphorus esters. *Biochem. J.* **185,** 739–747.

Clothier, B., Johnson, M. K., and Reiner, E. (1981). Interaction of some trialkyl phosphorothiolates with acetylcholinesterase. Characterization of inhibition, aging, and reactivation. *Biochim. Biophys. Acta* **660,** 306–316.

Davis, C. S., and Richardson, R. J. (1980). Organophosphorus compounds. *In* "Experimental and Clinical Neurotoxicology" (P. S. Spencer and H. H. Schaumburg, eds.), pp. 527–544. Williams & Wilkins, Baltimore, Maryland.

Davis, C. S., and Richardson, R. J. (1987). Neurotoxic esterase: Characterization of the solubilized enzyme and the conditions for its solubilization from chicken brain microsomal membranes with ionic, zwitterionic, or nonionic detergents. *Biochem. Pharmacol.* **36,** 1393–1399.

Davis, C. S., Johnson, M. K., and Richardson, R. J. (1985). Organophosphorus compounds. *In* "Neurotoxicity of Industrial and Commercial Chemicals" (J. L. O'Donoghue, ed.) Vol. II, pp. 1–23. CRC Press, Boca Raton, Florida.

Dudek, B. R., and Richardson, R. J. (1982). Evidence for the existence of neurotoxic esterase in neural and lymphatic tissue of the adult hen. *Biochem. Pharmacol.* **31,** 1117–1121.

Fournier, L., Fournier, E., and Lecorsier, A. (1987). Détermination de la neurotoxic-estérase in pathologie neurologique d'origine toxique. Mesure sur lymphocytes humains au cours d'intoxications par organophosphates, ciguatera, alcool éthylique. *Ann. Med. Interne* **138,** 169–172.

Fulton, M. H., and Chambers, J. E. (1985). Inhibition of neurotoxic esterase and acetylcholinesterase by organophosphorus compounds in selected ectothermic vertebrates. *Pestic. Biochem. Physiol.* **23**, 282–288.

Gurba, P. E., and Richardson, R. J. (1983). Partial characterization of neurotoxic esterase of human placenta. *Toxicol. Lett.* **15**, 13–17.

Hollingsworth, P. J., Dudek, B. R., Smith, C. B., and Richardson, R. J. (1984). Direct comparison of the distribution of neurotoxic esterase and acetylcholinesterase in rat and hen brain. *In* "Cholinesterases" (M. Brzin, E.A. Barnard and D. Sket, eds.), pp. 483–492, Walter de Gruyter, Berlin.

Johnson, M. K. (1969a). A phosphorylation site in brain and the delayed neurotoxic effect of some organophosphorus compounds. *Biochem. J.* **111**, 487–495.

Johnson, M. K. (1969b). The delayed neurotoxic effect of some organophosphorus compounds. Identification of the phosphorylation site as an esterase. *Biochem. J.* **114**, 711–717.

Johnson, M. K. (1970). Organophosphorus and other inhibitors of brain "neurotoxic esterase" and the development of delayed neurotoxicity in hens. *Biochem. J.* **120**, 523–531.

Johnson, M. K. (1974). The primary biochemical lesion leading to the delayed neurotoxic effects of some organophoshorus esters. *J. Neurochem.* **23**, 785–789.

Johnson, M. K. (1975a). The delayed neuropathy caused by some organophosphorus esters: Mechanism and challenge. *Crit. Rev. Toxicol.* **3**, 289–316.

Johnson, M. K. (1975b). Structure–activity relationships for substrates and inhibitors of hen brain neurotoxic esterase. *Biochem. Pharmacol.* **24**, 797–805.

Johnson, M. K. (1977). Improved assay of neurotoxic esterase for screening organophosphates for delayed neurotoxicity potential. *Arch. Toxicol.* **37**, 113–115.

Johnson, M. K. (1980). The mechanism of delayed neuropathy caused by some organophosphorus esters: Using the understanding to improve safety. *J. Environ. Sci. Health* **B15**, 823–829.

Johnson, M. K. (1982). The target for initiation of delayed neurotoxicity by organophosphorus esters: Biochemical studies and toxicological applications. *In* "Reviews in Biochemical Toxicology" (E. Hodgson, J. R. Bend, and R. M. Philpot, eds.), Vol. 4, pp. 141–212. Elsevier, New York.

Johnson, M. K. (1987). Receptor or enzyme: The puzzle of NTE and organophosphate-induced delayed polyneuropathy. *Trends Pharmacol. Sci.* **8**, 174–179.

Johnson, M. K. (1990). Organophosphates and delayed neuropathy—is NTE alive and well? *Toxicol. Appl. Pharmacol.* **102**, 385–399.

Johnson, M. K., and Lauwerys, R. (1969). Protection by some carbamates against the delayed neurotoxic effects of diisopropylphosphorofluoridate. *Nature (London)* **222**, 1066–1067.

Johnson, M. K., and Read, D. J. (1987). The influence of chirality on the delayed neuropathic potential of some organophosphorus esters: Neuropathic and prophylactic effects of stereoisomeric esters of ethylphenylphosphonic acid (EPN oxon and EPN) correlate with quantities of aged and unaged neuropathy target esterase *in vivo*. *Toxicol. Appl. Pharmacol.* **90**, 103–115.

Johnson, M. K., Read, D. J., and Yoshikawa, H. (1986). The effect of steric factors on the interaction of some phenylphosphonates with acetylcholinesterase and neuropathy target esterase of hen brain. *Pestic. Biochem. Physiol.* **25**, 133–142.

Kayyali, U. S., Moore, T. B., Randall, J. C., and Richardson, R. J. (1989). Neurotoxic esterase assay: Corrected wavelength and extinction coefficient. *Toxicologist* **9**, 73.

Kayyali, U. S., Moore, T. B., Randall, J. C., and Richardson, R. J. (1991). Neurotoxic esterase (NTE) assay: Optimized conditions based on detergent-induced shifts in the phenol/4-aminoantipyrine chromophore spectrum. *J. Analyt. Toxicol.* **15**, 86–89.

Klaassen, C. D. (1986). Principles of toxicology. *In "Casarett and Doull's Toxicology: The Basic Science of Poisons" 3rd Ed.* (C. D. Klaassen, M. O. Amdur, and J. Doull, eds.), pp. 11–32, Macmillan, New York.

Lotti, M. (1991). The pathogenesis of organophosphate polyneuropathy. *CRC Crit. Revs. Toxicol.* (in press).

Lotti, M., and Johnson, M. K. (1978). Neurotoxicity of organophosphorus pesticides: Predictions can be based on *in vitro* studies with hen and human enzymes. *Arch.Toxicol.* **41**, 215–221.

Lotti, M., and Johnson, M. K. (1980). Neurotoxic esterase in human nervous tissue. *J. Neurochem.* **34**, 747–749.

Lotti, M., Becker, C. E., Aminoff, M. J., Woodrow, J. E., Seiber, J. N., Talcott, R. E., and Richardson, R. J. (1983). Occupational exposure to the cotton defoliants DEF and merphos: A rational approach to monitoring organophosphorus-induced delayed neurotoxicity. *J. Occup. Med.* **25**, 517–522.

Lotti, M., Becker, C. E., and Aminoff, M. J. (1984). Organophosphate polyneuropathy: Pathogenesis and prevention. *Neurology* **34**, 658–662.

Lotti, M., Wei, E. T., Spear, R. C., and Becker, C. E. (1985). Neurotoxic esterase in rooster testis. *Toxicol. Appl. Pharmacol.* **77**, 175–180.

Lotti, M., Caroldi, S., Capodicasa, E., and Moretto, A. (1991). Promotion of organophosphate-induced delayed polyneuropathy by phenylmethanesulfonyl fluoride. *Toxicol. Appl. Pharmacol.* **108**, 234–241.

Main, A. R. (1980). Cholinesterase inhibitors. *In* "Introduction to Biochemical Toxicology" (E. Hodgson and F. E. Guthrie, eds.), pp. 193–223, Elsevier, New York.

Moretto, A., and Lotti, M. (1988). Organ distribution of neuropathy target esterase in man. *Biochem. Pharmacol.* **37**, 3041–3043.

Moretto, A., Capodicasa, E., Peraica, M., and Lotti, M. (1991a). Age sensitivity to organophosphate-induced delayed polyneuropathy: Biochemical and toxicological studies in developing chicks. *Biochem. Pharmacol.* **41**, 1497–1504.

Moretto, A., Bertolazzi, M., Capodicasa, E., Peraica, M. Richardson, R. J., Scapellato, M. L., and Lotti, M. (1991b). Phenylmethanesulfonyl fluoride elicits and intensifies the clinical expression of neuropathic insults. *Arch. Toxicol.* (in press).

Murphy, S. D. (1986). Toxic effects of pesticides. *In* "Casarett and Doull's Toxicology: The Basic Science of Poisons" 3rd Ed. (C. D. Klaassen, M. O. Amdur, and J. Doull, eds.), pp. 519–581. Macmillan, New York.

Novak, R., and Padilla, S. (1986). An *in vitro* comparison of rat and chicken brain neurotoxic esterase. *Fundam. Appl. Toxicol.* **6**, 464–471.

Peraica, M., Capodicasa, E., Scapellato, M. L., Bertolazzi, M., Moretto, A., and Lotti, M. (1991). Organophosphate-induced delayed polyneuropathy (OPIDP) in chicks: induction, promotion and recovery. *Toxicologist,* **11**, 306.

Pope, C. N., and Padilla, S. (1989a). "Potentiation of Mipafox-Induced Delayed Neurotoxicity by Phenylfluoride (PMSF)." Seventh International Neurotoxicology Conference, Little Rock, Arkansas, Abstract #21.

Pope, C. N., and Padilla, S. (1989b). Modulation of neurotoxic esterase activity *in vitro* by phospholipids. *Toxicol. Appl. Pharmacol.* **97**, 272–278.

Pope, C. N., and Padilla, S. (1989c). Chromatographic characterization of neurotoxic esterase. *Biochem. Pharmacol.* **38**, 181–188.

Pope, C. N., and Padilla, S. (1990). Potentiation of organophosphorus-induced delayed neurotoxicity by phenylmethylsulfonyl fluoride. *J. Tox. Environ. Health* **31**, 261–273.

Pope, C., Chapman, M., Arthun, D., Tanaka, D., and Padilla, S. (1991). Age and sensitivity to organophosphorus-induced delayed neurotoxicity (OPIDN): Effects of phenylmethylsulfonyl fluoride (PMSF). *Toxicologist,* **11**, 304.

Reiner, E., Davis, C. S., Schwab, B. W., Schopfer, L. M., and Richardson, R. J. (1987). Kinetics of heat inactivation of phenyl valerate hydrolases from hen and rat brain. *Biochem. Pharmacol.* **36,** 3181–3185.

Richardson, R. J. (1983). Neurotoxic esterase: Research trends and prospects. *Neurotoxicology,* **4,** 157–162.

Richardson, R. J. (1984). Neurotoxic esterase: Normal and pathogenic roles. *In* "Cellular and Molecular Neurotoxicology" (T. Narahashi, ed.), pp. 285–295. Raven Press, New York.

Richardson, R. J., and Dudek, B. R. (1983). Neurotoxic esterase: Characterization and potential for a predictive screen for exposure to neuropathic organophosphates. *In* "Pesticide Chemistry: Human Welfare and the Environment" (J. Miyamoto and P. C. Kearney, eds.),Vol. 3, pp. 491–495. Pergamon Press, Oxford, England.

Richardson, R. J., Davis, C. S., and Johnson, M. K. (1979). Subcellular distribution of marker enzymes and of neurotoxic esterase in adult hen brain. *J. Neurochem.* **32,** 607–615.

Schwab, B. W., Davis, C.-S. G., Miller, P. H., and Richardson, R. J. (1985). Solubilization of hen brain neurotoxic esterase in dimethylsulfoxide. *Biochem. Biophys. Res. Commun.* **132,** 81–87.

Thomas, T. C., Ishikawa, Y., McNamee, M. G., and Wilson, B. W. (1989). Correlation of neuropathy target esterase activity with specific tritiated diisopropylphosphorofluoridate-labeled proteins. *Biochem. J.* **257,** 109–116.

Vilanova, E., Barril, J., Carrera, V., and Pellin, M. C. (1990). Soluble and particulate forms of the organophosphorus neuropathy target esterase in hen sciatic nerve. *J. Neurochem.* **55,** 1258–1265.

Williams, D. G. (1983). Intramolecular group transfer is a characteristic of neurotoxic esterase and is independent of the tissue source of the enzyme. *Biochem. J.* **209,** 817–829.

Williams, D. G., and Johnson, M. K. (1981). Gel electrophoretic identification of hen brain neurotoxic esterase, labeled with tritiated diisopropylphosphorofluoridate. *Biochem. J.* **199,** 323–333.

Williams, G. M., and Weisberger, J. H. (1986). Chemical carcinogens. *In* Casarett and Doull's Toxicology: The Basic Science of Poisons" 3rd Ed. (C. D. Klaassen, M. O. Amdur, and J. Doull, eds.), pp. 99–173, Macmillan, New York.

Zech, R., and Chemnitius, J. M. (1987). Neurotoxicant-sensitive esterase. Enzymology and pathophysiology of organophosphorus ester-induced delayed neuropathy. *Prog. Neurobiol.* **29,** 193–218.

IV

Toxic Effects—Organismal

17

Triphenyl Phosphite: A Type II Organophosphorus Compound–Induced Delayed Neurotoxic Agent

Mohamed B. Abou-Donia
Department of Pharmacology
Duke University Medical Center
Durham, North Carolina

I. Introduction

A. Phosphorus Compounds

Phosphorus is essential to all forms of life, e.g., hereditary processes, growth, development, and maintenance. It does not exist in free form in nature and is found mostly in its fully oxidized phosphate state. Phosphates occur in all living organisms, soils, rocks, oceans, and in most foods. The normal phosphorus content in potable water is 0.001 to 0.1 ppm. Synthesis of orthophosphoric acid (PO_4H_3) utilizes more than 75% of the world's phosphate rock production, most of which is used to manufacture fertilizers, with less than

Organophosphates: Chemistry, Fate, and Effects

5% being used to synthesize other phosphorus compounds. Most of the naturally occurring inorganic phosphates are nontoxic. They are used as fertilizers and in soft drinks, toothpastes, detergents, and medicines. In contrast, synthetic organophosphorus (OP) compounds exhibit a very wide range of toxic effects.

B. Chemistry of Phosphorus Compounds

1. Properties of Phosphorus Atoms

The phosphorus atom, P, belongs to the Group V elements, which are sometimes referred to as pnicogens or pnictides, of the periodic table. It has the following properties:

Electronic structure	$1S^2\ 2S^2\ 2P^6\ 3S^2\ 3P^3$
Atomic number	15
Atomic weight	30.97
Ionization potentials, e.v.	
3rd	30.15
5th	65.00
Electronegativity	2.19
Radii, Å	
Van der Waal	1.50
3rd	2.12
5th	0.34
Natural Abundance	100.00%
Nuclear spin	0.5

2. Organophosphorus Compounds

Phosphorus-containing organic compounds may be divided into two major subgroups, the pentavalent phosphorus and the trivalent phosphorus-containing compounds.

a. Pentavalent Organophosphorus Compounds These compounds possess a pentavalent phosphorus atom that has a tetrahedral configuration. Pentavalent organophosphorus (OP) compounds are further subclassified according to the substituents attached to the phosphorus atoms shown below (See also Figs. 1 and 2 in Chapter 1 by Chambers in this volume for additional details and examples).

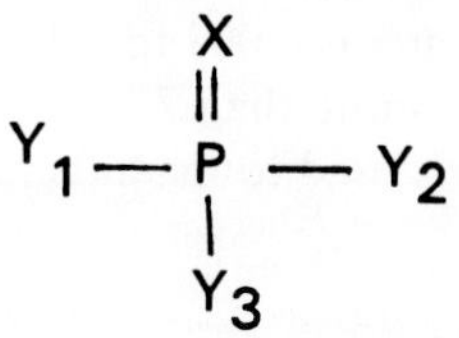

X
||
Y_1 — P — Y_2
|
Y_3

Subgroup	X	Y_1	Y_2	Y_3
Phosphate	O	OR	OR	OR
Phosphorothioate	S	OR	OR	OR
Phosphorothiolate (phosphorothioate)	O	SR	OR	OR
Phosphonate	O	OR	OR	R
Phosphonothioate	S	OR	OR	R
Phosphinite	O	OR	R	R
Phosphorotrithioate	O	SR	SR	SR
Phosphorofluoridate	O	OR	OR	F
Phosphonofluoridate	O	OR	R	F
Phosphinofluoridate	O	R	R	F
Phosphoroamidofluoridate	O	OR	HNR	F
Phosphorodiamidofluoridate	O	HNR	HNR	F
Phosphonocyanidate	O	OR	R	CN

Most synthetic OP compounds belong to the pentavalent group. These compounds may be used in agriculture as insecticides, acaricides, nematocides, veterinary pesticides, insect chemosterilants, fungicides, herbicides, rodenticides, and insecticide synergists, as insect repellants; in nerve gases; as pharmaceuticals; as flame retardants; and for other industrial uses (Abou-Donia, 1985).

b. Trivalent OP Compounds Trisubstituted phosphorus acid esters, i.e., triphosphites, contain a trivalent phosphorus atom that has a pyramidal configuration. The trivalent pyramidal phosphorus atom–containing molecule is less stable compared to the pentavalent tetrahedral phosphorus atom–containing molecule. Trivalent phosphorus compounds are further classified into subclasses based on the substituents attached to the phosphorus atom shown below.

$$Y_1 - \underset{\substack{|\\ Y_3}}{P} - Y_2$$

Subgroup	Y_1	Y_2	Y_3
Triaryl phosphite	OAr	OAr	OAr
Diaryl alkyl phosphite	OAr	OAr	OR
Aryl dialkyl phosphite	OAr	OR	OR
Trialkyl phosphite	OR	OR	OR
Phosphorotrithoite	SR	SR	SR

Trisubstituted aryl and alkyl phosphites, e.g., triphenyl phosphite and tri-*iso*-propyl phosphite, are used as antioxidants in many industries (U.S. EPA 1985a). Phosphorotrithioite, e.g., *S,S,S-tri-n*-butyl phosphorotrithioite (merphos) is used as a cotton defoliant.

II. Actions of Organophosphorus Esters

Organophosphorus esters are nervous system poisons. They adversely affect both the central (CNS) and peripheral (PNS) nervous systems. Most of these compounds have acute cholinergic toxicity, while some induce delayed neurotoxic action, which is referred to as OP compound–induced delayed neurotoxicity, OPIDN (Smith *et al.*, 1930; Abou-Donia, 1978, 1981).

A. Cholinergic Acute Toxicity

Organophosphorus compounds produce acute toxicity by inhibiting acetylcholinesterase (AChE), the enzyme responsible for hydrolyzing the neurotransmitter acetylcholine (ACh)(see discussion by Chambers in Chapter 1, this volume). ACh interacts with two types of ACh receptors: muscarinic and nicotinic in the CNS and PNS. Inhibition of AChE results in the accumulation of ACh at these receptor sites and leads to the development of signs of acute cholinergic effects (Abou-Donia, 1985).

Overstimulation of the muscarinic receptors, which are found primarily in smooth muscle, heart, and exocrine glands, results in the following signs: pupil constriction; increased lacrimation, salivation, and sweating; tightness in the chest; wheezing because of broncoconstriction and increased bronchial secretion; nausea; vomiting; abdominal pain; diarrhea and involuntary defecation resulting from increased gastrointestinal motility and gastric secretion; involuntary urination because of the constriction of urinary bladder smooth muscle; and bradycardia that may progress to heart attack.

Accumulation of ACh at the nicotinic receptors, which occur at the endings of motor nerves to the neuromuscular junction of the skeletal muscle and autonomic ganglia, leads to easy fatigability, and muscle weakness followed by involuntary twitching and cyanosis. Respiratory muscle weakness contributes to dyspnea, hypoxemia, and cyanosis. In severe poisoning, overstimulation of sympathetic ganglia overrides muscarinic action–induced bradycardia, and results in tachycardia, and leads to increased blood pressure and hyperglycemia.

Poisoning by OP compounds leads to the accumulation of ACh in the CNS resulting first in tension, anxiety, restlessness, insomnia, headache, emotional instability, excessive dreaming, and nightmares. Exposure to large amounts of OP compounds results in slurred speech, tremor, generalized weakness, ataxia, convulsions, depression of respiratory and circulatory centers, and coma.

Death due to OP compounds is produced by respiratory failure because of muscarinic effects resulting in bronchoconstriction and increased bronchial secretion; nicotinic action leading to respiratory muscle paralysis; and central nervous action because of depression and paralysis of respiratory centers.

B. Organophosphorus Compound–Induced Delayed Neurotoxicity

Although the immediate hazard of OP compounds is their inhibition of AChE, some of these compounds also produce a condition known as organophosphorus compound–induced delayed neurotoxicity, or OPIDN, in humans and sensitive animal species (Smith *et al.*, 1930; Abou-Donia, 1981; Abou-Donia and Lapadula, 1990). The earliest recorded cases of OPIDN in humans were attributed to the use of creosote oil for treatment of pulmonary tuberculosis in France in 1899 (Roger and Recordier, 1934). It was not until 1930, however, that Smith *et al.*, identified tri-*o*-cresyl phosphate (TOCP) as the agent responsible for OPIDN.

Delayed neurotoxicity is defined as a "delayed onset of prolonged locomotor ataxia resulting from a single or repeated exposure to an organophosphorus compound" (U.S. EPA, 1985b; Abou-Donia and Lapadula, 1990). It is characterized by a delay period of 6 to 14 days before onset of ataxia followed by paralysis. Neuropathologic changes are Wallerian-type degeneration of the axon and myelin of the large and long fiber tracts in the CNS and PNS, resulting in a sensory–motor distal neuropathy (Lillie and Smith, 1932; Cavanagh, 1964).

Smith *et al.*, (1930) and Aird *et al.*, (1940) both recognized that although triaryl phosphites produce delayed neurotoxicity, their action was different from that produced by TOCP and other phosphates. Recent studies have confirmed early findings that triphenyl phosphite–induced delayed neurotoxicity is different in several respects from that produced by pentavalent OP compounds (Veronesi *et al.*, 1986; Veronesi and Dvergsten, 1987; Padilla *et al.*, 1987; Carrington and Abou-Donia, 1988; Carrington *et al.*, 1988a,b; Abou-Donia and Brown, 1990; Katoh *et al.*, 1990; Tanaka *et al.*, 1990). Recently, delayed neurotoxicity induced by OP compounds has been classified into two groups: Type I delayed neurotoxicity that is induced by all OP compounds except the phosphites, which produce Type II (Abou-Donia and Lapadula, 1990). The characteristics of Type I and Type II OPIDN are listed in Table I.

TABLE I

Characteristics of Type I and Type II OPIDN

Characteristic	Type I	Type II
1. Chemical structure	Pentavalent phosphorus atom, e.g., TOCP, DFP	Trivalent phosphorus atom, e.g., TPP_i, $TOCP_i$
2. Species selectivity	Rodents are less sensitive	Rodents are sensitive
3. Clinical signs	Hen, flaccid paralysis Cat, flaccid paralysis Rat, no clinical signs	Hen, flaccid paralysis Cat, extensor rigidity Rat, partial flaccid paralysis, tail kinking, bidirectional circular motion
4. Length of latent period before onset of clinical signs	Hen, 6–14 days Cat, 14–21 days Ferret, 4 days Rat, no clinical signs	Hen, 4–6 days Cat, 4–7 days Ferret, 10–14 days Rat, 7 days
5. Age Sensitivity	Young chicks are insensitive	Young chicks are more sensitive
6. Neuropathological lesions	Wallerian-type degeneration of specific ascending and descending tracts of sensorimotor pathways of the brainstem and spinal cord and in peripheral nerves. No changes in nerve cell body or dorsal ganglia.	In the addition to the Wallerian-type degeneration, there is chromatolysis and necrosis of nerve cell body and ganglia; in addition to brainstem lesions, lesions also occur in corticol and thalamic regions.
7. Protection with phenylmethyl sulfonyl fluoride PMSF	Full protection against small doses, i.e., 125 mg TOCP per kg; partial protection against higher doses.	Protects against small doses, i.e., 250 mg TPP_i per kg, synergizes neurotoxicity of large doses, i.e., 1000 mg TPP_i per kg.
8. Inhibition of neurotoxic esterase NTE	Hen, 65–70% inhibition Rat, 65–70% inhibition	Hen, 70% inhibition Rat, 39% inhibition
9. Chromaffin cells	a. No effect on catecholamine secretion b. No effect on ^{45}Ca uptake evoked by 10 μM nicotine or 56 mM K^+ c. No effect on ATP synthesis via ^3H-adenosine incorporation d. No morphological changes	a. Inhibition of catecholamine secretion b. Inhibition of ^{45}Ca uptake c. Inhibits ATP synthesis d. Swollen and disrupted mitochondria

1. Chemical Structure–Delayed Neurotoxic Effect Relationship

Although numerous OP esters have been synthesized and screened for their ability to inhibit acetylcholinesterase, only a very limited number has been screened for OPIDN. Most of the chemicals tested belong to the Type I class, while only very few Type II chemicals (phosphites) have been tested for their potential to produce OPIDN.

a. The Potential for Type I Chemicals to Produce OPIDN By 1981 the results of studies that tested 237 OP compounds for the potential to produce OPIDN in chickens were reported in the literature (Abou-Donia, 1981). These chemicals have been classified according to their chemical structure.

Aliphatic Compounds Of the 68 chemicals tested, only 41 (60%) produced delayed neurotoxicity (Table II). This series of compounds contains 12 subclasses with varying delayed neurotoxicity potential. The descending order of delayed neurotoxicity of aliphatic phosphorus esters is phosphonates = phosphorofluoridates = phosphonofluoridates = phorphorodiamidofluoridates = phosphoroamidofluoridates > phosphates > phosphorotrithioates > phosphorothioates = phosphonothioate = phosphinofluoridates = phosphorochloridates.

Pyrophosphorus compounds This series of compounds includes phosphates, phosphonates, and phosphoroamidates. None of the nine compounds reported was shown to produce OPIDN (Table III).

Aliphatic aromatic compounds This group contains the largest number of chemicals studied for delayed neurotoxicity: 76 (Table III). Only 35 compounds (46%) of this group have the potential to produce OPIDN. The descending order of delayed neurotoxic potency is phosphorodiamidofluoridates > phosphonates > phosphonothioates > phosphates > phosphorothioates > phosphinates.

Triarylphosphates A total of 71 compounds belonging to this group were tested for delayed neurotoxicity (Table III). Only 25 chemicals (35%) were positive. Delayed neurotoxicity potential depends on the size, number, and position of the substituent. While the unsubstituted triphenyl phosphate does not produce delayed neurotoxicity, the potency of substituted alkyl phenyls decreases in the order CH_3 > C_2H_5 > *n*-C_3H_7 > *iso*-C_3H_7 > *sec*-butyl = *tert*-butyl. Most triaryl phosphate esters with one or more phenyl rings substituted in the 2-position (*ortho*) are generally neurotoxic. This may result from their metabolism *in vivo* in the more potent saligenin cyclic phosphate derivative in analogy with TOCP. The requirement that esters contain hydrogen on the α-carbon atom in order to be neurotoxic does not seem to be universal, since other compounds such as 2,5-dimethyl phenyl phosphate fail to cause delayed neurotoxicity. Compounds with an ethyl group in the *para* position produce OPIDN.

TABLE II

Chemical Structure–Delayed Neurotoxic Effect Relationship for Type I Compounds

Chemical Class	Number tested	Dose range (mg/kg)	Route of exposure[a]	Compounds exhibiting[b] +	−	±
Aliphatic compounds						
Phosphates	17	2–118	s.c./i.v.	7	8	2
Phosphorothioates	3	20–1000	s.c.	0	3	0
Phosphonates	3	100–200	s.c.	3	0	0
Phosphonothioates	2	40–75	p.o.	0	2	0
Phosphinate	1	5–20	i.v.	0	1	0
Phosphorotrithioate	6	1000–30000	i.p.	2	4	0
Phosphorofluoridates	11	0.3–30	i.m.	11	0	0
Phosphorofluoridates	6	1–5	i.m.	6	0	0
Phosphinofluoridates	4	2.5–5	i.m.	0	4	0
Phosphoroamidofluoridate	1	5	i.m.	1	0	0
Phosphorodiamidofluoridate	9	0.1–100	i.m.	9	0	0
Phosphorochloridates	3	20–100	i.m.	0	3	0
Pyrophosphorus compounds						
Phosphates	2	50–300	s.c.	0	2	0
Phosphonates	5	10	s.c.	0	5	0
Phosphoroamidates	2	160–300	p.o.	0	2	0
Aliphatic aromatic compounds						
Phosphates	22	12–3000	varies	5	17	0
Phosphorothioates	12	10–1600	s.c.	1	11	0
Phosphonates	16	5–500	s.c./p.o.	13	3	0
Phospho nothioates	20	40–1000	p.o.	13	6	0
Phosphorodia midofluoridates	3	10–100	i.m.	3	0	0
Phosphinates	3	10	s.c.	0	3	0
Triaryl phosphate compounds	71	25–3000	p.o.	25	44	2
Saligenin cyclic phosphorus	13	0.5–200	i.p.	8	5	0

[a]s.c., subcutaneous; p.o., oral; i.p., intraperitoneal; i.v., intravenous; i.m. intramuscular.
[b] +, produced OPIDN; −, did not produce OPIDN; ± reports showing positive and negative delayed neurotoxicity.

Saligenic cyclic phosphorus compounds Only 13 compounds of this group were tested, of which 8 (62%) produced OPIDN (Table III).

Table III summarizes the results of studies into the chemical structure–delayed neurotoxicity relationship for Type I compounds. Of 237 pentavalent phosphorus-containing organic compounds, only 109 (46%) produced delayed neurotoxicity (Fig. 1). Conflicting results were reported on the delayed neurotoxicity of five compounds (2%).

TABLE III

A Summary of Chemical Structure–Delayed Neurotoxic Effect Relationships for Type I Compounds

Chemical Class	Number tested	Dose range	Route of exposure[a]	Compounds exhibiting[b]		
				+	−	±
Aliphatic compounds	68	0.1–3000	varies	41 (60%)	25 (37%)	2 (3%)
Pyrophosphorus compounds	9	10–300	s.c./p.o.	0 (0%)	9 (100%)	0 (0%)
Aliphatic aromatic compounds	76	10–3000	varies	25 (46%)	40 (53%)	1(1%)
Triaryl phosphate compounds	71	25–3000	p.o.	25 (35%)	44 (62%)	2 (3%)
Saligenin cyclic phosphorus	13	0.5–200	i.p.	8 (62%)	5 (38%)	0 (0%)
Total	237 (100%)			109 (46%)	123 (52%)	5 (2%)

[a] s.c., subcutaneous; p.o., oral; i.p., intraperitoneal.
[b] +, produced OPIDN; −, did not produce OPIDN; ±, reports showing positive and negative effects.

b. The Potential for Type II Chemicals to Produce OPIDN Studies on delayed neurotoxicity potential have been reported on only four Type II compounds (Fig. 2). All compounds were able to produce OPIDN in chickens and in other species. These chemicals are triphenyl phosphite, *tri-o*-cresyl phosphite, *tri-m*-cresyl phosphite, and *tri-p*-cresyl phosphite (Fig. 1).

III. Triaryl Phosphites

A. Acute Convulsive Action of Triaryl Phosphites

Triphenyl phosphite (TPP_i) was used as a convulsive agent in experimental epilepsy (Cobb *et al.*, 1937, 1938). It produces convulsions in various species, e.g., rats, cats, chickens (Smith *et al.*, 1930, 1932, 1933) and dogs (Aird *et al.*, 1940). This action is produced with all triaryl phosphites tested, i.e., the *ortho*, *meta*, and *para* isomers of tri-cresyl phosphite. The signs of this acute effect develop within a few minutes of dosing, and if the animal does not die, they disappear within a few hours. These signs are characterized by fine or coarse generalized tremors involving mostly large muscles. *In vitro* and *in vivo* studies have demonstrated that the acute convulsive action of triaryl phosphites is produced by the phenol or substituted phenol, which is formed by the hydrolysis of these esters *in vivo*.

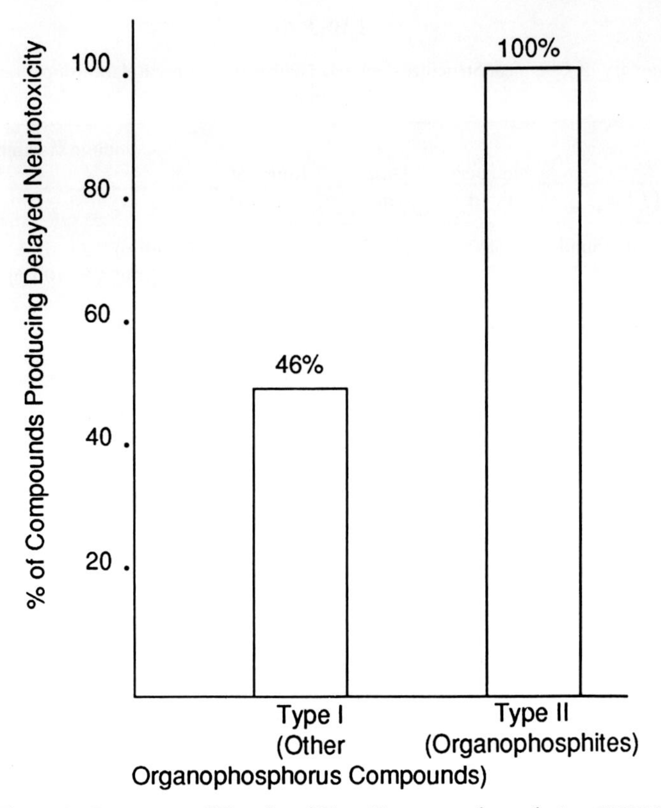

Figure 1 Percentage of Type I and Type II compounds producing OPIDN.

1. *In Vitro* Hydrolysis of Triaryl Phosphites

The hydrolysis of four triaryl phosphites was carried out in 0.01 N KOH in absolute alcohol (Smith *et al.*, 1933). The initial pseudo first-order rate constant for hydrolysis (min^{-1}) was in descending order: TPP_i, 2.77 > $TMCP_i$, 1.39 > $TOCP_i$, 0.69 > $TPCP_i$, 0.35. In contrast, the initial rate for the hydrolysis of TOCP in 0.1 N KOH (10 times as much as for phosphites) was 0.03 min^{-1}. On the other hand, all of the aryl phosphites and TOCP were more stable in absolute ethanol. The order of hydrolysis of triaryl phosphites in absolute ethanol was the opposite of that in KOH/absolute ethanol. Thus, the initial pseudo first-order rate constant for hydrolysis (min^{-1}) was in descending order: $TPCP_i$, 0.06 > $TOCP_i$, 0.006 > $TMCP_i$, 0.003 > TPP_i, 0.002, while TOCP showed no evidence of decomposition in 7 days.

The difference in the rate of hydrolysis of the phosphite by splitting of phenols, compared with that of the phosphates, correlates with the mode of convulsive action of these chemicals. Thus, TOCP, a Type I compound, is stable to hydrolysis, which results in the production of a cholinergic effect

TYPE I Other Organophosphorus Esters	OPIDN	TYPE II Organophosphites	OPIDN
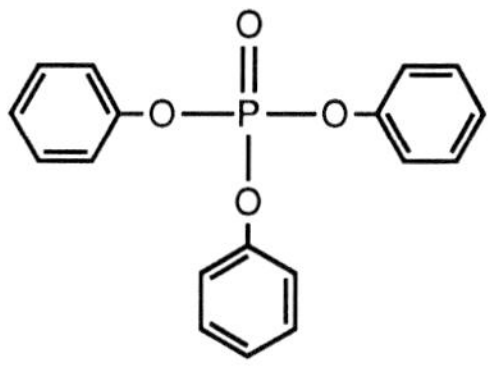 TPP	–	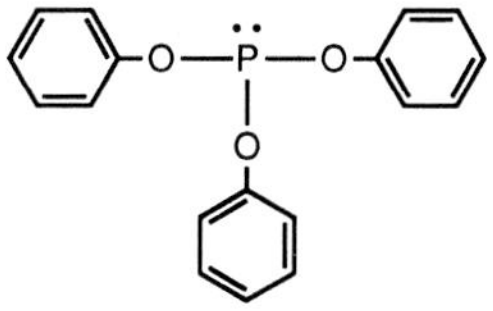 TPP_i	+
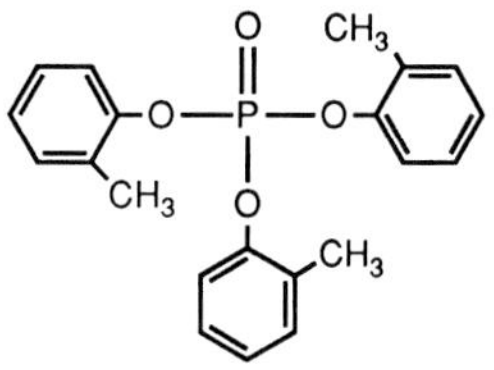 TOCP	+	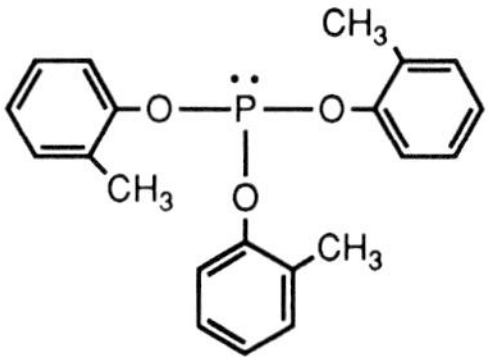 $TOCP_i$	+
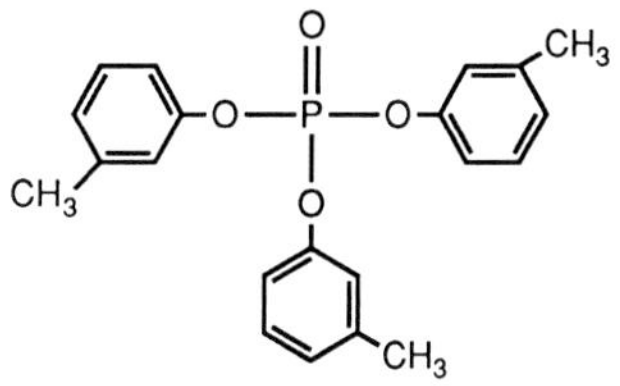 TMCP	–	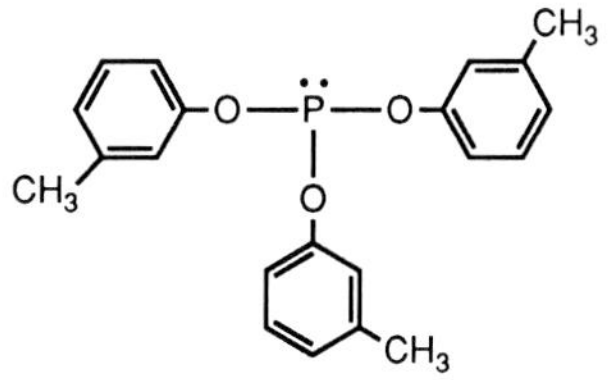 $TMCP_i$	+
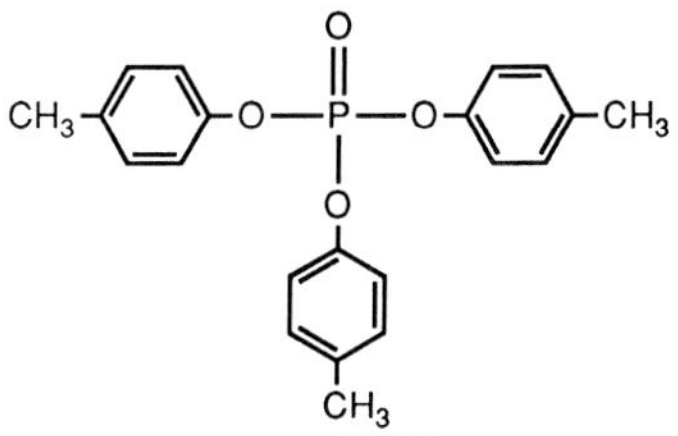 TPCP	–	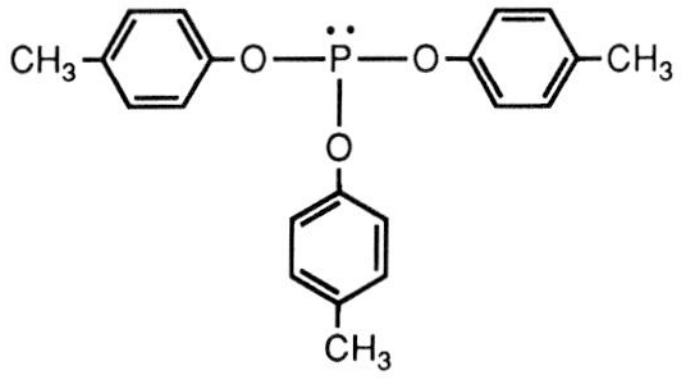 $TPCP_i$	+

Figure 2 Structures of some Type I and Type II compounds.

at large doses, with no convulsion since phenol is not released. On the other hand, triaryl phosphites are rapidly hydrolyzed resulting in the release of phenol, which produces convulsive action. Also, the ease with which tricresyl phosphite isomers were hydrolyzed correlated well with their acute convulsive toxicity. Thus, in rats the acute toxicity of these isomers followed similar order to that of their alkaline hydrolysis: $TPP_i > TPCP_i > TMCP_i > TOCP_i$. In the cat, $TPCP_i$, the least stable phosphite in absolute ethanol, was the most acutely toxic following subcutaneous injections.

2. *In Vitro* Hydrolysis of Triphenyl Phosphite

The aqueous hydrolysis of triphenyl phosphite (TPP_1) was studied in both buffered and unbuffered equivolume solutions of acetone and water using ^{31}P nuclear magnetic resonance (NMR) spectrometry (Carrington *et al.,* 1990). TPP_i was hydrolyzed to phenol and diphenyl phosphite that subsequently lost a second phenol to form phenyl phosphite at a rate faster than that of TPP_i hydrolysis. Finally, phenyl phosphite was hydrolyzed to phenol and phosphorus acid. It should be stressed that except for the triester (TPP_i), the di- and mono-substituted and free phosphorus acids exist mostly as phosphonic acid esters and were analyzed as such using ^{31}P NMR.

The di- and mono-substituted and free phosphorus acids all are trivalent phosphorus oxy acids that have P—OH bonds where the hydrogen atom is ionizable. All of these compounds undergo hydrogen transfer and form a very strong phosphoryl oxygen bond as outlined in Eq. (1).

$$\text{P—OH} \rightarrow \text{H—P=O} \tag{1}$$

The hydrogen atom in P—H bonds is not ionized. This process, which resembles the keto–enol shift in carbonyl compounds, is shown below for the conversion of diphenyl phosphite to diphenyl phosphonate (Fig. 3).

Diphenyl phosphite, which contains a trivalent phosphorus atom, can exist only as transitory species in concentration of 1 in 10^{12} (Cotton and Wilkinson, 1962), although in some of its reactions it behaves as a trivalent molecule. Similarly, monophenyl phosphite and phosphorus acid, each of which has a trivalent pyramidal molecule that is unstable in pyramidal form, exist in tautomeric equilibrium with tetrahedral phosphonic acid (Fig. 4).

The hydrolysis rate of TPP_i is dependent on the pH of the aqueous acetone solution. When the hydrolysis medium is buffered at pH 8, TPP_i hydrolysis takes place at a very slow rate, with only 10% hydrolyzed after 18 hr. In contrast, in unbuffered media, TPP_i is hydrolyzed rapidly, with almost 99% being hydrolyzed within 12 hr. Also, as the monophenyl phosphonate rises, the medium pH drops from about pH 4 to below pH 2.

3. Oxidation of Triphenyl Phosphite

The trivalent phosphorus atom in triphenyl phosphite contains a pair of electrons ready for reaction with oxygen and formation of a phosphoryl

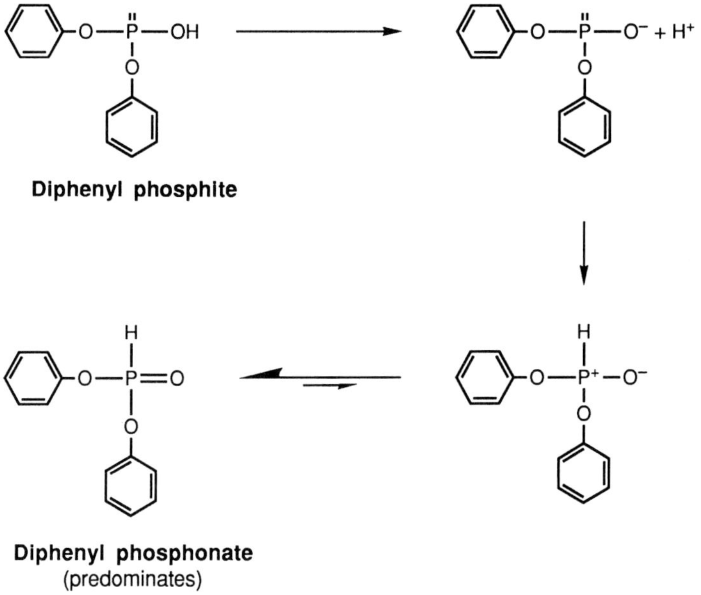

Diphenyl phosphite

Diphenyl phosphonate
(predominates)

Figure 3 Conversion of diphenyl phoshite to diphenyl phosphonate.

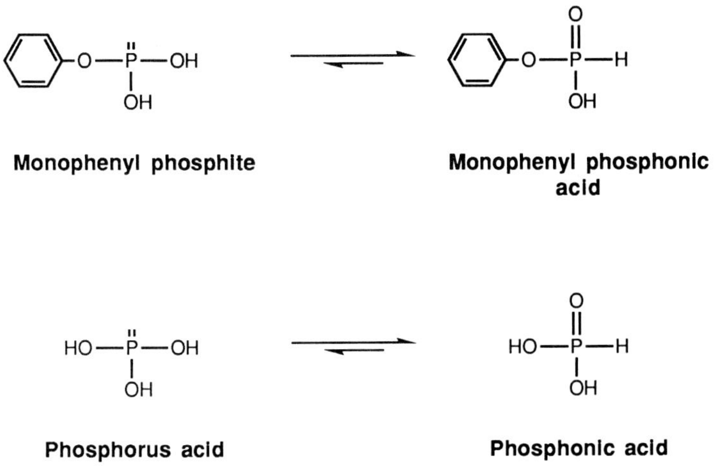

Monophenyl phosphite

Monophenyl phosphonic acid

Phosphorus acid

Phosphonic acid

Figure 4 Tautomeric equilibrium of trivalent monophenyl phosphite and phosphorus acid to pentavalent phosphonic acid.

oxygen bond. The formation of such a very strong bond is the driving force for many reactions, e.g., the oxidation of triphenyl phosphite to triphenyl phosphate (Fig. 5).

4. *In Vivo* Hydrolysis of Triphenyl Phosphite

Triphenyl phosphite produced convulsions in various species. In dogs, the severity of convulsive signs produced soon after a subcutaneous dose of TPP_i depended on the purity of the chemical used (Aird *et al.,* 1940). These studies demonstrated the correlation between the convulsive action of TPP_i and the degree of phenol contamination in the preparation used.

The hydrolysis of $[^{32}P]TPP_i$ (Fig. 6) was studied in cats following a single intraperitoneal injection (Aird *et al.,* 1940). One hour and 45 minutes after injection, only 8% of the radioactivity was unabsorbed in the peritoneal cavity. A small amount of ^{32}P-derived materials (TPP_i or its breakdown products) was found in the CNS after 1.5 hr, whereas the amount of phenol was considerably higher than normal levels in nervous tissues. This led to the conclusion that phenol produced via TPP_i hydrolysis was present in the brain in amounts sufficient to explain the early acute toxic action on the CNS that was manifested as convulsions. Of particular interest is the relatively higher concentration of ^{32}P-derived materials in the diencephalon and motor cortex, suggesting selective and preferential absorption of TPP_i by grey matter of the CNS. Triaryl phosphites were absorbed very slowly from the site of subcutaneous injection in chickens (Smith *et al.,* 1933). In hens, a subcutaneous injection of TPP_i persisted unhydrolyzed at the site of application for several weeks in six of eight treated hens (Carrington *et al.,* 1988b, 1990). TPP_i in the other two hens, however, was rapidly hydrolyzed to diphenyl phosphite (DPP_i) and phenol, as determined by ^{31}P NMR, within 4 to 6 hr. Both of these hens died of apparent phenol toxicity. Phenol toxicity may account for the all-or-none acute toxicity following the injection of TPP_i in which about 20% or more than 100 subcutaneously injected hens died with 24 hr of dosing, while the majority exhibited no signs of acute toxicity. It is noteworthy that subcutaneously injected DPP_i in the hen is hydrolyzed to monophenyl phosphite (MPP) and phenol faster than TPP_i.

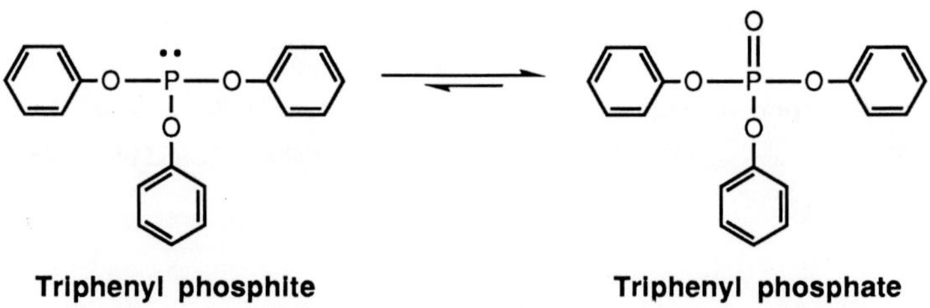

Triphenyl phosphite Triphenyl phosphate

Figure 5 Oxidation of triphenyl phosphite to triphenyl phosphate.

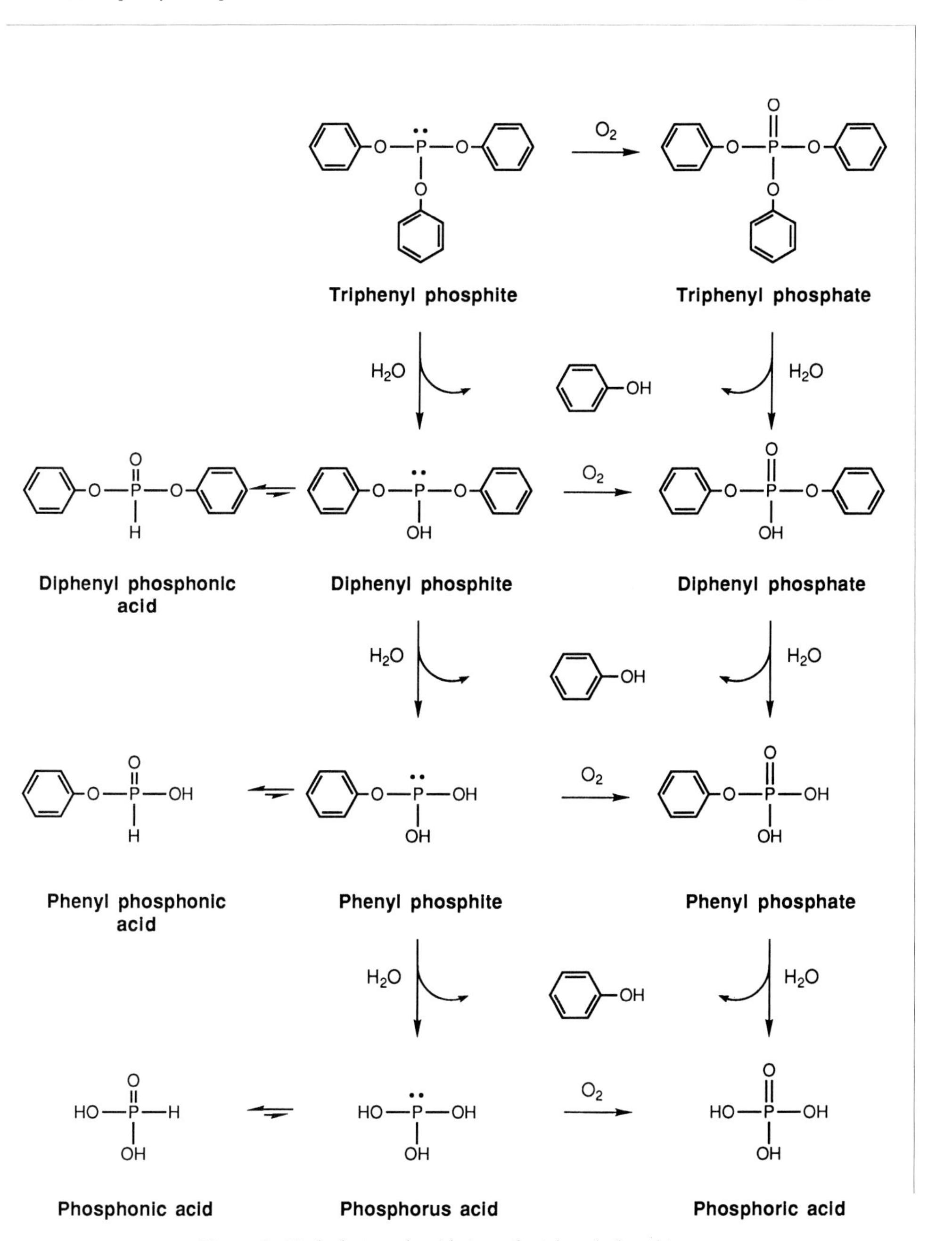

Figure 6 Hydrolysis and oxidation of triphenyl phosphite.

In rats, complete hydrolysis of TPP_i to DPP_i and phenol usually takes place in 6 hr but certainly within 24 hr. This rapid hydrolysis of TPP_i may account for the greater acute toxicity of TPP_i and greater incidence of dermal necrosis in the rat than in the hen.

These results are consistent with the hypothesis that the acute toxicity of TPP_i is caused by the phenol resulting from its hydrolysis *in vivo*. Differences in the rates of hydrolysis may account for the variability in the acute toxicity of TPP_i in the rat and the hen and among individual animals within the same species.

5. TPP-Induced Hyperthermia

In cats TPP_i produced hyperthermia, which ranged from 0.9 to 2.5°C in most animals that developed extensor rigidity (Smith *et al.,* 1933). The rise in temperature occurs at onset of ataxia and sometimes a day sooner. Hyperthermia usually persists during the stages of ataxia and paresis and occasionally through the period of extensor rigidity.

B. Differential Neurotoxic Actions of Triphenyl Phosphite Resulting from Differential Routes of Administration

1. Oral Administration

Of hens given a single oral dose of TPP_i at 100, 250, 500, or 1,000 mg/kg, only those given the highest dose developed acute phenol-like toxic effects within a few minutes of dosing. These effects were characterized by coarse tremors and weakness. These signs usually lasted a short time, approximately 3 hr. On day 21, all hens received a second dose (Abou-Donia and Brown, 1990).

A single oral dose of TPP_i at 100 to 1000 mg/kg did not produce delayed neurotoxicity in hens. On the other hand, a second oral dose of 250 mg/kg or higher produced clinical signs of delayed neurotoxicity. Hens given two doses of 250 mg/kg developed mild ataxia, while those given two doses of 500 or 1000 mg/kg progressed to gross and severe ataxia, respectively. Clinical signs of these hens were similar to those treated with TOCP (Abou-Donia and Brown, 1990).

2. Dermal Application

All hens given a daily dermal dose of TPP_i, 100 mg/kg, developed mild ataxia after four doses, and their condition progressed to paralysis after 23 topical applications (Abou-Donia and Brown, 1990).

The results indicate that dermal application of TPP_i is more effective in producing delayed neurotoxicity in hens than is the orally administered compound. Orally administered TPP_i was less effective in producing OPIDN because it was rapidly hydrolyzed in the aqueous environment in the gastrointestinal tract, as shown previously *in vivo* and *in vitro*. TPP_i in aqueous medium was hydrolyzed to phenol, resulting in acute phenol-like toxic effect when administered at high doses. On the other hand, percutaneously administered TPP_i remained intact long enough to reach the target for delayed neurotoxicity in the nervous system. This finding seems to be a characteristic of triaryl phosphites that produce Type II OPIDN in contrast to Type I OPIDN, which is induced following oral or dermal administration (Tables IV and V).

3. Neuropathological Changes

Hens that developed clinical signs of TPP_i-induced delayed neurotoxicity exhibited histopathological alterations in the CNS and PNS. The frequency and severity of lesions were dose dependent, as reflected in the degree of clinical signs. Following oral adminstration, all lesions were minimal except for one lesion that was slight. On the other hand, dermally treated hens that were paralyzed before termination showed minimal, slight, or moderate severity of histopathologic changes. TPP_i-induced lesions occurred in the white and/or grey matter of the CNS and PNS.

TABLE IV

Threshold Single Dose for Induction of Type I and II OPIDN in Hens

Type I (pentavalent organophosphorus esters)	Dose (mg/kg)	Route
TOCP	62.5	s.c.[a]
TOCP	250	Oral[a]
DFP	0.25	s.c.[a]
Cyanofenphos	5	Oral[c]
EPN	25	Oral[c]
Leptophos	100	Oral[c]
EPDP	800	Oral[c]
DEF	100	Oral[d]
DEF	250	Dermal[d]
Type II (Organophosphites)	**Dose (mg/kg)**	**Route**
TPP_i	250	s.c.[b]
$TOCP_i$	1919	s.c.[b]

[a] Carrington and Abou-Donia (1988a).
[b] Smith *et al.*, (1932).
[c] Abou-Donia (1979).
[d] Abou-Donia *et al.*, (1979b).

TABLE V

Threshold Daily Dose for Induction of Type I and II OPIDN in Hens

Type I (pentavalent organophosphorus esters)	Dose (mg/kg) [total]	Route
EPN	0.01 [0.2]	Dermal[a]
EPN	0.1 [1.7]	Oral[c]
Leptophos	0.5 [47]	Dermal[e]
DEF	0.5 [15]	Oral[f]
TOCP	0.5 [36]	Oral[g]
Leptophos	1.0 [62]	Oral[g]

Type II (Organophosphites)	Dose (mg/kg) [total]	Route
TPP_i	422 [1689]	s.c.[b]
TPP_i	100 [1200]	Dermal[d]

[a] Abou-Doniaet *al.*, (1983a).
[b] Smith *et al.*, (1930).
[c] Abou-Donia and Graham (1978a).
[d] Abou-Donia and Brown (1990).
[e] Abou-Donia and Graham (1978b).
[f] Abou-Donia *et al.*, (1979a).
[g] Abou-Donia and Graham (1979).

These results agree with previous studies in the chicken (Smith *et al.*, 1933; Carrington *et al.*, 1988a), the rat (Smith *et al.*, 1933; Veronesi and Dvergsten, 1987), and the cat (Smith *et al.*, 1933). In the spinal cord triaryl phosphites produce diffuse lesions in the ascending (sensory, affector) tracts, i.e., spinocerebellar and anterolateral. The decending (motor, effector) tracts involved were rubrospinal, vestibulospinal, tectospinal, lateral corticospinal, and anterolateral tracts. Cellular degeneration was characterized by cellular gliosis and decrease in number of motor cells in the anterior horn. Less frequently present were fatty degeneration, tigrolysis, and cellular necrosis. These changes were seen in the medulla, pons, brainstem, and cerebellar cortex (Smith *et al.*, 1933).

Studies in the ferret showed that TPP_i-induced neuropathologic changes were not confined to sensorimotor pathways and nuclei of the brainstem, but also included other auditory, visual, and higher-order sensory motor tracts (Tanaka *et al.,* 1990). TPP_i-induced lesions in cortical and thalamic regions indicated that Type II OPIDN is distinct from Type I, in which only brainstem, cerebellum, and spinal cord are affected. In contrast, in Type II, CNS degeneration is present in both forebrain and hindbrain areas.

Although hydrolyzed phenol seems to play a role in the acute convulsive action of TPP_i, its involvement in the mechanism of TPP_i-induced delayed neurotoxicity is doubtful. Delayed neurotoxicity induced by orally administered TPP_i is minimal. This might be attributed to the instability of TPP_i in aqueous media resulting in its hydrolysis to yield phenol in the gastrointestinal tract. Thus, the phenol existing partially in ionized, non-lipid-soluble form may not be able to reach the target for delayed neurotoxicity in the CNS and PNS.

Induction of delayed neurotoxicity by TPP_i requires that the intact molecule penetrate various biological membranes to reach the neurotoxicity target protein. This would necessitate a slow hydrolysis rate of circulated TPP_i following its absorption from the application site. It is assumed that TPP_i stability in the plasma would be similar to that under buffered conditions, resulting in a lower hydrolysis rate than seen in unbuffered aqueous solutions. This would allow enough of the intact TPP_i to reach the neurotoxicity target in the nervous system to result in delayed neurotoxicity. Such behavior is consistent with the hydrophobic character of TPP_i that results in its accumulation in the nervous tissue. Accumulated TPP_i at the neurotoxicity target results in the induction of both cell body and axonal degeneration characteristic of TPP_i-induced delayed neurotoxicity.

C. Effect of Age on Sensitivity of Chickens to TPP_i-Induced Delayed Neurotoxicity

Katoh *et al.,* (1990) studied the effect of age on susceptibility of chickens (groups of five) to TPP_i-induced delayed neurotoxicity following a single intravenous injection of 50 mg/kg in dimethylsulfoxide (DMSO), Tween 80 and saline, 1:1:3. Delayed neurotoxicity was assessed by observing gait and behavioral abnormalities during the 21-day experiment. Birds that were 41 or 60 days old did not exhibit neurological deficiency. On the other hand, one, five, and four chickens showed mild ataxia 10 to 12 days after dosing in the groups given 90, 120, and 180 mg/kg respectively. The same i.v. dose produced ataxia, which progressed to paralysis in one bird, when given to 23-month-old chickens. Age-related sensitivity to TPP_i correlated well with age-related clearance of TPP_i from the blood and most tissues examined. The biological half-life of TPP_i in the blood was 23.2 min and 30.5 min following an i.v. injection of TPP_i, 50 mg/kg, in 45- and 135-day-old chickens. Pharmacokinetics and disposition, however, do not explain the results that spinal cord and sciatic nerve concentrations of TPP_i were the same in both age groups.

In another study (Abou-Donia and Brown, 1990), three groups of 1-week-old chicks were treated with two subcutaneous doses at a 21-day interval of TPP_i, 1000 mg/kg, $TOCP_i$ per 1137 mg/kg, or TOCP. A group of untreated chicks was used as a control. Chicks were killed when they became moribund or 42 days after initial dose. Neither acute nor delayed neurotoxicity signs developed following a single subcutaneous dose of any compound. On the other hand, chicks given a second s.c. dose of TPP_i or $TOCP_i$ developed ataxia 4 to 5 days after the second dose, which progressed to ataxia with near paralysis. Chicks treated with two doses of TOCP remained normal.

TPP_i and $TOCP_i$ induced histopathologic lesions in the CNS and PNS of chicks similar to those seen in the hen. In the brain, lesions were seen in the brain stem and in the thalamus. Histologic alterations occurred in the grey and white matter of the spinal cord. Furthermore, there was a generalized atrophy and hypoplasia of the entire spinal cord, with neurons being smaller and fewer and more numerous glial aggregates diagnosed as gliosis. Sections of dorsal root ganglion showed central chromatolysis, satellitoses, and neuronophagia in ganglion cells. The sciatic nerve and its branches exhibited axonal swelling and fragmentation.

The effect of a single subcutaneous equimolar dose of 1000 mg TPP_i per kg, 1137 mg TOCP per kg, or 1187 mg $TOCP_i$ per kg, on the activity of chick brain neurotoxic esterase (NTE), AChE, and plasma BuChE were determined 24 hr after administration. The three compounds significantly inhibited brain NTE and plasma BuChE (more than 70% inhibition). None of these treatments had any significant effect on chick brain AChE. NTE activity is preferentially inhibited by delayed neurotoxic OP compounds but not by non-delayed neurotoxic compounds (Johnson, 1969, 1990). There is a good correlation between the inhibition and aging of hen brain NTE by OP compounds and their ability to produce OPIDN (Fig. 7). To produce OPIDN, an OP compound must result in 70% inhibition of NTE activity 24 to 48 hr after the administration of the unprotected LD_{50}.

These studies demonstrate the differential sensitivity of chicks to induction of delayed neurotoxicity by Type I and Type II OPIDN compounds. That chicks were not sensitive to TOCP-induced delayed neurotoxicity is in agreement with previous studies (Bondy *et al.*, 1961; Johnson and Barnes, 1970; Konno and Kinebuchi, 1978; Abou-Donia *et al.*, 1982; Katoh *et al.*, 1990). Insensitivity of chicks to a single dose of Type I compounds may be related to differences in toxicokinetics of these chemicals, akin to the differences in rates of accumulation and elimination of EPN (Type I) in sciatic nerves and spinal cord between 1-week-old and adult hens (Abou-Donia *et al.*, 1983a,b).

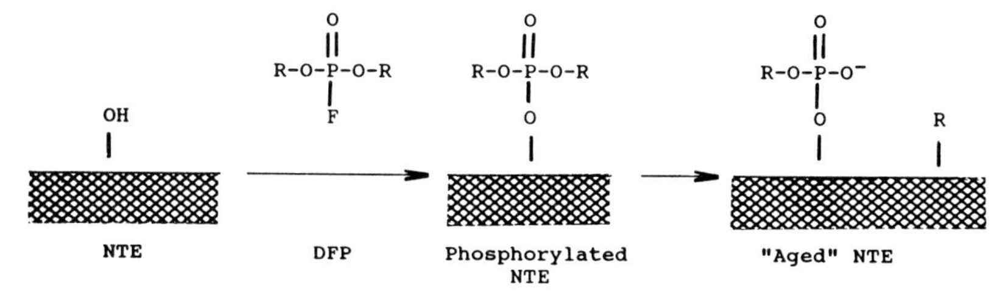

Figure 7 Binding and *aging* of OP compounds to NTE.

D. Biochemical Action of Triphenyl Phosphite

Intravenous injections of TPP_i adversely affected several hen skeletal muscle mitochondrial enzymes (Konno *et al.*, 1989). TPP_i significantly inhibited creatine kinase and succinate dehydrogenease activities in adductor magnus and soleus muscle mitochondria 24 to 48 hr after administration. Because creatine kinase is involved in energy transfer from mitochondria to myofibrils in skeletal muscle (Apple and Rogers, 1986), Konno *et al.*, (1989) suggested that the mitochondria might be the target for TPP_i toxicity.

We used primary cultures of bovine adrenomedullary cells to study the biochemical effect of TPP_i *in vitro*. These cells are derived from the neural crest stem cell and are considered truncated sypathetic neurons because they lack axonal-like projections. Because of this feature, these cells provide an ideal system for studying the neurotoxic action of TPP on the cell body in isolation. The differential effect of TPP_i and DFP on the exocytotic secretion of catecholamine from chromaffin cells has been studied (Abou-Donia and Knoth, 1989). TPP_i selectively inhibited catecholamine secretion irrespective of the secretagogue used. Concomitantly, TPP_i inhibited ^{45}Ca uptake into the cells. In contrast, DFP, a Type I OPIDN compound that does not affect the cell body, neither inhibited catecholamine secretion nor affected ^{45}Ca uptake into the cells (Abou-Donia *et al.*, 1990).

Further studies into the effect of TPP_i on chromaffin cells suggested that the mitochondria are the target for its delayed neurotoxicity (Anderson *et al.*, 1991). Early ultrastructural changes were seen as swollen or disrupted mitochondria. These findings correlated with the inhibitory action of TPP_i on the mitochondrial ability to synthesize ATP as reflected by the inhibition of [^{14}C]adenosine incorporation into ATP.

These results suggest that TPP_i may cause axonal swelling by inhibiting ATP synthesis, which leads to depletion of ATP stores and disruption of active transport. Such disruption breaks down ionic equilibrium normally maintained in the cell and subcellular organelles such as mitochondria and leads to the accumulation of intracellular sodium and water, resulting in the swelling of both the cell and its organelles, e.g., mitochondria (Schwertschlag *et al.*,

1986). Depletion of ATP synthesis by TPP_i may be responsible also for its inhibition of exocytotic process, since ATP is required for the exocytosis (Dunn and Holz, 1983).

TPP_i, an antioxidant, may produce neurotoxicity by oxidizing to phenyl phosphate and depleting oxygen from the mitochondria, resulting in anoxia. The driving force of this reaction is the formation of the very strong oxyphosphoryl bond. The resulting anoxia leads to mitochondrial swelling. The calcium ion has been proposed to play a key role in the mechanism of anoxia-induced mitochondrial swelling (Beatrice *et al.*, 1984; Tagawa *et al.*, 1985).

IV. Summary

1. All four tested triaryl phosphites were neurotoxic and showed some quantitative differences, but qualitatively, they were essentially similar.
2. In large enough doses, two stages of neurotoxic action may be elicited with all aryl phosphites.
3. An early stage of acute convulsive action characterized by transient tremors is produced by the phenol that results from the hydrolysis of these compounds.
4. The delayed neurotoxic action is characterized by bidirectional circling, primary ataxia, and later by extensor rigidity in the cat, and by flaccid paralysis of the extremities in chickens and rats.
5. It is unlikely, however, that phenol, which may be involved in the acute effect of TPP_i, is responsible for the delayed neurotoxicity induced by TPP_i. This is consistent with the finding that the phenol is rapidly detoxified by oxidation and conjugation, with the result that it does not cause degenerative changes in the CNS (Aird *et al.*, 1940).
6. Oral administration of aryl phosphites results in a very weak neurotoxic action, while subcutaneous and dermal administration are effective in producing neurotoxicity.
7. Young chicks are not sensitive to a single subcutaneous dose of TPP_i or $TOCP_i$, perhaps because of their rapid detoxication compared to that of adult chickens. Two subcutaneous doses of these aryl phosphites, however, unlike TOCP, produced delayed neurotoxicity in chicks.
8. Histopathologic lesions are characterized by Wallerian-type degeneration of the axon and myelin, and cell necrosis. In addition to lesions of the sensorimotor pathways of the brain stem and spinal cord, which cause ataxia and paralysis, lesions are also present in cortical and thalamic regions, resulting in impairment of cognitive functions.
9. Thus TPP_i, a Type II OPIDN, induces CNS degeneration in both forebrain and hindbrain, unlike Type I OPIDN, from the use of which only brain stem, cerebellum, and spinal cord are affected.

10. TPP_i-induced ultrastructural changes were characterized by swollen and degenerated mitochondria of chromaffin cells.

Acknowledgments

Supported in part by grants from the National Institute of Environmental Health Sciences No. ESO2717 and ESO5154 and from the National Institute for Occupational Safety and Health No. OHO 0823.

References

Abou-Donia, M. B. (1978). Role of acid phosphatase in delayed neurotoxicity induced by leptophos in hens. *Biochem. Pharmacol.* 27, 2055–2058.

Abou-Donia, M. B. (1979). Delayed neurotoxicity of phenylphosphonothioate esters. *Science* **205**, 713–715.

Abou-Donia, M. B. (1981). Organophosphorus ester–induced neurotoxicity. *Annu. Rev. Pharmacol. Toxicol.* **21**, 511–548.

Abou-Donia, M. B. (1985). Biochemical toxicology of organophosphorus compounds. *In* "Neurotoxicology" (K. Blum and L. Manzo, eds.), pp. 423–444. Marcel Dekker, New York.

Abou-Donia, M. B., and Brown, H. R. (1990). Triphenyl phosphite, a Type II OPIDN compound. *Proc. Spring Natl. Meet. Am. Chem. Soc., Boston, MA,* April 22–27, in press.

Abou-Donia, M. B., and Graham, D. G. (1978a). Delayed neurotoxicity of *o*-ethyl *o*-4-nitrophenyl phenylphosphonothioate: Subchronic (90 days) oral administration in hens. *Toxicol. Appl. Pharmacol.* **45**, 685–700.

Abou-Donia, M. B., and Graham, D. G. (1978b). Neurotoxicity produced by long-term low-level topical application of leptophos in the comb of hens. *Toxicol. Appl. Pharmacol.* **46**, 199–213.

Abou-Donia, M. B., and Graham, D. G. (1979). Delayed neurotoxicity of subchronic oral administration of leptophos: Recovery during four months after exposure. *J. Toxicol. Environ. Health* **5**, 1133–1147.

Abou-Donia, M. B., Graham, D. G., Abdo, K. M., and Komeil, A. A. (1979a). Delayed neurotoxic, late acute and cholinergic effects of *S,S,S*-tributyl phosphorotrithioate (DEF): Subchronic (90 days) administration in hens. *Toxicology* **14**, 229–243.

Abou-Donia, M. B., Graham, D. G., Makkawy, H. A., and Abdo, K. M. (1983a). Effect of a subchronic dermal application of *o*-ethyl *o*-4-nitrophenyl phenylphosphonothioate on producing delayed neurotoxicity in hens. *Neurotoxicology* **4**, 247–260.

Abou-Donia, M. B., Graham, D. G., Timmons, P. R., and Reichert, B. L. (1979b). Delayed neurotoxic and late acute effects of *S,S,S*-tributyl phosphorotrithioate on the hen: Effect of route of administration. *Neurotoxicology* **1**, 425–447.

Abou-Donia, M. B., Hernandez, Y. M., Ahmed, N. S., and Abou-Donia, S. A. (1983b). Distribution and metabolism of *o*-ethyl *o*-4-nitrophenyl phenylphosphonothioate after a single oral dose in one-week-old chicks. *Arch. Toxicol.* **54**, 83–96.

Abou-Donia, M. B., and Knoth, J. K. (1989). Differential effects of triphenyl phosphite and diisopropylphosphorofluoridate on catecholamine secretion from bovine adrenal medullary chromaffin cells. *Toxicologist* **9**, 74.

Abou-Donia, M. B., Lapadula, D. M., and Anderson, J. K. (1990). Selective inhibition of $^{45}Ca^{2+}$ uptake into synaptosomes and primary cell cultures by triphenyl phosphite, a Type II OPIDN. *Toxicologist* **10**, 106.

Abou-Donia, M. B., and Lapadula, D. M. (1990). Mechanisms of organophosphorus ester–induced delayed neurotoxicity: Type I and Type II. *Annu. Rev. Pharmacol. Toxicol.* **30,** 405–440.

Abou-Donia, M. B., Makkawy, H. M., Salama, A. E., and Graham, D. G. (1982). Effect of age of hens on their sensitivity to delayed neurotoxicity induced by a single dose of tri-*o* tolyl phosphate. *Toxicologist* **2,** 178.

Abou-Donia, M. B., Reichert, B. L., and Ashry, M. A. (1983). The absorption, distribution, excretion, and metabolism of a single oral dose of *o*-ethyl *o*-4-nitrophenyl phenylphosphonothioate in hens. *Toxicol. Appl. Pharmacol.* **70,** 18–28.

Aird, R. B., Cohn, W. E., Weiss, S. (1940). Convulsive action of triphenyl phosphite. *Proc. Soc. Exp. Biol. Med.* **45,** 306–309.

Anderson, J. K., Veronesi, B., Jones, K., Lapadula, D. M.,and Abou-Donia, M. B. (1991). Triphenyl phosphite-induced ultrastructural changes in bovine adrenomedullary chromaffin cells. *Toxicol. Appl. Pharmacol.* in press.

Apple, F. S., and Rogers, M. A. (1986). Mitochondrial creatine kinase activity alterations in skeletal muscle during long-distance running. *J. Appl. Physiol.* **62,** 482–486.

Beatrice, M. D., Stiers, D. L., and Pfeiffer, D. R. (1984). The role of glutathione in retention of Ca^{2+} by liver mitochondria. *J. Biol. Chem.* **259,** 1279–1287.

Bondy, H. F., Field, E. J., Worden, A. N., and Hughes, J. P. W. (1961). A study on the acute toxicity of the triaryl phosphates used as plasticizers. *Br. J. Ind. Med.* **17,** 190–200.

Carrington, C. D., and Abou-Donia, M. B. (1988). Triphenyl phosphite neurotoxicity in the hen: Inhibition of neurotoxic esterase and a lack of prophylaxis by phenylmethylsulfonyl fluoride. *Arch. Toxicol.* **62,** 375–380.

Carrington, C. D., Brown, H. R., and Abou-Donia, M. B. (1988a). Histopathological assessment of triphenyl phosphite neurotoxicity in the hen. *Neurotoxicology* **9,** 223–234.

Carrington, C. D., Burt, C. T., and Abou-Donia, M. B. (1988b). *In vivo* ^{31}P nuclear magnetic resonance studies on the absorption of triphenyl phosphite and tri-*o*-cresyl phosphate following subcutaneous administration in hens. *Drug Metab. Dispos.* **16,** 104–109.

Carrington, C. D., Burt, C. T., and Abou-Donia, M. B. (1990). Role of phenol in the toxicity of triphenyl phosphite (TPP). *Toxicologist* **10,** 167.

Cavanagh, J. B. (1964). The significance of the "dying back" process in experimental and human neurological disease. *Int. Rev. Exp. Pathol.* **3,** 219–267.

Cobb, S., Cohen, M. E., and Ney, J. (1937). Brilliant vital red as an anticonvulsant. *J. Nerv. Ment. Dis.* **85,** 438–444.

Cobb, S., Cohen, M. E., and Ney, J. (1938). Anticonvulsive action of vital dyes. *Arch. Neurol. Psychiat.* **40,** 1156–1177.

Cotton, F. A., and Wilkinson, G. (eds.) (1962). "Advanced Inorganic Chemistry. A Comprehensive Text," Interscience, New York.

Dunn, L. A., and Holz, R. W. (1983). Catecholamine secretion from digitonin-treated adrenomedullary chromaffin cells. *J. Biol. Chem.* **258,** 4989–4993.

Johnson, M. K. (1969). The delayed neurotoxic effect of some organophosphorus compounds. Identification of the phosphorylation site as an esterase. *Biochem. J.* **114,** 711–717.

Johnson, M. K. (1990). Organophosphates and delayed neuropathy—Is NTE alive and well? *Toxicol. Appl. Pharmacol.* **102,** 385–399.

Johnson, M. K., and Barnes, J. M. (1970). Age and sensitivity of chicks to the delayed neurotoxic effects of some organophosphorus compounds. *Biochem. Pharmacol.* **19,** 3045–3047.

Katoh, K., Konno, N., Yamauchi, T., and Fukushima, M. (1990). Effects of age of susceptibility of chickens to delayed neurotoxicity due to triphenyl phosphite. *Pharmacol. Toxicol.* **66,** 387-392.

Konno, N. K., Katoh, K., Yamauchi, T., and Fukushima, M. (1989). Delayed neurotoxicity of triphenyl phosphite in hens: Pharmacokinetics and biochemical studies. *Toxicol. Appl. Pharmcol.* **100,** 440–450.

Konno, N., and Kinebuchi, H. (1978). Residues of Phosvel in plasma and in adipose tissue of hens after single oral administration. *Toxicol. Appl. Pharmacol.* **45**, 541–547.

Lillie, R. D., and Smith, M. I., (1932). The histopathology of some neurotoxic phenol esters. *Natl. Inst. Health Bull.* **160**, 54–69.

Padilla, S. S., Grizzle, T. B., and Lylerty, D. (1987). Triphenyl phosphite: *In vivo* and *in vitro* inhibition of rat neurotoxic esterase. *Toxicol. Appl. Pharmacol.* **87**, 249–256.

Roger, H., and Recordier, M. (1934). Les polyneuritesphosphocreosotiques (phosphate de creosote, ginger paralysis, apiol). *Ann. Med. Paris* **35**, 44–63.

Schwertschlag, U., Schrier, R. W., and Wilson, P. (1986). Beneficial effects of calcium channel blockers and calmodulin-binding drugs on *in vitro* renal cell anoxia. *J. Pharmacol. Exp. Ther.* **238**, 119–124.

Smith, M. I., Elvove, E., Valer, P. J., Frazier, W. H., and Mallory, G. E. (1930). Pharmacological and chemical studies of the cause of so-called ginger paralysis. *U.S. Public Health Rep.* **45**, 1703–1716.

Smith, M. I., Engel, E. W., and Stohlman, E. F. (1932). Further studies on the pharmacology and certain phenol esters with special reference to the relation of chemical constitution and physiological action. *Nat. Inst. Health Bull. No. 160,* 1–53.

Smith, M. I., Lillie, R. D., Elove, E., and Stohlman, E. F. (1933). The pharmacologic action of the phosphorus acid esters of the phenols. *J. Pharmacol. Exp. Ther.* **49**, 78–99.

Tagawa, K., Nishida, T., Watanabe, F., and Koseki, M. (1985). Mechanism of anoxic damage of mitochondria: Depletion of intramitrochondrial ATP and concomitant release of free CA^{2+}. *Mol. Physiol.* **8**, 515–524.

Tanaka, D., Jr., and Bursean, S. J., Lehning, E. J., and Aulerich, R. J. (1990). Exposure to triphenyl phosphite results in widespread degeneration in the mammalian central nervous system. *Brain Res.* **53**, 250–298.

U.S. Environmental Protection Agency. (1985a). "Chemical Hazard Information Profile. Triphenyl Phosphite." Draft Report Washington, D.C., E.P.A.

U.S. Environmental Protection Agency. (1985b). Registration of pesitcides in the United States. Proposed guidelines. Subpart G. *Neurotoxicity Fed. Regist.* **50**, 39,458–39,470.

Veronesi, B., and Dvergsten, C. (1987). Triphenyl phosphite neuropathy differs from organophosphorus-induced delayed neuropathy in rats. *Neuropathol. Appl. Neurobiol.* **13**, 193–208.

Veronesi, B., Padilla, S., and Newland, D. (1986). Biochemical and neuropathological assessment of triphenyl phoshite in rats. *Toxicol. Appl. Pharmacol.* **83**, 203–210.

18

Rodent Models of Organophosphorus-Induced Delayed Neuropathy*

Bellina Veronesi
Stephanie Padilla
U.S. Environmental Protection Agency
Health Effects Research Laboratories
Neurotoxicology Division, MD-74B
Cellular and Molecular Toxicity Branch
Research Triangle Park, North Carolina

I. Introduction

The delayed neuropathy (OPIDN) that accompanies exposure to some organophosphorus (OP) compounds is poorly understood. Distinct and separate from the immediate physiological crisis (*viz.*, lacrimation, salivation, respiratory distress) associated with OP inhibition of acetylcholinesterase (AChE), OPIDN is a debilitating paralysis that occurs 2–3 weeks after acute or multiple exposures to *neuropathic* OP compounds. Neuropathologically, OPIDN affects the longest and largest fiber tracts of the spinal cord (CNS) and peripheral (PNS) nervous system in a dying-back pattern of degeneration (Cavanagh, 1963). Although OPIDN has occurred in epidemic proportions throughout this country and the rest of the world, little is known about its

*This document has been reviewed in accordance with U.S. Environmental Protection Agency policy and approved for publication. Mention of trade names and commercial products does not constitute endorsements or recommendation for use.

underlying mechanisms or treatment. This may be due in large part to the use of chickens as primary animal models to describe the pathological and biochemical changes associated with OPIDN. Few biochemical, pharmacological or neuroanatomical data are available for chickens, although they are highly sensitive to the ataxia associated with OPIDN. Nevertheless, this species is used exclusively to screen chemicals for possible OPIDN activity. Rodents are considered insensitive to OPIDN because of their failure to develop hind-limb paralysis after exposure (Abou-Donia, 1981). Resistance of rodents to OPIDN has been discussed in terms of neuroanatomy (Cavanagh, 1954, Johnson, 1975), pharmacokinetics and metabolism (Abou-Donia, 1983; Hansen, 1983), and qualitative differences in the target enzyme (Johnson, 1975; Hussain and Oloffs, 1979; Soliman *et al.*, 1982). Factors such as serum and brain carboxylesterase levels are other considerations. In reexamining the potential resistance of rodents to OPIDN, experiments from our laboratory have demonstrated that rats and mice exposed to single or multiple doses of OP compounds can develop neuropathological and biochemical changes typical of OPIDN in the absence of ataxia, suggesting that rodents may be viewed as viable animal models to study OPIDN.

II. Neuropathic Distribution

To develop a rodent model of the delayed neuropathy we initially used tri-*ortho*-cresyl phosphate (TOCP), a neuropathic OP compound that first came into notoriety in the 1920s as the culpable agent of the Ginger Jake epidemic, a toxic outbreak that ultimately paralyzed 60,000 victims in the United States during the prohibition era (Burley, 1930; Merritt and Moore, 1930; Smith and Lillie, 1931; Morgan, 1982). TOCP is considered a model OPIDN-producing chemical.

In our first series of experiments Sprague-Dawley (Veronesi and Abou-Donia, 1982), and Long Evans, male (Veronesi, 1984a) rats were pretreated with atropine sulfate and exposed by gavage to a biweekly high dose of TOCP (1160 mg/kg) or to daily low (116 mg/kg) doses. Rats were examined for pathology 2 weeks after exposure to the acute dose and every 6 weeks thereafter.

Histopathological examination of 1-micron epoxy sections indicated that severe spinal cord damage occurred after a single high (1160 mg/kg) dose of TOCP; the effect was localized in the dorsal columns (i.e., fasciculus gracilis) of the upper cervical cord (C2–5) region, an area that houses long sensory nerve fibers (Zemlan *et al.*, 1978). In the lumbar cord, only a scattered distribution of degeneration was found in the ventral–lateral and ventral columns, which house the distal ends of various descending tracts (Zemlan *et al.*, 1979). Microscopically, the degenerative changes consisted of myelin debris, astrocytic proliferation, hyaline bodies, and giant axonal swellings, which by electron microscopy were seen to contain accumulations of fragmented smooth endoplasmic reticulum (i.e., tubulovesicular profiles) and intra-axonal vacuoles (Figs. 1, 2). Such axonal lesions were ultrastructurally identical to those described in humans and other experimental models of OPIDN (Bischoff, 1967; Prineas, 1969; Bouldin and Cavanagh, 1979; Jortner *et al.*, 1990). The PNS at 2 weeks exposure was largely devoid of overt axonal degeneration. Instead, excessive axonal sprouting and other evidence of regeneration and remyelination could be seen. At later stages of intoxication, however, extensive degeneration was noted and consisted of swollen axons, collapsed myelin sheaths, myelin ellipsoids, and fragmented axons. Teased nerve-fiber preparations of the PNS revealed that the proximal length of affected fibers appeared relatively normal, with giant axonal swellings located between normal internodes and degenerated fiber lengths supporting a "dying-back" classification to the neuropathy.

Overt hind-limb dysfunction was not observed in rats until after 12 weeks of exposure to high doses of TOCP. At that time, affected rats developed hind-limb splay, a noticeable "heel-walk" and in some instance, "criss-crossing" of the hind-limbs when lifted by their tail. Such effects have been described in mercury poisoning and are related to severe degeneration of the

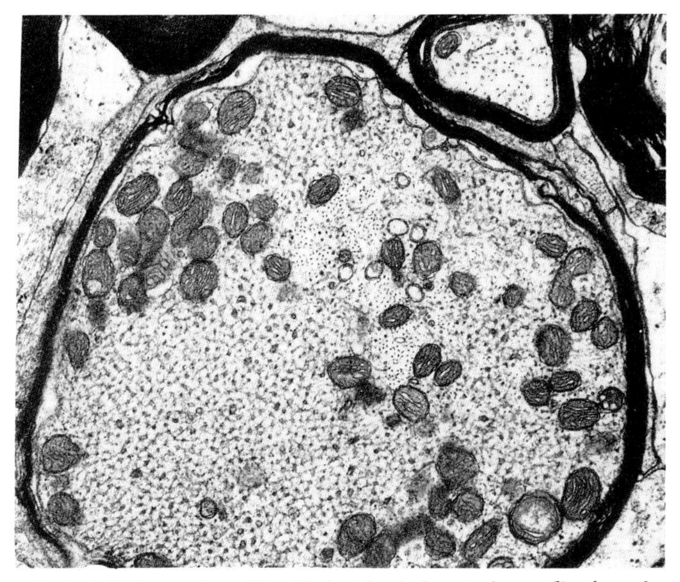

Figure 1 Typical CNS axonal swelling filled with tubulovesicular profiles from dorsal columns of rat dosed with TOCP, 1160 mg/kg, and sampled for electron microscopy 14 days later. × 12,500.

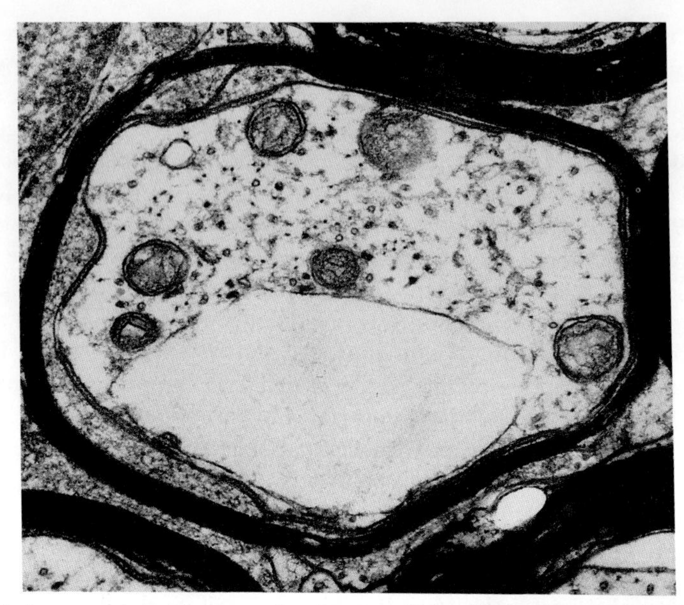

Figure 2 Intraaxonal vacuolation produced a type of axonal lesion characteristic of mipafox intoxication in the rat. Long Evans rat treated with mipafox 15 mg/kg and sacrificed 21 days later for electron microscopy. × 14,000.

dorsal column sensory fibers (Chang, 1980). The results of these experiments established that the rat, while being resistant to the neurological effect of TOCP, showed neuropathic degeneration similar to that of the conventional test species, the chicken.

III. Biochemical Index

OPIDN was next evaluated biochemically in the rat. Neurotoxic (or neuropathy-target) esterase (NTE) is an accepted biochemical marker of OPIDN (Johnson, 1969), although its purification and characterization has not yet been completed. Inhibition of NTE by ≥ 70% shortly after OP exposure generally predicts subsequent pathology in chickens, cats, farm livestock, and humans (Johnson, 1975; Lowndes *et al.*, 1974; Lotti and Johnson, 1980; Soliman *et al.*, 1982). To test this relationship in rodents, Long Evans rats were dosed (po) acutely with various levels of TOCP (i.e., 145–3480 mg/kg), and brain and spinal cord NTE activity was determined 24 and 44 hr later. Two weeks later, similarly treated animals were sacrificed and examined for neuropathic damage.

Severe spinal cord degeneration was observed in 90% of the rats dosed with TOCP (≥835 mg/kg) with only minimal pathology occurring at the lower doses. This dose (i.e., 835 mg/kg) resulted in a 66 and 72% mean inhibition of NTE activity in the brain and spinal cord, respectively (Padilla and Veronesi, 1985) (Table I). The relationship between NTE activity and pathology was further tested in rats treated with mipafox, another model neuropathic OP compound. Mipafox, unlike TOCP, does not require hepatic activation for neurotoxic potency. Again, severe cervical cord pathology was associated with mean NTE inhibition of 67 and 73% in the brain and spinal cord, respectively (Veronesi *et al.*, 1986a). These studies suggested that, for TOCP and mipafox, NTE inhibition was associated with neuropathic damage in rats, thus validating the rat as a biochemical model for OPIDN.

These studies suggested that there may be interspecies differences regarding the time course of NTE inhibition and recovery. For example, in rats dosed with TOCP (1160 mg/kg), brain NTE is depressed by only 25% 20 hr after exposure, whereas in chickens dosed with TOCP (1000 mg/kg), brain NTE is inhibited 90 to 95% 24 hr after exposure and remains depressed 45 to 50% 14 days after exposure (Ohkawa *et al.*, 1980). In rats, given TOCP at a dose (i.e., 3480 mg/kg) that produces 90% inhibition of NTE acutely, NTE activity returns to control values after 14 days. This faster recovery of NTE activity in the rat could be explained either by less *aging* of the inhibited enzyme (Clothier and Johnson, 1980) or by a more rapid resynthesis of NTE (Soliman *et al.*, 1982).

A subsequent study examined the qualitative similarities between chicken and rat brain NTE, by evaluating their *in vitro* sensitivities to inhibition by

TABLE I

The Relationship between NTE Inhibition and Cervical Cord Pathology

	NTE inhibition (%) (44 hr)		
TOCP (mg/kg)	Spinal cord	Brain	Spinal cord damage (2 weeks)
3480	87 + 5.7	89 + 2.4 (5)[a]	90%[b] (10)[a]
2320	91 + 6.8	98 + 1.0 (4)	100% (10)
1160	75 + 10.1	85 + 6.4 (5)	100% (6)
835	72 + 5.3	66 + 4.6 (15)	90% (10)
580	65 + 5.4	57 + 6.1 (12)	15% (13)
290	40 + 6.7	36 + 4.7 (11)	7.5% (13)
145	24 + 4.7	18 + 3.0 (9)	0 (5)
0	0	0 (18)	0 (12)

[a]Sample size.
[b]Percentage of animals showing cervical cord damage ≥ 3.

OP compounds (Novak and Padilla, 1986). These authors concluded that *in vitro* rat and chicken brain NTE were very similar with respect to inhibitor sensitivities, pH sensitivity, and molecular weight. A noted difference was that the specific activity of chicken brain NTE was approximately twice that found in rat brain.

An experiment was designed to manipulate the role of NTE in precipitating OPIDN in the rat (Veronesi and Padilla, 1985). Delayed neuropathy is thought to involve two separate events: (1) the inhibition of NTE activity due to the binding of the OP to the active site of NTE and (2) aging of the NTE–OP complex in which an alkyl substitution of the OP is hydrolyzed (Johnson, 1975; Clothier and Johnson, 1980; Williams, 1983). Both extensive inhibition and aging are steps necessary for neuropathy to develop (see Chapter 16 by Richardson, this volume, for discussion of interactions of OP compounds with NTE). Certain phosphinates, carbamates, and sulphonates can inhibit NTE over the critical 70%, but because they are unable to age they are nonneuropathic. Exposure to such chemicals before treatment with an ageable (i.e., neuropathic) OP compound will theoretically block the active site of NTE and protect against subsequent neuropathy (Johnson and Lauwerys, 1969; Lowndes *et al.*, 1974; Baker *et al.*, 1980; Caroldi *et al.*, 1984). To differentiate between these two events (i.e., enzyme inhibition and aging), rats were exposed first to a nonneuropathic nonaging, protecting agent phenylmethysulfonylfluoride (PMSF) (250 mg/kg, i.p.) and subsequently (4 hr later) to a neuropathic, ageable OP compound, mipafox (15 mg/kg, I.P.). Other animals were treated with both compounds but in the reverse order (i.e., mipafox, 4 hr, PMSF). A time course of brain NTE inhibition and recovery was defined in rats exposed to either PMSF or mipafox. A separate group of PMSF-pretreated rats was exposed to mipafox when brain NTE inhibition was 89%. Conversely, another group of rats, pretreated with mipafox, was dosed with PMSF when NTE inhibition was 90%. A third group of animals was treated with PMSF and exposed to mipafox 14 days later, when NTE activity had recovered to within 10% of control values.

Histopathology indicated severe cervical cord damage in the following percentages of the treated animals: PMSF, 0%; mipafox, 85%; PMSF, 4 hr, mipafox, 0%; mipafox, 4 hr PMSF, 100%; PMSF, 14 days, mipafox, 75%; vehicle controls, 0% (Fig. 3). It should be noted that a moderate amount of cervical cord damage was seen in rats treated with PMSF alone. In summary, these data indicated that PMSF pretreatment protected rats against mipafox-induced neurological damage, but that the timing of administration and the order of presentation were critical for this protection. These experiments demonstrated that the initial events of OPIDN were common among species and underscored the critical role of the inhibited and "aged" NTE for the neuropathic process of OPIDN in the rat as in other susceptible species.

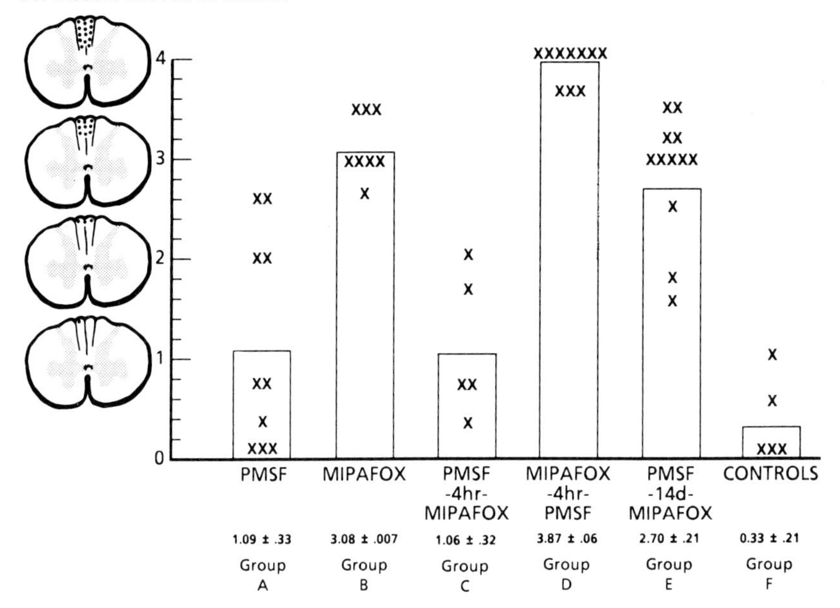

Figure 3 Semiquantification of cervical cord damage is represented in both scattergram and bar graph form and the mean and standard error for each group is noted. Neuropathy was scored 0–4 depending on the amount and location of axonal damage. Cord involvement is depicted diagrammatically along the vertical axis. Reprinted with permission from *Toxicol. Appl. Pharmacol.* 81, 262.

IV. Pharmacological Manipulation

The roles of metabolic activation and detoxification of OP compounds were also examined in rats (Veronesi, 1984b). The mixed function oxidases (MFO) system metabolizes a diverse array of xenobiotics and endogenous compounds. TOCP is activated by the MFO system to the neurotoxic metabolite saligenin *o*-tolyl-phosphate, which is subsequently hydrolyzed and detoxified by the MFO systems (Casida *et al.*, 1961). The effect of metabolic inhibition on the onset of OPIDN in rats was examined by pretreating Long Evans rats with the MFO inhibitor, piperonyl butoxide (PiPB) (50 mg/kg), 1 hr before administering TOCP (1160 mg/kg). The animals were killed after the last of three weekly treatments, and the spinal cord, various peripheral nerves, and the liver were examined microscopically.

Rats treated with PiPB in combination with TOCP showed significantly more damage in both the spinal cord and PNS then did those treated with TOCP alone. PiPB, which inhibits MFO when given in low doses shortly before exposure, is reported to also inhibit Phase II conjugation reactions

(James and Harbison, 1982; Lucier and Matthews, 1971). Since glutathione-dependent reactions (e.g., dealkylation, dearylation, demethylation) play important roles in the deactivation of various OP compounds (Morello *et al.*, 1968; Hollingsworth, 1969; Murphy, 1982), interference with the Phase II reactions presumably allowed the reactive intermediate of TOCP (saligenin-*o*-tolyl cyclic phosphate) a longer serum–tissue residency time to exert its neurotoxic effects. To support this thesis, an earlier experiment (Veronesi and Abou-Donia, 1982) pretreated Sprague-Dawley rats with PiPB for 4 days before TOCP administration. The exposure variables (i.e., multiple PiPB exposures) were designed to *induce* the synthesis of conjugating proteins, thereby improving the efficacy of TOCP deactivation. In these experiments, quantitatively less CNS and PNS degeneration developed in rats treated with PiPB and TOCP compared with TOCP alone. PiPB was used in this experiment to upset the biotransformational balance (i.e., activation and deactivation), an important determinant in the development of TOCP-induced neuropathy (Adou-Donia, 1983; Hansen, 1983). This study demonstrated that pharmacologic manipulation of metabolic pathways could affect both the onset and severity of TOCP neuropathy in rats and suggested that rodent resistance to OPIDN may have a hepatic component.

IV. Triphenyl Phosphite Neuropathy

More recently, the rat model was used to evaluate the neurotoxicity of triphenyl phosphite (TPP), an aryl phosphite structurally similar to TOCP. Like TOCP, this compound was investigated over 50 years ago and found to produce extensor rigidity in cats, in contrast to TOCP, which produced flaccid paralysis in the same species (Lillie and Smith, 1932). In spite of its increasing use in industry and agriculture as an antiplasticizer and as an insecticidal synergist (U.S. EPA, 1986), little research has been done to characterize the neurotoxicity of TPP. Recently, however, a British testing laboratory reported that chickens exposed to TPP developed the ataxia, pathology, and biochemical features of OPIDN (Roberts *et al.*, 1982). Because of the chemical resemblance of TPP to the classic OPIDN compound TOCP, we examined its neurotoxic effects in rats. Animals were exposed to two single doses (2 × 1.0 mg/kg, s.c.) and examined using neuropathological and biochemical criteria, pathognomonic of the delayed neuropathy in rats.

Our results failed to support TPP as an OPIDN-producing chemical for several reasons. Within 7 days after exposure, severe, TPP-induced, neurological dysfunction (hind- and fore-limb paralysis, dorsal flexion, and circling behavior), dissimilar to effects produced by TOCP, were observed (Fig. 4). In addition, spinal cord degeneration initiated in the lateral and ventral columns in a pattern of damage topographically different from the dorsal column

Figure 4 Triphenyl phosphite–treated rat displayed dorsal flexion and a distinct kink in the base of the tail as well as circling behavior and overt hind-limb dysfunction. Such features do not occur in rodent OPIDN.

damage produced in the rat by classic OPIDN compounds, such as TOCP and mipafox. In the TPP-treated animals, anterior horn necrosis in the spinal cord grey matter, damage of the spinal roots and dorsal root ganglion neurons, and degeneration and axonal swellings in the reticular formation were also noted. Last, both the neurological effects and neuropathic damage occurred in conjunction with only marginal (33%) inhibition of NTE, which is subthreshold to predict OPIDN (Veronesi *et al.*, 1986b).

To expand on our thesis that TPP did not produce OPIDN, we next exposed rats to TOCP, TPP, or a combination of both. In these rats, the neurological effects described in TPP treatment occurred within days of exposure, and neuropathic damage occurred in the dorsal, lateral, and ventral columns of the cervical cord. These effects were seen in addition to anterior horn cell damage and brain stem involvement. This final experiment (Veronesi and Dvegsten, 1987) showed that in rats, TPP and TOCP produce two distinct patterns of neuropathy. When TPP is administered to the chicken, the common test model of OPIDN, NTE inhibition (> 90%), ataxia, and spinal cord histological damage similar to OPIDN occurred (Roberts *et al.*, 1982; Carrington and Abou-Donia, 1986, 1988). Closer morphological examination of the histopathology, however, revealed that in addition to degeneration of the spinal cord tracts, TPP-dosed chickens exhibited anterior horn cell (i.e., motor neuron) necrosis and brainstem pathology (Carrington *et al.*, 1988), features not seen in OPIDN. To explain this difference in the neuropathic expression of mammals and avians, Abou-Donia and colleagues have suggested that in chickens, TPP produces both OPIDN and a neuronopathy, the latter being expressed only in mammals (Carrington *et al.*, 1988). However, recent tissue culture studies (Anderson *et al.*, 1991) using primary chromaffin cell cultures have clearly demonstrated that TPP produces ultrastructural and biochemical changes clearly distinct from OPIDN agents.

VI. Mouse Model of OPIDN

We have recently reported, using light and electron microscopy, that mice (CD-1 strain) are also neuropathically sensitive to single doses of TOCP, but that because of high intragroup variability to the neuropathic and NTE response, threshold NTE inhibition and spinal cord pathology can not be correlated. In this model the neuropathic damage is confined to the lateral and ventral column of the cervical cord rather than the dorsal columns, and NTE inhibition never exceeds ≥ 65% inhibition, in spite of increasing doses (Veronesi *et al.*, 1991).

VII. Interspecies Variations in OPIDN

Our data suggest several differences in responsiveness of rodents and chickens to OPIDN-producing agents. First, the exact topography of spinal cord damage seen in rats differs from that seen in hens. In OPIDN, the tracts containing the largest or longest nerve fibers are the most susceptible, regardless of the species, in keeping with the definition of a "dying-back" neuropathy. In humans and cats, the most severely damaged descending nerve fibers are in the pyramidal tracts, which are missing in the bird (Huber and Crosby, 1929). In the hen, the most vulnerable tracts are scattered throughout the lateral, ventral, and dorsal columns. In the rat, the most severely affected tracts appear to be the large-diameter sensory fibers terminating in the upper cervical cord dorsal columns. In the mouse, the most susceptible fibers are the lateral and ventral columns. It has been suggested that, in addition to topographic differences, PNS degeneration precedes CNS damage in the chicken (Cavanagh, 1954; Abou-Donia, 1981). In contrast, the rat expresses a more protracted onset of PNS degeneration (Veronesi, 1984a). The most striking difference in the response of chickens and rats to OP exposure is the latter's retention of hind-limb function in spite of CNS and PNS histopathology. Early studies have reported that even after 24 weeks of daily or intermittent exposure to OP compounds, rats demonstrate only minor motoric problems in spite of evidence of neuropathology (Majno and Karnovsky, 1961). Similarly, mice retain hind-limb function until 240 days of daily dosing with TOCP (Lapadula *et al.*, 1985). This is in contrast to the chicken, which becomes grossly ataxic after minor spinal cord pathology (Prentice and Roberts, 1983). This paradox is an engaging problem neurologically and may involve differences in neural compensatory reactions (i.e., regeneration) and tract sensitivities. In the rat, the dorsal columns, which house the sensory nerves, are almost exclusively damaged by OP compounds, whereas the descending tracts, which contain the small-diameter motor fibers, are relatively spared.

The prominent PNS regeneration seen in rodent OPIDN (Veronesi, 1984a,b) appears to play a key role in protecting the rat from functional debilitation. Although regeneration is often seen as a response to traumatic or experimental nerve injury, the predominance of this event, which would help to reestablish muscle strength and coordination throughout the early stages of TOCP damage may explain, in part, the rat's preservation of hind-limb function until later stages of intoxication. The low level of PNS damage in rodent OPIDN may in itself be sufficient to protect against detectable ataxia, since even in the chicken, when PMSF, administered in the sciatic artery, is used to protect the hen from subsequent DFP-induced PNS damage (but not cord degeneration), only ataxia occurs (Caroldi *et al.*, 1984).

VIII. Summary

This chapter describes our efforts to study OP-induced neurotoxicity in rodent species. To date, our results indicate that if morphological rather than functional endpoints are used, the rat is sensitive to the neuropathological effects of TOCP and mipafox. Although obvious cost and technical advantages exist in using the rat for studies of OPIDN, the major appeal is the extensive literature and data base that exists for this species.

References

Abou-Donia, M. B. (1981). Organophosphorus ester–induced delayed neurotoxicity. *Annu. Rev. Pharmacol. Toxicol.*, **21**, 511–548.

Abou-Donia, M. B. (1983). Toxicokinetics and metabolism of delayed neurotoxic organophosphorus esters. *Neurotoxicology* **4**(1), 113–130.

Anderson, J., Veronesi, B., Jones, K., Lapadula, D. M., and Abou-Donia, M. B. (1991). Triphenyl phosphite induced ultrastructure changes in bovine adrenomedullary chromaffin cells. *Toxic. Appl. Pharm.* (in press).

Baker, T., Lowndes, J. E., Johnson, M. K., and Sandborg, I. C. (1980). The effect of phenylmethanesulfonyl fluoride on delayed organophosphorus neuropathy. *Arch. Toxicol.*, **46**, 305–311.

Bischoff, A. (1967). The ultrastructure of tri-*ortho*-cresyl phosphate poisoning. I. Studies on myelin and axonal alterations in the sciatic nerve. *Acta Neuropathol. (Berl)* **9**, 158–174.

Bouldin, T. W., and Cavanagh, J. B. (1979). Organophosphorus neuropathy. II. A fine structural study of the early stages of axonal degeneration. *Am. J. Pathol.* **94**(2), 253–270.

Burley, B. T. (1930). The 1930 type of polyneuritis. *N. Engl. J. Med.* **202**, 1139–1142.

Caroldi, S., Lotti, M., and Masutti, A. (1984). Intraarterial injection of diisopropylfluro-phosphate or phenylmethanesulphonylfluoride produces unilateral neuropathy or protection, respectively, in hens. *Biochem. Pharmacol.* **33**, 3213–3217.

Carrington, C. D., and Abou-Donia, M. B. (1986). Delayed neurotoxicity of triphenyl phosphite (TPP) in the hen. *Toxicologist* **6**, 194 (Abstr.).

Carrington, C. D., and Abou-Donia, M. B. (1988). Triphenyl phosphate neurotoxicity in the hen: Inhibition of neurotoxic esterase and phophylaxis by phenylmethylsulfonyl fluoride. *Arch. Toxicol.* **62**, 375–380.

Carrington, C. D., Brown, H. R., and Abou-Donia, M. B. (1988). Histopathological assessment of triphenyl phosphite neurotoxicity in the hen. *Neurotoxicology* **9**, 223–234.

Casida, J. E., Eto, M., and Baron, R. L. (1961). Biological activity of tri-*o*-cresyl phosphate metabolite. *Nature* **191**, 1396–1397.

Cavanagh, J. B. (1954). The toxic effects of tri-*ortho*-cresyl phosphate on the nervous system: An experimental study in hens. *J. Neurol. Neurosurg. Psychiat.* **17**, 163–172.

Cavanagh, J. B. (1963). Organophosphorus neurotoxicity, a model "dying-back" process comparable to certain human neurological disorders. *Guy's Hospital Reports*, **17**, 163–172.

Chang, L. (1980). Methyl mercury. *In* "Experimental and Clinical Neurotoxicology." (P. S. Spencer and H. H. Schaumburg, eds.), pp. 508–526. Williams & Wilkins, Baltimore, Maryland.

Clothier, B., and Johnson, M. K. (1980). Reactivation and aging of neurotoxic esterase inhibited by a variety of organophosphorus esters. *Biochem. J.* **187**, 739–747.

Hansen, L. G. (1983). Biotransformation of organophosphorous compounds relative to delayed neurotoxicity. *Neurotoxicology* **4**, 97–111.

Hollingsworth, R. M. (1969). Dealkylation of organophosphorous esters by mouse liver enzymes *in vitro* and *in vivo*. *J. Agric. Food Chem.* **17**, 987–996.

Huber, G. C., and Crosby, E. C. (1929). The nuclei and fiber paths of the avian diencephalon with consideration of telencephalic and certain mesencephalic centers and connections. *J. Comp. Neurol.* **48**, 1–255.

Hussain, M. A., and Oloffs, P. C. (1979). Neurotoxic effects of leptophos (Phosvel) in chicken and rats following chronic low-level feeding. *J. Environ. Sci. Health* **B14**, 367–382.

James, R. C., and Harbison, R. D. (1982). Hepatic glutathione and hepatotoxicity. Effects of cytochrome P-450 complexing compounds SKF 525A L-a acetylmethadol (LAAM), nor-LAAM, and piperonyl butoxide. *Biochem. Pharmacol.* **31**, 1829–1835.

Johnson, M. K. (1969). The delayed neurotoxic effect of some organophosphorus compounds. Identification of the phosphorylation site as an esterase. *Biochem. J.* **114**, 711–717.

Johnson, M. K. (1970). Organophosphorous and other inhibitors of brain "neurotoxic esterase" and the development of delayed neuropathy in hens. *Biochem. J.* **120**, 523–531.

Johnson, M. K. (1975). The delayed neuropathy caused by some organophosphorus esters:

Mechanism and challenge. *CRC Crit. Rev. Toxicol.* 3, 289–316.

Johnson, M. K., and Lauwerys, R. (1969). Protection by some carbamates against the delayed neurotoxic effects of di-isopropylphosphorofluoridate. *Nature (London)* **222**, 1066–1067.

Jortner, B. S., Dyer, K. R., Shell, L. G., and Ehrich, M. (1990). Comparative studies of organophosphorus ester-induced delayed neuropathy (OPIDN) in rats and hens dosed with mipafox. Toxicologist, p. 340. (Abstr.)

Lapadula, D. M., Patton, S. E., Campbell, G. A., and Abou-Donia, M. B. (1985). Characterization of delayed neurotoxicity in the mouse following chronic oral administration of tri-*o*-cresyl phosphate. *Toxicol. Appl. Pharmacol.* **79**, 83–90.

Lillie, R. D., and Smith, M. I. (1932). The histopathology of some neurotoxic phenol esters. *Nat. Inst. Health Bull.* **160**, 54–64.

Lotti, M., and Johnson, M. K. (1980). Neurotoxic esterase in human nervous tissue. *J. Neurochem.* **34**(3), 747–749.

Lowndes, H. E., Baker, T., and Riker, W. F., Jr. (1974). Motor nerve dysfunction in delayed DFP neuropathy. *Eur. J. Pharmacol.* **29**, 66–73.

Lucier, G. W., and Matthews, H. B. (1971). Microsomal rat liver UDP glucoronyltransferase: Effect of piperonyl butoxide and other factors on enzyme activity. *Arch. Biochem. Biophys.* **145**, 520–530.

Majno, G., and Karnovsky, M. L. (1961). A biochemical and morphologic study of myelination and demyelination. III. Effect of an organophosphorous compound (mipafox) on the biosynthesis of lipid by nervous tissue of rats and hens. *J. Neurochem.* **8**, 1–16.

Merritt, H. H., and Moore, M. (1930). Peripheral neuritis associated with ginger extract ingestion. *N. Engl. J. Med.* **202**, 4–12.

Morello, A., Vardanis, A., and Spencer, E. Y. (1968). Mechanisms of detoxification of some organophosphorous compounds: The role of glutathione-dependent demethylation. *Can. J. Biochem.* **45**, 885–892.

Morgan, J. P. (1982). The Jamaica ginger paralysis. *J.A.M.A.* **248**, 1864–1867.

Murphy, S. D. (1982). Toxicity and hepatic metabolism of organophosphate insecticides in developing rats. *Environ. Fact. Hum. Growth. Dev.* **11**, 125–136.

Novak, R., and Padilla, S. (1986). An *in vitro* comparison of rat and chicken brain neurotoxic esterase. *Fundam. Appl. Toxicol.* **6**, 464–471.

Ohkawa, H., Oshita, H., and Miyamoto, J. (1980). Comparison of inhibitory activity of various organophosphorous compounds against acetylcholinesterase and neurotoxic esterase of hens with respect to delayed neurotoxicity. *Biochem. Pharmacol.* **29**, 2721–2727.

Padilla, S., and Veronesi, B. (1985). The relationship between neurological damage and neurotoxic esterase inhibition in rats acutely exposed to tri-*ortho*-cresyl phosphate. *Toxic. Appl. Pharmacol.* **78**, 78–87.

Prentice, D. E., and Roberts, N. L. (1983). Acute delayed neurotoxicity in hens dosed with tri-*ortho*-cresyl phosphate (TOCP): Correlation between clinical ataxia and neuropathological findings. *Neurotoxicology* **4**, 271–283.

Prineas, J. (1969). The pathogenesis of dying-back polyneuropathies. Part I. An ultrastructural study of experimental TOCP intoxication in the cat. *J. Neuropathol. Exp. Neurol.* **28**, 571–597.

Roberts, N. L., Prentice, D. E., and Cooke, L. (1982). "Screening Test for Neurotoxicity of Triphenyl Phosphite in the Chicken Following Oral Exposure." Huntingdon Research Centre, Huntingdon, Cambridgeshire, England.

Smith, M. I., and Lillie, R. R. (1931) The histopathology of tri-*ortho*-cresyl phosphate poisoning. The etiology of so-called ginger paralysis (third report). *Arch. Neurol. Psychiatry* **26**, 976–992.

Soliman, S. A., Linder, R., Farmer, J., and Curley, A. (1982). Species susceptibility of delayed toxic neuropathy in relation to *in vivo* inhibition of neurotoxic esterase by neurotoxic organophosphorous esters. *J. Toxicol. Environ. Health* **9**, 189–197.

U.S. Environmental Protection Agency. (1986). "Chemical hazard information profile (CHIP) on triphenyl phosphite." U.S. Environmental Agency. Office of Pesticides and Toxic Substances, Washington, D.C.

Veronesi, B. (1984a). A rodent-model of organophosphorous-induced delayed neuropathy: Distribution of central (spinal cord) and peripheral nerve damage. *Neuropathol. Appl. Neurobiol.* **10**(6), 357–368.

Veronesi, B. (1984b). The effect of metabolic inhibition with piperonyl butoxide on rodent sensitivity to tri-*ortho*-cresyl phosphate. *Exp. Neurol.* **85**, 651–660.

Veronesi, B., and Abou-Donia, M. D. (1982). Central and peripheral neuropathology induced in rats by tri-*ortho*-cresyl phosphate. *Vet. Hum. Toxicol.* **24**, 222. (Abstr.).

Veronesi, B., and Dvergsten, C. (1987). Triphenyl phosphite neuropathy differs from organophosphorous-induced delayed neuropathy in rats. *Neuropathol. Appl. Neurobiol.* **13**, 193–208.

Veronesi, B., and Padilla, S. (1985). Phenylmethylsufonyl fluoride protects rats from mipafox-induced delayed neuropathy. *Toxicol. Appl. Pharmacol.* **81**, 258–264.

Veronesi, B., Padilla, S., and Lylerly, D. (1986a). Biochemical and neuropathological correlates of mipafox-induced neuropathy in rats. *Neurotoxicology* **7**, 207–216.

Veronesi, B., Padilla, S., and Newland, D. (1986b). Biochemical and neuropathological assessment of triphenyl phosphite in rats. *Toxicol. Appl. Pharmacol.* **83**, 203–210.

Veronesi, B., Padilla, S., Blackmon, K., and Pope, C. (1991). A murine model of OPIDN: Neuropathic and biochemical description. *Toxicol. Appl. Pharmacol.* **107**, 311–324.

Williams, D. G. (1983). Intramolecular group transfer is a characteristic of neurotoxic esterase and is independent of the tissue source of the enzyme. *Biochem. J.* **209**, 817–829.

Zemlan, F. P., Leonard, C. M., Kow, L. M., and Pfaff, D. W. (1978). Ascending tracts of the lateral columns of the rat spinal cord: A study using the silver impregnation and horseradish peroxidase techniques. *Exp. Neurol.* **62**, 298–334.

Zemlan, F. P., Kow, L. M., Morrel, J. I., *et al.* (1979). Descending tracts of the lateral columns of the rat spinal cord: A study using the horseradish perioxidase and silver impregnation techniques. *J. Anat.* **123**, 489–510.

19

Immunotoxicity of Organophosphorus Compounds

Stephen B. Pruett
Department of Biological Sciences
Mississippi State University
Mississippi State, Mississippi

I. Overview of the Immune System

The immune system is a complex, interactive array of cells and molecules that acts to protect the body from foreign materials including microbes and neoplastic cells. The importance of this system is illustrated by the increased incidence of infectious disease and cancer in individuals with various types of immune dysfunctions. Substantial dysfunction of the type associated with severe combined immune deficiency (SCID, a congenital condition) or acquired immune deficiency syndrome (AIDS) invariably leads to lethal infection or neoplasia (Bortin and Rimm, 1977; Berkelman *et al.*, 1989). Mild to moderate immune dysfunction is associated with less severe, but still potentially life-threatening, consequences (Allen, 1976; Penn, 1988). Causes of this type of dysfunction include some of the drugs used to treat graft rejection and cancer (Penn, 1988), splenectomy (Wara, 1981), congenital conditions affecting individual immune defense mechanisms (Sneller and Strober, 1990), and some chemical toxicants (Luster *et al.*, 1988).

Organophosphates: Chemistry, Fate, and Effects

The protective mechanisms of the immune system can be classified in two broad categories, innate immunity and acquired immunity (for a general discussion of mechanisms of immunity, see Roitt *et al.*, 1985). Innate immunity requires little, if any, induction time; it is nonspecific; and it does not respond more effectively to second or subsequent encounters with a particular microbe than to the first encounter. The simplest of the innate defense mechanisms are physical or chemical barriers such as skin, mucous membranes, and stomach acidity. Internal innate defenses include phagocytic cells (e.g., monocytes, macrophages, and neutrophils) and natural killer cells. The former constitute a critical defense against bacteria, viruses, and fungi. The latter are lymphoid cells that are able to bind and kill susceptible tumor cells. Molecular systems involved in innate immunity include the complement system, interferons, acute phase proteins, and cytokines such as interleukins 1 and 6 and tumor necrosis factor. All of these cellular and molecular mechanisms are either present constitutively or rapidly induced by microbes or neoplasia and act to control or eliminate infection and cancer.

Because many microbes are able to circumvent one or more of the innate defense mechanisms, other defenses are needed. When microbes or neoplastic cells grow to sufficient numbers, their molecular components stimulate an acquired immune response. Unlike innate immunity, acquired immunity must be induced and is highly specific for the inducing material (antigen). In addition, second and subsequent encounters with this antigen elicit a more rapid and vigorous response than does the first encounter. An acquired immune response begins when the antigen is internalized and undergoes limited cleavage in an antigen-processing and -presenting cell (macrophages, dendritic cells, and B lymphocytes can perform this function) (Fig.1). Antigen fragments are then noncovalently associated with a major histocompatibility complex (MHC) protein and transported to the cell surface. Among the millions of lymphocytes circulating through the blood and lymphatics, those few that bear receptors complementary to this particular antigen–MHC complex will bind to the antigen-presenting cell, initiating a complex series of cellular and molecular interactions leading to the delivery of activation signals to antigen-specific B and T lymphocytes. This causes proliferation and differentiation to produce plasma cells that secrete specific antibodies and cytotoxic T cells, which are able to lyse antigen-bearing target cells. These effector systems mediate humoral and cellular immunity, respectively, and cause elimination of antigens by a variety of mechanisms. Cytokines and antibodies produced during acquired immune responses may function in part by enhancing the effectiveness of innate defenses. For example, antibodies specifically bind the antigen that stimulated their production and activate the complement system, which can lyse some bacteria and contribute to initiation of inflammation.

In summary, the immune system includes a number of different cell types at several anatomical locations. Cellular processes involved in immunity

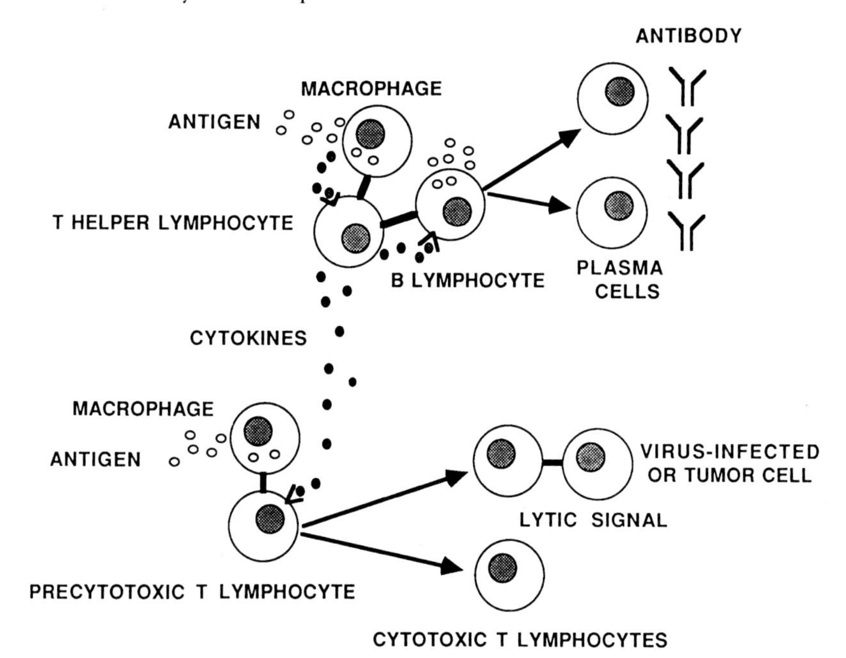

Figure 1 An acquired immune response begins when a foreign substance (antigen) is internalized by an antigen processing/presenting cell (illustrated here as a macrophage). Several types of cells and several molecular signals (cytokines) are involved in the generation of effector cells or molecules (such as cytotoxic T lymphocytes and antibodies), which act to localize and eliminate the foreign material. This sequence of events is described in more detail in the text.

include cellular signaling (including requisite second-messenger functions), cellular proliferation and differentiation, and endocytosis and secretion. In addition, immunity depends on intact barrier functions of the skin and mucous membranes and a number of other physiological functions not always considered part of the immune system (e.g., secretion of HCL by the stomach). It is not surprising that such a diverse system can be affected by many xenobiotics.

II. Overview of Immunotoxicology

The systematic use of animal studies to identify agents that may cause immunological problems is a relatively recent development (Luster *et al.*, 1988). Possible effects of drugs and chemicals on the immune system include suppression of immunity, initiation of autoimmunity (an acquired immune response to self-antigen), and generation of hypersensitivity (allergic) responses. Animal studies are particularly important in the identification of immunosuppressive agents, since exposure to a drug or chemical could easily be overlooked as a possible contributing factor in cases of infection or neoplasia in humans. Furthermore, the use of human epidemiological data to identify potentially immunotoxic agents is complicated by the fact that the adverse effects of immunosuppression are dependent not only on which immune function is affected and the degree of suppression, but also on the simultaneous occurrence of a challenge (exposure to microbes or appearance of neoplastic cells).

Several agents known to be immunosuppressive to humans are also immunosuppressive to mice, and the extensive data regarding the immune system of the mouse indicate that it is comparable in most important respects to the human immune system (Luster *et al.*, 1988). The use of a two-tiered system of validated immunological assays for immunotoxicity evaluation in the mouse has been developed under the sponsorship of the National Toxicology Program (Luster *et al.*, 1988) (Table I). A key feature of this system is that it includes assays for most major innate and acquired immune functions as well as host-resistance models to evaluate the composite, integrated performance of several immune functions. Because a different set of immune effec-

TABLE I

National Toxicology Program's Recommended Tests for Detecting Immune Alterations following Chemical or Drug Exposure in Rodents

Immunological parameter	Procedures
Screen (Tier I)	
Immunopathology	Hematology; weights: body, thymus, spleen, kidney, liver; cellularity: spleen; histology: spleen, thymus, lymph node.
Humoral immunity	Enumerate IgM antibody-forming cells following immunization with sheep red blood cells. Response to B-cell mitogen (lipopolysaccharide).
Cellular immunity	Mixed leukocyte response. Response to T-cell mitogen (concanavalin A).
Innate immunity	Natural killer (NK) cell activity.
Comprehensive (Tier II)	
Immunopathology	Quantitation of T cells and B cells.
Humoral immunity	Enumerate IgG antibody-forming cells.
Cellular immunity	Cytolytic T cell response. Delayed hypersensitivity response (DHR)
Innate immunity	Macrophage quantitation and functional assays.
Host resistance	Syngeneic tumors: PYB6 sarcoma, B16F10 melanoma. Bacterial models: *Listeria monocytogenes, Streptococcus* species. Viral models: Influenza Parasite models: *Plasmodium yoelii*

[a] Adapted from Luster *et al.* (1988).

tor mechanisms is involved in the response to each type of microbe and to neoplasia, several disease models are used (Luster *et al.*, 1988). This type of comprehensive evaluation is important, because some immune functions may be unaffected or even enhanced by agents that suppress other immune functions (Holsapple *et al.*, 1988).

The availability of validated assays for immune functions known to be important in host protection is also useful in investigations of the cellular targets and mechanisms of action of immunotoxic agents. Initial screening assays in such studies are generally done by administering the putative immunotoxic agent to mice and assessing particular immune functions in *ex vivo* assays. Some of these assays can be used to investigate the cellular target(s) of the immunotoxic agent. This is done by separating T cells, B cells, and macrophages from treated and control mice and reconstituting the three cell types in various combinations. Because acquired responses require all three cell types (Fig. 1), elimination or dysfunction of any of the three cell types caused by *in vivo* administration of an organophosphorus (OP) compound will suppress the response (Rodgers *et al.*, 1987; Dooley and Holsapple, 1988). In addition, some techniques are good representations of humoral or cellular acquired immune responses and can be induced and assayed entirely *in vitro* (Devens *et al.*, 1985; Luster *et al.*, 1988). This is useful in mechanistic studies because it allows assessment of the direct effects of putative immunotoxic agents on cells or molecules of the immune system in the absence of indirect effects that may occur *in vivo*. *In vitro* metabolic activation systems can be incorporated for assessment of agents that are not immunotoxic unless metabolically activated (White and Holsapple, 1984; Rodgers *et al.*, 1985a).

III. Immunotoxicity of Organophosphorus Compounds

A. Effects of *in Vivo* Administration

The most consistently reported OP-induced immunosuppression is associated with acute administration of neurotoxic dosages. All compounds tested by two laboratories (parathion, malathion, O,O-dimethyl-O-2,2-dichlorovinyl phosphate (DDVP), sarin, tabun, and soman) significantly inhibited the generation of antibody-forming cells (plasma cells) when administered at neurotoxic dosages (Casale *et al.*, 1983, 1984; Clement, 1985). In most cases, the immunosuppressive dosage killed some of the mice, and dose–response studies were not done. If further studies demonstrate that immunosuppression occurs at neurotoxic, but consistently nonlethal dosages, these findings could be relevant in the management of cases of human OP poisoning.

In contrast, subchronic oral administration of leptophos at dosages sufficient to produce significant inhibition of acetylcholinesterase activity, but not sufficient to produce neurotoxic signs, did not suppress generation of antibody-forming cells (Koller *et al.*, 1976). Similarly, subchronic oral administration of malathion at 0.1 LD_{50} per day did not affect the generation of antibody-forming cells or cytotoxic T lymphocytes (Rodgers *et al.*, 1986). However, Fan *et al.* (1978) noted suppressed serum antibody levels, suppressed host resistance to bacterial challenge, decreased thymus weight, and suppressed response to lymphocyte mitogens (an indicator of proliferative capacity) after chronic administration of methylparathion at 3.0 mg/kg/day. Desi and co-workers (1976) observed similar effects in rabbits that received malathion (100 mg/kg/day) or DDVP (2.5 mg/kg/day) in the diet for several months.

The most thoroughly studied OP compound with regard to immunotoxicity is O,O,*S*-trimethylphosphorthioate (O,O,*S*-TMP). This compound is a contaminant of a number of commercial insecticide preparations, and it may be the only OP compound for which there is convincing evidence of selective immunotoxicity. Devens and colleagues (1985) have shown that acute administration of O,O,*S*-TMP suppresses humoral and cellular immunity when administered orally at 10 mg/kg. This dosage does not decrease body weight, cause histopathological changes, inhibit serum acetylcholinesterase, or cause neurotoxic signs.

Because of the relatively small number of OP compounds studied, as well as differences in experimental animals, immunological assays, schedules of administration, and dosages, general conclusions regarding the immunotoxicity of sub-neurotoxic levels of OP compounds cannot be reached at present. It should also be noted that most of the compounds that did not affect the immune functions examined have not been comprehensively tested. Thus, it is possible that immunosuppression of important immunological functions has been overlooked.

B. Effects of *in Vitro* Exposure

Several immune functions are affected by *in vitro* exposure to OP compounds. Since most OP insecticides are rapidly metabolically activated *in vivo*, the issue of metabolic activation is important in such studies. Some investigators have incorporated an *in vitro* activation system (isolated hepatocytes or S9 fraction) (White and Holsapple, 1984; Yang *et al.*, 1986), whereas others have added the active metabolite directly to cultures (Pruett and Chambers, 1988). It should be noted, however, that spleen cells are able to metabolize some xenobiotics (White and Holsapple, 1984). Therefore, effects reported in studies that do not include a metabolic activation procedure may be caused by metabolites produced *in situ*. It is also possible that the parent compound acts directly and that metabolic activation is not required. This seems to be the case with malathion, which suppresses lymphocyte activation by antigen or mitogens only if it has not been exposed to a metabolic activation system (Rodgers *et al.*, 1985a; Rodgers and Ellefson, 1990). In contrast, *O,O,S*-TMP suppresses the generation of cytotoxic T lymphocytes *in vitro* only if it is first metabolically activated (Rodgers *et al.*, 1985a).

Immune functions that can be altered by exposure to OP compounds *in vitro* include antigen processing and presentation, mitogen-induced lymphocyte proliferation and cytokine production, the generation and function of cytotoxic T lymphocytes, production of hydrogen peroxide by macrophages, and the activity of the complement system (Esa *et al.*, 1988; Rodgers *et al.*, 1985a; Rodgers *et al.*, 1985b; Pruett and Chambers, 1988; Rodgers *et al.*, 1987; Casale *et al.*, 1989; Rodgers and Ellefson, 1990). Most of these effects occur at OP concentrations that could theoretically be obtained *in vivo*. We observed significant suppression of the generation of antibody-forming cells in Mishell-Dutton cultures treated with parathion or paraoxon, but not methyl parathion or diazinon, over a broad range of concentrations (Figs. 2, 3). The apparent enhancement noted with methylparathion is not unprecedented. We have previously reported enhancement of lymphocyte proliferation by phenyl phosorothioates (Pruett *et al.*, 1989) that is related to the sulfur moieties in these compounds.

As with *in vivo* administration, none of the immune functions tested was affected *in vitro* by all OP compounds. The most consistently affected functions involve macrophages. For example, several OP compounds used in flame retardant and lubricant preparations are potent inhibitors of antigen processing and presentation (Esa *et al.*, 1988). This effect was also noted with *O,O,S*-TMP following *in vivo* administration (Rodgers *et al.*, 1985b). Mouse macrophages exposed to metabolized malathion are able to produce more H_2O_2 than control macrophages following an appropriate triggering stimulus (Rodgers and Ellefson, 1990). Generally, increased production of H_2O_2 is characteristic of activated macrophages, which are more potent antimicrobial and antitumor effectors than are unactivated macrophages (Adams and Hamilton, 1987). Increased thiol production is also an indicator of macrophage activation (Watanabe and Bannai, 1987), and thiol production is increased by exposure of macrophages to OP compounds *in vitro* (Fig. 4). Since activated macrophages can suppress the generation of acquired immune responses (Lee *et al.*, 1985), it is not surprising that suppressive macrophages have been isolated from *O,O,S*-TMP–treated mice (Rodgers *et al.*, 1987). The impact of these effects on host resistance to microbes has not been examined. Interestingly, treatment of mice with known macrophage activators enhances resistance to *Salmonella typhimurium* infections, even though generation of acquired immunity is suppressed (Lee *et al.*, 1985). It is likely that the enhanced innate antimicrobial function of activated macrophages compensates for the decreased generation of acquired immunity in these mice. Whether this occurs also in mice exposed to OP compounds and whether it similarly

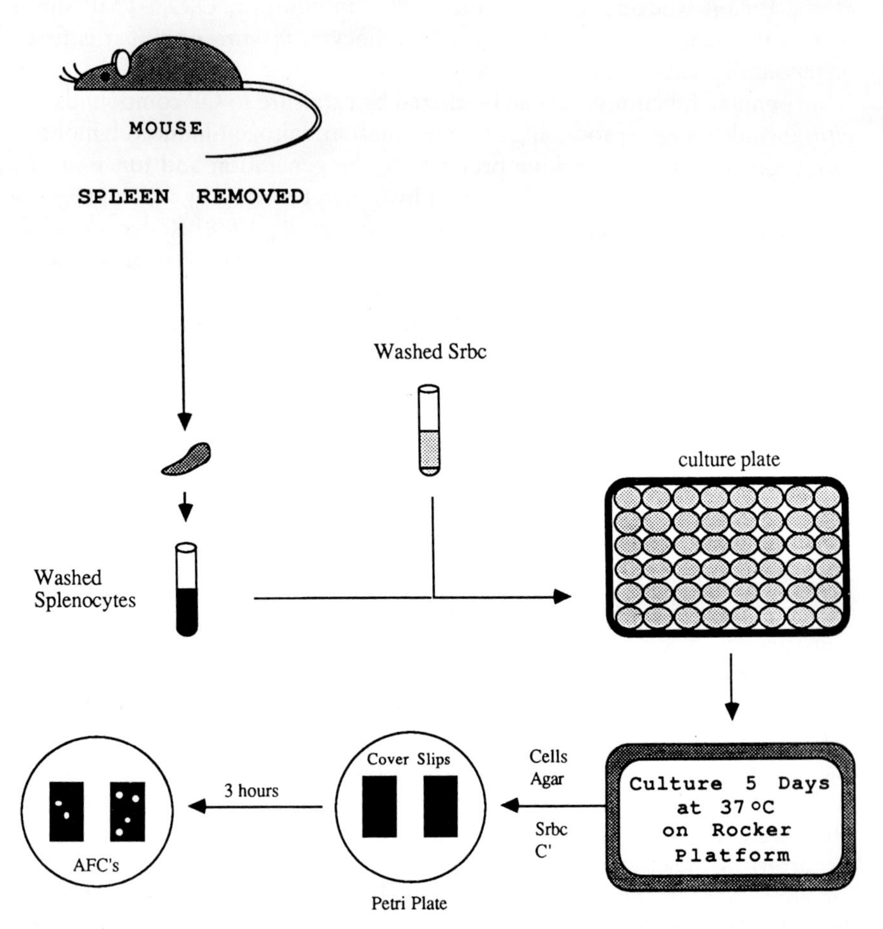

Figure 2 Experimental design of the Mishell Dutton assay for *in vitro* generation and enumeration of antibody-forming cells (AFC). Antibody-forming cells are induced by antigen (sheep red blood cells, S_{rbc}) and measured by the plaque assay. In this assay, antibodies from the AFCs bind to S_{rbc} and are lysed by complement (c′), producing a clear zone in the opaque lawn of S_{rbc}. This method is especially useful for assessing direct effects of immunotoxic agents on cells or molecules involved in humoral immunity. This can be done by including the agent in the culture, or by exposing the splenocytes to the agent *in vitro* and washing it away before initiating cultures. This method can also be used following *in vivo* exposure to immunotoxic agents, to determine whether any effects on the immune system persist when splenocytes are removed from the host and stimulated in culture.

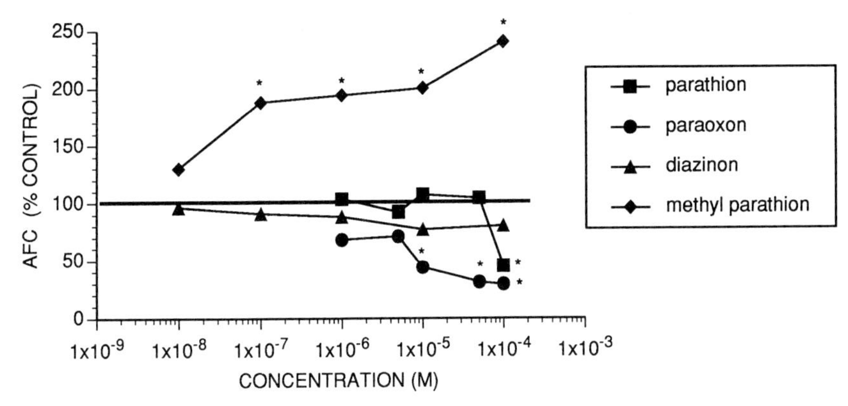

Figure 3 Direct effects of selected OP compounds on the generation of antibody-forming cells in Mishell Dutton cultures. The OP compounds were dissolved in ethanol and added to Mishell Dutton cultures (see Fig. 2) at the indicated concentrations. Response was defined as 100% for vehicle controls, and results shown are based on comparison to these controls. Results are means of triplicate cultures in a single experiment and are representative of at least two experiments, except for diazinon which was only tested once. The assay was performed as described by Luster *et al.*, (1988). Values for AFC/10^6 splenocytes ranged from 1048 to 2009 in the experiments shown. This is within the range typically reported for this assay (Luster *et al.*, 1988). Values that differ significantly from vehicle controls ($P < 0.05$ by Dunnett's test) are indicated by *.

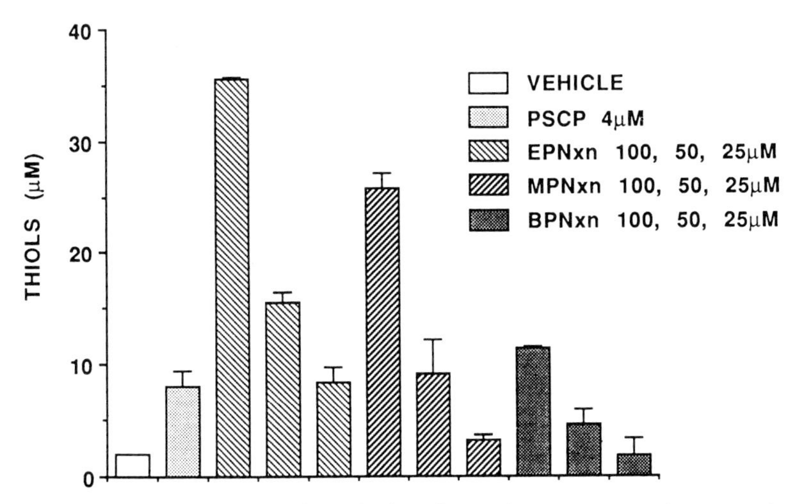

Figure 4 Effects of OP compounds on thiol production by mouse macrophages in culture. Protease peptone–elicited mouse macrophages were cultured at 3×10^6 cells/ml in RPMI 1640 medium with 5% fetal bovine serum. OP compounds were added at the indicated concentrations, and thiol concentration was determined after 20 hr in culture using our modification of the DTNB method (Pruett and Kiel, 1988). PSCP, phenyl saligenin cyclic phosphate; EPN_{XN}, the oxon form of EPN; MPN_{XN} and BPN_{XN}, the methyl and butyl analogs of EPN_{XN}. Results shown are means ± standard error for triplicate samples. In other experiments (not shown), polymyxin B was used to competitively inhibit any LPS contamination. This did not affect the results, indicating that stimulation by endotoxin is not responsible for these results. Furthermore, no endotoxin was detected in OP dilutions used in these studies by the Limulus assay.

affects resistance to other microbes would be revealed by comprehensive immunotoxicological evaluation of selected OP compounds. Such studies would also permit more definitive conclusions regarding the biological significance of the immunological effects of *in vitro* exposure to OP compounds.

IV. Cellular Targets and Mechanisms of Action

A. Mechanisms of Action

As noted by Casale *et al.* (1983), three classes of mechanisms might explain the immunosuppression caused by neurotoxic dosages of OP compounds. The compounds or their metabolites might act directly on cells of the immune system. Alternatively, excessive cholinergic stimulation, which is the prominent feature of OP-induced neurotoxicity, could also cause immunotoxicity. Although there are reports of cholinergic receptors on lymphocytes, the reported binding affinity of these receptors is generally much less than that of cholinergic receptors in other tissues, and some investigators have not detected receptors (Strom *et al.*, 1981; Costa *et al.*, 1988; Maslinski, 1989; Wazer and Rotrosen, 1984; Maloteaux *et al.*, 1982). The basis for these conflicting reports is unclear, but the possibility that cholinergic agonists or antagonists might affect lymphocyte or macrophage function cannot be excluded (Richman and Arnason, 1979; Strom *et al.*, 1981). Finally, it is possible that the immunosuppressive effects of high dosages of OP compounds are the result of indirect mechanisms such as the stress response or hypothermia caused by neurotoxic dosages (Clement, 1985).

Parathion and its active metabolite paraoxon at concentrations equal to or greater than obtainable *in vivo* significantly decrease the generation of antibody-forming cells in *in vitro* cultures (Fig. 3). This suggests that the suppression of antibody-forming cell generation *in vivo*, which has been reported following administration of a single neurotoxic dose of parathion (Casale *et al.*, 1983, 1984), may be caused by direct action on cells of the immune system. In contrast, we found no evidence of suppression of antibody-forming cell generation in cultures exposed *in vitro* to the cholinergic agonist, carbachol (Fig. 5). The enhancement of the generation of antibody-forming cells noted in some of these experiments could be mediated by cholinergic receptors on lymphocytes or macrophages. Interestingly, we also noted enhancement by methylparathion (Fig. 3), and some OP compounds can act directly on cholinergic receptors (see Chapter 13 in this volume). In any case, there is no evidence of immunosuppression by cholinergic agonists, and these results do not support direct cholinergic stimulation of cells of the immune system as a mechanism of OP-induced immunosuppression. Studies are in progress to simultaneously assess the *in vitro* and *in vivo* effects of parathion and paraoxon as well as selected cholinergic agonists and an-

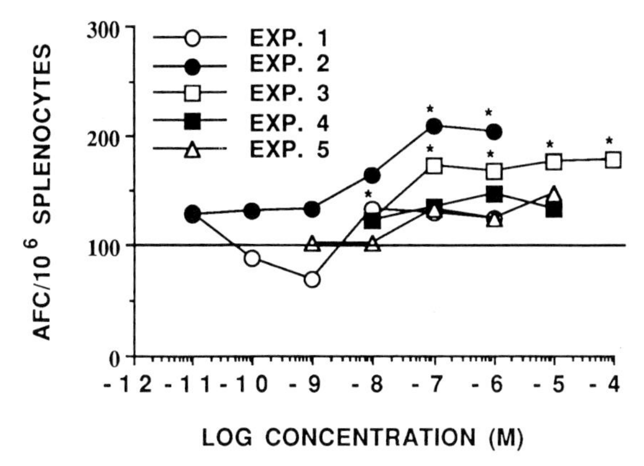

Figure 5 Effect of the cholinergic agonist carbachol on generation of antibody-forming cells (AFC) in Mishell Dutton cultures. Vehicle (culture medium) controls are defined as 100%, and all other values are in comparison to these controls. AFC/10^6 splenocytes for vehicle controls ranged from 1128 to 1498 in the five experiments shown. This is within the range of values normally reported for this assay (Luster *et al.*, 1988). The values shown are means for triplicate cultures. Similar patterns were noted in multiple experiments with bethanecol and nicotine (data not shown). Values significantly different from vehicle controls ($P < 0.05$ by Dunnett's test) are indicated by *.

tagonists. These studies should clarify the roles of direct action of OP compounds and direct action of cholinergic stimulation in OP-induced immunosuppression. Further studies regarding the identity of indirect effects and their possible involvement in immunosuppression by OP compounds (Casale *et al.*, 1983) will be of interest.

In contrast with immunosuppression by neurotoxic doses of OP compounds, there is no indication that indirect effects are responsible for the action of low dosages. In fact, direct effects on cells of the immune system have been demonstrated for *O,O,S*-TMP, which is immunotoxic at low dosages (Rodgers *et al.*, 1987). The cellular targets and possible mechanisms of action of this and other OP compounds are considered in the following sections.

B. Cellular Targets

As noted in sections III, A and III, B, many of the effects of *in vitro* exposure to OP compounds seem to involve macrophages; however, few studies have been specifically designed to examine the cellular targets of immunotoxic OP compounds. Rodgers *et al.* (1987) have shown that treatment of mice with an immunotoxic dose of *O,O,S*-TMP (which is not neurotoxic) generates a population of macrophages that will not support mitogen- or antigen-driven lymphocyte proliferation. This was determined by isolating T cells, B cells, and macrophages from control and treated mice and mixing them in various combinations to reconstitute the necessary mixture of all three cell types. When T cells or B cells obtained from treated mice were mixed with macrophages from control mice, the immune response was decreased only slightly. However, when macrophages from treated mice were mixed with T cells and B cells from control mice, the immune response was substantially suppressed. Direct assessment of antigen processing and presentation by macrophages from mice treated with O,O,*S*-TMP indicates impairment of this function (Rodgers *et al.*, 1985b). Macrophages from treated mice are not dysfunctional merely in terms of antigen processing and presentation, but also actively suppress responses by lymphocytes and macrophages from control mice (Rodgers *et al.*, 1987). As expected, these macrophages have the phenotypic characteristics of highly activated macrophages (Rodgers *et al.*, 1985b).

In vitro exposure of human monocytes to triphenyl phosphate and several related compounds decreases their antigen-processing and -presenting functions, and this seems to account for the ability of these compounds to suppress antigen-driven lymphocyte proliferation *in vitro* (Esa *et al.*, 1988). The effects of *in vivo* administration of these compounds on the immune system have not been examined.

In vitro exposure of mouse splenocytes (T cells, B cells, and macrophages) to metabolized malathion does not affect the lymphocyte response to mitogens, but increases production of H_2O_2 by splenic macrophages (Rodgers and Ellefson, 1990). *In vivo* administration of malathion does not suppress humoral or cellular immunity (Rodgers *et al.*, 1986). This demonstrates that activation of macrophages by OP compounds *in vitro* (as indicated by increased production of H_2O_2) does not necessarily correlate with immunosuppression *in vivo*. It should be noted, however, that macrophage activation proceeds along a continuum, and it is likely that only the most highly activated macrophages suppress lymphocyte proliferation (Adams and Hamilton, 1987). Lower levels of activation may be sufficient to increase the ability of macrophages to produce H_2O_2 but insufficient to cause immunosuppression. Analysis of the activation status of macrophages treated *in vitro* with metabolically activated malathion and macrophages obtained from malathion-treated mice are needed to clarify this matter.

Although there is no direct evidence that T or B lymphocytes are the predominant targets of any immunosuppressive OP compound, only a few compounds have been examined; it remains possible that the primary target for some OP compounds will be a cell type other than macrophages.

C. Molecular Targets and Mechanisms

The molecular targets that mediate the effects of OP compounds on macrophages are not known. Two reports indicate that the nonspecific esterase activity of monocytes (antigen-processing and -presenting cells in the blood that can differentiate into macrophages) is decreased in workers exposed to OP compounds (Wysocki *et al.*, 1987; Mandel *et al.*, 1989). However, the normal physiological function of this esterase is not known, and the functional status of these cells was not examined. The observation that *in vivo* or *in vitro* exposure to OP compounds can activate macrophages (Rodgers *et al.*, 1987, Fig. 4), suggests transmembrane signaling or second-messenger components as possible sites of action. In particular, it is possible that OP compounds could phosphorylate some of the same proteins phosphorylated by protein kinases involved in cellular activation.

In view of the ability of many OP compounds to phosphorylate and inhibit serine proteases, any such enzymes involved in immune functions may be considered potential target molecules. The most obvious component of the immune system in this regard is the complement system. Activation of this important defense system involves the action of serine proteases in a cascade of proteolytic cleavages. Complement activation is inhibited *in vitro* by dichlorvos at concentrations of 1 or 3 m*M*, but not at lower concentrations, which would be nonlethal *in vivo* (Casale *et al.*, 1989). In similar experiments, we noted no inhibition of complement activity *in vitro* by paraoxon, methyl paraoxon, butyl paraoxon, or phenyl saligenin cyclic phosphate at concentrations up to 0.1 m*M* (unpublished results, 1988). These results do not exclude the possibility that some OP compounds or their metabolites may inhibit the complement system, but complement is not a major target of the compounds tested.

O,O,S-TMP inhibits an early post recognition event in the process leading to the lysis of target cells by cytotoxic T lymphocytes and natural killer cells (Rodgers *et al.*, 1988). The molecules involved in this previously unrecognized step are not known. This study illustrates the utility of OP compounds as probes in mechanistic studies of immunological functions. However, the high concentration of *O,O,S*-TMP required to suppress cytolytic activity (50 μ*M*) indicates that this is not an important mode of action of this compound *in vivo*.

A trypsin-like protease, which is involved in the early events leading to the proliferation of T lymphocytes (Wong *et al.*, 1987), is another potential target of OP compounds. This enzyme is inhibited by the relatively specific trypsin inhibitor, N-α-*p*-tosyl-L-lysine chloromethyl ketone (TLCK), and by less-specific inhibitors of serine proteases such as aprotinin, but not by soybean trypsin inhibitor. Inhibition of T-cell proliferation and cytokine production and responsiveness by *in vitro* exposure to OP compounds has been reported and might be explained by inhibition of this protease (Pruett and Chambers, 1988). To test this hypothesis we compared the inhibition of T-lymphocyte proliferation, trypsin activity, and acetylcholinesterase (AChE) activity by several structurally related OP compounds. The oxon forms of the OP compounds were used to avoid the need for a metabolic activation system.

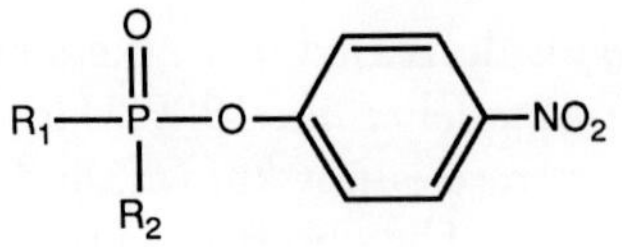

Basic Structure of All Compounds except PSCP and SSCP

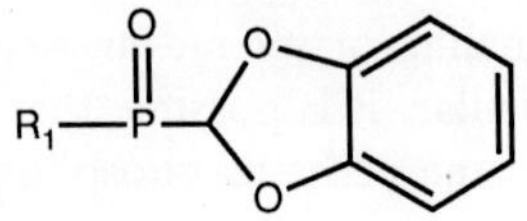

PSCP and SSCP

Figure 6 Structures and abbreviations of OP compounds used to obtain the results shown in Table II.

The results indicate little correlation between inhibition of trypsin and inhibition of T-lymphocyte proliferation (Fig. 6, Table II). This does not support the hypothesis that a trypsin-like serine protease is the molecular target of OP compounds that inhibit T cell proliferation, unless this protease differs considerably from trypsin in its susceptibility to OP inhibition. In addition, most of the OP compounds tested were one or two orders of magnitude less potent in inhibition of T-cell proliferation as compared to inhibition of AChE activity. Only phenyl saligenin cyclic phosphate (an analog of the active metabolite of tri-*o*-cresyl phosphate) exhibited similar potencies for both of these functions. Thus, it is unlikely that any of the compounds examined would be selectively immunotoxic with regard to their effects on T cells. A dose that could prevent T-lymphocyte proliferation would very likely produce fatal inhibition of AChE. It remains possible, however, that other immunological functions are more sensitive than T-cell proliferation to inhibition by these compounds.

The chemical reactivity of OP compounds that is responsible for their immunotoxicity is not known. In the preceding discussion of molecular targets, there was the tacit assumption that phosphorylation of immunologically important proteins is involved in immunosuppression just as phosphorylation of AChE is responsible for the acute neurotoxicity of these compounds. This has not been formally demonstrated. We observed decreased inhibition of T-lymphocyte proliferation *in vitro* by ethyl *p*-nitrophenylthiobenzene phosphonate (EPN) oxon if the compound was first incubated overnight in culture medium. Spontaneous hydrolysis during this incubation was demonstrated colorimetrically by the presence of the leaving group, *p*-nitrophenol (unpublished data, 1989). This is consistent with a mechanism involving phosphorylation of a critical T-cell protein, but other explanations are possible. Because the oxon forms of OP compounds are more active in protein phosphorylation that are the parent compounds (Chapter 1 by Chambers, this volume), the requirement for metabolic activation in the immunotoxicity of *O,O,S*-TMP (Rodgers *et al.*, 1985a) is also consistent with an important role

TABLE II

Effect of Selected OP Compounds

Lymphocyte activation IC_{50}[a] (μM)	AChE activity IC_{50} (nM)	Trypsin activity IC_{50} (μM)	R_1	R_2
C[b]	C	D	Me—O	Me—O
B	B	D	Et—O	Et—O
B	B	D	nPr—O	nPr—O
B	B	B	nBu—O	nBu—O
B	B	C	nPe—O	nPe—O
C	B	B	Me—O	iPr—O
C	D	D	iPr—O	iPr—O
B	B	B	Me—O	Ph
B	B	A	Et—O	Ph
B	ND	ND	nPr—O	pH
B	ND	ND	nBu—O	nBu—O
B	ND	ND	nPe—O	nPe—O
B	C	A	PNP—O	Ph
B	B	B	nPr—S	Me—O
D	D	A	nPr—S	Ph
D	D	A	nBu—S	pH
D	C	A	nBu—S	nBu—S
C	B	B	Me	Ph
A	B	A	Et	Ph
A	D	A	Ph	—
A	D	A	O-Ph	—

[a]IC_{50}, concentration that causes 50% inhibition; ND, not determined, O, oxygen; Me, methyl; Bu, butly; Et, ethyl; Pr, propyl; Pe, pentyl; Ph, phenyl; S, sulfur; PNP, *p*-nitrophenol; n, normal; i, iso.
[b]A, 0–5; B, 6–50; C, 51–150; D, >151.

for phosphorylation. However, more direct evidence with additional OP compounds is needed to confirm the importance and generality of phosphorylation as a mechanism of OP immunotoxicity.

V. Summary and Conclusions

There is ample evidence that some OP compounds can suppress immunological functions. The most consistent suppression has been reported following administration of a single neurotoxic dose. Because there are numerous cases of human OP poisoning each year (Murphy, 1986), this matter warrants further investigation. If the generality of this phenomenon is confirmed for a number of OP compounds, and suppression of host resistance to infection is demonstrated, it may be appropriate to consider prophylactic use of antibiotics in cases of OP poisoning.

A wide range of immunological functions can be suppressed by OP compounds, and many of them involve macrophages. Macrophages have been identified as a major cellular target of OP compounds in all studies in which this issue has been addressed. Only a few compounds have been examined, however, and the conclusion that macrophages are the major cellular target of all immunotoxic OP compounds would be premature at this point.

Only limited data are available regarding the molecular targets and mechanisms of action of OP compounds in the immune system. The working hypothesis of a number of investigators has been that OP compounds suppress immune functions by phosphorylating and inhibiting critical proteins involved in immune functions. There is indirect evidence that this is the case, but the potency of OP compounds with regard to inhibition of immune functions is generally at least an order of magnitude less than their potency with regard to inhibition of AChE activity. It remains possible, however, that some OP compounds are potent inhibitors of immune functions. Candidate compounds include phenyl saligenin cyclic phosphate and several compounds structurally related to tri-phenyl phosphate (Pruett and Chambers, 1988; Esa *et al.*, 1988). These compounds inhibit immune functions *in vitro* at concentrations of 1 μM or less. The vast majority of OP compounds have not been tested for immunotoxicity, and until more is known about molecular targets, mechanisms of action, and structure–activity relationships, the only reliable way to identify those that may be immunotoxic is through the use of screening tests (Luster *et al.*, 1988).

Probably the most productive direction for future research on the immunotoxicity of OP compounds would be comprehensive studies of a few compounds that have been implicated as immunotoxicants. This would include assays of the major immune functions and of host resistance to microbes and cancer. Such studies would allow subsequent investigations to focus on those functions most affected by a particular compound. *In vitro* representations of these relevant functions could then be used to examine the cellular and molecular targets and mechanisms of action of the compounds. In time, this approach could yield the information needed to begin assessment of the possible impact of OP-induced immunosuppression on human health.

Acknowledgments

This work was supported in part by NIH Biomedical research support grant RR07215, NIH Grant ES05371-01, and NIH Senior Fellowship ES05499-01. The author wishes to thank Yun Cheng Han for his excellent technical assistance.

References

Adams, D. O., and Hamilton, T. A. (1987). Molecular bases of signal transduction in macrophage activation induced by interferon gamma and second signals. *Immunol. Rev.* **97**, 5–28.

Allen, J. C. (1976). Infection complicating neoplastic disease and cytotoxic therapy. *In* "Infection and the Compromised Host" (J. C. Allen, ed.), pp. 151–171, Williams & Wilkins, Baltimore, Maryland.

Berkelman, R. L., Heyward, W. L., Stehr-Green, J. K., and Curran, J. W. (1989). Epidemiology of human immunodeficiency virus infection and acquired immunodeficiency syndrome. *Am. J. Med.* **86**, 761–770.

Bortin, M. M., and Rimm, A. A. (1977). Severe combined immunodeficiency disease. Characteristics of the disease and results of bone marrow transplantation. *JAMA* **238**, 591–600.

Casale, G. P., Cohen, S. D., and DiCapua, R. A. (1983). The effects of organophosphate-induced cholinergic stimulation on the antibody response to sheep erythrocytes in inbred mice. *Toxicol. Appl. Pharmacol.* **68**, 198–205.

Casale, G. P., Cohen, S. D., and DiCapua, R. A. (1984). Parathion-induced suppression of humoral immunity in inbred mice. *Toxicol. Lett.* **23**, 239–247.

Casale, G. P., Bavari, S., and Connolly, J. J. (1989). Inhibition of human serum complement activity by diisopropylfluorophosphate and selected anticholinesterase insecticides. *Fundam. Appl. Toxicol.* **12**, 460–468.

Clement, J. G. (1985). Hormonal consequences of organophosphate poisoning. *Fundam. Appl. Toxicol.* **5**, 561–577.

Costa, L. G., Kaylor, G., and Murphy, S. D. (1988). Muscarinic cholinergic binding sites on rat lymphocytes. *Immunopharmacol.* **16**, 139–149.

Desi, I., Farkas, I., Varga, G., Gonczi, C., Szlobodnyik, J., and Kneffel, Z. (1976). Immunosuppressive effects of hydrocarbon and organophosphate pesticide administration. *Egeszegtudomany* **20**, 358–368. (English abstract obtained through BIOSIS).

Devens, B. H., Grayson, M. H., Imamura, I. K., and Rodgers, K. E. (1985). *O,O,S*-trimethyl phosphorothioate effect on immunocompetence. *Pestic. Biochem. Physiol.* **24**, 251–259.

Dooley, R. K., and Holsapple, M. P. (1988). Elucidation of cellular targets responsible for tetrachlorodibenzo-*p*-dioxin (TCDD)-induced suppression of antibody responses: I. The role of the B lymphocyte. *Immunopharmacol.* **16**, 167–180.

Esa, A. H., Warr, G. A., and Newcombe, D. S. (1988). Immunotoxicity of organophosphorus compounds. Modulation of cell-mediated immune responses by inhibition of monocyte accessory functions. *Clin. Exp. Immunol.* **49**, 41–52.

Fan, A., Street, J. C., and Nelson, R. M. (1978). Immunosuppression in mice administered methyl parathion and carbofuran by diet. *Toxicol. Appl. Pharmacol.* **45**, 235.

Holsapple, M. P., White, K. L., McCay, J. A., Bradley, G. S., and Munson, A. E. (1988). An immunotoxicological evaluation of 4,4′-thio-*bis*-(6-*t*-butyl-*m*-cresol) in female B6C3F1 mice. 2. Humoral and cell-mediated immunity, macrophage function, and host resistance. *Fundam. Appl. Toxicol.* **10**, 701–716.

Koller, L. D., Exon, J. H., and Roan, J. G. (1976). Immunological surveillance and toxicity in mice exposed to the organophosphate, leptophos. *Environ. Res.* **12**, 238–242.

Lee, J.-C., Gibson, C. W., and Eisenstein, T. K. (1985). Macrophage-mediated mitogenic suppression induced in mice of the C3H lineage by a vaccine strain of *Salmonella typhimurium*. *Cell. Immunol.* **91**, 75–91.

Luster, M. I., Munson, A. E., Thomas, P. T., Holsapple, M. P., Fenters, J. D., White, K. L., Lauer, L. D., Germolac, D. R., Rosenthal, G. J., and Dean, J. H. (1988). Methods evaluation. Development of a testing battery to assess chemical-induced immunotoxicity: National Toxicology Program's guidelines for immunotoxicity evaluation in mice. *Fundam. Appl. Toxicol.* **10**, 2–19.

Maloteaux, J. M., Waterkein, C., and Laduron, P. M. (1982). Absence of dopamine and muscarinic receptors on human lymphocytes. *Arch. Int. Pharmacodyn.* **258**, 174–176.

Mandel, J. S., Berlinger, N. T., Kay, N., Connett, N., and Reape, M. (1989). Organophosphate exposure inhibits non-specific esterase staining in human blood monocytes. *Am. J. Ind. Med.* **15**, 207–212.

Maslinski, W. (1989). Cholinergic receptors of lymphocytes. *Brain Behavior and Immunity* **3**, 1–14.

Murphy, S. D. (1986). *In* "Casarett and Doull's Toxicology" (C. D. Klaassen, M. O. Amdur, and J. Doull, eds.), pp. 527–539. Macmillan, New York.

Penn, I. (1988). Tumors of immunocompromised patients. *Annu. Rev. Med.* **39**, 63–73.

Pruett, S. B., and Chambers, J. E. (1988). Effects of paraoxon, *p*-nitrophenol, phenyl saligenin cyclic phosphate, and phenol on the rat interleukin-2 system. *Toxicol. Lett.* **40**, 11–20.

Pruett, S. B., and Kiel, J. L. (1988). Quantitative aspects of the feeder cell phenomenon: Mechanistic implications. *Biochem. Biophys. Res. Comm.* **150**, 1037–1043.

Pruett, S. B., Chambers, J. E., and Chambers, H. W. (1989). Potential immunomodulatory activity of phenylphosphonothioates. *Int. J. Immunopharmacol.* **11**, 385–393.

Richman, D. P., and Arnason, B. G. W. (1979). Nicotinic acetylcholine receptor. Evidence for a functionally distinct receptor on human lymphocytes. *Proc. Natl. Acad. Sci. U.S.A.* **76**, 4632–4635.

Rodgers, K. E., and Ellefson, D. D. (1990. Modulation of respiratory burst and mitogenic response of human peripheral blood mononuclear cells and murine splenocytes by malathion. *Fundam. Appl. Toxicol.* **14**, 309–317.

Rodgers, K. E., Grayson, M. H., Imamura, T., and Devens, B. H. (1985a). *In vitro* effects of malathion and *O,O,S*-trimethyl phosphorothioate on cytotoxic T-lymphocyte responses. *Pestic. Biochem. Physiol.* **24**, 260–266.

Rodgers, K. E., Imamura, T., and Devens, B. H. (1985b). Investigations into the mechanism of immunosuppression caused by acute treatment with *O,O,S*-trimethyl phosphorothioate. II. Effect on the ability of murine macrophages to present antigen. *Immunopharmacol.* **10**, 181–189.

Rodgers, K. E., Leung, N., Ware, C. F., Devens, B. H., and Imamura, T. (1986). Lack of immunosuppressive effects of acute and subacute administration of malathion on murine cellular and humoral immune responses. *Pestic. Biochem. Physiol.* **25**, 358–365.

Rodgers, K. E., Imamura, T., and Devens, B. H. (1987). Investigations into the mechanism of immunosuppression caused by acute treatment with *O,O,S*-trimethyl phosphorothioate: Generation of suppressive macrophages from treated animals. *Toxicol. Appl. Pharmacol.* **88**, 270–281.

Rodgers, K. E., Grayson, M. H., and Ware, C. F. (1988). Inhibition of cytotoxic T lymphocyte and natural killer cell-mediated lysis by *O,O,S*-trimethyl phosphorodithioate is at an early post-recognition step. *J. Immunol.* **140**, 564–570.

Roitt, I. M., Brostoff, J., and Male, D. K. (1985). "Immunology," C. V. Mosby, St. Louis, Missouri.

Sneller, M. C., and Strober, W. (1990). Abnormalities of lymphokine gene expression in patients with common variable immunodeficiency. *J. Immunol.* **144**, 3762–3769.

Strom, T. B., Lane, M. A., and George, K. (1981). The parallel, time-dependent, bimodal change in lymphocyte cholinergic binding activity and cholinergic influence upon lymphocyte-mediated cytotoxicity after lymphocyte activation. *J. Immunol.* **127**, 705–710.

Wara, D. W. (1981). Host defense against *Streptococcus pneumoniae*: The role of the spleen. *Rev. Infect. Dis.* **3**, 299–309.

Watanabe, H., and Bannai, S. (1987). Induction of cystine transport activity in mouse peritoneal macrophages. *J. Exp. Med.* **165**, 628–637.

Wazer, D. E., and Rotrosen, J. (1984). Murine lymphocytes lack clearly defined receptors for muscarinic and dopaminergic ligands. *J. Pharm. Pharmacol.* **36**, 853–854.

White, K. L., and Holsapple, M. P. (1984). Direct suppression of *in vitro* antibody production by mouse spleen cells by the carcinogen benzo(*a*)pyrene but not by the noncarcinogenic congener benzo(*e*)pyrene. *Cancer Res.* **44**, 3388–3393.

Wong, R. L., Gutowski, J. K., Katz, M., Goldfarb, R. H., and Cohen, S. (1987). Induction of DNA synthesis in isolated nuclei by cytoplasmic factors: Inhibition by protease inhibitors. *Proc. Natl. Acad. Sci. U.S.A.* **84**, 241–245.

Wysocki, J., Kalina, Z., and Owczarzy, I. (1987). Effect of organophosphoric pesticides on the behavior of NBT-dye reduction and E-rosette formation tests in human blood. *Int. Arch. Occup. Environ. Health* 59, 63–71.

Yang, K. H., Kim, B. S., Munson, A. E., and Holsapple, M. P. (1986). Immunosuppression induced by chemicals requiring metabolic activation in mixed cultures of rat hepatocytes and murine splenocytes. *Toxicol. Appl. Pharmacol.* 83, 420–429.

20

Teratogenic Effects of Organophosphorus Compounds

Paul A. Kitos and Oranart Suntornwat[*]
Department of Biochemistry
The University of Kansas
Lawrence, Kansas

I. Introduction

A. Teratogenic Considerations

The toxic, mutagenic, carcinogenic, and teratogenic effects of organophosphorus (OP) compounds are important considerations in evaluating their suitability for agricultural, industrial, and domestic use. Although OP compounds have many different applications, they are used in especially large amounts in pest control where they frequently come into contact with people and animals. This chapter deals with their roles in causing birth defects.

A teratogen is an agent (chemical, physical, or biological) that causes congenital malformations. It adversely affects processes of differentiation and development, causing structural and/or functional abnormalities in the orga-

*Current Address: Department of Chemistry, Silpakorn University, Nakorn Pathom, Thailand.

Organophosphates: Chemistry, Fate, and Effects

nism. The biological defects are initiated either during or before embryogenesis and usually do not involve mutagenic or carcinogenic events (Wilson, 1972). In general, the cells of immature tissues, somatic or germinal, are more susceptible to teratogens than are those of mature tissues. The production of abnormalities is a result of failures in the ontogeny of the organism: aberrations in cell mortality, mitotic rate, schedule of differentiation, cell motility, or any of a number of other developmentally important processes. The consequences of these actions may become apparent even within hours of the insult (Fishbein, 1976).

The biological actions of a teratogen depend on a multitude of factors: the type of organism (oviparous, viviparous); the route of administration; the stage of development of the organism at the time of the challenge; the magnitude of the challenge (duration, amount of the agent); agent accessibility to the susceptible tissue; the ability of the organism to metabolize the agent to its teratogenically active form (if the agent, as administered, is not the active species); the ability of the organism to destroy the teratogen; etc. The literature is replete with apparently conflicting results about the teratogenic actions of OP compounds, the basis of much of which can be attributed to variations in the test conditions.

B. Early Studies

The earliest studies of OP-induced teratogenesis were carried out mainly on birds. That the OP insecticides are embryotoxic and/or teratogenic for the chick embryo was reported by McLaughlin *et al.* (1963), Baron and Johnson (1964), Marliac (1964), and Marliac *et al.* (1965). OP-induced abnormal neurological and skeletal responses in either chicken or duck embryos were described by Roger *et al.* (1964, 1969), Khera (1966), Khera and LaHam (1965), Khera *et al.* (1965), Khera and Bedok (1967), Walker (1967), and Greenberg and LaHam (1970). Foremost among the observed teratogenic problems were growth retardation, micromelia, beak and plumage defects, and axial skeletal problems (Walker, 1968; Roger *et al.*, 1969, Greenberg and LaHam, 1970). Figure 1 shows a normal, day 15 chick embryo and a day 15 embryo from an egg that had been injected on day 4 by the intravitelline route with 0.2 mg diazinon. In this picture the treated embryo exhibits all of the above-mentioned deformities (Henderson and Kitos, 1982).

Prior to 1980 there were relatively few reports of teratogenic effects of the OP insecticides on mammals, and most of these were conducted on rats (Marliac *et al.*, 1965; Fish, 1966; Tanimura *et al.*, 1967; Dobbins, 1967; Kimbrough and Gaines, 1968) and mice (Tanimura *et al.*, 1967; Budreau and Singh, 1973). Fish and amphibians received even less attention in these regards than mammals.

Figure 1 Diazinon treated and untreated chick embryos. Embryonated chicken eggs were injected by the intravitelline route on day 4 with corn oil (CON) or 0.2 mg of diazinon in corn oil (DZN) and then incubated further to day 15. The embryos were then photographed.

II. Teratogenic Organophosphorus Compounds

A. Pesticides, Chemotherapeutic Agents, and Other Substances

A very large number of OP compounds are currently, or have been recently, in widespread use. Some of them are confirmed teratogens and the others can be considered potential teratogens. It would be presumptuous to assume that any OP compound is nonteratogenic. OP pesticides include insecticides and acaracides, nematocides, insect chemosterilants, fungicides, herbicides and plant growth regulators, rodenticides, insecticide synergists, and insect repellants (Eto, 1974; Toy and Walsh, 1987). OP compounds, however, are not used just as pesticides. Cyclophosphamide (Fig. 2) and some of its derivatives, which are nitrogen mustards as well as phosphotriester derivatives, are used as cancer chemotherapeutic agents and have been used as defleecing agents for sheep. Dialkyl phosphonates are used as wetting agents, metal extractants, and oil additives. Trialkyl and triaryl phosphites are used as polymer stabilizers in the manufacture of vinyl plastics and synthetic rubber. Phenylphosphonous dichloride is used in making nylon. Triarylphosphates are used as

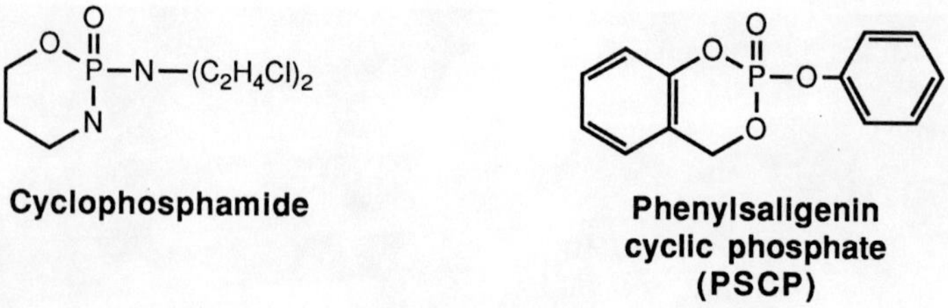

Cyclophosphamide

Phenylsaligenin cyclic phosphate (PSCP)

Figure 2 Molecular structures of cyclophosphamide and phenylsaligenin cyclic phosphate.

plasticizers in the production of cellulose acetate and vinyl polymers and are also effective in these products as flame retardants. They have also been used in leaded gasolines to combat misfiring. Tributylphosphate is used to purify uranium and as a nonflammable component of the hydraulic systems of large commercial aircraft. Tributyl trithiophosphate is used as a defoliant for cotton plants, to facilitate the cotton harvest without killing the plants (Toy and Walsh, 1987). In fact, there are numerous OP products with which we interact in our homes and workplaces, and many of these chemicals present known and unknown hazards to the unborn.

B. Biological Actions

1. Agent Administration

In order to study the teratogenic actions of foreign substances, one must deliver the putative teratogen to the embryonic system (fertile egg, pregnant animal, etc.) at the appropriate stage of development and by a route that will facilitate interaction of the substance with the target. Several modes of delivery have been used for viviparous animals: *ad libitum* in the food or water of the mother, oral gavage, intraperitoneal injection, etc. Likewise, several routes have been used for birds: immersing the eggs in the test solution (Meiniel, 1974), spraying the shell with a solution of the compound (Hoffman and Eastin, 1981), exposing the egg in a closed box to a volatile insecticide (dichlorvos) (Lutz-Ostertag and Bruel, 1981); cutting a window in the shell and pipetting the agent onto the extraembryonic membrane (Hall, 1977; Hodach *et al.*, 1974) or injecting it into the extraembryonic coelom (Goel and Jurand, 1976; Overman *et al.*, 1976) or directly beneath the embryo (Grubb and Montiegel, 1975); injecting it into the air chamber (Gebhardt, 1968; van Steenis and van Logten, 1971); and, most commonly, injecting it into the yolk (McLaughlin *et al.*, 1963; Clegg, 1964; Gebhardt, 1972; Wyttenbach *et al.*, 1981; Kitos *et al.*, 1981a), with or without prior mixing with yolk contents (Walker, 1967). The foreign substance has been administered in a variety of carriers: water (Roger *et al.*, 1969; Meiniel, 1976a); ethanol (Flockhart and Casida, 1972); propylene glycol (Gebhardt, 1972); methoxytriglycol (Proctor *et al.*, 1976; Moscioni *et al.*, 1977; Seifert and Casida, 1978); vegetable oil (Walker, 1968; Greenberg and LaHam, 1970; Meiniel, 1976a; Wyttenbach *et al.*, 1981; Kitos *et al.*, 1981a). Each solvent system has its advantages and limitations.

2. Avian Subjects

The most commonly used birds for teratogenic studies are white leghorn chickens (*Gallus domesticus*), mallard ducks (*Anas platyrhynchos*), Japanese quail (*Coturnix coturnix Japonicus*), bobwhite quail (*Colinus virginianus*), and chukar partridge (*Alectoris chukar*). Very young avian embryos have been reported to be less susceptible than older embryos to the actions of injected substances (Landauer, 1975a; Roger *et al.*, 1969; Schom and Abbott, 1977). A possible reason for this apparent lack of susceptibility of very young embryos to OP compounds is that because of the small size of the embryo and the uncertainty of its location in the egg, it is difficult to deliver the test substance reproducibly. Using a precision delivery system, Wyttenbach *et al.* (1981) and Wyttenbach and Thompson (1985) showed that up to day 4 of incubation, the toxicity and teratogenicity of malathion for the chick embryo is inversely proportional to the embryo's age, that younger embryos are, in fact, more vulnerable to the adverse effects of the exogenous agent. In the very early chick embryo, the most obvious OP-induced abnormalities are slower development, folding or undulation of the notochord and neural tube (Meiniel, 1977b; Wyttenbach and Hwang, 1984; Garrison and Wyttenbach, 1985a), and unilateral retardation of the cranial sense organs (Garrison and Wyttenbach, 1985b).

a. Types of Deformities There are striking similarities between the teratogenic actions of OP and alkylcarbamate (AC) compounds. Within the AC group, eserine (physostigmine), carbaryl, and carbofuran are potent teratogens (Proctor *et al.*, 1976). When embryonated chicken eggs are exposed early in development (days 3 to 7) to OP or AC compounds they cause either or both of two classes of deformities, identified as Types 1 and 2 (Moscioni *et al.*, 1977).

b. Type 1 Deformities In avian embryos the Type 1 deformities include micromelia (straightened and shortened legs), fused appendages, abnormal feather and beak formation, and frequently, gross abdominal edema (Eto, 1981), and can be prevented or ameliorated by concurrent administration of niacin and some of its derivatives and metabolic precursors, including tryptophan (Roger *et al.*, 1964; Upshall *et al.*, 1968; Greenberg and LaHam, 1970; Wenger and Wenger, 1973; Proctor and Casida, 1975; Proctor *et al.*, 1976; Meiniel, 1976a and 1976b; Meiniel and Autissier-Navarro, 1980; Moscioni *et al.*, 1977; Misawa *et al.*, 1981; Kitos *et al.*, 1981b; Henderson and Kitos, 1982). The production of Type 1 teratisms is preceded by a lowering of the tissue concentrations of pyridine nucleotides; the addition of niacin analogs or tryptophan restores the NAD concentrations to or above normal and prevents or lessens the severity of the Type 1 lesions. Shown in Fig. 3 are the skeletal elements of a day 15 control chick embryo and those of embryos exposed from day 4 to either 0.5 mg per egg of diazinon, a highly teratogenic insecticide, or 2 mg per egg of EPN, a weakly teratogenic insecticide.

c. Type 2 Deformities Type 2 deformities include problems of the axial skeleton (e.g., wry and short neck and *rumplessness*) (Meiniel *et al.*, 1970; Meiniel, 1974, 1976a, 1977a, b), arthrogryposis (i.e., retention of a joint in a flexed position), and muscular hypoplasia, and are not antagonized by niacin derivatives or tryptophan (Moscioni *et al.*, 1977). The deformities are antagonized by simultaneously administered, relatively nontoxic oximes such as pralidoxime (2-pyridinealdoxime methonium ion, 2-PAM) (Meiniel, 1974, 1976a, b, 1978a; Landauer, 1977; Moscioni *et al.*, 1977; Misawa *et al.*, 1981; Kitos *et al.*, 1981b).

d. Other Effects Some effects of OP compounds on avian embryos may not be identified as Type 1 or 2 anomalies, including lowered hatch rate,

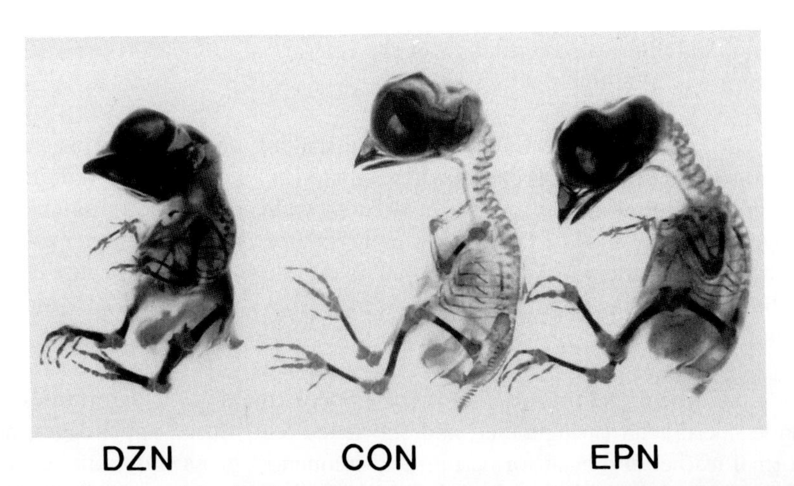

Figure 3 Skeletons of day 15 chick embryos after exposure to OP insecticides. Embryonated chicken eggs were injected on day 4 with corn oil (CON), 0.5 mg diazinon (DZN) in corn oil, or 2 mg EPN in corn oil. At day 15 the embryos were processed by the method of Dingerkus and Uhler (1977), clarifying the soft tissues and staining the cartilage with alcian blue and the calcified bone with alizarin red. In this black-and-white reproduction, the regions of calcification of the long bones appear dark and the cartilaginous regions, grey. Severe skeletal deformities are evident in the DZN embryo, and minor skeletal deformities, in the EPN embryo.

retarded growth, hypoglycemia (Arsenault *et al.*, 1975; Laley and Gibson, 1977), abnormal thyroid function (Richert and Prahlad, 1972), hypoactivity, shivering, and sleeping syndrome (Khera *et al.*, 1965), muscle weakness (Schom *et al.*, 1979), etc.

Varnagy (1981) and Varnagy *et al.* (1982) examined the embryotoxic and teratogenic effects of several OP insecticides on Japanese quail and pheasant (*Phasianus colchicus*). Methyl parathion was found to be about 150 times more teratogenic for pheasant than for quail, producing lordoscoliosis, umbilical hernia, and microphthalmia to varying degrees. Heptenophos was embryotoxic at days 9 and 10 of incubation, but neither it nor mevinphos was found to be teratogenic to quail or pheasant. In a similar study, Somlyay and Varnagy (1986) observed that parathion, methyl parathion, and phosmet were more toxic and teratogenic for pheasant embryos than for chick embryos, and that of these three compounds, parathion was the most active.

Farage-Elawar and Francis (1988) injected chicken eggs with fenthion (FEN), fenitrothion (FTR) or desbromoleptophos (DBL) on day 15 and looked for posthatch responses, particularly changes in gait and acetylcholinesterase (AChE) and neuropathy target esterase (NTE) activities. Day 15 embryos were used because they were considered to be beyond the time of OP-induced teratogenesis and before the period of sensitivity to organophosphate-induced delayed neuropathy (OPIDN). The three agents have different OPIDN potentials: DBL, strong; FEN, mixed; FTR, none. In each case, growth of the OP-exposed chicks was the same as that of the controls. Neither FEN nor FTR inhibited NTE, whereas DBL did. The gait of both the DBL and FEN chicks was affected for 6 weeks posthatch, whereas that of the FTR chicks was not. They concluded that DBL and FEN affected the gait of the chicks irreversibly by a process that involved neither AChE nor NTE.

Hanafy and El-Din (1986) showed that up to 2 mg methamidophos per 50 g chicken egg, administered on day 4 into the air chamber, produced serious reproductive problems: a few anomalies (open umbilicus and thin toes and feet), retarded growth, and a significant level of lethality.

Simulating field conditions (external exposure; 153 liter aqueous/ha or 16.8 liter oil base/ha), Hoffman and Eastin (1981) showed that, if the OP compound was applied at any time from day 3 to 8 of incubation, the order of embryotoxicity for the mallard duck was parathion (PTN) > diazinon (DZN) > malathion (MLN). At agriculture levels of application, embryo growth was stunted, and there was a high frequency of malformations, especially distortions in the axial skeleton at the cervical level. All three compounds decreased the levels of plasma and brain cholinesterase (ChE), and the decreases were still apparent at hatching. Ethyl *p*-nitrophenylthiobenzene phosphonate (EPN) administered at day 3 caused high mortality, impaired growth, axial scoliosis, and edema (Hoffman and Sileo, 1984). Brain weights, NTE and AChE, plasma ChE, and alkaline phosphatase were all decreased at the time of hatching. The hatchlings were weak and had difficulties righting themselves. In contrast to OPIDN, there was no demyelination or axonopathy of the spinal cord.

Early administration (day 2 or 3) of DZN or PTN to bob white quail eggs did not produce notochordal folding or vascular enlargement as it did in the chick embryo (Meneely and Wyttenbach, 1989). It did produce a short, contorted, vertebral axis and tibiotarsal, rib, and sternum defects. PTN was more potent than DZN in causing the skeletal defects and did not produce any Type 1 defects.

3. Nonavian Subjects

The biological effects of OP compounds on mammalian and amphibian embryos have not been so well characterized as those on avian embryos, and the findings seem less consistent.

a. Amphibians The effects of phenylsaligenin cyclic phosphate (PSCP)(see Fig. 3), leptophosoxon (LPTO), tri-*o*-tolylphosphate (TOTP), and paraoxon, all AChE inhibitors or potential inhibitors, were determined using the gray treefrog, narrow-mouthed frog, and leopard frog (Fulton and Chambers, 1985). Neither TOTP (10 ppm) nor PXN (100 ppm) was toxic or teratogenic. PXN probably does not survive in the aqueous milieu long enough to have an effect. PSCP (0.5 ppm) was both toxic and teratogenic, producing abdominal edema, blisters, and spinal abnormalities in all 3 species. LPTO (2.2 ppm) was toxic only to the gray treefrog but was not teratogenic. MLN and PTN and their oxygen analogs and monocrotophos and dicrotophos were tested on South African clawed frog embryos (Snawder and Chambers, 1989). The first four of these compounds produced severe defects: reduced growth, abnormal pigmentation and gut development, and notochordal defects. The latter two compounds had milder effects. All compounds caused a reduction in the NAD level, suggesting that the lowered NAD had nothing to do with the developmental problems. MLN (up to 10 mg/liter) and malaoxon (up to 1 mg/liter) were administered in the bathing fluid to *Xenopus laevis* during the first 4 days of development and had teratogenic and NAD effects similar to those noted above (Snawder and Chambers, 1990). Simultaneous administration of tryptophan restored the NAD level to normal but did not alleviate the deformities. The agents were most effective in the post-organogenesis period (48–96 hr).

Tadpoles (*Rana catesbeiana*) have been shown to concentrate pesticides from their bathing fluid (Hall and Kolbe, 1980). A single meal of tadpoles that had been exposed to 1 ppm PTN or 5 ppm fenthion was lethal to mallard ducks. Not enough dicrotophos, MLN, or acephate was accumulated by tadpoles to be lethal as a single meal. The duck brain ChE activity correlated well with the dose of insecticide and the lethal effect. Clearly, the OP-concentrating ability of the amphibian embryos can have a major effect on life forms higher in the food chain.

b. Mammals Embryotoxicity and teratogenicity resulting from exposure to OP compounds are not so readily produced in mammals as in birds. There is a low incidence of abnormalities, and at least some of the responses seem to be anecdotal. Part of the basis for the difference is the maternal presence, the capacity of the mother to absorb, detoxify, and excrete foreign substances, and the existence of the placenta as a barrier between the mother and the fetus. Another reason for lower sensitivity of the mammalian fetus to toxic and teratogenic OP compounds may be that the mammal does not synthesize its NAD from tryptophan, eliminating kynurenine formamidase (KFase) as a possible metabolic target. Rather, it uses preformed niacin, a vitamin, for the synthesis of this coenzyme (Seifert and Casida, 1981). In the present discussion, the mammalian responses will be considered by species.

Mouse In one of the earliest studies of the effects of OP compounds on mouse embryos, Budreau and Singh (1973) showed that demeton (up to 10 mg/kg) and fenthion (up to 80 mg/kg), given between days 7 and 10, were embryotoxic and mildly teratogenic, causing digestive, nervous, and skeletal problems, malformed digits and vertebrae and cleft palate. If the compounds were administered later, they had much less teratogenic effect and little or no effect on litter size and birth weight. They found no correlation between ChE inhibition and teratogenesis. Tanimura *et al.* (1967) showed that methyl PTN was teratogenic to mice but not to rats. Courtney *et al.* (1986) administered trichlorfon (200 mg/kg) daily by oral gavage to mice and found it to be teratogenic, fetotoxic and, at the higher levels, maternally lethal. They observed fewer calcified centers in the fetal paws and delayed fetal maturation. Nehez *et al.* (1987) examined the cytogenetic, genetic, and embryotoxic effects of a hemiacetal analog (a hypothetical impurity in technical preparations) of trichlorfon and of trichlorfon, itself, and found that a single dose of 81 mg/kg of either compound cause chromosomal aberrations in the maternal bone marrow cells. However, four consecutive 81 mg/kg doses of either compound (given on days 2 through 5) was only weakly embryotoxic. Deacon *et al.* (1980) found that chlorpyrifos in daily (day 6 through 15) oral gavage doses of less than 25 mg/kg was not teratogenic, but that 25 mg/kg was fetotoxic and produced minor skeletal variants.

The genotoxicity of a single 5 mg/kg dose (acute) or five 1 mg/kg doses (chronic) of phosphamidon in mice was studied using three different assays: bone marrow cell chromosomal assay, micronucleus formation, and sperm shape changes (Behera and Bhunya, 1987). The results indicated that the compound was genetoxic, especially if given as a single large dose. In similar studies, acephate (Behera and Bhunya, 1989) and ethion (Bhunya and Behera, 1989) were also implicated as mutagens. Phosphamidon was found to be more embryotoxic than teratogenic on Swiss Albino mice (Bhatnagar and Soni, 1988). It was most effective during the post-implantation (day 7) and late organogenesis (day 13) periods but relatively ineffective during the early organogenesis period (day 10). In another study with mice, phosphamidon was administered at 35 ppm in the drinking water before and during gestation and effects on the embryos were noted much later (Soni and Bhatnagar, 1989). The results were mixed: if the treatment was initiated 30 days before mating, there were fewer implantations, smaller litter sizes and fetal weights, and more resorptions than in the controls. If it was initiated 60 days before mating there were no effects. The authors attribute the latter results to a time-dependent development of resistance to the OP compound by the Swiss mice.

Rat The rat embryo seems to be less susceptible to OP compounds than the mouse embryo. PTN, methyl parathion and diisopropylfluorophosphate (DFP) caused decreases in AChE activity of rat embryos but produced no obvious teratisms (Fish, 1966). MLN, PTN, DZN, and dichlorvos, when fed to the dam, gave only slight indications of teratogenicity (Dobbins, 1967). Tanimura *et al.* (1967) and Kimbrough and Gaines (1968) likewise reported little or no teratogenic effects of OP insecticides on rats. Two alkylaryl phosphate esters, (2-ethylhexyldiphenyl phosphate and isodecyldiphenyl phosphate), that are used as plasticizers were administered at up to 300 mg/kg/day by oral gavage to pregnant rats, beginning at day 6 of gestation (Robinson *et al.*, 1986). There were no teratogenic responses even though 2-ethylhexyldiphenyl phosphate decreased the weight gains. Welsh *et al.* (1987) fed up to 1% triphenylphosphate (i.e., up to 690 mg/kg) to rats, beginning at 4 weeks postweaning and continuing for 91 days through mating and gestation. There were no toxic effects on the mother or offspring and no increases in anomalies in the fetuses. Daily oral doses of oxydemetonmethyl (up to 4.5 mg/kg) were given to pregnant rats from day 6 to 15 of gestation, and the mother and fetuses were examined at days 16, 20, and 21 (postpartum) (Clemens *et al.*, 1990). There were dose-related decreases in maternal plasma, red blood cell and brain ChE, food consumption, and body weight. The dam also developed tremors. However, there was no evident fetotoxicity or teratogenicity. There was a small decrease in fetal brain AChE, but neonatal survival, growth, and development seemed unaffected. There were also no changes in sensory reflexes, learning ability, or open field activity of the young animals.

There are, however, reports of teratogenic actions of OP compounds on rats. A single 200 mg dose of chloracetophon given on day 13 of gestation resulted in some bone disorders. Repeated doses at one fifth that level resulted in no embryotoxicity or teratogenicity (Khadzhitodorova and Andreev, 1984). Methamidophos [up to 2 mg/kg (0.1 LD_{50})], given orally to pregnant rats, resulted in some embryo lethality, growth retardation, anencephaly, and anotia (Hanafy *et al.*, 1986). Larger doses had more severe effects and caused cyanosis. Trichlorfon by oral gavage to CD rats at up to 200 mg/kg is maternally lethal. At lower doses, the ribs, urinary system, and skull of the embryos were affected and the agent was fetotoxic (Courtney *et al.*, 1986). Clearly, rat embryos can be affected by OP compounds, but the range and severity of the problems are much less than those for mice.

Rabbit Fenchlorfos (up to 50 mg/kg), given to rabbits at day 6 of gestation, did not affect implantation, fetal weight, and the proportion of live fetuses (Nafstad *et al.*, 1983). However, there was an increase in the incidence of cardiovascular, brain, and skeletal anomalies. The cardiovascular malformations and cerebellar hypoplasia were dose dependent. Machin and McBride (1989) administered MLN (100 mg/kg) at days 7 to 10 to rabbits and found no indications of teratogenesis.

Guinea pig Trichlorfon (100 mg/kg) was administered by oral gavage to pregnant guinea pigs at approximately days 37 and 52 of gestation (Berge *et al.*, 1986). The pups developed locomotory problems and had decreased brain weights. There were weight reductions of the cerebellum, medulla oblongata, hippocampus, thalamus, and colliculi. The cerebellum had reduced external granular and molecular layers, regional absences of Purkinje cells, and reduced cholineacetyltransferase and glutamate decarboxylase.

Humans The possibility that OP compounds could be fetotoxic and/or teratogenic in humans is, of course, a major question for which there are no good answers. Most studies on the subject are, of necessity, epidemiological. There are some reports, however, that give us reason to be concerned. OP pesticides are known to inhibit ChE in humans as well as in other animals (Shellenberger, 1980). Phosphine gas, while not actually an OP compound, is used to fumigate grain and as a reagent in the microchip industry. It has been shown to cause chromosomal rearrangements, mainly translocations, in lymphocytes of fumigant applicators (Garry *et al.*, 1989). Lindhout and Hageman (1987) reported an amyoplasia-like condition in the case of maternal exposure to MLN during the 11th and 12th week of pregnancy. A study of birth defects and low birth weight was made from vital records following aerial application of a low dose of MLN over a 13,000-square-mile area of the San Francisco Bay area (Grether *et al.*, 1987). No biologically plausible pattern of association was found from the data.

Romero *et al.* (1989) reported an incident of possible human teratogenesis by a mixture of three pesticides: oxydemetonmethyl, mevinphos and the carbamate methomyl. Thirty-five field workers, including one who was 4 weeks pregnant, entered a field contaminated with residues of these three insecticides. The child was born weighing 3.2 kg and having multiple defects: cardiac, bilateral optic nerve colobomas, microphthalmia of the left eye, cerebral and cerebellar atrophy, and facial anomalies. The cardiac defects included ventricular and atrial septal defects, stenosis of the pulmonary artery, and a patent ductus arteriosus. The child lived 14 days. While this information is episodal and does not provide a cause-and-effect relationship, it does raise our awareness of the possible hazards for humans of OP insecticides, alone or in combinations.

C. Structural Determinants

Some of the molecular determinants of toxicity and teratogenicity of OP compounds have been described by Casida *et al.* (1963), Eto (1974), and Ohkawa (1982). In addition, Upshall *et al.* (1968), Proctor *et al.* (1976) and Moscioni *et al.* (1977) surveyed a large number of OP compounds for their abilities to produce physical deformities in chick embryos. The test compounds included triesters of phosphate, phosphothioate, phosphothiolate, and phosphothiolothioate, and diesters of phosphonothioates. The frequency and severity of the teratogenic lesions were found to correlate with a few structural themes in the OP molecules (Eto *et al.*, 1980; Eto, 1981). The teratogenicities of the OP compounds occurred with both thioate and oxoate structures, and structures that have either aliphatic or aromatic leaving groups (Table I, R_3), and were usually only slightly influenced by the nature of the simple phosphate ester substituents (Table I, $R_{1\text{ and }2}$).

1. Crotonamide Derivatives

The leaving groups (Table I, R_3) of most of the commercially available, strongly teratogenic OP compounds are either crotonamide congeners or substituted pyrimidines (Table I). Within each of these categories there are some structural consistencies. The teratogenic form of dicrotophos (DCP) contains the cis, but not the trans, isomer of crotonamide (Roger *et al.*, 1969). Compounds (2) and (3) can be formed metabolically from DCP and are also very teratogenic (Roger *et al.*, 1969). Compound (5), a strong teratogen, contains an α-pyridone ring which is a structural analog of *cis*-crotonamide. Crotoxyphos (8) and mevinphos (9) contain crotonic acid esters, rather than amides, and are poor teratogens. The accumulated data imply that the amide nitrogen and its position in the crotonamide moiety are important teratogenic determinants of these compounds in the chick embryo (Eto, 1981).

2. Pyrimidine Derivatives

DZN, a commonly used pyrimidyl insecticide, is among the most teratogenic of OP compounds for avian embryos (Eto *et al.*, 1980; Eto, 1981). Pyrimidine-containing esters of both phosphate and phosphothioate are active, but most of the members of this group that are actually used as insecticides are phosphothioates. The phosphates are relatively unstable in aqueous milieu and may be rendered inactive *in vivo* before they can cause developmental problems (Eto *et al.*, 1980). The phosphothioates are slowly converted to their oxo counterparts in animal tissues, as discussed in Chapters 1, 10, and

TABLE I

OP Insecticide Structures and Teratogenic Signs in Avian Embryos

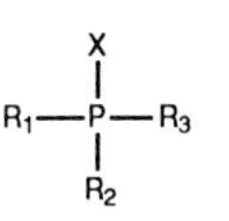

Parent Compound

Compound	R_1, R_2 [a]	X [b]	R_3 [c]	Teratogenicity [d]
(1) Dicrotophos (DCP)	1	O	–O–C(C)=C–C(=O)–N(C)(C)	++
(2) Monocrotophos (MCP)	1	O	–O–C(C)=C–C(=O)–N(C)(H)	++
(3) *N*-Demethyl-MCP	1	O	–O–C(C)=C–C(=O)–N(H)(H)	++
(4) SD 5911	1	O	–O–C(C)=C–C(=O)–N(C–C)(C–C)	++
(5) 2-Pyridon-4-yl diethylphosphorothioate	2	O	–O–(2-pyridone ring: C=O, NH)	++
(6) SD 5562	1	O	–O–C(C)=C(Cl)–C(=O)–N(C)(C)	++
(7) Phosphamidon	1	O	–O–C(C)=C(Cl)–C(=O)–N(C–C)(C–C)	+
(8) Crotoxyphos	1	O	–O–C(C)=C–C(=O)–O–C(C)–(phenyl)	–
(9) Mevinphos	1	O	–O–C(C)=C–C(=O)–O–C	–
(10) Dichlorvos	1	O	–O–C=C–Cl_2	–

(continued)

TABLE I *(cont.)*

Compound	R_1, R_2[a]	X[b]	R_3[c]	Teratogenicity[d]
(11) 4-Pyrimidyldiethyl phosphothioate	2	S	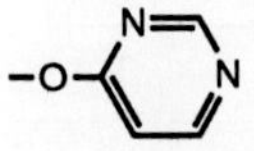	++
(12) Diazinon (DZN)	2	S	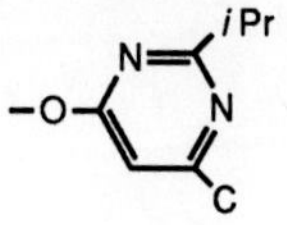	++
(13) Etrimfos (ETF)	1	S	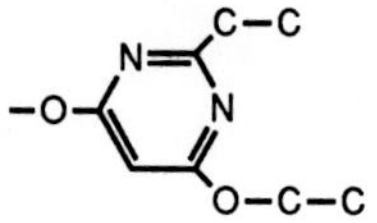	++
(14) Pyrimiphos, methyl	1	S	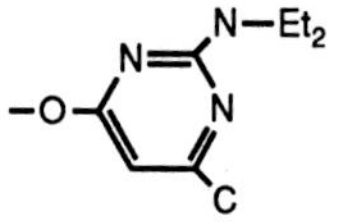	++
(15) Pyrimiphos, ethyl	2	S	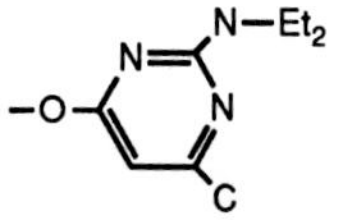	++
(16) Parathion (PTN)	2	S	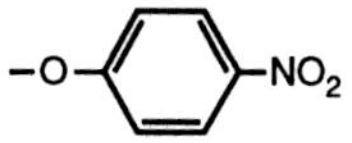	++
(17) PTN-methyl	1	S	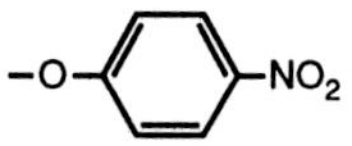	–
(18) 3,5-Dimethylphenyl-diethylphosphorothioate	2	S	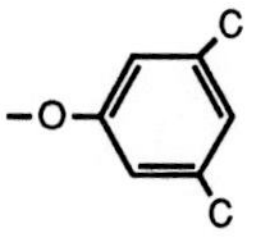	–
(19) Coumaphos	2	S	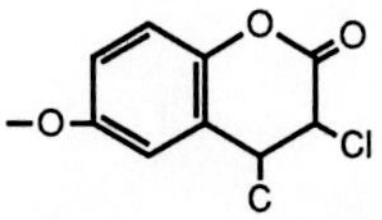	–

[a]1, methoxy; 2, ethoxy.
[b]O, oxygen; S, sulfur.
[c]All H atoms have been omitted except where necessary to show structure.
[d]++ is very teratogenic; + is mildly teratogenic; – is not teratogenic.

11, making them continuously available in small concentrations. Many of the *in vitro* teratogenic studies of the pyrimidyl OP compounds have been done with the oxo rather than the thio forms because they are considered to be the principal reactive species *in situ*.

The pyrimidyl compounds that have diethyl or di-*n*-propyl simple phosphate ester substituents (R_1 and R_2) have been found to be slightly more teratogenic than those containing dimethyl, diisopropyl or di-*n*-butyl. If R_1 is -*S*-alkyl or if the compound is a phenylphosphonate or phenylthiophosphonate, the teratogenicity is very much reduced (Eto, 1981).

The teratogenic activities of the pyrimidine-containing OP compounds are not specifically the result of the presence of an aromatic R_3 function, because similar OP compounds with nonpyrimidyl, aromatic leaving groups are usually not teratogenic (Table I). PTN (16) (R_3 = 4-nitrophenyl) is an exception. It is very teratogenic but does not cause so broad a range of deformities as DZN, ETF, etc., and PTN methyl (17) is much less teratogenic than PTN (Proctor *et al.*, 1976; Moscioni *et al.*, 1977). Examples of non-teratogenic OP compounds with aromatic leaving groups are 3,5-dimethylphenyl-diethylphosphothioate (18) and coumaphos (19) (Proctor *et al.*, 1976).

3. Other *N*-Heterocyclic Derivatives

The teratogenicities of many OP compounds containing other *N*-heterocyclic R_3 functions have been determined. For example, consider DZN analogs in which the pyrimidyl moiety is replaced by a pyridyl, linked at its 2, 3, or 4 position to the phosphorus atom. The 2-linked compound [Table II, compound (20)] is not teratogenic, while the 3- and 4-linked compounds [(22) and (24), respectively] are strongly teratogenic, causing mainly Type 2 defects. Methylation of the pyridine ring at positions adjacent to the N atom [(21), (23) and (25)] does not greatly alter the teratogenic intensity, but does change the defect from mainly Type 2 to mainly Type 1. Pyrimidine-containing OP compounds (11) and (26) have the same general structure–activity relationships as their pyridine-containing counterparts, the unalkylated forms causing mainly Type 2 teratisms and the alkylated causing mainly Type 1. If unsubstituted pyrazine (27) is the leaving group, the compound is also highly teratogenic, with strong Type 2 effects. Replacing the heterocyclic substituent with a homocyclic, aromatic residue, either with (29) or without (28), an exocyclic nitrogen-containing substituent, results in virtual loss of teratogenicity (Eto *et al.*, 1980).

4. Generalities

OP insecticides differ greatly in their embryotoxic and teratogenic effects. All are insecticidal, presumably because they or their active metabolic derivatives are strong inhibitors of ChE. They are also anticholinergic but not always lethal in their unintended targets (birds, mammals, etc.).

TABLE II

Aromatic Leaving Groups of OP Compounds and Teratogenic Signs in Avian Embryos

$$C{-}C{-}O{-}\overset{\overset{S}{\|}}{P}({-}O{-}C{-}C){-}R_3 \quad \text{Parent Compound}$$

Compound	R_3	Teratogenicity Intensity	Teratogenicity Type
(20) Pyridine		–	–
(21) Pyridine		–	–
(22) Pyridine		++	2
(23) Pyridine		++	1
(24) Pyridine		++	2
(25) Pyridine		++	1
(11) Pyrimidine		++	2
(26) Pyrimidine		++	1
(27) Pyrazine		++	2
(28) Phenyl		–	–
(29) *N*-substituted phenyl		–	–

A summary of some of the important structural determinants of teratogenic OP compounds on chick embryos is presented in Fig. 4. The compounds can be divided into two groups: the *cis*-crotonamide phosphates and the *N*-heterocyclic phosphates and phosphothioates (Eto, 1981). In both instances, R_1 and R_2 are usually simple ester or thiolester functions. The thiolesters are generally much less teratogenic than the corresponding oxygen esters. In the *cis*-crotonamide category, R_3 can be a methyl group, R_4, an H or Cl atom, and R_5 and R_6, H or small alkyl groups. Crucial for teratogenic activity are the cis configuration of the crotonamide and the presence of the amide nitrogen.

For strong teratogenicity among *N*-heterocyclic phosphates or phosphothioates, there should be at least one ring nitrogen in the leaving group, usually either meta or para to the phosphate ester bond. The heterocyclic moiety can be a pyridine, pyrimidine, or pyrazine. Alkyl substituents on the ring influence the biological nature of the deformity, emphasizing Type 1. There are many exceptions to the structure–activity relationships that have been outlined here, an important example of which is PTN [Table I, (16)], which does not contain a ring nitrogen but is highly teratogenic. There are also many other kinds of OP compounds that are not even considered in this survey, the teratogenic effects of which need to be determined on a case-by-case basis.

III. Mechanisms of Teratogenesis

The initial step of any mechanism of OP-induced teratogenesis must be an interaction between the toxicant and one or more components of the biological system. The OP insecticides or their metabolic derivatives react with AChE, which is a serine esterase. However, in most living organisms there are many different serine esterases, all having essentially the same mechanism of

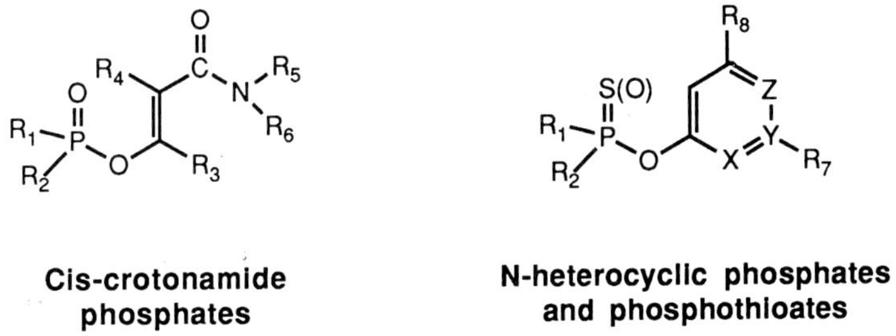

Cis-crotonamide phosphates

N-heterocyclic phosphates and phosphothioates

Figure 4 Salient molecular features of OP compounds that influence their teratogenicities. X, Y, and Z, C or N; R_1 and R_2, simply alkyl esters or thiol esters; R_3, H or methyl group; R_4, H or Cl; R_5 and R_6, H or small alkyl groups; R_7 and R_8, small alkyl groups.

action and, therefore, being candidates for inhibition. The number and OP-sensitivity of these targets and their importance to the functional integrity of the organism are unknown. Suntornwat and Kitos (1990) sought to determine the number and relative abundance of OP insecticide sensitive protein targets in the day 10 chick yolk sac membrane (YSM). Any protein that reacts with an OP insecticide should also be able to react with diisopropylfluorophosphate (DFP) (Fig. 5).

Chicken eggs at day 4 were injected by the intravitelline route with OP insecticides and then incubated further to day 10. Cell-free extracts of the YSMs were prepared and incubated at 0°C for 10 min with [3H]DFP. The proteins of the reaction mixtures were resolved by sodium dodecyl sulfate polyacrylamide gel electrophoresis (SDS PAGE), and the radioactive bands were detected by autofluorography and quantified by gel slicing and scintillation spectrometry. In the control system (no insecticide) there were three major radioactive protein bands (Fig. 6, A) with M_r of 93, 83, and 72 kDa. The 72-kDa peak was by far the most abundant, accounting for 88% of the radioactivity. If DZN (0.5 mg/egg), a strong teratogen, was injected at day 4, the radioactivity in all three peaks was decreased, but especially in peaks 1 and 2 (Fig. 6, B). If EPN (2 mg/egg), a weak teratogen, was injected, the radioactivity in peak 3 disappeared completely, while that in the other two peaks was only marginally lower (Fig. 6, C). These findings establish that inhibition of the 72-kDa protein of the YSM has nothing to do with the major teratogenic anomalies, while inhibition of the other two could be determinants in

HO—(E) + $(C_2H_5O)_2$—P(=S(O))—O—[pyrimidinyl: $CH(CH_3)_2$, CH_3] ⟶ $(C_2H_5O)_2$—P(=S(O))—O—(E) + HO—[pyrimidinyl: $CH(CH_3)_2$, CH_3]

Serine esterase **Diazinon (diazoxon)** **Inactive serine esterase**

HO—(E) + $[(CH^{*}_3)_2CH^{*}O]_2$—P(=O)—F ⟶ $[(CH^{*}_3)_2CH^{*}O]_2$—P(=O)—O—(E) + HF

$[^3H]$DFP **Radioactive serine esterase-DIP**

Figure 5 The $[^3H]$DFP-binding assay. (1) The OP compound (e.g., diazinon or diazoxon) binds covalently *in ovo* to the seryl hydroxyl at the active site of a susceptible YSM serine esterase. (2) The unbound fraction of the serine esterase in the YSM cell-free extract of the treated or untreated embryo then reacts with $[^3H]$DFP *in vitro* to form ^3H-labeled enzyme in proportion to the amount of the free serine esterase.

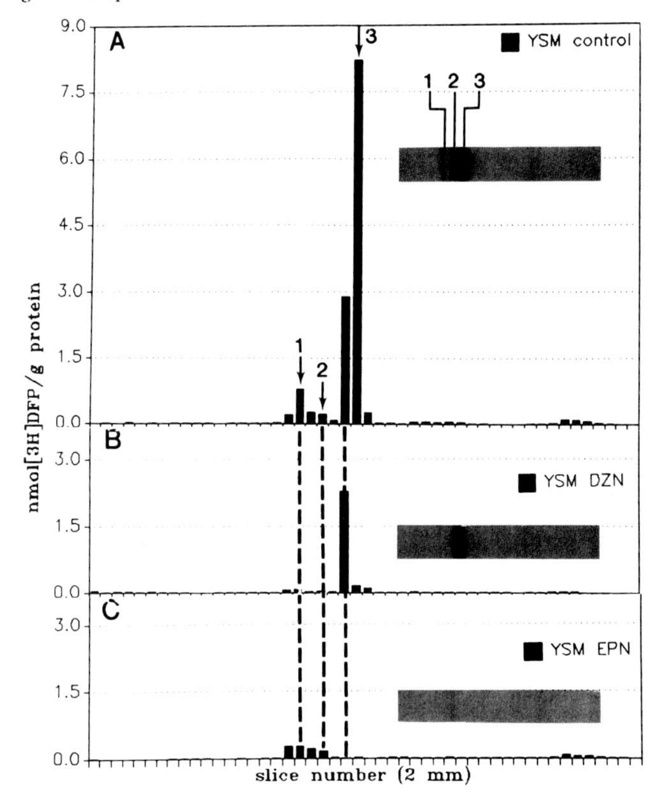

Figure 6 The effect of OP insecticides on target proteins of the YSM. The proteins in the cell-free extracts of day 10 chick YSMs, treated as explained in Fig. 5, were resolved by SDS PAGE. The DFP-labeled protein bands were visualized by autofluorography and quantified by gel slicing and scintillation spectrometry. (A) radioactivity profile from eggs injected with corn oil on day 4; (B) profile from eggs injected with 0.5 mg of diazinon in corn oil on day 4; (C) profile from eggs injected with 2 mg of EPN in corn oil on day 4. The amounts of radioactivity in the labeled peaks (ordinate) were converted to their molar equivalents. The inserts are the corresponding autofluorograms.

the DZN-induced malformations. These studies do not provide any information about the specific functions of the three DFP-sensitive proteins. Unfortunately, this [3H]DFP-labeling method is not sensitive enough to detect many of the minor DFP-reactive proteins in the extract. Since kynurenine formamidase has a 60-kDa M_r, it must not be one of the observed inhibited proteins.

Flockhart and Casida (1972) prepared starch-gel zymograms of the proteins of extracts of YSMs from OP-treated and untreated chicken eggs. They found 6 major, DFP-sensitive, α-naphthylphenylacetate hydrolyzing bands in the control preparations, and fewer bands in the extracts from dicrotophos- and EPN-treated eggs. However, they did not find good correlation between the bands that were inhibited and the teratogenic potencies of the OP compounds.

A. Type 1 Teratogenesis

Much of our current knowledge about the mechanisms of OP- and AC-induced Type 1 teratogenesis in birds comes from the work of Casida and his colleagues (Roger *et al.*, 1964, 1969; Upshall *et al.*, 1968; Flockhard and Casida, 1972; Proctor and Casida, 1975; Proctor *et al.*, 1976; Moscioni *et al.*, 1977; Seifert and Casida, 1978, 1979, 1982; Eto *et al.*, 1980). The more recent of these studies suggest that severe NAD deficits, brought on by many of the OP and AC compounds, are responsible for the deformities. They showed that there is excellent correlation between the teratogenicity of the OP compound, the NAD content of the embryonic tissues and the inhibition of kynurenine formamidase (KFase, EC 3.5.1.9) of the YSM (Proctor *et al.*, 1976; Moscioni *et al.*, 1977). There was not a corresponding change in the concentration of the other nucleotides, nor was there an effect on the reduced: oxidized ratio of either NAD or NADP (Kushaba-Rugaaju and Kitos, 1985). KFase is a target of the teratogenic OP compounds, and its inhibition prevents the synthesis of NAD from tryptophan (Moscioni *et al.*, 1977; Seifert and Casida, 1978, 1979, 1982; Eto *et al.*, 1980; Eto, 1981). As already mentioned, nicotinamide and several of its congeners and tryptophan ameliorate the Type 1 anomalies (Roger *et al.*, 1964; Greenberg and LaHam, 1970 Wenger and Wenger, 1973; Proctor *et al.*, 1976; Moscioni *et al.*, 1977; Kitos *et al.*, 1981b; Henderson and Kitos, 1982).

1. The Role of Kynurenine Formamidase Inhibition

KFase catalyzes the hydrolysis of *N*-formylkynurenine to kynurenine and formate. There are two major forms (A and B) of KFase in the YSM and chicken liver (Seifert and Casida, 1979). Only the B form (M_r 60 kDa) is inhibited by OP and AC compounds and by phenylmethanesulfonylfluoride (PMSF); the A form catalyzes the hydrolysis of N^1,N^α-diformylkynurenine but not of N^1-formylkynurenine, the natural substrate in the tryptophan to NAD pathway (Seifert and Casida, 1978, 1981). Therefore, only the B form is relevant to the *de novo* synthesis of NAD. Paradoxically, the inhibition kinetics of B-KFase do not correlate well with the teratogenic potencies of the OP and AC compounds. Seifert and Casida (1978) injected 21 OP compounds into the yolks of day 4 embryonated eggs and measured the inhibition of the YSM B-KFase 5 days later. The inhibition correlated with the lowering of the NAD content of the embryo and the intensity of the Type 1 teratogenic signs (Seifert and Casida, 1979). In agreement with these findings, Moscioni *et al.* (1977) showed that DZN inhibits the conversion of [^{14}C]tryptophan to NAD *in ovo*, as would be expected if KFase were inhibited.

There are pronounced differences between the effects of some OP compounds on KFase *in vivo* and *in vitro*. They can generally be explained by the fact that the oxoate is much more reactive than the thioate. The thioate must be converted *in vivo* to its more active oxoate form. For example, Eto *et al.* (1981) showed that some phosphothioates (DZN and several of its analogs) were active *in vivo* but inactive *in vitro*, while the corresponding oxoates were very active *in vitro*. In some cases the oxoates were inactive *in vivo*, probably because of their intrinsic instability in aqueous systems.

2. The Role of Serine Proteases and Tryptophan

There is reason to question the singularity of the role of KFase in causing avian Type 1 teratogenesis. DZN and dicrotophos can both cause Type 1 defects even if the NAD content of the embryo remains normal (Kitos *et al.*, 1981b). This happens when 2-PAM is administered to the early embryo along with the insecticide. As already indicated, 2-PAM prevents only the Type 2 deformities. Furthermore, tryptophan, administered *in ovo*, is effective in preventing both the NAD deficit in insecticide-treated embryos and Type 1 teratisms (Greenberg and LaHam, 1970; Wenger and Wenger, 1973, 1980; Henderson and Kitos, 1982). Since tryptophan precedes the KFase step in the NAD synthetic pathway, it would not be expected to suppress the deformities if KFase were the only OP target in causing Type 1 teratogenesis. However, tryptophan is essentially as effective in preventing the Type 1 deformities as several post-KFase intermediates (L-kynurenine, 3-hydroxyanthranilic acid, and quinolinic acid) of the tryptophan-to-NAD pathway (Henderson and Kitos, 1982).

Tryptophan is one of only a few free amino acids whose concentration in the chick embryo is reduced by the administration of a teratogenic dose of DZN (Kushaba-Rugaaju and Kitos, 1985). Larger concentrations of tryptophan were needed to support chondrogenesis (measured as sulfate incorporation into proteoglycans) than to support the proliferation of chick limb bud cells in micromass culture (Byrne and Kitos, 1983). Consequently, at least part of the underlying basis of Type 1 teratogenesis could be a limitation in the availability of free tryptophan, which, in turn, could affect the synthesis of NAD, proteins, and proteoglycans in the developing embryo (Greenberg and LaHam, 1970; Kitos *et al.*, 1987). This would be consistent with the report of Wenger (1974), which provided evidence for protease inhibition in the yolk sac membrane as a mechanism for OP-induced teratogenesis. It would also be consistent with the results of Freeman and Lloyd (1983, 1985) and Freeman and Brown (1985), who showed that nutritional deprivation in rats due to decreased protease activity could cause congenital deformities.

In studying the teratogenic effects of OP compounds on frog embryos, Snawder and Chambers (1989, 1990) noted that there was a reduction in the tissue NAD content of the subjects. The NAD level could be restored to normal by supplementation of the bathing fluid with tryptophan or niacin, but in neither case did the supplement decrease the severity of the deformity. In these situations, therefore, the NAD deficiency would not seem to be involved in the teratogenesis.

3. The Role of Poly(ADP ribose)

Caplan and Rosenberg (1975) reported that the NAD level of embryonic chick limb mesenchymal cells influences their chondrogenic and myogenic tendencies. Nicotinamide is both a substrate for NAD synthesis and a product inhibitor of poly(ADP ribose) synthesis. It can also inhibit chondrogenic expression of the chick limb mesenchymal cells. They suggested that poly(ADP ribose) synthesis is inhibited by high cellular levels of NAD and that this, in turn, programs the cells away from chondrogenesis; that low tissue NAD levels favor chondrogenic differentiation. Hwang *et al.* (1988) were unable to confirm that chondrogenic differentiation was promoted by low tissue NAD levels. Proctor *et al.* (1976), Moscioni *et al.* (1977) and Kitos *et al.* (1981b) observed that low tissue NAD levels of the chick limb bud, caused by the teratogenic OP compounds, resulted in micromelia, achondroplasia of the long bones of the legs. Benzamide is not an NAD precursor but is an inhibitor of poly(ADP ribose) polymerase. It was shown to augment chondrocytic differentiation (Nishio *et al.*, 1983) and increase the NAD levels (Nakanishi *et al.*, 1984; Nakanishi and Uyeki, 1985) of chick limb bud cells in *micromass* culture. These findings point to a possible involvement of poly(ADP ribose) polymerase activity in the generation of Type 1 teratisms but, unfortunately, they are difficult to interpret in terms of mechanism of action.

B. Type 2 Teratogenesis

1. The Role of Acetylcholinesterase Inhibition

OP-induced Type 2 teratogenesis, like the actions of OP compounds on insects, has been attributed to disruption of neurotransmission at the cholinergic neuromuscular junction (Upshall *et al.*, 1968; Eto, 1974; Meiniel, 1976b, 1977a, 1978b, 1981; Landauer, 1977; Misawa *et al.*, 1981; Meneely and Wyttenbach, 1989). It appears to be because of the inhibition of AChE and results in localized accumulations of ACh and the loss of regulation of cholinergic neurotransmission. The acute toxic and Type 2 teratogenic effects are not alleviated by niacin, but are by nontoxic oximes (Meiniel, 1974; Moscioni *et al.*, 1977). They include skeletal dysplasias, much like those produced by carbamylcholine and neostigmine, both of which are carbamate analogs of acetylcholine, and the actions of which are antagonized by 2-PAM (Landauer, 1977). Three kinds of findings support cholinergic involvement in Type 2 teratogenesis (Seifert and Casida, 1981): (1) AChE activity is decreased and tissue ACh concentration is increased by the teratogens (e.g., PTN, dicrotophos, eserine) (Khera, 1966; Upshall *et al.*, 1968; Greenberg and LaHam, 1970; Walker, 1971; Meiniel, 1977a); (2) certain pyridine aldoximes decrease the Type 2 teratisms (Meiniel, 1974, 1975, 1976a,b,c; Moscioni *et al.*, 1977; Landauer, 1977), probably by dephosphorylating the inhibited AChE (Aldridge and Reiner, 1972; Kiffer and Delamanche, 1983); and (3) several cholinergic agonists (e.g., gallamine, succinylcholine) cause axial skeletal defects similar to those produced by the Type 2 OP teratogens (Upshall *et al.*, 1968; Roger *et al.*, 1969; Landauer, 1975a,b; Meiniel, 1978b).

2. The Possible Involvement of Proteoglycans

The OP-induced Type 2 axial skeletal deformities in the chick embryo appear early in development and can be generated at any time up to day 11 (Meiniel and Autissier-Navarro, 1980). Agents that favor muscle membrane depolarization (cholinergic agonists and ChE inhibitors) cause Type 2 teratisms. Meiniel (1981) suggested that the spinal scoliosis and vertebral fusion are the result of sustained muscle contraction. Garrison and Wyttenbach (1985a) reported strong cervical notochordal folding, deformities of the adjacent spinal cord and distention of the major blood vessels beginning as early as 24 hr after the start of incubation and diminishing with time thereafter. They postulated that failures in the development of the supportive sheath of these morphological structures cause the deformations. In support of this contention are the findings of Ho and Gibson (1972) that MLN causes undersulfation of chondroitin sulfate, the principal glycosaminoglycan of the sheath. Frederickson and Low (1971) had shown that chondroitin sulfates were important components in linking collagen fibrils in the notochordal sheath. However, caution should be exercised in interpreting the correlation since Ho and Gibson were examining the OP effect on the chick embryo tibiotarsus, for which niacin is an antagonist, while Garrison and Wyttenbach (1985a) were considering defects in the notochord and spinal cord, for which niacin is not an antagonist. Using DZN and dicrotophos, injected at day 3, Misawa *et al.*, (1982) observed undulating notochord and fused cervical rings as early as day 6. They considered the neck deformities attributable to early changes in the processes of differentiation.

3. Other Possible Contributing Factors

There are other ways in which the OP compounds can affect biological systems, including embryos. For example, sister chromatic exchange (SCE), a possible index of mutagenicity, was induced in hamster and human cell lines by several OP insecticides (methyl PTN, demeton, trichlorfon, dimethoate, MLN, and methidathion) but not by DZN or disyston (Chen *et al.*, 1981). All of these OP compounds caused cell-cycle delays, especially in hamster cells. Nishio and Uyeki (1981) examined dichlorvos, dicrotophos, MLN, PTN, leptophos, DZN, malaoxon, paraoxon, leptophosoxon, and diazoxon at concentrations up to 1 m*M* for their abilities to produce SCE in Chinese hamster cells. None was as effective as the standard mutagen ethylmethanesulfonate, but all except DZN caused an increase in SCE. The oxoates were more effective antiproliferative and SCE-producing agents than the thioates. MLN-induced chromosomal aberrations in bone marrow cells of mice have also been reported (Dulout *et al.*, 1983). The antiproliferative IC_{50}s of methyl PTN, DZN, paraoxon, chlorpyrifos and methylchlorpyrifos, and leptophos on chick ganglion cultures ranged from 10^{-6} *M* to 10^{-2} *M* (Sharma and Obersteiner, 1981). The cytotoxic changes included decreased migration of cells from the tissue explant, varicosities in and shortening of cells, vacuolization and rounding of neuroglial cells, pigment degeneration, and abolition of cell growth. There was no correlation of nerve fiber or glial cell cytotoxicity with the LD_{50} in animals, ChE inhibition, or lipophilicity of the agents.

Kiermayer and Fedtke (1977) reported that amiprophosmethyl has a strong antimicrotubule action, inhibiting postmitotic migration of the nucleus. PTN has been shown to inactivate rat liver cytochrome P_{450} even more effectively than does CO (Halpert and Neal, 1979). Nichol *et al.* (1983) and Nichol and Angel (1984) reported the inhibition of porphyrin biosynthesis in rats by isodiazinon. It causes decreases in the activities of liver ferrochelatase and coproporphyrinogen oxidase but not in glutamic dehydrogenase, succinic dehydrogenase, or kynurenine hydroxylase, suggesting that the action is porphyrin biosynthesis rather than the result of mitochondrial damage.

V. Overview

It should not be in the least astonishing that OP compounds are serious potential hazards to the born and unborn, especially as they are the chemical kin of the lethal family of nerve gases. By design, the OP insecticides, acaracides, nematocides, and rodenticides focus their destructive actions on targets within organisms that are fundamentally similar to those in humans and other higher animals and thus, lacking discretion, they challenge all of animal life. In this realm, the most insidious of their consequences can be upon the least mature, the most formative, the unborn. On the other hand, with their destructive potential, it is remarkable that the OP compounds are as innocuous to most animals as they seem to be. However, vigilance and caution in their use is the wisdom that gives these compounds their current acceptance.

Acknowledgments

For the support of these studies, the authors thank the Fulbright Foundation (for its support of O.S.), the KUGRF (Grant No. 3243) and the Wesley Foundation of Wichita, Kansas (Grant No.

8812007). The Wesley Foundation is an independent, nonprofit organization whose mission is to improve the quality of health in Kansas.

References

Aldridge, W. N., and Reiner, E. (1972). "Enzyme Inhibitors as Substrates." American Elsevier, New York.

Arsenault, A. L., Gibson, M. A., and Mader, M. E. (1975). Hypoglycemia in malathion-treated chick embryos. *Can. J. Zool.* 53, 1055–1057.

Baron, R. L., and Johnson, H. (1964). Neurological disruption prolonged in hens by two organophosphate esters. *Br. J. Pharmacol.* 23, 295–304.

Behera, B. C., and Bhunya, S. P. (1987). Genetoxic potential of an organophosphate insecticide, phosphamidon (Dimecron): An *in vivo* study in mice. *Toxicol. Lett.* 37, 269–277.

Behera, B. C., and Bhunya, S. P. (1989). Studies on the genotoxicity of Asataf (acephate), an organophosphate insecticide, in a mammalian *in vivo* system. *Mutat. Res.* **223**, 287–293.

Berge, G. N., Nafstad, I., and Fonnum, F. (1986). Prenatal effects of trichlorfon on the guinea pig brain. *Arch. Toxicol.* **59**, 30–35.

Bhatnagar, P., and Soni, I. (1988). Evaluation of the teratogenic potential of phosphamidon in mice by gavage. *Toxicol. Lett.* **42**, 101–107.

Bhunya, S. P., and Behera, B. C. (1989). Evaluation of mutagenicity of a commercial organophosphate insecticide, Tafethion, in mice tested *in vivo*. *Caryologia* **42**, 139–145.

Budreau, C. H., and Singh, R. P. (1973). Teratogenicity and embryotoxicity of demeton and fenthion in CF #1 mouse embryos. *Toxicol. Appl. Pharmacol.* **24**, 324–332.

Byrne, D. H., and Kitos, P. A. (1983). Teratogenic effects of cholinergic insecticides in chick embryos. IV. The role of tryptophan in protecting against limb deformities. *Biochem. Pharmacol.* **32**, 2881–2890.

Casida, J. E., Baron, R. L., Eto, M., and Engel, J. L. (1963). Potentiation and neurotoxicity induced by certain organophosphates. *Biochem. Pharmacol.* **12**, 73–83.

Caplan, A. I., and Rosenberg, M. J. (1975). Interrelationship between poly(ADP-ribose) synthesis, intracellular NAD levels, and muscle or cartilage differentiation from mesodermal cells of embryonic chick limb. *Proc. Natl. Acad. Sci. U.S.A.* **72**, 1852–1857.

Chen, H. H., Hsueh, J., Sirianni, S. R., and Huang, C. C. (1981). Induction of sister chromatid exchanges and cell cycle delay in cultured mammalian cells treated with eight organophosphorus pesticides. *Mut. Res.* **88**, 307–316.

Clegg, D. J. (1964). The hen egg in toxicity and teratogenicity studies. *Food Cosmet. Toxicol.* **2**, 717–727.

Clemens, G. R., Hatnagel, R. E., Bare, J. J., and Thyssen, J. H. (1990). Teratological, neurochemical, and postnatal neurobehavioural assessment of Metasystox-R, an organophosphate pesticide in the rat. *Fundam. Appl. Toxicol.* **14**, 131–143.

Courtney, K. D., Andrews, J. E., and Springer, J. (1986). Assessment of teratogenic potential of trichlorfon in mice and rats. *J. Environ. Sci. Health*, part B. **21**, 207–227.

Deacon, M. M., Murray, J. S., Pilny, M. K., Rao, K. S., Dittenber, D. A., Hanley, T. R., and John, J. A. (1980). Embryotoxicity and fetotoxicity of orally administered chlorpyrifos in mice. *Toxicol. Appl. Pharmacol.* **54**, 31–40.

Dingerkus, G., and Uhler, L. D. (1977). Enzyme clearing of Alcian blue–stained whole small vertebrates for demonstration of cartilage. *Stain Technol.* **52**, 229–231.

Dobbins, P. K. (1967). Organic phosphate insecticides as teratogens in the rat. *J. Fla. Med. Assoc.*54, 452–456.

Dulout, F. N., Pastori, M. C., and Olivero, O. A. (1983). Malathion-induced chromosomal aberrations in bone-marrow cells of mice: Dose–response relationships. *Mut. Res. Lett.* **122**, 163–167.

Eto, M. (1974). "Organophosphorus Pesticides: Organic and Biological Chemistry." CRC Press, Cleveland, Ohio.

Eto, M. (1981). Structure and avian teratogenicity of organophosphorus compounds. *J. Pestic. Sci.* **6**, 95–106.

Eto, M., Seifert, J., Engel, J. L., and Casida, J. E. (1980). Organophosphorus and methylcarbamate teratogens: Structural requirements for inducing embryonic abnormalities in chickens and kynurenine formamidase inhibition in mouse liver. *Toxicol. Appl. Pharmacol.* **54**, 20–30.

Farage-Elawar, M., and Francis, B. M. (1988). Effects of fenthion, fenitrothion, and desbromoleptophos on gait, acetylcholinesterase, and neurotoxic esterase in young chicks after *in ovo* exposure. *Toxicology* **49**, 253–261.

Fish, S. A. (1966). Organophosphorus cholinesterase inhibitors and fetal development. *Am. J. Obstet. Gynecol.* **96**, 1148–1154.

Fishbein, L. (1976). Teratogenic, mutagenic, and carcinogenic effects of insecticides. *In* "Insecticide Biochemistry and Physiology" (C. F. Wilkinson, ed.), pp. 555–604. Plenum Press, New York.

Flockhart, I. R., and Casida, J. E. (1972). Relationship of the acylation of membrane esterases and proteins to the teratogenic action of organophosphorus insecticides and eserine in developing hen eggs. *Biochem. Pharmacol.* **21**, 2591–2603.

Frederickson, R. G., and Low, F. N. (1971). The fine structure of perinotochordal microfibrils in control and enzyme-treated chick embryos. *Am. J. Anat.* **130**, 347–375.

Freeman, S. J., and Brown, N. A. (1985). Comparative effects of cathepsin inhibitors on rat embryonic development *in vitro*. Evidence that cathepsin D is unimportant in the proteolytic function of yolk sac. *J. Embryol. Exp. Morphol.* **86**, 271–281.

Freeman, S. J., and Lloyd, J. B. (1983). Inhibition of proteolysis in rat yolk sacs as a cause of teratogenesis. Effects of leupeptin *in vitro* and *in vivo*. *J. Embryol. Exp. Morphol.* **78**, 196–197.

Freeman, S. J., and Lloyd, J. B. (1985). Interference with embryonic nutrition as the teratogenic mechanism of action of suramin and aurothiomalate in rats. *Biochem. Soc. Trans.* **13**, 196–197.

Fulton, M. H., and Chambers, J. E. (1985). The toxic and teratogenic effects of selected organophosphorus compounds on the embryos of three species of amphibians. *Toxicol. Lett.* **26**, 175–180.

Garrison, J. C., and Wyttenbach, C. R. (1985a). Notochordal development as influenced by the insecticide dicrotophos (Bidrin). *J. Exp. Zool.* **234**, 243–250.–Garrison, J. C., and Wyttenbach, C. R. (1985b). Teratogenic effects of the organophosphate insecticide dicrotophos (Bidrin): Histological characterization of defects. *Anat. Rec.* **213**, 464–472.

Garry, V. F., Griffith, J., Danzl, T. J., Nelson, R. L., Whorton, E. B., Krueger, L. A., and Cervenka, J. (1989). Human genotoxicity: Pesticide applicators and phosphine. *Science* **246**, 251–255.

Gebhardt, D. O. E. (1968). The teratogenic action of propyleneglycol (propanediol-1,2) and propanediol-1,3 in the chick embryo. *Teratology* **1**, 153–161.

Gebhardt, D. O. E. (1972). The use of the chick embryo in applied teratology. *Adv. Teratol.* **5**, 97–111.

Goel, S. C., and Jurand, A. (1976). Effects of hydrocortisone acetate on the development of chicken embryos. *Teratology*, **13**, 139–149.

Greenberg, J., and LaHam, Q. N. (1970). Malathion-induced teratism in the developing chick. *Can. J. Zool.* **47**, 539–542.

Grether, J. K., Harris, J. A., Neutra, R., and Kizer, K. W. (1987). Exposure to aerial malathion application and the occurrence of congenital anomalies and low birthweight. *Am. J. Pub. Health* 77, 1009–1010.

Grubb, R. B., and Montiegel, E. C. (1975). The teratogenic effects of 6-mercaptopurine on chick embryos *in ovo*. *Teratology* **11**, 179–185.

Hall, B. K. (1977). Thallium-induced achondroplasia in chicken embryos and the concept of critical periods during development. *Teratology* **15**, 1–16.

Hall, R. J., and Kolbe, E. (1980). Bioconcentration of organophosphorus pesticides to hazardous levels by amphibians. *J. Toxicol. Environ. Health* **6**, 853–860.

Halpert, J. R., and Neal, R. A. (1980). Mechanism of the inactivation of rat liver cytochrome P450 by parathion. *In* "Microsomes, Drug Oxidation, and Chemical Carcinogenesis" (M. J. Coon, A. H. Conney, and R. W. Estabrook, eds.), Vol. 1, pp. 323–326. Academic Press, New York.

Hanafy, M. S. M., and El-Din, A. W. K. (1986). The effect of an organophosphorus insecticide Tamaron on the developing chick embryo. *Vet. Med. J.* **34**, 347–356.

Hanafy, M. S. M., Atta, A. H., and Hashim, M. M. (1986). Studies on the teratogenic effects of Tamaron (an organophosphorus pesticide). *Vet. Med. J.* **34**, 357–363.

Henderson, M., and Kitos, P. A. (1982). Do organophosphate insecticides inhibit the conversion of tryptophan to NAD *in ovo*? *Teratology* **26**, 173–181.

Ho, M., and Gibson, M. A. (1972). A histochemical study of the developing tibiotarsus in malathion-treated chick embryos. *Can. J. Zool.* **50**, 1293–1298.

Hodach, R. J., Gilbert, E. F., and Fallon, J. F. (1974). Aortic arch anomalies associated with the administration of epinephrine in chick embryos. *Teratology* **9**, 203–209.

Hoffman, D. J., and Eastin, W. C. (1981). Effects of malathion, diazinon, and parathion on mallard embryo development and cholinesterase activity. *Environ. Res.* **26**, 472–485.

Hoffman, D. J., and Sileo, L. (1984). Neurotoxic and teratogenic effects of an organophosphorus insecticide [phenylphosphonothioic acid *o*-ethyl *o*-(4-nitrophenyl) ester] on mallard development. *Toxicol. Appl. Pharmacol.* **73**, 284–294.

Hwang, P. M., Byrne, D. H., and Kitos, P. A. (1988). Effects of molecular oxygen on chick limb bud chondrogenesis. *Differentiation* **37**, 14–19.

Khadzhitodorova, E., and Andreev, A. (1984). Effect of the organophosphate insecticide chloracetophon on ossification of the skeleton and development of the internal organs of fetuses of white rats. *Eksp. Med. Morfol.* **23**, 201–205.

Khera, K. S. (1966). Toxic and teratogenic effects of insecticides in duck and chick embryos. *Toxicol. Appl. Pharmacol.* **8**, 345 (Abstr.).

Khera, K. S., and Bedok, S. (1967). Effects of thiol phosphates on mitochondrial and vertebral morphogenesis in chick and duck embryos. *Food Cosmet. Toxicol.* **5**, 359–365.

Khera, K. S., and LaHam, Q. N. (1965). Cholinesterases and motor end-plates in developing duck skeletal muscle. *J. Histochem. Cytochem.* **13**, 559–562.

Khera, K. S., LaHam, W. N., and Grice, H. C. (1965). Toxic effects in ducklings hatched from embryos inoculated with EPN or systox. *Food. Cosmet. Toxicol.* **3**, 581–586.

Kiermayer, O., and Fedtke, C. (1977). Strong antimicrotubule action of amiprophos methyl (APM) in *Micrasterias*. *Protoplasma* **92**, 163–166.

Kiffer, D., and Delamanche, I. S. (1983). *In vitro* study of organophosphorus inactivators of membrane acetylcholinesterase and of reactivating pyridinium oximes using rat brain slices. *Biochimie* **65**, 477–484.

Kimbrough, R. D., and Gaines, T. B. (1968). Effect of organic phosphorus compounds and alkylating agents on the rat fetus. *Arch. Environ. Health* **16**, 805–808.

Kitos, P. A., Wyttenbach, C. R., Olson, K., and Uyeki, E. M. (1981a). Precision delivery of small volumes of liquids to very young avian embryos. 2. Description of the injection system. *Toxicol. Appl. Pharmacol.* **59**, 49–53.

Kitos, P. A., Anderson, D. S., Uyeki, E. M., Misawa, M., and Wyttenbach, C. R. (1981b). Teratogenic effects of cholinergic insecticides in chick embryos. 2. Effects on the NAD content of early embryos. *Biochem. Pharmacol.* **30**, 2225–2235.

Kitos, P. A., Tennant, K. D., and Kim, H- Y. (1987). Tryptophan deficiency: Role in organophosphorus teratogenesis. *In* "Progress in Tryptophan and Serotonin Research, 1986" (D. A. Bender, M. H. Joseph, W. Kochen, and H. Steinhart, eds.), Walter de Gruyter, New York.

Kushaba-Rugaaju, S., and Kitos, P. A. (1985). Effects of diazinon on nucleotide and amino acid contents of chick embryos. Teratogenic considerations. *Biochem. Pharmacol.* **34**, 1937–1943.

Laley, B. O., and Gibson, M. A. (1977). Association of hypoglycemia and pancreatic islet tissue with micromelia in malathion-treated chick embryos. *Can. J. Zool.* **55**, 261–264.

Landauer, W. (1975a). Cholinomimetic teratogens: Studies with chicken embryos. *Teratology* **12**, 125–145.

Landauer, W. (1975b). Cholinomimetic teratogens: II. Interaction with inorganic ions. *Teratology*, **12**, 271–276.

Landauer, W. (1977). Cholinomimetic teratogens: V. The effect of oximes and related cholinesterase reactivators. *Teratology* **15**, 33–42.

Lindhout, D., and Hageman, G. (1987). Amyoplasia congenita-like condition and maternal malathion exposure. *Teratology* **36**, 7–9.

Lutz-Ostertag, Y., and Bruel, M. T. (1981). Embryotoxic and teratogenous action of dichlorvos (organophosphate insecticide) on quail embryo development. *C. R. Seances Acad. Sci. Ser.* 3, **292**(18), 1051–1054.

Machin, M. G., and McBride, W. G. (1989). Teratological study of malathion in the rabbit. *J. Toxicol. Environ. Health* **26**, 249–253.

Marliac, J. P. (1964). Toxic and teratogenic effects of 12 pesticides in the chick embryo. *Fed. Proc.* **23**, 105 (Abstr.).

Marliac, J. P., Verrett, M. J., McLaughlin, J., Jr., and Fitzhugh, O. G. (1965). A comparison of toxicity data obtained from 21 pesticides by the chicken embryo technique for acute, oral LD_{50}s in rats. *Toxicol. Appl. Pharmacol.* **7**, 490 (Abstr.).

McLaughlin, J., Marliac, J. P., Verret, M. J., Mutchler, M. K., and Fitzhugh, O. G. (1963). The injection of chemicals into the yolk sac of fertile eggs prior to incubation as a toxicity test. *Toxicol. Appl. Pharmacol.* **5**, 760–771.

Meiniel, R. (1974). Action protectrice de la pralidoxime vis-à-vis des effets tératogenes du parathion sur le squélette axial de l'embryon de caille. *C. R. Acad. Sci.* Sec. D, **279**, 603–606.

Meiniel, R. (1975). Praladoxime, specific antiteratogen compound for bidrin-induced axial deformities in quail embryos. *C. R. Hebd. Seances Acad. Sci.* Ser. D **280**, 1019–1022.

Meiniel, R. (1976a). Prevention des anomalies induities par deux insecticides organophosphores (parathion et bidrin) chez l'embryon de caille. *Arch. Anat. Microsc. Morphol. Exp.* **65**, 1–15.

Meiniel, R. (1976b). Plurality in the determinism of organophosphorus teratogenic signs. *Experientia* **32**, 920–922.

Mieniel, R. (1976c). Expression of parathion axial teratogenesis after giving various compounds known to hve antiteratogenic and antitoxic action in the adult or embryo of vertebrates after exposure to organic phosphates. Study on quail embryo (*Coturnix coturnix japonica*). *C. R. Hebd. Seances Acad. Sci.* Ser. D **283**, 1085–1087.

Meiniel, R. (1977a). Cholinesterase activities and expression of axial teratogenesis in the quail embryo exposed to organophosphates. *C. R. Hebd. Seances Acad. Sci.* Ser. D **285**, 401–404.

Meiniel, R. (1977b). Tératogenie des anomalies axiales induities par un insecticide organophosphore (le parathion) chez l'embryon d'oiseau. *Wilhelm Roux's Arch. Dev. Biol.* **181**, 41–63.

Meiniel, R. (1978a). Agents anticholinesterasiques et tératogenese axiale chez l'embryon de caille. *Wilhelm Roux's Arch. Dev. Biol.* **185**, 209–225.

Meiniel, R. (1978b). Neuroactive compounds and vertebral teratogenesis in the bird embryo. *Experientia* **34**, 394–396.

Meiniel, R. (1981). Neuromuscular blocking agents and axial teratogenesis in the avian embryo. Can axial morphogenetic disorders be explained by pharmacological action upon muscle tissue? *Teratology* **23**, 259–271.

Meiniel, R., and Autissier-Navarro, C. (1980). Teratogenic activity of organophosphate pesticide in chick embryos. *Acta Embryol. Morphol. Exp.* **1**, 33–41.

Meiniel, R., Lutz-Ostertag, Y., and Lutz, H. (1970). Effets tératogenes du parathion (insecticide organophosphore) sur le squélette embryonnaire de la caille Japonaise (*Coturnix coturnix japonica*). *Arch. Anat. Micr. Morph. Exp.* **59**, 167–183.

Meneely, G. A., and Wyttenbach, C. R. (1989). Effects of the organophosphate insecticides diazinon and parathion on bob-white quail embryos: Skeletal defects and acetylcholinesterase activity. *J. Exp. Zool.* **252**, 60–70.

Misawa, M., Doull, J., Kitos, P. A., and Uyeki, E. M. (1981). Teratogenic effects of cholinergic insecticides in chick embryos. 1. Diazinon treatment on acetylcholinesterase and choline acetyltransferase activities. *Toxicol. Appl. Pharmacol.* **57**, 20–29.

Misawa, M., Doull, J., and Uyeki, E. M. (1982). Teratogenic effects of cholinergic insecticides in chick embryos. 3. Development of cartilage and bone. *J. Toxicol. Environ. Health* **10**, 551–563.

Moscioni, A. D., Engel, J. L., and Casida, J. E. (1977). Kynurenine formamidase inhibition as a possible mechanism for certain teratogenic effects of organophosphorus and methylcarbamate insecticides in chicken embryos. *Biochem. Pharmacol.* **26**, 2251–2258.

Nafstad, I., Berge, G., Sannes, E., and Lyngset, A. (1983). Teratogenic effects of the organophosphorus compound fenchlorphos in rabbits. *Acta Vet. Scand.* **24**, 295–304.

Nakanishi, S., Nishio, A., and Uyeki, E. M. (1984). Effect of benzamide on cell growth, NAD and ATP levels in cultured chick limb bud cells. *Biochem. Biophys. Res. Commun.* **121**, 710–716.

Nakanishi, S., and Uyeki, E. M. (1985). Benzamide on chondrocytic differentiation in chick limb bud cell culture. *J. Embryol. Exp. Morphol.* **85**, 163–175.

Nehez, M., Huszta, E., Mazzag, E., Scheufler, H., Schneider, P., and Fischer, G. W. (1987). Cytogenetic, genetic, and embryotoxicity studies with dimethyl-(2,2,2-trichloro-1-hydroxyethoxy)-phosphonate, a hypothetical impurity in technical grade trichlorfon. *Ecotoxicol. Environ. Safety* **13**, 216–224.

Nichol, A. W., Elsbury, S., Angel, L. A., and Elder, G. H. (1983). The site of inhibition of porphyrin biosynthesis by an isomer of diazinon in rats. *Biochem. Pharmacol.* **32**, 2653–2658.

Nichol, A. W., and Angel, L. A. (1984). A comparative study of porphyrin accumulation in tissue cultures of chicken embryo hepatocytes treated with organophosphorus insecticides. *Biochem. Pharmacol.* **33**, 2511–2515.

Nishio, A., and Uyeki, E. M. (1981). Induction of sister chromatid exchanges in Chinese hamster ovary cells by organophosphate insecticides and their oxygen analogs. *J. Toxicol. Environ. Health* **8**, 939–946.

Nishio, A., Nakanishi, S., Doull, J., and Uyeki, E. M. (1983). Enhanced chondrocytic differentiation in chick limb bud cell cultures by inhibitors of poly(ADP-ribose) synthetase. *Biochem. Biophys. Res. Commun.* **111**, 750–759.

Ohkawa, H. (1982). Stereoselectivity of organophosphorus insecticides. *In* "Insecticide Mode of Action," pp. 163–185. Academic Press, New York.

Overman, D. O., Graham, M. N., and Roy, W. A. (1976). Ascorbate inhibition of 6-aminonicotinamide teratogenesis in chicken embryos. *Teratology* **13**, 85–93.

'roctor, N. H., and Casida, J. E. (1975). Organophosphorus and methylcarbamate insecticide teratogenesis: Diminished NAD in chicken embryos. *Science* **190**, 85–93.

Proctor, N. H., Moscioni, A. D., and Casida, J. E. (1976). Chicken embryo NAD levels lowered by teratogenic organophosphorus and methylcarbamate insecticides. *Biochem. Pharmacol.* 25, 757–762.

Richert, E. P., and Prahlad, K. V. (1972). Effect of the organophosphate *o,o*-diethyl *S*-[(ethylthio)methyl] phosphorodithioate on the chick. *Poult. Sci.* 51, 613–619.

Robinson, E. C., Hammond, B. G., Johannsen, F. R., Levinskas, G. J., and Rodwell, D. E. (1986). Teratogenicity studies of alkylaryl phosphate ester plasticizers in rats. *Fundam. Appl. Toxicol.* 7, 138–143.

Roger, J.-C., Chambers, H., and Casida, J. E. (1964). Nicotinic acid analogs: Effects on response of chick embryos and hens to organophosphate toxicants. *Science* **144**, 539–540.

Roger, J.-C., Upshall, D. G., and Casida, J. E. (1969). Structure–activity and metabolism studies on organophosphate teratogens and their alleviating agents in developing hen eggs with special emphasis on bidrin. *Biochem. Pharmacol.* 18, 373–392.

Romero, P., Barnett, P. G., and Midtling, J. E. (1989). Congenital anomalies associated with maternal exposure to oxydemeton methyl. *Environ. Res.* 50, 256–261.

Schom, C. B., and Abbot, U. K. (1977). Temporal, morphological, and genetic responses of avian embryos to azodrin, an organophosphate insecticide. *Teratology* 15, 81–87.

Schom, C. B., Abbot, U. K., and Walker, N. E. (1979). Adult and embryo responses to organophosphate pesteicide: Azodrin. *Poult. Sci.* 58, 60–66.

Seifert, J., and Casida, J. E. (1978). Relation of yolk sac membrane kynurenine formamidase inhibition to certain teratogenic effects of organophosphorus insecticides and of carbaryl and eserine in chicken embryos. *Biochem. Pharmacol.* 27, 2611–2615.

Seifert, J., and Casida, J. E. (1979). Inhibition and reactivation of chicken kynurenine formamidase: *In vitro* studies with organophosphorates, *N*-alkylcarbamates and phenylmethanesulfonyl fluoride. *Pestic. Biochem. Physiol.* 12, 273–279.

Seifert, J., and Casida, J. E. (1981). Mechanisms of teratogenesis induced by organophosphorus and methylcarbamate insecticides. *In* "Progress in Pesticide Biochemistry" (D. H. Hutson and T. R. Roberts, eds.), Vol. 1, pp. 219–246. John Wiley, New York.

Sharma, R. P., and Obersteiner, E. J. (1981). Cytotoxic responses of selected insecticides in chick ganglia cultures. *Can. J. Comp. Med.* 45, 60–69.

Shellenberger, T. E. (1980). Organophosphate pesticide inhibition of cholinesterase in laboratory animals and man and effects of oxime reactivators. *J. Environ. Sci. Health* part B 15, 795–822.

Snawder, J. E., and Chambers, J. E. (1989). Toxic and developmental effects of organophosphorus insecticides in embryos of the South African clawed frog. *J. Environ. Sci. Health*, **B24**, 205–218.

Snawder, J. E., and Chambers, J. E. (1990). Critical time periods and the effect of tryptophan in malathion-induced developmental defects in Xenopus embryos. *Life Sci.* 46, 1635–1642.

Somlyay, I. M., and Varnagy, L. E. (1986). The avian embryo as test model in first-line pesticide screening for teratology and embryotoxicology. *Meded. Fac. Landbouwwet, Rijksuniv. Gent* 51(2a), 219–225.

Soni, I., and Bhatnagar, P. (1989). Embryotoxic and teratogenic studies of phosphamidon in Swiss albino mice. *Teratogen. Carcinogen. Mutagen.* 9, 253–257.

Suntornwat, O., and Kitos, P. A. (1990). Yolk sac membrane targets of the organophosphorus insecticides. *FASEB J.* 4(7), A2051.

Tanimura, T., Katsuya, T., and Nishimura, H. (1967). Embryotoxicity of acute exposure to methyl parathion in rats and mice. *Arch. Environ. Health* 15, 613–619.

Toy, A. D. F., and Walsh, E. N. (1987). "Phosphorus Chemistry in Everyday Living," 2d Ed. American Chemical Society, Washington, D.C.

Upshall, D. G., Roger, J. -C., and Casida, J. E. (1968). Biochemical studies on the teratogenic action of bidrin and other neuroactive agents in developing hen eggs. *Biochem. Pharmacol.* 17, 1529–1542.

van Steenis, G., and van Logten, M. J. (1971). Neurotoxic effect of the dithiocarbamate Tecoram on the chick embryo. *Toxicol. Appl. Pharmacol.* **19**, 675–686.

Varnagy, L. (1981). Teratological examination of agricultural pesticides on Japanese quail (*Coturnix coturnix japonica*) eggs. *Acta Vet. Acad. Sci. Hung.* **29**, 77–83.

Varnagy, L., Imre, R., Fancsi, T., and Hadhazy, A. (1982). Teratogenicity of methyl parathion 18 WP and Wofatox 50 EC in Japanese quail and pheasant embryos with particular reference to osteal and muscular systems. *Acta Vet. Acad. Sci. Hung.* **30**, 135–146.

Walker, N. E. (1967). Distribution of chemicals injected into fertile eggs and its effect upon apparent toxicity. *Toxicol. Appl. Pharmacol.* **10**, 290–299.

Walker, N. E. (1968). Use of yolk-chemical mixtures to replace hen egg yolk in toxicity and teratogenicity studies. *Toxicol. Appl. Pharmacol.* **12**, 94–104.

Walker, N. E. (1971). The effect of malathion and malaoxon on esterases and gross development in the chick embryo. *Toxicol. Appl. Pharmacol.* **19**, 590–601.

Welsh, J. J., Collins, T. F. Whitby, K. E., Black, T. N., and Arnold, A. (1987). Teratogenic potential of triphenylphosphate in Sprague-Dawley (Spartan) rats. *Toxicol. Indust. Health* **3**, 357–369.

Wenger, B. S. (1974). Protease inhibition as a teratogenic mechanism. *Am. Zool.* **14**, 1305 (Abstr.).

Wenger, B. S., and Wenger, E. (1973). Prevention of malathion-induced malformations in chick embryos by nicotinamide and tryptophan. *Proc. Can. Fed. Biol. Soc.* **16**, 61 (Abstr.).

Wenger, B. S., and Wenger, E. (1980). Pyridine nucleotides in drug teratogenesis. *Teratology* **21**, 74a (Abstr.).

Wilson, J. G. (1972). Interrelationships between carcinogenicity, mutagenicity, and teratogenicity. *In* "Mutagenic Effects of Environmental Contaminants" (H. E. Sutton and M. I. Harris, eds.), pp. 185–195. Academic Press, New York.

Wyttenbach, C. R., and Hwang, J. D. (1984). Relationship between insecticide-induced short and wry neck and cervical defects visible histologically shortly after treatment of chick embryos. *J. Exp. Zool.* **229**, 437–446.

Wyttenbach, C. R., and Thompson, S. C. (1985). The effects of the organophosphate insecticide malathion on very young chick embryos: Malformations detected by histological examination. *Am. J. Anat.* **174**, 187–202.

Wyttenbach, C. R., Thompson, S. C., Garrison, J. C., and Kitos, P. A. (1981). Precision delivery of small volumes of liquids to very young avian embryos. 1. Locating and positioning the embryo *in ovo*. *Toxicol. Appl. Pharmacol.* **59**, 40–48.

21

Neurobehavioral Effects of Organophosphorous Compounds

Zoltan Annau
Department of Environmental Health Sciences
The Johns Hopkins University
Baltimore, Maryland

I. Introduction

Long before organophosphorous (OP) compounds became associated with chemical warfare, they had been involved in a massive outbreak of poisoning in the United States. During the prohibition era, an alcoholic extract of Jamaican ginger that was contaminated by tri-ortho-cresyl phosphate (TOCP) was sold as a liquor substitute. In this well-publicized and well-studied episode, approximately 20,000 people developed various degrees of paralysis and other symptoms of poisoning in what became popularly known as Ginger Jake paralysis (Kidd and Langworthy, 1933). This large-scale tragedy became immortalized even in the popular music of the time (Morgan and Tullos, 1976). The symptoms consisted of a slowly developing paralysis, particularly in the legs, accompanied by tremors. Follow-up of many of these patients revealed that even after 6 years, there was little recovery, although the symptoms of muscle weakness were replaced by spasticity and hyperreflexia as well as abnormal reflexes (Aring, 1942).

Fortunately the OP compounds developed during World War II as nerve gases were never used on human populations so that large-scale exposures did not occur after the TOCP incident. Nevertheless, as the potent insecticidal properties of these compounds became recognized, their wide-

Organophosphates: Chemistry, Fate, and Effects

spread use resulted in exposures of farm workers and others associated with handling of the chemicals. Most of the human behavioral data obtained on the neurotoxicity of OP compounds has been recorded from occupational exposures, with the exception of some studies in which human volunteers were given nerve gas agents. These studies will be reviewed in some detail because they provide clear evidence of the behavioral effects of these agents at low doses under carefully controlled conditions.

II. Human Studies

Grob and Harvey (1953, 1958) described the central nervous system (CNS) effects of human subjects exposed to the nerve agent sarin (isopropyl methyl phosphonofluoridate). Sarin was administered in a water or propylene glycol solution orally, daily for periods up to 5 days, to 10 subjects. Symptoms in the subjects appeared usually within 20 min after drug administration as signs of muscarinic poisoning. These symptoms consisted (at the lower doses) of anorexia, nausea, and tightness of the chest. At higher doses, or repeated administration, abdominal cramps, vomiting, diarrhea, salivation, and lacrimation were reported. The CNS effects were described (Grob and Harvey, 1953) as consisting of tension, anxiety, emotional lability, and insomnia. With more prolonged exposure, headache, drowsiness, mental confusion, and slowness of recall were additional symptoms recorded. Changes in the EEG were seen also, consisting of slow waves and increased amplitude. Bowers *et al.* (1964) studied another agent, identified only as EA-1701, a classified nerve agent similar to sarin. The behavioral effects seen in this experiment were very similar to those described above. Subjects had difficulty concentrating, remembering tasks they had to perform, and were somewhat irritable. Thought processes seemed to fade away continually during the exposure and, when present, were exceedingly slow.

Exposure of human subjects to the insecticide parathion (*p*-nitrophenyl diethyl thionophosphate), or DFP (diisopropyl fluorophosphate), (Grob *et al.*, 1947, 1950), produced very similar although not identical effects to those described for sarin. An interesting difference was noted between DFP, which produced nightmares, and parathion, which did not.

Unfortunately, because these experiments were performed in a clinical setting, no systematic methodologies were used for recording behavioral responses. The descriptions, while extremely valuable since such human experimentation is for obvious reasons no longer permitted, are clinical in nature and difficult to quantitate. The main advantage of these human experiments is that they have served as a guide for animal experimentation in focusing on behavioral functions in animals that might subserve similar CNS processes.

Additional evidence from human exposure studies also points in the same direction as the clinical studies. These studies are clinical and epidemiological surveys of agricultural workers exposed to OP compounds, usually repeatedly and having or having had obvious symptoms of OP intoxication. Metcalf and Holmes (1969) tested industrial and agricultural workers with both behavioral and electrophysiological techniques. The most obvious signs of intoxication were disturbed memory and difficulty in maintaining alertness and attention. The EEG showed waveforms suggestive of narcolepsy, perhaps corroborating the inability to maintain alertness. Levin and Rodnitzky (1976) reviewed the effects of OP compounds in humans, both in experimental and industrial settings, and came to the conclusion that the most important signs of intoxication were memory deficits, linguistic disturbances, depression, anxiety, and irritability. Duffy *et al.* (1979) showed that when EEG measures were taken, even 1 year after workers had been exposed to OP compounds, significant alterations could be seen in beta activity, as well as in several other frequencies. The long persistence of symptoms has also been reported by Coye *et al.* (1986) and Savage *et al.* (1988), even after serum cholinesterase levels had returned to normal. Headache, giddiness, paresthesia, and ocular symptoms were most commonly observed in workers exposed to fenthion (O,O-dimethyl-O-(4-methylmercapto-3-methylphenyl)-phosphorothioate). These workers also had significantly reduced serum cholinesterase levels (Misra *et al.*, 1985).

These studies suggest that the repeated exposure of human subjects to OP compounds can have long-lasting effects, sometimes even after the usual biochemical indices of exposure, such as serum cholinesterase, have returned to normal. As will be discussed in the section on animal studies, behavioral tolerance to OP exposure develops rapidly, but this tolerance may hide to some extent the real intoxication that has already taken place.

III. Animal Studies

Because of their extreme neurotoxicity, OP compounds were widely studied in their capacity as toxic pesticides, as well as in their capacity as probes of cholinergic function (Bignami *et al.*, 1975). In this section, the acute treatment effects in studies using animals will be reviewed first. A second section describing subacute and chronic studies will follow. Finally, the last section will deal with the effect of OP compounds on the developing organism.

A. Acute Effects

The general effect of OP compounds on behavior in animals is a disruption of the behavior, once sufficiently toxic doses are administered. A variety of experimental compounds and insecticides have been studied in this context. Raslear and Kaufman (1983) have shown that a single dose of DFP (1.75 mg/kg) given to a rat disrupts running in an activity wheel, where a lever was also available for food reinforcement, within 24 hr of administration. Lever pressing and food intake recovered in 3 to 4 days, but wheel running remained depressed for the 21 days of the experiment. In addition, there were significant phase shifts in the running pattern during the light–dark cycle of the treated rats, suggesting a disruption of circadian rhythms. These circadian disruptions extended to all three behaviors. In a subsequent study, Raslear *et al.* (1986) showed that this disruption was not attributable to motor deficiencies caused by DFP, but was likely to be a direct effect on central mechanisms controlling circadian behaviors.

Using the conditioned taste-aversion paradigm, Roney *et al.* (1986) showed that this paradigm was more sensitive than more traditional operant tests for detecting toxicity. Doses of parathion, dichlorvos, or DFP that did not produce signs of toxicity were able to reduce water intake in rats that had been previously paired with the chemicals. The taste-aversion paradigm has been used by many investigators to demonstrate the potential toxicity of compounds (Kallman *et al.*, 1983). There is some question, however, whether this test is particularly relevant to rodents only, or whether it has general biological significance as a test for toxicity. It is certainly not a specific test for neurotoxicity.

While most experimental procedures have employed animals trained to perform a task before OP exposure, Geller *et al.* (1985) evaluated the effects of prior exposure on subsequent learning. Rats were injected with 0.31 or 0.46 μg/kg of the nerve agent soman on day 1 and 3 of the experiment. On day 6 the animals were trained to lever press in the presence of a tone to avoid an electric shock. None of the high-dose soman animals learned to avoid the shock, even after 16 weeks of training. Five of eight of the control animals and four of seven of the low dose soman animals were successful in acquiring this obviously difficult task during the same period. The high-dose soman was 75% of the LD_{50} and 13 of 20 animals died prior to training. Only three animals died in the low-dose group.

A somewhat similar approach was taken by McDonough *et al.* (1986) who trained rats first on a continuous reinforcement schedule to lever press for milk reinforcement. The animals were subsequently administered either saline or 100–110 μg/kg of soman. Three weeks later, the rats were trained on a differential reinforcement of low rates schedule (DRL) of 20 sec. On this schedule the animal has to wait 20 sec between lever presses in order to obtain the milk. As before, the soman exposed subjects were unable to learn the DRL schedule, although the lever-pressing response remained unaltered. Nine of 24 subjects died following the soman exposure. The surviving animals had significant brain pathology in the dorsal thalamus, the pyriform cortex, and the amygdala. The degree of neuropathology and behavioral deficits appeared to be directly correlated.

This high-dose treatment with a very toxic chemical that causes brain lesions poses problems in the interpretation of the nervous system/behavioral effects of OP compounds. It is impossible to determine from the results whether the behavioral effects were due to the soman or the lesions or a combination of both. The complexity of the lesions makes it impossible to control for the location of the lesion effect, and thus, the results of this study, while valuable in terms of near lethal dose studies, do not help us understand the consequences of exposure to more typical doses encountered by humans.

Chambers and Chambers (1989) administered lethal (2.0 mg/kg) and sublethal (1.3 mg/kg) doses of paraoxon to rats performing on a fixed ratio 10 (FR 10) operant schedule. These lethal doses were followed by the centrally acting antidote atropine sulfate or the peripherally acting methyl bromide atropine and methyl nitrate atropine. The behavior of the rats was evaluated for 2 days after the treatments. These high doses of paraoxon depressed both brain cholinesterase and behavioral performance severely. The centrally acting atropine sulfate at high doses (90 mg/kg) was sufficient to substantially counteract the disruptive effects of the OP compound by day 2. Neither peripherally acting atropine compound was effective on the first day in restoring the behavioral performance although there was partial recovery by day 2. Similar findings were reported in chronic studies at lower doses by Russell *et al.* (1971) and Chambers *et al.* (1988), using different behavioral paradigms.

B. Chronic Treatment and the Development of Tolerance

A series of studies by Glow and co-workers established some of the early parameters of the behavioral effects of OP compounds. Food-deprived rats were trained to lever press in order to obtain a food pellet. Daily sessions were restricted to each rat's obtaining a total of 20 pellets. In one of the earliest papers, Glow and Rose (1965) followed this training by extinction sessions, during which lever presses were no longer reinforced with food pellets. During extinction, the animals were divided into four groups. Group 1 received injections of water. Group 2 was given DFP plus an oxime as a peripheral cholinesterase activator. Group 3 received DFP plus water, and group 4, water plus the oxime. Group 3 took significantly longer to extinguish than the other groups and, in addition, the animals made significantly fewer lever presses on day one of extinction than animals in the other groups. These results indicated that when cholinesterase was inhibited both centrally and peripherally, significant alterations in behavior resulted, particularly on the first day following treatment.

As a follow-up to this study, Glow and Richardson (1967) measured the effects of continued treatment with DFP on the operant response. Rats were trained to make 20 responses per day in order to obtain food pellets. When response rates had stabilized, the animals were administered a dose of DFP (1.0 mg/kg) on the first day and 0.5 mg/kg every third day thereafter. The effect of treatment was a significant disruption on the first day in the time it took the animals to obtain the 20 pellets. By the ninth day of the experiment, however, the DFP-treated animals were not significantly different from controls. This clear-cut demonstration of behavioral tolerance established the pattern for many of the subsequent studies with animals in other laboratories.

In an attempt to determine whether the type of reinforcement used would influence the outcome of the DFP treatment, Richardson and Glow (1967) trained rats on a visual discrimination task. Rats had to discriminate between vertical and horizontal stripes in a runway. Half the animals received a food pellet for a correct response (positive reinforcement), and half the animals received both a food pellet for the correct response and an electric shock for an incorrect response (negative reinforcement). Training started after the animals received their first DFP injection, and drug treatment continued as in the previously described study. The DFP-treated animals made significantly more errors than controls, as would be expected, during training. An interesting aspect of this study, however, was that the animals that received both positive and negative reinforcements made fewer errors than the group receiving positive reinforcement only. These results indicated that while OP treatment could disrupt a learning task significantly, the extent of the disruption depended on the reinforcement contingencies. Painful electric shocks reduced the disruption, while food pellets alone were not sufficiently motivating to the animal to reduce errors. The authors point to the potential interaction between the toxic effects of a chemical and motivating factors in determining the behavioral outcome of exposure to chemicals.

Russell and his co-workers, during the late 1960s, started a series of experiments to investigate the phenomenon of *behavioral tolerance* in animals exposed chronically to OP compounds and other cholinesterase inhibitors. Using the treatment regimen described earlier by Glow *et al.* (1966), they injected rats intramuscularly with an initial dose of 1.0 mg/kg of DFP, followed by 0.5 mg/kg every third day (Russell *et al.*, 1969). A variety of conditioned and unconditioned behaviors were chosen to investigate the disruptive effects of treatment. Rats were deprived of either food or water or 23 hr and then were allowed ad lib access to either for 1 hr. The initial effect of the DFP treatment was to disrupt both feeding and drinking. This disruption however, was most severe following the first treatment and was followed by rapid recovery toward baseline until the next DFP injection 3 days later. A series of diminishing oscillatory behavioral depressions followed each injection with an increasing tendency toward baseline as treatment continued. Food intake recovered, despite continued treatment after 10 days, and water intake, after 22 days. Concurrent measurements of brain cholinesterase in identically treated animals showed these levels to be depressed throughout the experiment by approximately 70%.

In order to determine the effects of DFP treatment on motor function as well as an operant response maintained by negative reinforcement, other animals were trained on a continuous avoidance schedule. Animals had to press a lever every 40 sec in order to postpone receiving an electric shock. Failure to respond resulted in the delivery of a shock every 20 sec. Following the establishment of stable baseline behavior, animals were given DFP. Treatment resulted in a decrease in the number of avoidances and an increase in the number of shocks received. Tolerance to the DFP developed by the tenth day of treatment. Two other conditioned behaviors were evaluated in these experiments, a discrete trial and single alternation task. On both of these tasks, the animals developed tolerance within 10 days.

These experiments indicated that the development of tolerance was not due to the debilitating (perhaps peripheral or neuromuscular) effects of cholinesterase inhibition, but rather to some interaction between these peripheral and central effects. In the central effects, the particular pathway involved may have been important in determining the duration of tolerance development. Thus, drinking behavior which is more directly involved in cholinergic pathways (Grossman, 1961), was affected more than eating and the other behaviors that are less directly linked to cholinergic systems.

The importance of central mechanisms in the development of tolerance was tested further by Russell *et al.* (1971) by evaluating the effects of cholinolytic agents on tolerant animals. Rats were trained to lever press in a chamber in order to obtain their entire daily water supply during 1 hour sessions. When response rates had stabilized, they were subjected to the previously described DFP treatment. After the ninth injection of DFP, behavioral tolerance was evidenced by a recovery of the response rate of the treated animals. Both DFP-treated and untreated animals were then challenged by several doses of atropine (1.0 to 15.0 mg/kg). In addition, methylatropine was used in order to compare central versus peripheral effects of these challenges. Atropine at the two highest doses (10.0 and 15.0 mg/kg) depressed the response rates of the DFP-treated animals significantly more than it did the response rates of controls. At the lower doses of atropine, there were no differences between the groups, and mehtylatropine challenge did not differentiate the two groups of animals. These results indicated that DFP treatment altered central rather than peripheral mechanisms in the development of tolerance and that this could be demonstrated by relatively high-dose challenges with atropine. This was interpreted in terms of atropine competing for receptors that had been desensitized by the DFP treatment.

Overstreet *et al.* (1972) explored the possibility of decreased sensitivity of the cholinergic system by challenging animals with pilocarpine, a cholinergic agonist. In this study rats were trained to eat powdered food out of containers that could be measured for total consumption. After the third DFP treatment, food consumption had returned to normal and the animals were challenged with both amphetamine (1.0 and 2.0 mg/kg) and pilocarpine (2.0 and 4.0 mg/kg). Amphetamine reduced the food intake of both control and DFP-treated groups equally. Pilocarpine at 4.0 mg/kg reduced the food intake of the control animals by 77% and of the DFP treated animals by only 38%, a highly significant difference. This test of receptor sensitivity confirmed the results of the Russell *et al.* (1971) study by indicating that the behavioral changes seen during tolerance involved the cholinergic receptors rather than the adrenergic system.

A rather interesting extension of these studies was published by Chippendale *et al.* (1972). Measuring drinking volume in water deprived rats, they treated them with a range of DFP doses (0.2, 0.4, 0.5, and 1.0 mg/kg) and then challenged them with scopolamine, methyl scopolamine, and alpha-methyl-para-tyrosine. The lowest dose of DFP (0.2 mg/kg) had no effect on water intake; the two intermediate doses (0.4 and 0.5 mg/kg) disrupted water intake temporarily; while the highest dose (1.0 mg/kg) disrupted drinking totally after the third injection, with no recovery, i.e., no tolerance. Despite the lack of effect at the 0.2 mg/kg dose, when the animals were challenged with scopolamine, their water intake was significantly more depressed than the water intake of the untreated animals. Neither methyl scopolamine nor alpha-methyl-para-tyrosine differentiated control from treated animals.

This experiment demonstrated two interesting principles. First, it showed that doses of OP compounds low enough not to disrupt behavior could still alter the underlying neurochemical substrate as revealed by pharmacologic challenge. Second, if the dose of the OP compound was high enough, it disrupted behavior, and, at least in the confines of this experimental design, no tolerance could be seen.

In subsequent experiments Overstreet *et al.* (1974) showed that both muscarinic and nicotinic receptor sensitivity is reduced by DFP treatment and that the muscarinic receptors are more prone than the nicotinic receptors to disruption. These authors also reported that behavioral disruption did not occur until cholinesterase levels were lowered below 46% of control. Russell *et al.* (1986) in a complex series of experiments with the nerve agent soman, demonstrated that tolerance was not a unitary phenomenon, but was restricted to some biological and behavioral systems but not to others. More specifically, using low-level exposures to soman (35 μg/kg for 3 days followed by three injections per week) in rats, they showed that there was no effect on water intake. There was an initial period of hypothermia that lasted 4 days; there was an increase in pain threshold that persisted throughout the experiment; locomotor activity decreased for 5 days and then returned to baseline; there was no difference between control and soman-treated animals in the number of trials required to reach criterion in a shock avoidance learning task;

and performance on a fixed-interval 30-second operant task was disrupted for the first 4 days. This experiment, using the considerably more toxic nerve agent soman, showed again that subsymptomatic exposures to OP compounds can cause significant alterations in a variety of biobehavioral functions.

Bignami and his colleagues (Bignami *et al.*, 1985; Giardini *et al.*, 1981, 1982) were able to demonstrate that the development of tolerance was not only under the control of the biological mechanisms postulated by Russell and co-workers, but was also influenced by the experimental conditions that controlled the behavioral events. Specifically, three groups of animals were trained on a shock-avoidance schedule. Following stable performance, group A was given paraoxon (0.125 mg/kg) daily, 2 hours before the behavioral session; group B was given paraoxon immediately after the termination of the avoidance session or approximately 23.5 hr before the beginning of the next session; group C was treated with paraoxon, but not tested on the avoidance task. After 28 days, group B was switched to the same regimen as group A, namely, injected 2 hr before testing, and group C resumed behavioral testing.

The results indicated that group A was severely disrupted by paraoxon in terms of the number of avoidances made by the rats during the first 16 days of treatment. By day 28 however, this group had developed tolerance and avoidance performance had returned to baseline. Group B was less disrupted than group A, and its performance also returned to baseline on day 18. When group B was shifted to treatment 2 hr before testing, its performance deteriorated to the same degree as the initial performance of group A, in effect showing no transfer of the previously demonstrated tolerance. Group C, which had not been allowed to perform in the avoidance task for 28 days, was similarly disrupted. Both groups subsequently showed the development of behavioral tolerance during continued treatment.

These results clearly demonstrate that behavioral tolerance is situation specific and that it does not develop unless an animal receives the chemical exposure in association with the behavioral performance. The results also imply that attempts to associate changes in enzyme and receptor levels in the brain with the behavioral changes may be misleading, in that while under some circumstances they may be parallel processes, under other circumstances, such as the above experiments, they may be uncoupled.

The idea that repeated exposure to OP compounds and the appearance of behavioral tolerance does not mean that the organism has recovered its normal function has been explored by others as well. Costa and Murphy (1982) treated mice chronically with the insecticide disulfoton (O,O-diethyl *S*-[2(ethylthio)ethyl]phosphorodithionate). After about 5 days, the mice no longer exhibited typical signs of OP poisoning. When tested in a passive avoidance task, treated mice were not different from untreated controls. When challenged with scopolamine, however, in a retention test, the disulfoton exposed mice were significantly disrupted as compared to controls.

This type of *pharmacological* challenge has often been found useful in revealing latent expressions of neurotoxicity (Annau, 1987). Using a somewhat similar approach, Raffaele *et al.* (1990) showed that rats trained on a serial discrimination reversal procedure and then exposed to DFP at low doses were not disrupted by the drug treatment. When challenged with scopolamine, however, these animals also showed impairment as compared to controls.

From these studies of the effect of OP compounds, a picture emerges that is somewhat different from expectations after the Jamaican ginger episode described at the beginning. It is quite clear that repeated exposure to moderate doses of OP compounds can lead initially to behavioral symptoms of poisoning, followed by the disappearance of these symptoms and thus behavioral tolerance. Alternatively, very low-dose exposure to OP compounds may not have any discernible symptoms at all. In both cases though, challenging these asymptomatic animals results in the emergence of behavioral responses to the pharmacological agent, suggestive of an altered biological substrate. The conclusion must be drawn, therefore, that despite the rather large volume of research and human literature on OP poisoning, it is still not clear what level of exposure does not result in latent toxicity. In the next section, this concept is extended, when the problem of exposure of the developing organism is described.

C. Perinatal Exposures

Spyker and Avery (1977) investigated the effects of prenatal exposure of mice to diazinon, an OP compound commonly used in agriculture. This compound is metabolized to yield the potent cholinesterase-inhibiting compounds diazoxon and tetraethylmonothiopyrophosphate. Pregnant mice were exposed through diet from day one of gestation to 0, 0.18, and 9.0 mg/kg of body weight throughout pregnancy. All females gave birth to normal-appearing offspring. The high-dose offspring showed a reduction in growth up to 1 month of age. Behavioral testing of all offspring showed that there were abnormalities in endurance and coordination on a rod-clinging test and the inclined plane test. The higher dose offspring also showed abnormalities in a Lashley maze and a swimming test. At this dose, examination of the brains of the offspring at 101 days of age revealed neuropathological alterations.

Gupta *et al.* (1985) treated pregnant rats from day 6 to day 20 of gestation with methylparathion at either 1.0 or 1.5 mg/kg. In a somewhat puzzling outcome, the results indicated that on a series of behavioral measures, cage emergence, locomotor activity, and an operant task, the low-dose offspring were significantly different from control animals, but the high-dose offspring were not. Neurochemical measures indicated that the high dose was more effective in altering cholinesterase levels in the brain of both mothers and offspring than was the low dose. The results of this experiment remain somewhat unclear, given the contradictory results between the chemistry and the behavioral measures and the lack of a dose-response effect.

In order to determine whether postnatal exposure has similar effects on subsequent behavioral outcomes, Stamper *et al.* (1988) exposed rat pups from day 5 to day 20 postnatally to parathion at 1.3 or 1.9 mg/kg. At the higher dose, the pups exhibited signs of intoxication for the first few days of treatment. Both doses decreased growth in the pups up to 20 days postnatally, the last time point of these measures. While most early developmental behavioral measures showed no effects of treatment, there were significant effects on spontaneous alternation in a T-maze on postnatal day 24 and a significant reduction in working memory in a radial arm maze on postnatal day 36 with both doses.

IV. Conclusions

This review of the behavioral effects of OP intoxication has shown that in all species examined and at all ages, exposure to these compounds can have deleterious and long-lasting, perhaps irreversible consequences. What has not emerged in the literature is whether the seemingly long-lasting effects of these compounds will change with time. It is also unknown whether an interaction between ageing and toxicity will exacerbate the symptoms. This is not an inconsequential problem, since farm workers who are exposed to these compounds throughout life are likely to suffer the most consequences during ageing. It is also unclear whether the prenatal exposures leading to abnormal behavioral responses will dissipate with time, remain unaltered, or whether asymptomatic organisms will remain altered in their response to pharmacological agents.

Despite the numerous research reports, as seen in this review of the effects of OP compounds on behavior, the literature shows very little cohesion, and systematic exploration of these issues. In order to answer the questions relevant to human health, it is important to approach the field with a more systematic series of experiments that use the same organism, the same behavioral tasks, and attempt to determine the long-term health consequences of OP exposure.

References

Annau, Z. (1987). The use of pharmacological challenges in behavioral toxicology. *Zentralbl. Bakteriol. Mikrobiol. Hyg.* 185, 61–64.

Aring, C. D. (1942). The systemic nervous affinity of triorthocresyl phosphate (Jamaica ginger palsy). *Brain*, 63, 34–47.

Bignami, G., Giardini, V., and Scorrano, S. (1985). Behaviorally augmented versus other components in organophosphate tolerance: The role of reinforcement and response factors. *Fundam. Appl. Toxicol.* 5, S213–S224.

Bignami, G., Rosic, N., Michalek, H., Milosevic, M., and Gatti, G. L. (1975). Behavioral toxicity of anticholinesterase agents: Methodological, neurochemical and neuropsychological aspects. In "Behavioral Toxicology" (B. Weiss and V. G. Laties, eds.), pp. 155–215. Plenum, New York.

Bowers, M. B., Jr., Goodman, E., and Sim, V. M. (1964). Some behavioral changes in man following anticholinesterase administration. *J. Nerv. Ment. Dis.* **138**, 383–389.

Chambers, J. E., and Chambers, H. W. (1989). Short-term effects of paraoxon and atropine on schedule-controlled behavior in rats. *Neurotoxicol. Teratol.* **11**, 427–432.

Chambers, J. E., Wiygul, S. M., Harkness, J. E., and Chambers, H. W. (1988). Effects of acute paraoxon and antropine exposures on retention of shuttle avoidance behavior in rats. *Neurosci. Res. Commun.* 3, 85–92.

Chippendale, T. J., Zawolkow, G. A., Russell, R. W., and Overstreet, D. H. (1972). Tolerance to low acetylcholinesterase levels: Modification of behavior without acute behavioral change. *Psychopharmacologia (Berl.)* **26**, 127–139.

Costa, L. G., and Murphy, S. D. (1982). Passive avoidance retention in mice tolerant to the organophosphorus insecticide disulfoton. *Toxicol. Appl. Pharmacol.* **65**, 451–458.

Coye, M. J., Barnett, P. G., Midtiling, J. E., Velasco, A. R., Romero, P., Clemments, C. L. O'Malley, M., and Tobin, M. W. (1986). Clinical confirmation of organophosphate poisoning of agricultural workers. *Am. J. Indust. Med.* **10**, 399–409.

Duffy, F. D., Burchfiel, J. L., Bartels, P. H., Goan, M., and Sim, V. M. (1979). Long-term effects of an organophosphate upon the human electroencephalogram. *Toxicol. Appl. Pharm.* **47**, 161–176.

Geller, I., Hartmann, R. J., and Gause, E. M. (1985). Effects of subchronic administration of soman on acquisition of avoidance–escape behavior by laboratory rats. *Pharmacol. Biochem. Behav.* **23**, 224–230.

Giardini, V., de Acetis, L., Amorico, L., and Bignami, G. (1981). Test factors affecting the time course of avoidance depressions after DFP and paraoxon. *Neurobehav. Toxicol. Teratol.*3, 331–338.

Giardini, V., Meneguz, A., Amorico, L., De Acetis, L., and Bignami, G. (1982). Behaviorally augmented tolerance during chronic cholinesterase reduction by paraoxon. *Neurobehav. Toxicol. Teratol.* **4**, 335–345.

Glow, P. H., and Richardson. (1967). Control of a response after chronic reduction of cholinesterase. *Nature* **214**, 629–630.

Glow, P. H., and Rose, S. (1965). Effects of reduced acetylcholinesterase levels on extinction of a conditioned response. *Nature* **206**, 475–477.

Glow, P. H., Richardson, A., and Rose, S. (1966). Effect of acute and chronic inhibition of cholinesterase upon body weight, food intake and water intake in the rat. *J. Comp. Physiol. Psychol.* **61**, 295–299.

Grob, D., and Harvey, A. M. (1953). The effects and treatment of nerve gas poisoning. *Am. J. Med.* **14**, 52–63.

Grob, D., and Harvey, J. C. (1958). Effect in man of the anticholinesterase compound sarin (isopropyl methyl phosphonofluoridate). *J. Clin. Invest.* **37**, 350–368.

Grob, D., Harvey, A. M., Langworthy, O. R., and Lilienthal, J. L., Jr. (1947). The effect on the central nervous system with special reference to the electrical activity of the brain. *Bull. Johns Hopkins Hosp.* **81**, 257–266.

Grob, D., Garlick, W. L., and Harvey, A. M. (1950). The toxic effects in man of the anticholinesterase insecticide parathion (*p*-nitrophenyl diethyl thionophosphate). *Bull. Johns Hopkins Hosp.* **87**, 106–129.

Grossman, S. P. (1962). Direct adrenergic and cholinergic stimulation of hypothalamic mechanisms. *Amer. J. Physiol.* **202**, 872–882.

Gupta, R. C., Rech, R. H., Lovell, K. L., Welsch, F., and Thornburg, J. E. (1985). Brain cholinergic, behavioral, and morphological development in rats exposed *in utero* to methylparathion. *Toxicol. Appl. Pharmacol.* 77, 405–413.

Kallman, M. J., Lynch, R. R., and Landauer, M. R. (1983). Taste aversions to several hydrogenated hydrocarbons. *Neurobehav. Toxicol. Teratol.* 5, 23–27.

Kidd, J. G., and Langworthy, O. R. (1933). Jake paralysis. Paralysis following the ingestion of Jamaica ginger extract adulterated with tri-ortho-cresyl-phosphate. *Bull. Johns Hopkins Hosp.* **52**, 39–66.

Levin, H. S., and Rodnitky, R. L. (1976). Behavioral effects of organophosphate pesticide in man. *Clin. Toxicol.* 9, 391–405.

McDonough, J. H., Smith, R. F., and Smith, C. D. (1986). Behavioral correlates of soman-induced neuropathology: Deficits in DRL acquisition. *Neurobehav. Toxicol. Teratol.* 8, 179–187.

Metcalf, D. R., and Holmes, J. H. (1969). EED, psychological and neurological alterations in humans with organophosphate exposure. *Ann. N. Y. Acad. Sci.* **160**, 357–365.

Michalek, H., Pintor, A., Fortuna, S., and Bisso, G. M. (1985). Effects of diisopropyfluorophosphate on brain cholinergic systems of rats at early developmental stages. *Fundam. Appl. Toxicol.* 5, S204–S215.

Misra, U. K., Nag, D., Bhushan, V., and Ray, P. K. (1985). Clinical and biochemical changes in chronically exposed organophosphate workers. *Toxicol. Let.* **24**, 187–193.

Morgan, J. P., and Tullos, T. C. (1976). The Jake Walk Blues. A toxicologic tragedy mirrored in American popular music. *Ann. Intern. Med.* 85, 804–808.

Overstreet, D. H., Hadick, D. G., and Russell, R. W. (1972). Effects of amphetamine and pilocarpine on eating behavior in rats with chronically low acetylcholinesterase levels. *Behav. Bio.* 7, 217–226.

Overstreet, D. H., Russell, R. W., Vasquez, B. J., and Dalglish, F. A. (1974). Involvement of muscarinic and nicotinic receptors in behavioral tolerance to DFP. *Pharmacol. Biochem. Behav.* 2, 45–54.

Raffaele, K., Olton, D., and Annau, Z. (1990). Repeated exposure to diisopropylfluorophosphate (DFP) produces increased sensitivity to cholinergic antagonists in discrimination retention and reversal. *Psychopharmacology* **100**, 267–274.

Raslear, T. G., and Kaufman, L. W. (1983). Diisopropyl phosphorofluoridate (DFP) disrupts circadian activity patterns. *Neurobehav. Toxicol. Teratol.* 5, 407–411.

Raslear, T. G., Leu, J. R., and Simmons, L. (1986). The effects of diisopropyl phosphorofluoridate (DFP) on inter-response time and circadian patterns of lever-pressing in rats. *Neurobehav. Toxicol. Teratol.* 8, 655–658.

Richardson, A. J., and Glow, P. H. (1967). Discrimination behavior in rats with reduced cholinesterase activity. *J. Comp. Physiol. Psychol.* 63, 240–246.

Roney, P. L., Jr., Costa, L. G., and Murphy, S. D. (1986). Conditioned taste aversion induced by organophosphate compounds in rats. *Pharmacol. Biochem. Behav.* **24**, 737–742.

Russell, R. W., Warburton, D. M., and Segal, D. S. (1969). Behavioral tolerance during chronic changes in the cholinergic system. *Commun. Behav. Bio.* 4, 121–128.

Russell, R. W., Vasquez, B. J., Overstreet, D. H., and Dalglish, F. W. (1971). Effects of cholinolytic agents on behavior following development of tolerance to low cholinesterase activity. *Psychopharmacologia* **20**, 32–41.

Russell, R. W., Booth, R. A., Lauretz, S. D., Smith, C. A., and Jenden, D. J. (1986). Behavioral, neurochemical and physiological effects of repeated exposures to subsymtomatic levels of the anticholinesterase, soman. *Neurobehav. Toxicol. Teratol.* 8, 675–685.

Savage, E. P., Keefe, T. J., Mounce, L. M., Heaton, R. K., Lewis, J. A., and Burcar, P. J. (1988). Chronic neurologic sequelae of acute organophosphate pesticide poisoning. *Arch. Environ. Health* 43, 38–45.

Spyker, J. M., and Avery, D. L. (1977). Neurobehavioral effects of prenatal exposure to the organophosphate diazinon in mice. *J. Toxicol. Environ. Health* 3, 989–1002.

Stamper, C. R., Baluini, W., Murphy, S. D., and Costa, L. G. (1988). Behavioral and biochemical effects of postnatal parathion exposure in the rat. *Neurotoxicol. Teratol.* 10, 261–266.

V

Summary and Conclusions

22
Summary and Conclusions

Janice E. Chambers
College of Veterinary Medicine
Mississippi State University
Mississippi State, Mississippi

The organophosphorus (OP) compounds represent a large and diverse group of anthropogenic chemicals that have enjoyed a long and colorful history dating from 1820, with the anticholinesterases, which are of primary emphasis in this volume, dating from the 1930s (H. Chambers, Chapter 1). While all are based on organic phosphorus, they display wide diversity in their structures, chemical reactivities, chemical stabilities, biological reactivities, disposition, metabolism, toxicological effects, and toxicological potencies. The chemistry of the OP compounds can be quite complex (Abou-Donia, Chapter 17); this is illustrated with the synthesis, analysis, and stereochemical considerations for one group of OP compounds, the phosphorothiolates (Thompson, Chapter 2). Although clearly there is much commonality among such parameters as potential metabolic pathways or biochemical target molecules, there is also a tremendous amount of individuality in the types and potencies of toxicological effects elicited by individual compounds, and also uniqueness in the responses of individual species or taxonomic groups.

One of the most important reasons for the long-standing popularity of the OP compounds as insecticides is the fact that, unlike the organochlorine insecticides which preceded them, the OP insecticides are usually nonpersistent in the environment and they do not typically bioaccumulate (Racke, Chapter 3). The environmental entry of OP insecticides by the indirect routes of drift, volatilization, leaching, and runoff has been low and for the most part insignificant. Since most OP insecticides tend to sorb well to soil, the likelihood of ground water contamination is low. The environmental reactions are frequently similar to metabolic reactions. Thus, the chemical and biological lability of these compounds has made them attractive replacements for the persistent organochlorine insecticides, and their efficacy as insecticides has afforded them great utility even today after over 30 years of use. Frequently, however, they have high acute toxicity levels, and therefore do pose a threat of accidental poisoning during occupational handling.

Organophosphates: Chemistry, Fate, and Effects

For most OP compounds, neurotoxicity is the most serious toxicological concern. The primary mechanism of acute toxicity in vertebrates and invertebrates of both the insecticidal OP compounds (or their activated metabolites) or the OP chemical warfare agents is usually considered to be inhibition of the critical enzyme acetylcholinesterase (AChE) in cholinergic synapses and neuromuscular junctions, with resultant hyperactivity of cholinergic pathways (H. Chambers, Chapter 1; Wallace, Chapter 4). These anticholinesterases are very reactive, and readily phosphorylate the active site serine in serine esterases and serine proteases; this phosphorylation leads to a persistent inhibition of the enzyme activity, which can be life-threatening or result in severe signs of cholinergic system toxicity. If used promptly, therapy from many of the insecticidal OP compounds (but not the nerve agents) can be assisted by the use of oxime reactivators (Wilson *et al.*, Chapter 5). (The oximes are useful experimental tools and may also be useful in assessing field exposure to OP insecticides.) This anticholinesterase action, long believed to be nondisputable, is currently more controversial in light of recent evidence of OP interaction with cholinergic receptors at very low OP concentrations (Eldefrawi *et al.*, Chapter 13; Costa, Chapter 14). Thus through direct interaction with the receptor, the OP compound could mediate the same intracellular events caused by endogenous acetylcholine. As a result the primary effects on AChE may be exacerbated or attenuated by direct effects on cholinergic receptors, depending on the location (presynaptic or postsynaptic) and associated function of these receptors.

Of course, species selectivity is clearly a goal in the discovery of new OP insecticides such that nontarget species sensitivity is as low as possible. While the *in vitro* metabolic activities for activation and detoxication are not in themselves good predictors of species sensitivities, the AChE sensitivity as determined by molecular features may be related to species sensitivities and may yield a more useful approach to predicting species selectivity (Wallace, Chapter 4). Nevertheless, an understanding of disposition and metabolism will also be required in order to predict a species' or strain's *in vivo* response, where such factors as the differences among species in the amounts, locations, and sensitivities of protective molecules such as carboxylesterases (Maxwell, Chapter 9) or the activities of detoxication enzymes such as phosphorotriester hydrolases in resistant strains (Kasai *et al.*, Chapter 8) are important in modulating the toxicity of an OP compound.

While these documented effects on AChE are clearly lethal or life-threatening, sublethal levels of exposure have the potential of altering behavior because of the importance of cholinergic pathways in a variety of behaviors including memory (Annau, Chapter 21). Thus, cholinergic effects of the OP compounds would be expected to threaten normal behavioral performance, and, indeed, a number of behavioral effects have been documented in a variety of laboratory situations. Also implications have been generated of persistent OP-induced effects from behavioral abnormalities and memory deficits in humans accidentally exposed to high levels of OP compounds. Even subsymptomatic doses of OP compounds have resulted in behavioral effects, as have prenatal and early postnatal exposures. Again, with behavioral responses, individual OP compounds seem to induce different types of responses, and the nature of the experimental paradigm seems to be a confounding factor in the magnitude of the effects observed. Thus, a more systematic approach to behavioral effects is warranted to understand the nature, persistence, and severity of OP-induced effects.

Despite the fact that the basic metabolic pathways for OP compounds have been studied for many years, the relevance of these in the *in vivo* situation is still unclear. Many of the OP insecticides are phosphorothionates and must be metabolically activated, in most cases by the cytochrome P450-dependent monooxygenases (Nakatsugawa, Chapter 10) and in some cases by the flavin-containing monooxygenases (Levi and Hodgson, Chapter 6). The latter enzymes, the flavin-containing monooxygenases, have a number of similarities and a number of dissimilarities to the more intensively studied P450-dependent monooxygenases, and certainly warrant further investigation for their role in the metabolism of OP compounds as well as other pesticides and xenobiotics.

A variety of detoxication pathways exist which include several P450-mediated reactions, A-esterase (phosphorotriester hydrolase)-mediated hydrolysis, carboxylesterase-mediated detoxication (both catalytic for some compounds, and stoichiometric for all phosphates) and glutathione-mediated conjugation (Sultatos, Chapter 7; Kasai *et al.*, Chapter 8; Maxwell, Chapter 9; Nakatsugawa, Chapter 10). For toxicity to occur, the delicate balance between bioactivation and detoxication must favor activation to the extent required for oxons to escape detoxication, sequestration, nonspecific binding and any other mechanisms that exist to prevent oxon from reaching molecular targets. Predicting these *in vivo* metabolic relationships and ultimately toxicity from *in vitro* data is very difficult (Sultatos, Chapter 7; Nakatsugawa, Chapter 10) because relative enzyme affinities, intracellular OP concentrations, blood and tissue protein and membrane binding characteristics, as well as numerous other potential confounders can come into play to affect the ultimate amount of OP compound which reaches critical target molecules. The *in vivo* significance of some of these pathways, such as glutathione-mediated detoxication (Sultatos, Chapter 7) is still quite controversial. Such paradoxical effects as an attenuation of phosphorothionate acute toxicity following animal pretreatment with chemicals such as phenobarbital (which induce higher levels of phosphorothionate activation) are complex concepts to explain, and require an understanding of numerous factors involving metabolism and disposition. Thus, *in vivo* and *in vitro* data for a number of different OP compounds displaying different properties and toxicities will need to be correlated before *in vitro* data can be used effectively in a predictive sense.

The mammalian liver is clearly extremely important because of its high activities of xenobiotic metabolizing enzymes and because, by both catalytic and/or noncatalytic mechanisms, the amount of an OP compound in the blood exiting from the liver is reduced (Nakatsugawa, Chapter 10). The fact that toxic effects can occur at dosages where activated insecticide metabolites (oxons) fail to exit the liver, suggests that extrahepatic activation can be significant. Since many occupational exposures to insecticidal OP compounds will be dermal, the ability to penetrate the skin and the metabolic potential, either activation or detoxication, of the skin may be extremely important factors in many accidental poisonings. Because this absorptive surface interfaces more directly with the environment than respiratory or digestive tract epithelia do, environmental factors such as temperature and humidity can have a great influence on degree of absorption. Compounds absorbed through the skin can be sequestered by the subcutaneous fat for slow release. Sophisticated experimental techniques have been developed to study these intricate relationships (Riviere and Chang, Chapter 12).

Both the fact that oxons are highly reactive as well as the fact that so many protective (catalytic and noncatalytic) mechanisms exist throughout the organism, most notably the carboxylesterases in the liver and bloodstream, suggest the importance of target site activation. In fact, the brain does have a low but significant phosphorothionate activation potential and the *in vitro* activities are supportive of *in vivo* relevance. Additionally, the brain has been shown to be able to substantially bioactivate a phosphorothionate at realistic concentrations in the intact organism (J. Chambers, Chapter 11). Target site activation may not be significant at very high phosphorothionate dosages where the protective capacity of the liver is exceeded, but may be significant at lower oral exposure levels where first pass extraction by the liver is operational or with dermal exposures where the phosphorothionate can circulate directly to the brain.

Several critical defense systems, such as the variety of detoxication enzymes already mentioned, enable the organism to survive the toxic insults resulting from OP compound exposure. In addition, a neurochemical defense against repeated exposures to OP compounds has been documented, i.e., the down-regulation of muscarinic cholinergic receptors so that fewer receptors are available to respond to the excess acetylcholine and therefore an attenuated toxic response occurs (Hoskins and Ho, Chapter 15). This receptor down-regulation corresponds in time to some of the behavioral tolerance displayed in animals receiving multiple exposures to OP compounds (Annau, Chapter 21). Receptor down-regulation, possibly in conjunction with other adaptive responses, is undoubtedly an important homeostatic mechanism which allows the organism to adapt to an altered environment.

While inhibition of acetylcholinesterase, possibly in concert with direct effects on cholinergic receptors, is the critical mode of action of OP compounds in acute toxicity and in mediation of behavioral effects, there are noncholinergic actions of OP compounds which have yielded some very profound effects on various groups of organisms. One of the most serious noncholinergic toxicities is organophosphate-induced delayed neurotoxicity (OPIDN), a peripheral neuropathy caused by certain OP compounds leading to paralysis and ataxia in highly sensitive species. Although the exact mechanism of action is still controversial, there is an excellent correspondence between ageable OP compounds which can inhibit a curious enzyme activity termed neurotoxic esterase (neuropathy target enzyme; NTE) and delayed neuropathic potential (Richardson, Chapter 16). The endogenous function of NTE is unknown, as is its relationship to OPIDN. Therefore, at this point, NTE is, at the least, a prediction of delayed neuropathic potential as well as a biomarker of exposure to a delayed neurotoxicant with useful implications in regulation and public health, and, at the most, a direct monitor of target interaction. With OPIDN, also, great species selectivity exists, with chickens and humans being among the most sensitive and rodents classically considered among the least sensitive or insensitive. However, recent developments have indicated that rodents indeed do respond to delayed neurotoxicants and develop notable nervous system degeneration in response to exposure (Veronesi and Padilla, Chapter 18). While they are less likely to develop severe ataxia, their responsiveness indicates that rodent species, which are clearly more accessible and convenient than chickens, may well be useful experimental models in the study of OPIDN.

Recent work with noninsecticidal/nonnerve agent trivalent OP compounds such as triphenyl phosphite, have indicated that there is a second type of OP-induced delayed neurotoxicity (Type II) which can be induced (Abou-Donia, Chapter 17). This seems to involve neither AChE nor NTE, although it does result in nerve fiber degeneration. The histopathologic lesions, ataxia, paralysis, and impairment of cognitive function are more extensive than those effects resulting from Type I delayed neurotoxicants. This Type II toxicity may result from toxicity to mitochondria.

Finally, systems other than the nervous system also seem to be targets for OP compounds. The immune system can be adversely affected by OP compounds (Pruett, Chapter 19). While immunosuppression at neurotoxic, near lethal exposures has been reported, the likelihood of immunotoxicity from chronic OP exposures is unknown, with both positive and negative results occurring in the literature. Many aspects of the immune system can be affected by OP compounds *in vitro*, although such effects may not be the result of phosphorylation. However, the literature base on this area is not comprehensive enough to draw concrete conclusions on the immunotoxic potential of OP compounds.

Another nonneurotoxic effect displayed by some of the OP compounds is that of developmental defects, documented most strongly in chicken embryos (Kitos and Suntornwat, Chapter 20). Severe structural defects have resulted. One of the most likely targets is kynurenine formamidase, although effects on acetylcholinesterase and poly(ADP ribose) polymerase may also contribute to the outcome. One of the reasons for structural abnormalities is the occurrence of incorrect proteoglycans. Lower vertebrates appear to be more sensitive to teratogenic effects than mammals largely because of the protection afforded to the embryo/fetus by the mother, as well as a lack of one of the primary targets, kynurenine formamidase.

Lastly, a delayed pulmonary toxicity has also been documented, which is not the result of phosphorylation (Thompson, Chapter 2). This toxicity is elicited by some phosphorothiolates, a class of compounds which has limited utility as insecticides but which can form relatively easily by rearrangement of commercial insecticides. Additionally, some of these phosphorothiolates are potent anticholinesterases.

In summary, the OP insecticides have displayed a long history of use because of their efficacy and general lack of environmental concerns, but they are responsible for an acute toxicity hazard, as is also true for the OP nerve agents. Because of the complexity of their biological reactivity in terms of target enzyme sensitivity, disposition, and metabolism, it is at present extremely difficult, if not impossible, to predict *in vivo* toxicity levels with accuracy from the available *in vitro* data. Also, numerous questions still exist as to the biochemical mechanisms responsible for delayed toxicities, teratogenicity, immunotoxicity, or behavioral effects induced by OP compounds. Clearly, the existing data base on OP compounds is tremendous, but more information is still required to understand OP toxicity, both acute and chronic, such that meaningful assessments of OP risks can be ascertained.

Index

Page entries in bold indicate chemical structures.